Fermentation
Effects on
Food Properties

Chemical and Functional Properties of Food Components Series

SERIES EDITOR
Zdzisław E. Sikorski

Fermentation: Effects on Food Properties
Edited by Bhavbhuti M. Mehta, Afaf Kamal-Eldin and Robert Z. Iwanski

Methods of Analysis of Food Components and Additives, Second Edition
Edited by Semih Otles

Food Flavors: Chemical, Sensory and Technological Properties
Edited By Henryk Jelen

Environmental Effects on Seafood Availability, Safety, and Quality
Edited by E. Grazyna Daczkowska-Kozon and Bonnie Sun Pan

Chemical and Biological Properties of Food Allergens
Edited By Lucjan Jedrychowski and Harry J. Wichers

Chemical, Biological, and Functional Aspects of Food Lipids, Second Edition
Edited by Zdzisław E. Sikorski and Anna Kołakowska

Food Colorants: Chemical and Functional Properties
Edited by Carmen Socaciu

Mineral Components in Foods
Edited by Piotr Szefer and Jerome O. Nriagu

Chemical and Functional Properties of Food Components, Third Edition
Edited by Zdzisław E. Sikorski

Carcinogenic and Anticarcinogenic Food Components
Edited by Wanda Baer-Dubowska, Agnieszka Bartoszek and Danuta Malejka-Giganti

Toxins in Food
Edited by Waldemar M. Dąbrowski and Zdzisław E. Sikorski

Chemical and Functional Properties of Food Saccharides
Edited by Piotr Tomasik

Chemical and Functional Properties of Food Proteins
Edited by Zdzisław E. Sikorski

Fermentation
Effects on
Food Properties

EDITED BY

Bhavbhuti M. Mehta
Afaf Kamal-Eldin
Robert Z. Iwanski

CRC Press
Taylor & Francis Group
Boca Raton   London   New York

CRC Press is an imprint of the
Taylor & Francis Group, an **informa** business

CRC Press
Taylor & Francis Group
6000 Broken Sound Parkway NW, Suite 300
Boca Raton, FL 33487-2742

First issued in paperback 2016

Library of Congress Cataloging-in-Publication Data

Fermentation : effects on food properties / editors, Bhavbhuti M. Mehta, Afaf
    Kamal-Eldin, Robert Z. Iwanski.
        p. ; cm. -- (Chemical and functional properties of food components series)
    Includes bibliographical references and index.
    ISBN 978-1-4398-5334-4 (hardcover : alk. paper)
        I. Mehta, Bhavbhuti M. II. Kamal-Eldin, Afaf. III. Iwanski, Robert Z. IV. Series:
Chemical and functional properties of food components series.
        [DNLM: 1. Fermentation. 2. Food Microbiology. 3. Food Contamination--prevention
& control. 4. Food Preservation--methods. 5. Foods, Specialized. QW 85]

    664.001'57246--dc23   2012001491

Visit the Taylor & Francis Web site at
http://www.taylorandfrancis.com

and the CRC Press Web site at
http://www.crcpress.com

Contents

Preface

This book is composed of monographic chapters discussing the role of fermentation reactions in modifications of the contents, chemical, functional, and sensory properties, as well as biological activity of food components. The large variety of food products obtained by fermentation of different raw materials all over the world is presented. Special emphasis is placed on the effect of processing conditions on the enzymatic, fermentative reactions leading to the generation of various products and the development of desirable sensory, functional, and biological properties. Also, the role of different fermentation products and various nutritional habits in supplying the organism in many nutrients and other constituents that increase the human well-being is presented.

Unique, concise chapters are contributed by experts on food biochemistry and microbiology, food technology, and nutritionists having a sound background and personal experience in research and academic teaching. The information available in current world literature is critically evaluated and presented in a very concise and user-friendly form in one medium-sized book.

The text is based on the research and teaching experience of the authors. Moreover, they have also critically evaluated current literature, which they have cited in their chapters. The book is primarily addressed to food science graduate students, as well as to food technologists in the industry and in food quality control organizations who participate in continuing education systems, to the teaching staff specializing in food science and technology, researchers in food chemistry and technology, chemists, technical personnel in food processing plants, and nutritionists. Many topics are also interesting for students of chemistry and biology. Some sections of the book can be used by other educated readers interested in the quality of food, such as journalists and politicians interested in food, nutrition, and health issues. The book is useful as a concise, valuable source in any university course on food chemistry. Such a course is a must in teaching programs of food science and technology in any university in the world. However, the book is not written as a textbook for a specific course with a fixed number of teaching hours or credits. It is also useful as a resource for students of nutrition.

A very strict editorial control in all stages of preparation of the material has been exercised. In preparing the book, I have had the opportunity of working with a large group of colleagues from several universities and research institutions. Their ready acceptance of editorial suggestions and the timely preparation of high-quality manuscripts are sincerely appreciated. I am very glad to work with my co-editor, Prof. Afaf Kamal-Eldin, who has greatly helped in editing various chapters and providing guidance, and without whose kind support, it would have been very difficult to complete the book project. I also extend my sincere thanks to Prof. Robert S. Iwanski for his help. I cannot forget Prof. Zdzisław E. Sikorski, series editor of the Chemical and Functional Properties of Food Components Series who has provided me with the opportunity to work as an editor of the book. He helped me in preparation of the table contents and guided me when required. Words cannot express my sincere thanks to our all family members, friends, and colleagues who have helped and provided moral support to us during the entire project.

Bhavbhuti M. Mehta
Editor
bhavbhuti5@yahoo.co.in

Editor

Bhavbhuti M. Mehta is an assistant professor in the Dairy Chemistry Department, Sheth M.C. College of Dairy Science at Anand Agricultural University, Anand Gujarat, India. He earned his Bachelor of Technology. (dairy technology) and Masters of Science (dairying) from Sheth M.C. College of Dairy Science, Gujarat Agricultural University. Presently, he is pursuing his Ph.D. in the field of dairy chemistry. He teaches various subjects on dairy and food chemistry at the undergraduate as well as at the post graduate levels. His major specialty is various occurring physico-chemical changes during milk and milk product processing, and food chemistry in general. He is an associate editor of the *International Journal of Dairy Technology* and referee/reviewer of a number of journals. He has published 25 technical/research/review papers/chapters/booklets/abstracts/monographs in national as well as international journals, seminars, and conferences.

Associate Editors

Afaf Kamal-Eldin is a professor of food science at the United Arab Emirates University in Al-Ain, UAE. Her major specialty is chemistry, biochemistry and nutrition related to bioactive compounds and to food chemistry in general. She is a member of the editorial boards of a number of journals and has edited/co-edited four books published by the American Oil Chemists' Society Press. She has published about 150 original publications and 30 reviews and book chapters in addition to a large number of conference abstracts. Afaf is conducting research and teaching in the area of food for health and has supervised a large number of M.Sc. and Ph.D. theses.

Robert Z. Iwański is an assistant professor in the Department of Food Technology at the West Pomeranian University of Technology in Szczecin, Poland. His major research specialization is food microbiology, but he also holds a professional degree in bakery technology. His main research interest regards the effect of lactic acid bacteria and fermentation processes on the rheological properties of various species of bread baked from conventional and unconventional types of flour. Dr. Iwanski is the author of 15 original publications and has presented numerous conference papers. As an expert of the Polish bakery industry he prepared several opinions regarding industrial bakery processes. He teaches mainly on fermentation and cereal technology at the undergraduate and post graduate levels and participates as an instructor in courses organized for Polish farmers, supported by the European Union.

Contributors

Javier Carballo
Food Technology Department
University of Vigo
Ourense, Spain

Maricê Nogueira de Oliveira
Department of Biochemical and
    Pharmaceutical Technology
São Paulo University
São Paulo, Brazil

Nathalie Desmasures
Unité des Micro-organismes d'Intérêt
    Laitier et Alimentaire
Université de Caen Basse-Normandie
Caen, France

Izabela Dmytrów
Department of Dairy Technology and
    Food Storage
West Pomeranian University of
    Technology
Szczecin, Poland

Juana Fernández-López
IPOA Research Group, AgroFood
    Technology Department
Escuela Politécnica Superior de
    Orihuela
Miguel Hernández University
Orihuela-Alicante, Spain

Muhammad Imran
Department of Microbiology
Quaid-i-Azam
Islamabad, Pakistan

Robert Z. Iwański
Department of Food Technology
West Pomeranian University of
    Technology
Szczecin, Poland

Sam Jennings
Berry Ottaway & Associates Ltd.
Hereford, United Kingdom

Nilesh H. Joshi
Fisheries Research Station
Junagadh Agricultural University
Okha, India

Afaf Kamal-Eldin
Department of Food Science
United Arab Emirates University
Al-Ain, United Arab Emirates

Krzysztof Kryża
Laboratory of Food Storage
West Pomeranian University of
    Technology
Szczecin, Poland

Kazimierz Lachowicz
Meat Science Department
West Pomeranian University of
    Technology
Szczecin, Poland

Edyta Malinowska-Pańczyk
Department of Food Chemistry,
    Technology and Biotechnology
Gdansk University of Technology
Gdansk, Poland

Bhavbhuti M. Mehta
Dairy Chemistry Department
Sheth M.C. College of Dairy Science
Anand Agricultural University
Anand Gujarat, India

Peter Berry Ottaway
Berry Ottaway & Associates Ltd.
Hereford, United Kingdom

Jose A. Pérez-Alvarez
IPOA Research Group, AgroFood
    Technology Department
Escuela Politécnica Superior de
    Orihuela
Miguel Hernández University
Orihuela-Alicante, Spain

Zulema Coppes Petricorena
Department of Biochemistry
University of the Republic of Uruguay
Montevideo, Uruguay

Aly Savadogo
Département de Biochimie-Microbiologie
Universite de Ouagadougou
Ouagadougou, Burkina Faso

Maria E. Sayas-Barberá
IPOA Research Group,
    AgroFood Technology
    Department
Escuela Politécnica Superior de
    Orihuela
Miguel Hernández University
Orihuela-Alicante, Spain

Esther Sendra
IPOA Research Group, AgroFood
    Technology Department
Escuela Politécnica Superior de Orihuela
Miguel Hernández University
Orihuela-Alicante, Spain

Malgorzata Sobczak
Meat Science Department
West Pomeranian University of
    Technology
Szczecin, Poland

Robert Tylingo
Department of Food Chemistry
Gdansk University of Technology
Gdansk, Poland

Jean-Paul Vernoux
Unité des Micro-organismes d'Intérêt
    Laitier et Alimentaire
Université de Caen Basse-Normandie
Caen, France

Marek Wianecki
Department of Food Technology
West Pomeranian University of
    Technology
Szczecin, Poland

Joanna Żochowska-Kujawska
Meat Science Department
West Pomeranian University of
    Technology
Szczecin, Poland

1 Introduction

Bhavbhuti M. Mehta, Robert Z. Iwański, and Afaf Kamal-Eldin

Fermentation and drying are the oldest methods of food preservation and preparation. Unlike drying, fermentation gives food a variety of flavors, tastes, textures, sensory attributes, and nutritional and therapeutic values. The history of fermented foods is discussed by Prajapati and Nair (2003) and Hutkins (2006). The art of fermentation originated in the Middle East, the Indian subcontinent, and the Far East. As early as 4000 and 3000 BC, fermented bread and beer were known in Pharaonic Egypt and Babylonia. In the sacred book of the Hindus, Rigveda (ca. 1500 BC), it was mentioned that fermentation technology started to develop after observations of fermentative changes in fruits and juices (Upadhyay 1967; Prajapati and Nair 2003). Despite the long history of fermentations, the understanding of the sciences behind these arts came quite late and are not yet well achieved. In 1857, Louis Pasteur was the first to show that bacteria is involved in milk fermentation, a first step in elucidating the chemistry of fermentation. He described the process by the term *la vie sans air*, or life without air. Fermentation carried out without oxygen is an anaerobic process, and organisms that must live without air are called *obligate anaerobes*, while those that can live with or without air are called *facultative aerobes*. The understanding of the role of enzymes in fermentation reactions followed the experiments carried out in 1896 by the German chemists Hans and Eduard Buchner. In 1907, *Lactobacillus* was isolated from fermented milk by the Russian microbiologist Ellie Metchnikoff, and in 1930, Hans von Euler obtained the Nobel Prize for his work on the fermentation of sugars and fermentation enzymes.

The term *fermentation* is derived from the Latin word *fermentum*, meaning to boil. Prescott and Dunn (1957) defined fermentation in a broad sense as

a process in which chemical changes are brought about in an organic substrate, whether carbohydrate or protein or fat or some other type of organic material, through the action of biochemical catalysts known as "enzymes" elaborated by specific types of living microorganisms.

Campbell-Platt (1987) defined fermented foods as

those foods that have been subjected to the action of micro organisms or enzymes so that desirable biochemical changes cause significant modification in the food.

Although fermentation is actually an anaerobic process, it was extended to include aerobic processes and nonmicrobial processes such as those affected by

1

isolated enzymes. Thus, fermentation describes all of the processes via which complex organic foods are converted into simpler compounds with the production of chemical energy in the form of adenosine triphosphate (ATP). Upon glycolysis, the pyruvate that is produced is oxidized and generates additional ATP and NADH (the reduced form of NAD, nicotinamide adenine dinucleotide) in the tricarboxylic acid (TCA) cycle and by oxidative phosphorylation. This is known as aerobic respiration, which requires the presence of oxygen. On the other hand, oxygen is not required for the growth of facultative anaerobic organisms. In the absence of oxygen, NAD^+ is generated, and this is one of the fermentation pathways that occurs during lactic acid fermentation.

The fermentation process is divided into several categories, depending on the end products obtained. In the alcoholic fermentation process, enzymes produced by yeast degrade carbohydrate into ethanol and carbon dioxide (Reaction 1). Similarly, lactic acid is produced from carbohydrates in lactic acid fermentation (Reaction 2). In heterolactic fermentation, one molecule of glucose is converted into one molecule of lactic acid, one of ethanol, and one of carbon dioxide (Reaction 3).

$$C_6H_{12}O_6 \rightarrow 2C_2H_5OH + 2CO_2 \qquad \text{(Reaction 1)}$$

$$C_6H_{12}O_6 \rightarrow 2CH_3CHOHCOOH + 22.5 \text{ kcal} \qquad \text{(Reaction 2)}$$

$$C_6H_{12}O_6 \rightarrow CH_3CHOHCOOH + C_2H_5OH + CO_2 \qquad \text{(Reaction 3)}$$

Knowledge of microorganisms is essential to understand the process of fermentation, as it mainly involves the growth and various activities of microorganisms that produce a wide range of desirable and undesirable substances. Fermentation can be achieved by encouraging growth of the right microorganisms and discouraging the growth of the undesirable microorganisms that cause spoilage. The microorganisms (cultures) for fermentations are selected primarily on the basis of their ability to produce desirable products and to preserve or stabilize the food. Frazier (1958) has reported most of the bacterial cultures that are used in fermentation processes. These include bacteria (lactic acid cultures, propionic culture, acetic acid bacteria, cheese smear organisms, other bacterial cultures, etc.), yeasts (wine yeasts, baker's yeasts, yeasts for malt beverages, etc.), and molds (*Penicillium, Aspergillus*, etc.). The lactic acid bacteria are a heterogeneous bacterial group that includes species of the genera *Lactobacillus, Lactococcus, Leuconostoc, Streptococcus, Carnobacterium, Enterococcus, Oenococcus, Pediococcus, Tetragenococcus, Vagococcus, Weissella*, etc. The complex microbial community in fermented foods is described in Chapter 2 of this book.

Different fermented foods and the associated microorganisms, bacteria (lactic acid bacteria), and fungi are known for a wide diversity of matrices. For example, fermented cereal products include sorghum beer (*amgba/bili-bili/burukutu*), and fermented legume products include soybean sauce, tempeh, tofu, miso, nattō, *cheonggukjang, amriti, dhokla, dosa, idli, papad, wadi*, etc. The fermented vegetable products includes kimchi, sauerkraut, etc.; fermented dairy products include yogurt, *dahi* (curd), *shrikhand*, Bulgarian butter milk, acidophilus milk, kefir, koumiss, and varieties of cheese, etc.; fermented meat products include various types of dry sausages; and fermented fish is in the form of sauces, pastes, etc.

During fermentation, various chemical/biochemical changes take place. These changes primarily depend on the quality of raw materials, the various processing steps, the microorganisms used, and the type of products manufactured. Various compounds are formed from the food components (carbohydrates, proteins, and lipids) during fermentation. These compounds have an unequal contribution to the flavor and aroma of the fermented products, depending on the chemical structure and properties of the molecules. Compounds formed during fermentation include lactic acid, acetic acid, propionic acid, diacetyl, carbon dioxide, ethyl alcohol, exopolysaccharides, bacteriocins, etc., which affect the flavor, texture, and consistency of the product and inhibit spoilage and pathogenic microorganisms (Walstra, Wouters, and Geurts 2006). The generation of the flavor and aroma compounds in fermented foods is discussed in Chapter 3.

Whitaker (1978) describes changes in texture as one of the fundamental objectives of the fermentation process. For example, fermentative changes directly affect the basic measurable rheological properties like hardness, consistency, adhesiveness, viscosity, etc., in the products. Chapter 4 covers the rheological properties of such foods, and Chapter 5 thoroughly describes the role of fermentation processes in changing the color of foods. *Bioactive compounds* is a general term used to describe food constituents with extra nutritional value, usually in connection with reducing the effects of aging and degenerative diseases. The physiological effects of bioactive compounds can relate to signaling, cholesterol lowering, lipid modulation, immunity, hypotensivity, or other beneficial physiological effects. Fermentation induces different types of bioactivities caused by the hydrolysis and further metabolism of carbohydrates and proteins as well as the biosynthesis of vitamins and other compounds with potent bioactivities. All of these aspects are covered in Chapter 6.

Many foods in raw state contain toxins and antinutritional compounds that must be eliminated before consumption. The action of microorganisms during fermentation can remove or detoxify such compounds, and the details of these processes are discussed in Chapter 7. In addition, Chapter 8 covers the fortification of products derived from fermentation processes and various technical issues in the production and distribution of such foods.

Cereals (wheat, rye, rice, maize, barley, oats, sorghum, and millet) and legume (all types of beans and peas) are widely consumed throughout the world in different types of fermentations. While cereal grains are important sources of carbohydrates and energy, legumes are important sources of proteins. Fermentation is performed traditionally with different grades of grains and with mixed starter cultures, leading to variability in product quality (appearance, taste, flavor, nutritional value, and safety), and these issues are addressed in Chapter 9. Fermented vegetable products are high in nutritive value; are rich sources of vitamin C, dietary fiber, mineral salts, and antioxidants; and have a positive influence on human health. Chapter 10 looks into the details of fermenting cucumbers, cabbage (sauerkraut and kimchi), and olives. Chapter 11 thoroughly describes the fermentation of milk and milk products, as these products are highly nutritious, therapeutic, and healthy foods, as proven by Ayurveda, the old science of medicine. The various chemical changes that take place before and during fermentation are covered in this chapter. Chapter 12 discusses the fermented seafood products, especially fermented fish sauces and pastes. The various chemical changes that take place during fermentation of meat products is described in Chapter 13. For any food product, the safety of human beings is a must, and the production of safe foods is only achieved by following good manufacturing practices—the Hazard Analysis and Critical Control Points (HACCP) principles—which are discussed in Chapter 14.

Nowadays, one or more types of fermentation are practiced at home and by the different food manufacturers. Mastering today's art of fermentation processes requires detailed knowledge of food raw materials, microbiology, enzymology, chemistry/biochemistry, and physics/engineering/technology, which makes it a very challenging undertaking. This book has attempted to focus mainly on the various chemical changes that take place during processing, both pre- and post-fermentation, that ultimately affect food properties and the quality of the finished products. The microbiology parts are briefly mentioned here but are discussed more thoroughly in a number of recent books. For example, *Principles of Fermentation Technology*, 2nd edition (by P. F. Stanbury, S. Hall, and A. Whitaker, 1995, Elsevier Sci. Ltd., ISBN 0-7506-4501-6) focuses on fermentation technologies and bioprocess engineering and includes information on fermentation media, sterilization procedures, inocula, recombinant DNA techniques of industrial microorganisms, and fermenter design. Other useful books in this field include

Fermentation Microbiology and Biotechnology (by El-Mansi, 1999, Taylor & Francis Ltd., ISBN 0-7484-0734-0)

Fermentation and Food Safety (by Martin Adams, Rob Nout, and M. J. R. Nout, 2001, Springer US, ISBN 0834218437)

Wild Fermentation: The Flavor, Nutrition, and Craft of Live-Culture Foods (by Sandor E. Katz, 2003, Chelsea Green Publishing, ISBN 1-931498-23-7)

Food Fermentation (by Rob M. J. Nout, 2005, Wageningen Academic Publishers, ISBN 9076998833)

Fermentation: Vital or Chemical Process? (by Joseph S. Fruton, 2006, Brill Academic Publishers, ISBN 9004152687)

On Fermentation (by Paul Schutzenberger, 2008, Bibliolife, ISBN 055931597X)

Industrial Fermentation Food Microbiology and Metabolism (by Meenakshi Jindal, 2010, ISBN 9380013206)

There are also a number of older yet very valuable books to which the reader is kindly referred for complementary information.

REFERENCES

Campbell-Platt, G. 1987. *Fermented foods of the world: A dictionary and guide*. London: Butterworth.

Frazier, W. C. 1958. *Food microbiology*. New York: McGraw-Hill.

Hutkins, R. W. 2006. *Microbiology and technology of fermented foods*, 3–14. Ames, IA: Blackwell Publishing Professional.

Prajapati, J. B., and B. M. Nair. 2003. The history of fermented foods. In *Handbook of fermented functional foods*, ed. E. D. Farnworth, 1–25. Boca Raton, FL: CRC Press.

Prescott, S. C., and C. G. Dunn. 1957. *Industrial microbiology*. New York: McGraw-Hill.

Upadhyay, B. 1967. *Vedic sahitya aur sanskruti*, 3rd ed., 436–450. Varanasi, India: Sharda Mandir.

Walstra, P., J. T. M. Wouters, and T. J. Geurts. 2006. Lactic fermentation. In *Dairy science and technology*, 357–398. Boca Raton, FL: Taylor & Francis Group, LLC (CRC Press).

Whitaker, J. R. 1978. Biochemical changes occurring during the fermentation of high-protein foods. *Food Technology* 5: 175–180.

2 Complex Microbial Communities as Part of Fermented Food Ecosystems and Beneficial Properties

Muhammad Imran, Nathalie Desmasures, and Jean-Paul Vernoux

CONTENTS

2.1 INTRODUCTION

2.1.1 MICROBIAL CLASSIFICATION

Complex microbial ecosystems are composed of different microorganisms that can be classified based on the inferred evolutionary relationships among microorganisms, i.e., upon similarities in their genetic characteristics. Using such an approach and by comparing rRNA sequences, most microorganisms appear in the universal phylogenetic tree of life in three domains: Bacteria, Archaea, and Eukarya. Microbial classification can also be based on global similarity (morphological, physiological, ecological, molecular, metabolic, and genetic characteristics) between microorganisms. Classification of prokaryotic organisms starts from two domains (e.g., Bacteria and Archaea), which are then divided into phyla (29 and 5 for Bacteria and Archaea, respectively) (Euzéby 2011). Further subdivision is organized from the *class* rank to the *species* rank, sometimes to the *subspecies* rank (Figure 2.1). The smallest level of differentiation is the *strain* level, which could be more or less compared to the *individual* at the human scale. The main difference remains that, even at the strain level, microorganisms, as opposed to the individual, are always considered as populations of cells.

Classification of fungi (eukaryotic organisms) follows the same rules. Yeasts and molds are members of fungi distinguished according to specific morphology, respectively a predominant monocellular or pluricellular state; they are not true taxonomical groups. According to the rules of the International Committee on Taxonomy of Viruses, virus classification starts at the *order* level, but they are excluded from the phylogenetic tree of life.

Besides the scientific classification of microorganisms, a technological classification is often used, based on general microbial properties or activities of groups of microorganisms expressed along the food chain, from agricultural and fishery resources to the gastrointestinal tract of the consumer. Classification can be based—sometimes disregarding taxonomical proximity—on properties positively or negatively linked with food processing, on fermentation activity (e.g., lactic acid bacteria, butyric acid bacteria, acetic acid bacteria), or on the role during maturation

Domain → Phylum → Class → Subclass → Order → Suborder → Family → Subfamily →

Tribe → Subtribe →Genus → Species → Subspecies

FIGURE 2.1   Organization of microbial classification from largest to smallest ranks

of fermented food (e.g., ripening bacteria). It can also be based on temperature or salt tolerance/requirement for growth, e.g., psychrotrophic, thermophilic/thermoduric, or halophilic microorganisms. Microorganisms can also be classified as risk indicators regarding hygiene of the process or health of the consumer (e.g., coliform bacteria, anaerobic spore-forming bacteria, total plate counts or aerobic counts, and coagulase-negative and coagulase-positive staphylococci). None of these classifications is completely satisfying. Firstly, members of a given taxonomic rank are susceptible to fall within different technological categories and vice versa. Secondly, the technological classification is often true in a given food but not transposable to another. Thirdly, such grouping is done at the genus or species level, while many food-related properties are strain specific.

2.1.2 COMPLEX MICROBIAL ECOSYSTEMS

A complex microbial ecosystem is composed of a microbial community living in a matrix. It is defined as a multispecies assemblage in which microorganisms live together in a contiguous environment and interact with each other (Konopka 2009). A microbial community can be defined as a group of microorganisms that has different functions and activities distinguishing it from any other. From an ecological point of view, the study of a single isolated microorganism of one species is far from the reality of a microbial ecosystem. To approach the ecological reality, behavior of strains should be analyzed in the corresponding community. This point is critical to anticipate correctly the functionality of strains in a complex ecosystem.

Microbial communities have an impact on human well-being. Evolution, disease, corrosion, degradation, bioremediation, and global cycling are just a few of the many thousands of ways that microbial communities affect our lives. The microbial layers on the insides of household water pipes and in the rolling tanks of bioreactor grains, the microorganisms that extract nutrients from streams, soil microbial communities, ocean plankton microbes, gastrointestinal tract biofilms, oral-cavity microbial population, and food-related microbial communities are just a few examples. Complex microbial communities, such as biofilms in the oral cavity and gastrointestinal tract, play an important role in health and diseases (Potera 1999). Microbial communities in humans contain 100 times as many genes as the human genome (Versalovic and Relman 2006). Microbes in the form of biofilms are being linked to common human diseases ranging from tooth decay to prostatitis and kidney infections (Potera 1999).

Microbial communities are major constituents of fermented foods, and they contribute to the preservation of nutrients and to increasing the shelf life of food. Different fermented foods and the associated microorganisms, bacteria, and fungi are known for a wide diversity of matrices, e.g., milk, yogurt, cheese, beer, wine, sauerkraut, bread, cider, etc. (Table 2.1). The microbial communities related to fermented food have a direct effect on human life regarding safety and health effects. Food products as well as the processing surfaces are sites of multispecies biofilm communities that could include pathogenic microbes (Kumar and Anand 1998). The role of microbial communities in human health, industrial processes, and ecological functions is under discussion, and special attention is given to these microbial

TABLE 2.1

Fermentation and Fermented Products

Fermentation type	Microorganisms	Food products
Alcoholic	*Saccharomyces*	Beer, wine, cider, bread, naan, kefir
Lactic	Lactic acid bacteria	
Homolactic	*Lactococcus* and certain *Lactobacillus*	Fermented milks (*karmdinska*), cheeses, dry sausages
Heterolactic	*Leuconostoc* and certain *Lactobacillus*	Fermented milks (kefir), sauerkraut, green olives, bread
Malolactic	*Oenococcus*	Wine
Propionic	*Propionibacterium*	Hard cheese
Acetic	*Acetobacter aceti*	Vinegar

Source: Micheline Guéguen, personal communication.

ecosystems, which are evolving ecosystems. In fact, they modify progressively with time and in space due to changes in available metabolites and in the physicochemical parameters of their own *in situ* environment. To illustrate this statement, cheeses—especially smear- and mold surface-ripened cheeses—were chosen as examples of such evolutive complex microbial ecosystem and are described in Section 2.3.

2.2   MICROORGANISMS INVOLVED IN MANUFACTURING OF FERMENTED FOOD

Microorganisms involved in fermented food correspond mainly to bacteria and fungi (yeasts and molds) but also to viruses, including bacteriophages, which can have a strong negative impact on the fermentation process by destroying a specific strain. These bacteriophages could also contribute to microbial ecology and succession of lactic acid bacteria (LAB) species in vegetable fermentations (Lu et al. 2003). Among bacteria and fungi encountered in fermented food, most are chemoorganotrophic organisms, which can be either thermophilic, mesophilic, psychrotrophic, or psychrophilic. A nonexhaustive list of food-borne microorganisms in fermented food is shown in Table 2.2.

2.3   COMPOSITION OF COMPLEX MICROBIAL COMMUNITIES: EXAMPLES OF SMEAR- AND SOFT-CHEESE MICROFLORA

Cheese making began about 6000–9000 years ago, originating from the Middle East (Fox, Cogan, and McSweeney 2000), and now there are about 1,400 cheese varieties manufactured worldwide (Beresford et al. 2001). The primary objective of cheese manufacturing was to extend the shelf life and to conserve the nutritious components of milk. Manufacturing of most varieties of cheese involves the combination of four components/ingredients: milk, rennet, microorganisms, and salt. These are processed in the following steps during fermentation: acid production, gel

TABLE 2.2
Some Food-borne Bacteria and Fungi Reported in Fermented Food

Bacteria		Fungi
Acetobacter	*Macrococcus*	**Yeasts**
Acinetobacter	*Marinilactibacillus*	*Brettanomyces/Dekkera*
Aeromonas	*Marinomonas*	*Candida*
Agrococcus	*Microbacterium*	*Cryptococcus*
Alcaligenes	*Micrococcus*	*Geotrichum/Galactomyces*
Arthrobacter	*Morganella*	*Hanseniaspora*
Bacillus	*Mycetocola*	*Issatchenkia*
Brachybacterium	*Natrinema*	*Kluyveromyces*
Brevibacterium	*Paenibacillus*	*Metschnikowia*
Brevundimonas	*Pantoea*	*Pichia*
Brochothrix	*Pediococcus*	*Rhodotorula*
Carnobacterium	*Proteus*	*Saccharomyces*
Chryseobacterium	*Providencia*	*Schizosaccharomyces*
Citrobacter	*Pseudoalteromonas*	*Torulaspora*
Clostridium	*Pseudomonas*	*Trichosporon*
Corynebacterium	*Psychrobacter*	*Yarrowia*
Enterobacter	*Raoultella*	*Zygosaccharomyces*
Enterococcus	*Salmonella*	
Erwinia	*Serratia*	**Molds**
Escherichia	*Shewanella*	*Alternaria*
Flavobacterium	*Sphingobacterium*	*Aspergillus*
Hafnia	*Sphingomonas*	*Aureobasidium*
Halalkalibacillus	*Staphylococcus*	*Botrytis*
Halobacterium	*Streptococcus*	*Byssochlamys/Chrysosporium*
Halococcus	*Stenotrophomonas*	*Cladosporium*
Halomonas	*Tetragenococcus*	*Fusarium*
Halorubrum	*Tetrathiobacter*	*Monilia*
Klebsiella	*Vibrio*	*Mucor*
Kocuria	*Weissella*	*Penicillium*
Lactobacillus	*Yersinia*	*Rhizomucor*
Lactococcus		*Rhizopus*
Leucobacter		*Scopulariopsis/Trichothecium*
Leuconostoc		
Listeria		

formation, whey expulsion, and salt addition followed by a specific period of ripening (Figure 2.2). Fermented dairy food products like cheeses are examples of complex microbial communities, involving many strains of different species and genera grown together. Some of their roles are presented in Table 2.3.

Cheese microflora can be divided into two main functional groups. Lactic acid bacteria contribute to acid production, bringing on the curd-making and ripening microbiota. Ripening of smear- and mold-ripened cheeses starts with the growth of a large number of yeasts, which increase surface pH. As a result, a salt-tolerant, usually very

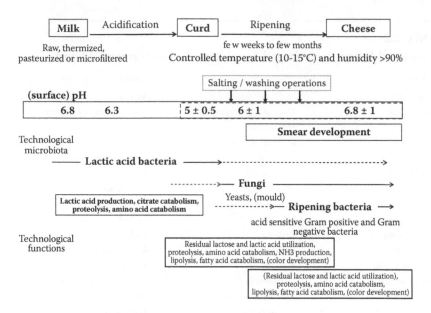

FIGURE 2.2   Schematic overview of microbial succession and functions of the different microbial groups involved during smear-ripened cheese making

complex and undefined bacterial consortium begins to develop and eventually covers the entire surface of the cheese, including coryneform bacteria, i.e., *Arthrobacter*, *Brevibacterium* (Cogan et al. 2011), and some gram-negatives (Larpin-Laborde et al. 2011), depending on cheese varieties and on dairies (Figure 2.2).

2.3.1   LACTIC ACID BACTERIA

Lactic acid bacteria are present in a variety of foods and participate in the development of texture, flavor, and safety quality of many fermented products, including cheeses. They have a common characteristic of lactic acid production from lactose. The starter bacteria encountered most often in cheese technology are members of the genera *Lactococcus*, *Lactobacillus*, *Streptococcus*, *Leuconostoc*, *Pediococcus*, and *Enterococcus* (Beresford et al. 2001). These microorganisms are gram-positive, catalase-negative, nonspore forming, microaerophilic or facultative anaerobic bacteria. The DNA of lactic acid bacteria has less than 55% G+C contents (Stiles and Holzapfel 1997). The use of starter cultures for cheese and sour milk production was introduced by Weigmann in 1890 (Stiles and Holzapfel 1997). The primary function of lactic acid bacteria is to produce acid during the fermentation process; however, they also contribute to cheese ripening in which their enzymes are involved in proteolysis and conversion of amino acids into flavor compounds (Fox and Wallace 1997). Lactic acid bacteria are either added deliberately at the beginning of cheese making, or they may be naturally present in raw milk and are called nonstarter lactic acid bacteria (NSLAB). Generally, starter bacteria can be distinguished into two groups: (a) mesophilic starters (i.e., *Lactococcus*, *Leuconostoc*) used for the cheese types

TABLE 2.3

Role in Cheese Making of Main Milk Microbial Groups

Microorganisms	Principal roles	Other possible roles	Examples
Lactic acid bacteria	Acidification of milk and curd; involvement in taste formation (proteolysis, aroma production) and in texture and opening of paste	Production of biogenic amines Antipathogenic activity (bacteriocins, lactic acid)	*Lactococcus, Leuconostoc, Pediococcus,* thermophilic *Streptococcus, Lactobacillus, Enterococcus*
Propionic acid bacteria	Participation in the formation of taste and opening of cooked pressed cheeses (Emmental, Comté, Gruyère) by propionic acid fermentation of lactate to give propionic and acetic acids and carbon dioxide	Antipathogenic activity (bacteriocins)	*Propionibacterium acidipropionici, Propionibacterium freudenreichii, Propionibacterium jensenii, Propionibacterium thoenii*
Surface bacteria	Constituents of the flora of surface-ripened cheeses; essential role in texture and color formation and in production of flavors (sulfur compounds) of washed rind, bloomy rind, or mixed cheeses (Munster, Camembert, Pont l'Evêque)	Antipathogenic activity (bacteriocins)	Nonpathogenic *Staphylococcus* (*S. equorum, S. xylosus, S. lentus*), coryneform bacteria (*Micrococcus, Brevibacterium, Arthrobacter*)
Yeasts	Curd deacidification at start of ripening, which allows subsequent implantation of acid-sensitive microorganisms such as coryneform bacteria; also involved in the formation of taste and texture of cheeses	*Geotrichum candidum* may be responsible for spoilage (slippery rind) in soft-cheese technology	*Kluyveromyces lactis, K. marxianus, Geotrichum candidum, Debaryomyces hansenii, Yarrowia lipolytica*
Molds	Ripening agents present at the surface (*P. camemberti* for soft cheeses, *Rhizomucor* for Tomme de Savoie and Saint Nectaire) or inside (*P. roqueforti* for blue-veined cheeses); role in the formation of sensory characteristics of cheeses; may cause spoilage: as *Rhizomucor* for "cat's fur" in soft cheese, characterized by a defect in cheese appearance, associated with a bad taste	Production of mycotoxins	*Chrysosporium sulfureum, Cladosporium herbarum, Penicillium camemberti, Penicillium roqueforti, Rhizomucor fuscus, Rhizomucor plumbeus, Scopulariopsis brevicaulis, Trichothecium domesticum* (formerly *Cylindrocarpon*)

(Continued)

TABLE 2.3 (CONTINUED)
Role in Cheese Making of Main Milk Microbial Groups

Microorganisms	Principal roles	Other possible roles	Examples
Butyric acid	Butyric acid fermentation produces butyric acid and hydrogen; consequences are defects of taste and openness ("late blowing"), which can grow in the cheese (cooked or uncooked pressed)		*Clostridium tyrobutyricum, Clostridium butyricum, Clostridium beijerinckii, Clostridium sporogenes*
Coliform bacteria (coliforms)	Some may be present in the digestive tract and feces; all these bacteria are grouped under the term *total coliforms*; some (thermotolerant coliforms) are often used with more or less success as indicators of hygienic conditions in production	Early swelling (soft cheese with a spongy appearance); production of aroma compounds of cheese interest	*Escherichia coli, Enterobacter cloacae, Hafnia alvei, Klebsiella oxytoca*
Psychrotrophic bacteria	Can grow at refrigeration temperature and produce thermoresistant lipases and proteases, causing off-flavors (rancidity, bitterness) in cheeses; some may be the cause of pigmentation defects or stickiness on the surface of cheese	Production of aroma compounds of cheese interest	Strains of *Acinetobacter, Bacillus, Flavobacterium, Pseudomonas putida, Ps. fluorescens, Xanthomonas*
Thermoresistant bacteria	Can survive pasteurization and cause deterioration after treatment; some can produce thermoresistant lipases and proteases; they include all sporulating bacteria and some particularly heat-resistant bacteria among nonsporulating bacteria		*Bacillus, Clostridium, Enterococcus,* strains of lactobacilli and coryneform bacteria

in which temperature of the curd is not raised more than 40°C during acidification and (b) thermophilic starters (*Lactobacillus helveticus*, *Lactobacillus delbrueckii*, *Streptococcus salivarius* ssp. *thermophilus*) mostly used for cheese types where curd temperature may rise above 40°C (Fleet 1999). Both mesophilic and thermophilic cultures can be subdivided into (a) mixed (undefined) cultures in which the number of strains is unknown and (b) defined cultures with known microbial composition.

Starters are normally added to milk at initial population of between 10^5 and 10^7 cfu/ml, but these bacteria develop rapidly and are concentrated by whey expulsion to attain a concentration of 10^9–10^{10} cfu/g in the curd after one day of inoculation. Regarding diversity of lactic acid bacteria in soft cheeses, several studies have been done on the Camembert of Normandy, France. *Lactococcus* sp. is the dominant microflora, with about 10^9 cfu/g during ripening with predominance of *Lactococcus lactis* (Richard 1984; Desmasures, Bazin, and Guéguen 1997). *Lactobacillus* is the second dominant group of lactic acid bacteria found in Camembert, with up to 3 ´ 10^7 cfu/g during ripening. *Lactobacillus paracasei* and *L. plantarum* were the two species more frequently found in Camembert cheese (Henri-Dubernet, Desmasures, and Guéguen 2004).

2.3.2 Ripening Microflora

2.3.2.1 Yeasts and Molds

Fungi contribute to the organoleptic quality of cheese mainly by the phenomena of proteolysis and lipolysis and by consumption of lactic acid and production of alkaline metabolites, such as ammonia, and also a few yeast species by fermentation of lactose (Addis et al. 2001). Their role in surface deacidification is an important phenomenon that allows further growth of acid-sensitive bacteria (Lenoir et al. 1985; Guéguen and Schmidt 1992).

Yeasts have ubiquitous characteristics and can be found in different ecological habitats. Their acid-tolerant characteristics justify their presence in soft cheeses, where low pH, moisture content, and temperature and high salinity favor their growth, with their numbers on the surface rapidly reaching 10^5–10^8 cfu/g (Fleet 1999; Larpin et al. 2006). Yeasts are found in a wide variety of cheeses, but diversity is particularly high in those made from raw milk. They are capable of degradation of different organic substances, and their role in curd deacidification and in formation of metabolites such as ethanol, acetaldehyde, and CO_2 is beneficial. Nearly 1,500 yeast species from about 100 genera are documented (some are described in Barnett, Payne, and Yarrow 2000); among them, about 50 species have been described in ripened cheeses. The yeast genera frequently isolated from different cheese types include: *Candida*, *Debaryomyces*, *Geotrichum*, *Kluyveromyces*, *Pichia*, *Rhodotorula*, *Saccharomyces*, *Trichosporon*, *Torulaspora*, *Yarrowia*, and *Zygosaccharomyces* spp. (Beresford and Williams 2004). Investigations on microbial diversity of the surface of Livarot, Limburger, and Muenster cheeses have shown that *G. candidum*, *D. hansenii*, *Kluyveromyces lactis*, and *K. marxianus* are the yeast species more frequently present and added as fungal starter. *Geotrichum candidum* and *D. hansenii* develop at the start of ripening on the surface of a number of soft cheeses, including Camembert, Pont l'Evêque, Tilsit, Limburger, Reblochon, and Livarot (Guéguen and Schmidt 1992; Eliskases-Lechner and Ginzinger 1995b; Larpin et al. 2006; Goerges et al. 2008). Some species of *Candida*, like *C. natalensis*,

C. catenulata, C. intermedia, C. anglica, C. deformans, and *C. parapsilosis* are also found on the surface of either Livarot, Reblochon, or gubbeen cheeses (Mounier et al. 2009; Cogan et al. 2011; Larpin-Laborde et al. 2011).

Some molds are also found as technological agents in several cheese varieties, including *Penicillium camemberti, Penicillium roqueforti, Trichothecium domesticum* (Lenoir, Lamberet, and Schmidt 1983; Choisy et al. 1997), but generally these are not used in smear-cheese production. In other cheese varieties, *P. roqueforti* or other molds like *P. expansum, P. janthinellum, P. viridicatum,* and *Rhizomucor* spp. can affect the cheese characteristics by acting as spoilage agents (Bockelmann et al. 1999). Fungi are transferred to cheese generally from fabrication premises (air, soil, walls, humans; raw milk, water brine solution) (Baroiller and Schmidt 1990; Viljoen, Khoury, and Hattingh 2003; Mounier et al. 2005, 2006a).

2.3.2.2 Surface Bacterial Flora

Technological nonlactic acid bacteria have been found on the surface of different cheeses. This flora is usually aerobic, mesophilic, and halotolerant but is acid sensitive. The main groups are coryneform bacteria and Staphylococcaceae (Maoz, Mayr, and Scherer 2003; Mounier et al. 2005; Goerges et al. 2008; Larpin et al. Forthcoming), and these two groups represent 61% to 71% of bacterial isolates (Mounier et al. 2008). Gram-negative bacteria can also develop on the surface of smear-ripened cheeses (Bockelmann et al. 1997; Feurer et al. 2004a, 2004b; Bockelmann et al. 2005; Rea et al. 2007; Larpin et al. Forthcoming). For example, it has been shown that Livarot surface flora consisted of about 34% gram-negative isolates (Larpin et al. Forthcoming).

2.3.2.2.1 Gram-Positive Bacteria

2.3.2.2.1.1 *Coryneform bacteria*   The term *coryneform* is dedicated to bacteria whose characteristic feature is their tendency to arrange themselves in a V-like pattern or lined up much like logs stacked one against the other. These are gram positive, nonmobile, and mostly aerobic. They are a branch of Actinobacteria (G+C contents > 50%). In the last 15 years, these have been deeply studied and are grouped into the suborders Micrococcineae and Corynebacterineae, composed of nine and six families, respectively (Stackebrandt, Rainey, and Ward-Rainey 1997). The coryneform genera found in cheese more frequently include *Arthrobacter, Brevibacterium, Brachybacterium, Corynebacterium, Microbacterium,* and *Micrococcus.*

Arthrobacter is a dominant genus on the surface of certain cheeses like Tilsit, Ardrahan, Durrus, and Milleen. Many *Arthrobacter* species, including *A. nicotianae, A. citreus, A. globiformis, A. variabilis,* and *A. mysorense,* have been isolated from cheese (Eliskases-Lechner and Ginzinger 1995a; Feurer et al. 2004b; Mounier et al. 2005). *Arthrobacter arilaitensis* and *A. bergerei* have been described on the surface of French cheeses (Feurer et al. 2004b; Irlinger et al. 2005; Larpin et al. 2006; Rea et al. 2007). These bacteria, especially *A. arilaitensis,* can come from commercial ripening culture and the environment (Goerges et al. 2008). The source of these bacteria is not yet clearly understood, but *A. arilaitensis* has been isolated from raw milk (Mallet, Guéguen, and Desmasures 2010).

Brevibacterium is the unique genus of the family Brevibacteriaceae. The most frequently described species in cheese was *B. linens* (Rattray and Fox 1999). It has been isolated from the surface of various cheeses including Gruyère (Kollöffel, Meile, and Teuber 1999) and Gubbeen (Brennan et al. 2002). In 2002, by using molecular techniques, a large diversity was observed in the *B. linens* species (Alves et al. 2002). By using DNA/DNA hybridization, *B. linens* was divided into the three new species: *B. aurantiacum*, *B. antiquum*, and *B. permense* (Gavrish et al. 2004). The reference strain *B. linens* ATCC9175, which was frequently found in cheese, was very similar to *B. aurantiacum*. So, a large number of strains isolated from cheese identified as *B. linens* would be now reclassified in *B. aurantiacum*.

The *Brachybacterium* genus contains 12 species, including *Br. nesterenkovii*, *Br. faecium*, *Br. alimentarium*, and *Br. Tyrofermentans*, that have been isolated from many cheeses such as Salers (Duthoit, Godon, and Montel 2003), Livarot (Larpin et al. Forthcoming), Gruyère (Schubert et al. 1996), Beaufort (Ogier et al. 2004), and Saint Nectaire (Delbès, Ali-Mandjee, and Montel 2007).

Corynebacterium is the dominant genus on the surface of many cheeses. Five species have been isolated from smear-ripened cheeses. *Corynebacterium variabile*, *C. casei*, *C. flavescens*, and *C. ammoniagenes* represented about 32% of all coryneform isolated from Brick (Valdès-Stauber, Scherer, and Seiler 1997). In Gubbeen, *C. casei* represented about 50% of isolates (Brennan et al. 2002), and *C. mooreparkense*—subsequently identified by Gelsomino et al. (2005) as a later synonym of *C. variabile*—was also described. Most *Corynebacterium* isolates were obtained after three weeks of ripening (Larpin et al. Forthcoming; Rea et al. 2007).

The genus *Microbacterium* is found in very low number on the surface of smear-ripened cheeses (Eliskases-Lechner and Ginzinger 1995a; Valdès-Stauber, Scherer, and Seiler 1997). In Gubbeen, 12% coryneform bacteria were identified as new species, one being *Microbacterium gubbeenense* (Brennan et al. 2001, 2002). Later, this species was again found in Gubbeen (Mounier et al. 2005) and Domiati (El-Baradei, Delacroix-Buchet, and Ogier 2007) cheeses.

Micrococcus is a genus of the Micrococcaceae family, phylogenetically very close to *Arthrobacter* but with identical morphology to *Staphylococcus*. The only species of this genus detected in cheese was *Micrococcus luteus* (Bockelmann et al. 1997; Mounier et al. 2005). Various other coryneform bacteria have been described recently on the surface of smear cheeses, such as *Leucobacter* spp. (Larpin et al. Forthcoming), *Mycetocola reblochoni* (Bora et al. 2008), and *Agrococcus casei* (Bora et al. 2007).

2.3.2.2.1.2 Staphylococcaceae Among the five genera included in the Staphylococcaceae family, two have been associated with cheese: *Macrococcus* and mainly *Staphylococcus*, while *Jeotgalicoccus* and *Salinicoccus* were detected in raw milk (Callon et al. 2007; Mallet, Guéguen, and Desmasures 2010). For a long time, *Staphylococcus* has remained associated to *Micrococcus* from a taxonomical point of view. Since 1997, a new classification of *Actinobacteria* has been proposed in order to redefine the family Micrococcaceae, so *Staphylococcus* was classified separately (Stackebrandt, Rainey, and Ward-Rainey 1997). This genus is a subbranch of *Clostridium-Bacillus*, which consists of gram-positive bacteria having G+C < 50%.

Some strains produce coagulase and/or enterotoxins and are potentially pathogenic or opportunistic pathogens (like *S. aureus*), contrary to the coagulase-negative species found in the cheese ecosystem: *Staphylococcus equorum, S. vitulinus*, and *S. xylosus*, which are the main species found in smear cheese (Irlinger et al. 1997; Hoppe-Seyler et al. 2004). In Gubbeen cheese, the *Staphylococcus* strains represented about 2.5% of total bacterial isolates (Brennan et al. 2002), while in Livarot about 9.5% of bacterial isolates were *Staphylococcus* spp. (Larpin et al. Forthcoming).

2.3.2.2.2 Gram-Negative Bacteria

While literature data is abundant regarding the presence of yeasts/molds or gram-positive bacteria in cheese, gram-negative bacteria (GNB) have been studied rarely. However, a wide diversity of GNB can be found at relatively high population levels in raw milk (Desmasures, Bazin, and Guéguen 1997; Lafarge et al. 2004) and in various cheeses including smear cheeses (Maoz et al. 2003; Larpin et al. Forthcoming). GNB usually represent from 18% to 60% of the bacteria isolated from the surface of European smear cheeses (Maoz et al. 2003; Mounier et al. 2005; Larpin et al. Forthcoming). GNB present on the surface of ripened soft cheese belong mainly to the Moraxellaceae, Pseudomonadaceae, and Enterobacteriaceae families (Tornadijo et al. 1993; Maoz, Mayr, and Scherer 2003; Bockelmann et al. 2005; Mounier et al. 2005). Recently, a study of the GNB associated with French milk and cheeses indicated the existence of a large biodiversity of at least 26 different genera, represented by 68 species, including potential new species identified among the 173 studied isolates (Coton et al. in press). *Pseudomonas, Chryseobacterium, Enterobacter*, and *Stenotrophomonas* were the genera most frequently found in cheese core and milk samples, while *Proteus, Psychrobacter, Halomonas*, and *Serratia* were the most frequent genera among surface samples. *Alcaligenes* sp., *Hafnia alvei*, *Marinomonas* sp., *Raoultella planticola*, and *Ewingella americana* were also described on the surface of Livarot cheese (Larpin et al. Forthcoming).

Until now, the presence of gram-negative bacteria, and particularly coliform bacteria, in food was considered as an indicator of bad handling, which can spoil the product. There is now evidence of their positive contribution to cheese organoleptic qualities, as demonstrated for *Proteus vulgaris* (Deetae et al. 2007) or by the use of *H. alvei* as commercial ripening culture (Alonso-Calleja et al. 2002). *Pseudomonas* spp. is also able to produce a variety of volatile compounds that may contribute positively to the sensory qualities of cheese (Morales, Fernandez-Garcia, and Nunez 2005a), including sulfur compounds as demonstrated for *P. putida* (Jay, Loessner, and Golden 2005).

2.3.3 PATHOGENIC MICROFLORA IN CHEESE

Cheeses are currently considered to be safe foods for consumers, as they have been implicated in only 1.8% of verified food-borne outbreaks due to zoonotic agents in the EU in 2008 (EFSA 2010). Historically there have been outbreaks of diseases associated with the consumption of cheeses, and the predominant organisms responsible have included *Staphylococcus aureus* and zoonotic bacteria such as *Salmonella*, *Listeria monocytogenes*, and verocytotoxin-producing *Escherichia coli* (VTEC)

(Zottola and Smith 1991; De Buyser et al. 2001; Little et al. 2008; EFSA 2010). Nearly 2,500 serovars have been identified inside the *Salmonella* genus. Among them, most (about 2,000) belong to the subspecies *Salmonella enterica* ssp. *enterica*. The two main ubiquitous serovars involved in food-borne outbreaks worldwide are *Salmonella enteritidis* and *S. typhimurium*, which are responsible for gastroenteritis.

Staphylococcus aureus is the bacteria most frequently associated with food-borne outbreaks from dairy products, and 85.5% of these outbreaks between 1992 and 1997 were due to *S. aureus*, according to De Buyser et al. (2001). Enterotoxin production by some strains can be responsible for diarrhea and sometimes causes vomiting. About 20 different thermostable enterotoxins have been described (Hennekinne et al. 2003).

While raw milk was reported as an important food vehicle in food-borne *Campylobacter* outbreaks in 2007 (EFSA 2009), cheeses are rarely associated with campylobacteriosis. *Listeria monocytogenes* is the causative agent of listeriosis. It is a rare pathology in Europe; its incidence in 2008 was 0.3 per 100,000 compared to campylobacteriosis at 40.7 per 100,000 (EFSA 2010). However, the pathogenicity of *L. monocytogenes* (with around 20%–30% fatalities) is an important public health concern.

Food ecosystems like cheese are composed of biotic and abiotic components. These components are determining factors in the growth of *L. monocytogenes* in cheese. The presence of *L. monocytogenes* in a food environment is favored by its ability to grow at refrigeration temperatures (2°C to 4°C) with a survival range of 0°C–45°C and a tolerance to pH values as low as 4.5 and to high sodium chloride concentration (up to 10%). Thus, it is very difficult to control *L. monocytogenes* in a cheese environment during ripening (Farber and Peterkin 1991).

During ripening of bacterial surface-ripened cheeses (red smear cheeses), the increase in pH on the surface creates a favorable environment for the growth of microorganisms, including contaminants such as *L. monocytogenes*. The rind of these cheeses is usually considered edible; the accidental presence of *L. monocytogenes* on surface-ripened cheeses can pose a potential health risk for certain consumers. Because some outbreaks occurred where cheese was found to be the source, several investigations have been conducted for testing the occurrence of *L. monocytogenes* in different smear-ripened cheeses (Terplan et al. 1986; Beckers, Soentoro, and Delgou-van Asch 1987; Breer and Schopfer 1988; Pini and Gilbert 1988; Eppert et al. 1995; Loncarevic, Danielsson-Tham, and Tham 1995; Loncarevic et al. 1998; Rudolf and Scherer 2001).

While *Listeria* was the emerging food-borne pathogen of the 1980s and has gained public attention, recent investigations on the hygienic status of red smear cheese are rarely available. It is therefore unknown whether the lack of outbreaks in recent years is related to the improved hygienic status of these cheeses or due to other factors. The presence of *L. monocytogenes* in soft and semisoft cheese made from raw or low-heat-treated cow's milk was detected in three out of seven qualitative investigations. For those investigations with positive findings, the proportions of positive samples ranged from 0.5% to 3.6%. Findings of levels above 100 cfu/g, which is the upper limit in cheese according the European regulation (EC 2073/2005), were not reported. For the batch-based sampling at retail, about 2.8% noncompliance was reported for soft and semisoft cheese (EFSA 2009). It is generally considered on the basis of investigations that *L. monocytogenes* was found more frequently in high-moisture than in low-moisture cheese (Ryser 1999). Incidence of *Listeria monocytogenes* in soft cheeses was found to be surprisingly higher in those made from pasteurized milk (8%) than from raw milk (4.8%) (Rudolf and Scherer 2001). Comparable data have been recently obtained by the European Food Safety Agency for European cheeses (EFSA 2009) with, respectively, 4.2%–5.2% for pasteurized and 0.3%–0.4% for raw milk cheeses, and this tendency was confirmed recently (EFSA 2010). Low incidence of *L. monocytogenes* in raw milk cheeses may be due to various factors. One of them is the strong monitoring of raw milk quality encountered in raw milk processing dairies. Another one is the presence of some natural factors in raw milk and thereafter in raw milk cheeses, which might be destroyed during pasteurization (Gay and Amgar 2005), e.g., the lactoperoxidase system and raw milk microbial communities.

2.4 IMPORTANT MICROBIAL METABOLIC PATHWAYS IN CHEESE RIPENING

Many studies have been done on the physicochemical parameters of smear-ripened cheeses (Lenoir et al. 1985; Choisy et al. 1997; Fox and Wallace 1997; Leclercq-Perlat, Corrieu, and Spinnler 2004a). Many biochemical reactions coexist or occur successively in cheese, the major components of the curd being lactose, lactate, proteins, fat, and their derivatives. The summary of all processes and their succession is explained in Figure 2.2, and the main metabolites are shown in Figure 2.3.

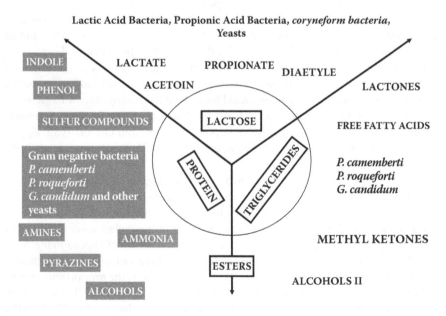

FIGURE 2.3   Main roles of microorganisms in the degradation of curd constituents (according to Micheline Guéguen, personal communication)

2.4.1 DEGRADATION OF LACTOSE AND LACTIC ACID

The lactose is transformed into D- and, principally, L-lactate by lactic acid bacteria during fabrication of curd, and part of the lactose is eliminated with the whey. Some yeasts also have the capacity of degrading lactose at the start of the ripening process (Leclercq-Perlat et al. 1999, 2000; Corsetti, Rossi, and Gobbetti 2001) by the action of a β–galactosidase that has been identified in *Kluyveromyces lactis*, *K. marxianus*, and *Debaryomyces hansenii* (Roostita and Fleet 1996; Fleet 1999). Lactic acid in cheese is usually found in the form of lactate, which is the principal carbon source for yeasts and molds. The degradation of lactate is achieved on the cheese surface during ripening, mainly by *G. candidum* and/or *D. hansenii* up to negligible concentration (<0.02%) at the end of ripening (Choisy et al. 1997; Gripon 1993; McSweeney and Sousa 2000). The mechanism of lactate degradation in yeasts has received little study. The presence of lactate oxidase enzymes in *G. candidum* that degrade the lactate in pyruvate and hydrogen peroxide in the presence of oxygen has been reported (Sztajer et al. 1996). Another lactate-degrading enzyme, NAD-dependent lactate dehydrogenase, was also reported in *G. candidum* (Hang and Woodams 1992). *Geotrichum candidum* and other yeasts degrade lactate and liberate ammonia; these two phenomena contribute to pH increase, which promotes the implantation of acid-sensitive microorganisms like ripening bacteria (Lenoir 1984; Choisy et al. 1997).

Lactose and lactate concentration and pH were estimated in five European smear-ripened cheeses along the ripening period (Cogan et al. 2011). Lactose was not detected in any cheeses except in the curd of one batch of Reblochon, which contained a low amount (3 g/kg dry matter). The lowest increase of pH levels during ripening was in Gubbeen (from 5.0 to 5.7) and the highest was in Livarot and Tilsit, which reached final pH values of 7.8 and 7.5, respectively. Except in Tilsit, the tendency was toward complete utilization of the lactate, which was below 2 g/kg dry matter in Limburger cheese from the seventh ripening day.

2.4.2 PROTEOLYSIS IN CHEESE RIPENING

The degradation of proteins is the major event in cheese ripening. This degradation modifies the textural aspects of cheese and promotes flavor development. By rupture of protein interlinks, the curd is transformed into a soft mass that contains a number of peptides and amino acids that are precursors of a number of aromatic compounds (Table 2.2). This is executed by a number of microbial enzymes and plasmine (milk enzyme) (Choisy et al. 1997; Rattray and Fox 1999; McSweeney and Sousa 2000). An intracellular enzyme was found in *G. candidum*, whose activity was very high in an exponential growth phase (Hannan and Guéguen 1985). Another study proved the extracellular proteolytic activity of *G. candidum* (Guéguen and Lenoir 1975). *Geotrichum candidum* is able to contribute to primary proteolysis by degrading α_{s1}- and βa^2-caseins (Auberger et al. 1997; Boutrou, Kerriou, and Gassi 2006). It also possesses aminopeptidase and carboxypeptidase activities to degrade small peptides.

Other microorganisms present in smear cheese also exhibit proteolytic activity, like *K. lactis*, *D. hansenii*, and *B. linens* (Frings, Holtz, and Kunz 1993; Rattray and Fox 1999; Klein, Zourari, and Lortal 2002). *Arthrobacter nicotianae* play a role in the proteolytic process by possessing two extracellular enzymes that can hydrolyze the α- and β-caseins (Smacchi, Fox, and Gobbetti 1999). *Corynebacterium* spp. (particularly *C. casei*) and *Brachybacterium* spp. express aminopeptidase activity (Curtin, Gobbetti, and McSweeney 2002). Some *Micrococcus* spp. and *Staphylococcus* spp. isolated from soft cheese had proteolytic activity (Addis et al. 2001). Gram-negative bacteria found in cheese as *Pseudomonas* spp. (*P. fluorescens, P. aeruginosa, P. chlororaphis, P. putida*) had a strain-dependent proteolytic activity (Sablé et al. 1997). Various genera of Enterobacteriaceae including *Hafnia, Serratia, Enterobacter*, and *Escherichia* of dairy origin also had strain-dependent proteolytic activity (Sablé et al. 1997; Morales, Fernández-García, and Nuñez 2003).

Analyzing five European smear-ripened cheeses, Cogan et al. 2011 demonstrated that all cheeses appeared to be different, proteolysis at the cheese surface being highest in Gubbeen and Livarot and lowest in Reblochon. Some differences were observed between the amino soluble nitrogen (ASN) and the nonprotein nitrogen (NPN). For example, in Tilsit the ASN was relatively low and the NPN relatively high. Both of these indicators of proteolysis were low in Reblochon and high in Gubbeen. These differences may have implications for flavor development.

2.4.3 LIPOLYTIC ACTIVITY IN CHEESE RIPENING

Triacylglycerols represent about 98% of fat content in curd. The hydrolytic action of microbial lipases generates free fatty acids and di- or monoglycerides and seldom glycerol. In smear cheeses, the degradation of triglycerides is usually carried out by yeasts, e.g., *Y. lipolytica* (Pereira-Meirelles, Rocha-Leão, and Sant'Anna 2000) and *G. candidum* (Choisy et al. 1997), but also by ripening bacteria including *B. linens* and *S. equorum* (Rattray and Fox 1999; Curtin, Gobbetti, and McSweeney 2002). *Corynebacterium* spp. (especially *C. casei*) and *Brachybacterium* spp. possess esterase activity (Curtin, Gobbetti, and McSweeney 2002). *Micrococcus* spp. and *Staphylococcus* spp. isolated from soft cheese have lipolytic and esterase activities (Sablé et al. 1997; Rattray and Fox 1999; Addis et al. 2001; Curtin, Gobbetti, and McSweeney 2002). The degradation of fatty acids produces methyl ketones by β-oxidation. The methyl ketones then might be reduced by action of reductases. Esterification of short- to medium-chain fatty acids could also be possible by combination with alcohols or thiols to give esters and thioesters. This might be a detoxification reaction that allows the elimination of large quantities of alcohol and fatty acids. The production of esters in Camembert cheese is associated with yeasts like *G. candidum* (Molimard and Spinnler 1996) as well as the production of ketones by this pathway (Jollivet et al. 1994).

2.4.4 PRODUCTION OF AROMATIC COMPOUNDS

Yeasts contribute to the flavor of smear-ripened cheeses by production of a large variety of volatile compounds (Molimard and Spinnler 1996), either directly or indirectly, by proteolytic and lipolytic activities (Lenoir 1984; Choisy et al. 1997; Leclercq-Perlat, Corrieu, and Spinnler 2004a). For example *D. hansenii, K. lactis*, and *K. marxianus* strains produce numbers of esters (Leclercq-Perlat, Corrieu, and Spinnler 2004a), e.g., ethyl acetate. This production is correlated with growth of these yeasts and their ability to metabolize ethanol and lactose (Leclercq-Perlat, Corrieu, and Spinnler 2004a, 2004b). The amino acids are also precursors of a number of aromatic compounds. Catabolism of branched-chain amino acids produces alcohols and aldehydes. Aromatic amino acids, degradation of which is initiated by a transaminase, are also precursors of various volatile aromatic compounds.

Aromatic-amino-acid transferases have been described in various LAB and coryneform bacteria (Deetae 2009). *In vitro* studies proved that *B. linens, Brevibacterium* spp., and *Microbacterium* spp. are capable of producing a variety of aromatic compounds, but this ability is strain dependent (Jollivet et al. 1994; Bonnarme et al. 2001). A phenylalanine dehydrogenase has been identified in *G. candidum* (Hemme et al. 1982). Catabolism of sulfur-containing amino acids, mainly methionine, leads (under the action of lyase or transaminase) to the production of various sulfur compounds that are strongly involved in the organoleptic properties of smear cheeses. *Geotrichum candidum* has been shown to produce flavor sulfides (Demarigny et al. 2000).

Various coryneform bacteria like *C. casei* and *M. luteus* produce sulfur volatile compounds but in low quantity (Brennan et al. 2002). This was demonstrated with *C. casei, C. mooreparkense*, and *M. gubbenense*, which were able to produce methanethiol from L-methionine (Brennan et al. 2002). Enterobacteriaceae strains of dairy origin, as *Proteus vulgaris*, were shown to be able to produce high amounts of volatile sulfur compounds in red smear cheese (Deetae 2009). Other gram-negative bacteria (e.g., *Pseudomonas putida, Psychrobacter celer*) are known for producing in food (Jay, Loessner, and Golden 2005; Morales, Fernandez-Garcia, and Nunez 2005b), including cheese (Deetae 2009), volatile sulfur compounds such as dimethyl disulfide, which significantly contribute to the aroma of cheese (Kagkli et al. 2006). Such activities may have a positive effect during cheese ripening.

2.5   FUNCTIONAL PROPERTIES OF MICROORGANISMS

2.5.1   Interactions in Dairy Microbial Communities
and against *Listeria monocytogenes*

Microbial interactions in communities occur either (a) directly via physical contact and/or signaling molecules and include predation and parasitism or (b) indirectly by changing environmental conditions, and include commensalism, protocooperation and mutualism, and competition and amensalism (Viljoen 2001; Sieuwerts et al. 2008). In complex microbial ecosystems, all these types of interactions can coexist. The functionality of a microbial community is determined by the number, types, and biochemical properties of the microorganisms, which mostly depend on the ecosystem in which they exist (Young et al. 2008).

2.5.1.1   Positive Interactions between Microorganisms
Among positive interactions, commensalism involves partners of different species, one being benefited without any effect on the second. Regarding relationships implying benefit for both partners, mutualism is characterized by the necessity for the microorganisms implicated to interact, while protocooperation is a facultative relation. In dairy products, the most widely reviewed positive interaction (protocooperation) is

between *Streptococcus salivarius* ssp. *thermophilus* and *L. delbrueckii* ssp. *bulgaricus* (Robinson, Tamime, and Wszolek 2002). Regarding cheese, it was demonstrated that *D. hansenii* and *Y. lipolytica*, when used as adjunct starter in cheddar cheese, interact positively with lactic acid bacteria, resulting in the acceleration of ripening and flavor production (A. Ferreira and Viljoen 2003). Positive growth interactions between *G. candidum* and other yeast species like *K. lactis*, *K. marxianus*, *D. hansenii*, *S. cerevisiae*, and *Zygosaccharomyces rouxii* have been observed (Guéguen and Schmidt 1992). *Debaryomyces hansenii* and *G. candidum*, by consumption of the lactate produced by lactic acid bacteria in curd, and secondly by secretion metabolites obtained by degradation of amino acids, play an important role in the deacidification of cheese, which gives acid-sensitive bacteria the chance to grow (Mounier et al. 2008).

2.5.1.2   Negative Interactions between Microorganisms

Amensalism describes a unidirectional process in which one microorganism has a negative impact on a second, by the production of a specific compound (antibiotic, bacteriocin). Competition takes place in a microbial community when various microorganisms need the same resource (space, specific nutrient). These two kinds of interaction relate to the general concept of antagonism, which have been extensively studied in food ecosystems because it can be applied as a natural biocontrol strategy (biopreservation) to enhance food quality and safety (Fleet 1999). Biopreservation involves multiple mechanisms, including colonization/space competition, synthesis of antimicrobials, production of lytic enzymes, detoxification of toxins, and degradation of virulence factors (Compant et al. 2005) as well as exploitation of bacteriophages/bacteria interactions (e.g., phage biosanitation and phage biocontrol) (Garcia et al. 2010).

Industrial application of antagonistic activity of the food microflora has resulted in the production of protective cultures (PCs), which are used because of production of antagonistic metabolites such as bacteriocin, antimicrobial peptides and enzymes, and low-molecular-weight nonproteinaceous compounds including organic acids and fatty acids (Holzapfel, Geisen, Schillinger 1995). Some examples of the use of PCs are shown in Table 2.4. Nevertheless, *in situ* application of metabolites produced by many food-related microflora against pathogens proved to be less effective than *in vitro*, possibly due to degradation, cross reactions, or higher concentration requirements in food systems (Dieuleveux and Guéguen 1998).

Likewise, the effective application of bacteriocins or bacteriocin-producing strains is scarce, as long-term effectiveness during food processing and storage has not yet been achieved because of their instability and reduced efficacy in complex food matrices (Chen and Hoover 2003; Gálvez et al. 2007). Moreover, the use of these single antimicrobial factors can lead to the development of resistance in the target pathogens, especially for bacteriocins (Vadyvaloo et al. 2004; Nilsson et al. 2005; Peschel and Sahl 2006). This fact makes the development of PCs a challenging job (Grattepanche et al. 2008). An effective microbiological safety insurance in food, especially with milk and cheese, has not yet fully been achieved, as 2% of cheeses made from raw, thermized, and pasteurized milk have a higher level of *Escherichia coli*, *Staphylococcus aureus*, and/or *Listeria monocytogenes* than authorized by European Commission regulations (Little et al. 2008).

TABLE 2.4
Target of Protective Cultures Used in Some Food

Protective cultures	Inoculum (cfu/g)	Spoiling or pathogenic agent	Food products
Pediococcus cerevisiae[a]	10^9	*Pseudomonas* sp.	Raw poultry
Lactococcus lactis ssp. *Lactis*	5.10^7	*E. coli* O157:H7	Raw chicken meat
Enterococcus faecium	10^4, 10^7	*Listeria monocytogenes*	Vacuum-packaged fish
Lactobacillus plantarum	10^8	*Salmonella typhimurium*	Ready-to-eat vegetables
	10^7	*Yersinia enterocolitica*	Shrimp extracts
Carnobacterium divergens (formerly *C. piscicola*)	5.10^6, 10^8	H_2S-producing bacteria, *Listeria monocytogenes*	Smoked salmon

Source: Rodgers S., *Trends in Food Science & Technology,* 19, 2008.
[a] (*reclassified as Lactobacillus dextrinicus*), Haakensen et al. 2009.

An interesting fact, cited previously, was the higher incidence of *L. monocytogenes* found in soft and semisoft cheeses manufactured from pasteurized milk than in raw milk cheeses (Rudolf and Scherer 2001). This finding opened new fields of investigation. The presence of diversified microbial communities in raw milk could be one of the keys to explaining the differences observed between raw and pasteurized milk cheeses. Indeed, as many as 150 microbial species—including lactic acid bacteria; coryneform bacteria; α, β, and γ proteobacteria; yeasts; and molds—have been described in raw milk from cows (Mallet, Guéguen, and Desmasures 2010) and goats (Callon et al. 2007). While *L. monocytogenes* growth was shown on smear cheeses when defined ripening cultures containing *Debaryomyces hansenii, Geotrichum candidum,* and *Brevibacterium linens* were used (Eppert et al. 1997; Loessner et al. 2003), some complex microbial consortia developed on the cheese surface and inhibited listerial growth.

Antilisterial activity of complex microbial communities of cheese has been the focus of several studies in the past 15 years (Eppert et al. 1997; Maoz, Mayr, and Scherer 2003; Mayr et al. 2004; Saubusse et al. 2007; Imran, Desmasures, and Vernoux 2010a, 2010b; Monnet et al. 2010; Retureau et al. 2010; Roth et al. 2010). Despite all these studies, no inhibitory substance has been identified in isolated strains, and inhibition does not correlate to pH or acid content in cheese. In some studies, antilisterial activity was attributed to the production of antibiotics or bacteriocins by isolated strains (Ryser et al. 1994; Carnio, Eppert, and Scherer 1999). However, full restoration of inhibition was not achieved by bacteriocin-producing strains reinoculated on artificially contaminated cheeses (Eppert et al. 1997).

The study of diversity–function relationships into complex microbial communities is a challenge for research (Waide et al. 1999). Diversity–function relationships in microorganisms have not been well studied yet, possibly because of conceptual and methodological difficulties, since microbial communities usually consist of many microbial species (Torsvik and Øvreås 2002; Gans, Wolinsky, and Dunbar 2005). This level of microbial diversity could result in functional redundancy (different species performing the same functional role in ecosystems so that changes in species diversity does not affect ecosystem functioning), so function could have a greater relationship with the environment than microbial diversity (Boyle et al. 2006). Methodologically, it is not easy to prove the relation between diversity and functionality. Since diversity and other factors vary simultaneously, it is difficult to study the impact of the transfer of the microbial communities from one condition to another. Complex microbial diversity is the source of an abundance of genes and functionalities. The relationship between functionality and species diversity can be influenced by disturbance (Huston 1994; Morin 2000) or niche specialization (Kassen et al. 2000). Evolution of microbial assemblages also influences the productivity–diversity pattern (Fukami and Morin 2003).

Antilisterial activity was observed as a stable function of complex microbial consortia when they were analyzed at intervals of six months in two independent experiments (Maoz, Mayr, and Scherer 2003), and antilisterial activity was linked with microbial diversity. In the more inhibitory community, *C. variabile* constituted 80% to 83% of the microbial isolates. An increase in the population of *B. linens* decreased the antilisterial activity significantly in the second microbial community (Maoz, Mayr, and Scherer 2003). Some inhibitory cheeses were characterized by the presence of *Enterococcus faecium* and *E. saccharominimus*, *Corynebacterium* spp. and *C. flavescens*, *Lactococcus garvieae*, and *L. lactis*, but no substance was identified (Saubusse et al. 2007). In another study, the antilisterial activity of inhibitory consortia was also attributed to the presence of high level of lactic acid bacteria at the beginning of the ripening period. Moreover, it was concluded that inhibitory consortia were more diverse than noninhibitory ones and contained *Marinilactibacillus psychrotolerans*, *Carnobacterium mobile*, *Arthrobacter nicotianae* or *A. arilaitensis*, *A. ardleyensis* or *A. bergerei*, and *Brachybacterium* sp. (Retureau et al. 2010).

Roth et al. (2010) have identified and studied the antilisterial properties of two complex microbial communities. Both communities had similar sequential development of eight dominant microbial species, i.e., *Lactococcus lactis*, *Staphylococcus equorum*, *Alkalibacterium kapii*, *Corynebacterium casei*, *Brevibacterium linens*, *Corynebacterium variabile*, *Microbacterium gubbeenense*, and *Agrococcus casei*. It was concluded that listerial inhibition was a result of the competition for nutrients between *Listeria* spp. and marine LAB, because growth of *M. psychrotolerans* and *Al. kapii* occurred simultaneously with a decrease of the listerial count for both communities. In another study by Monnet et al. (2010), it was demonstrated that succession of microbial populations results in an increase in antilisterial activity. After four propagations, *Listeria* counts decreased by four logs. In noninhibitory consortia, three yeasts were dominant: *D. hansenii*, *Y. lipolytica*, and *G. candidum*, with 10 groups of bacteria with dominance of *Corynebacterium*, *Leucobacter*, and *Brachybacterium*. In inhibitory cheese, all three yeasts were found, but the dominant one was *Y. lipolytica*, while *G. candidum* was codominant in the initial consortium, but not after propagation. Bacteria of the *Vagococcus–Carnobacterium–Enterococcus* group became dominant from 3.3% to 78.6% by eliminating other bacterial genera found in cheese. The propagations of complex microbial consortia can bias the dominance of specific flora on a specific model cheese, and in this study, the microbial population had totally changed.

More recently, minimal model microbial communities showing antilisterial activity have been constructed from an inhibitory complex undefined smear. Representative members of the inhibitory surface consortia were identified, and then minimal microbial communities with similar antilisterial activity as that of the initial smear were prepared. The communities were quite simple so that interaction could be easily tested. Two minimal microbial communities with different microbial composition were constructed. Community A 1.1.1.1 consisted of *G. candidum, Candida natalensis, Serratia liquefaciens, Paenibacillus* sp., *A. arilaitensis,* and *Marinomonas;* community E 1.1.1.1 consisted of *G. candidum, D. hansenii, A. arilaitensis, Marinomonas* sp., *Pseudomonas putida,* and *Staphylococcus equorum.* The antilisterial activity of these two minimal microbial communities was not significantly different from that of the complex inhibitory cheese smear over the course of ripening. There was no correlation between pH and *L. monocytogenes* growth (Imran, Desmasures, and Vernoux 2010b). It has also been observed that substantial changes in the flora composition do not necessarily lead to the loss of antilisterial activity of protective cultures. In another work, it was indicated that nonculturable species have no role in antilisterial activity (Carnio, Eppert, and Scherer 1999).

From all of the previous discussion regarding the composition of inhibitory communities, it can be inferred that microbial inhibition does not depend strictly on the microbial composition, but it may be specific for a particular ecological habitat, as for each of the cases discussed here. According to studies of Retureau et al. (2010) and Roth et al. (2010), it can be established that if two communities with different composition, originating from the same cheese, show similar antilisterial activity, it is definitely the microbial balance that plays an important role in antilisterial activity. Stable microbial composition is not mandatory for stable antilisterial activity of a red smear cheese consortium (Mayr et al. 2004; Monnet et al. 2010), but ecological conditions can interfere in expression of activity. An understanding of this balance is crucial for the safety of cheese.

While studying the antilisterial activity capacity of three microbial consortia from cheese surfaces at different ripening temperatures, it was found that two communities (one dominated by *C. casei, B. linens,* and *M. gubbeenense* and a second dominated by *A. nicotianae* and *C. casei*) totally inhibited *L. monocytogenes* at ripening temperatures of 12°C, 13°C, and 16°C. But for a third consortium dominated by *A. nicotianae* and *B. linens,* inhibition was observed only at 13°C (Mayr et al. 2004). It can be inferred from this study that inhibition potential depends not only on the microbial composition of consortia, but also on abiotic factors like temperature. Temperature is a major factor in microbial growth and enzyme activity, and therefore it greatly influences the course of cheese ripening (Weber and Ramet 1986). Besides direct impact on growth rates, it also influences the microbes indirectly, together with relative humidity, and it controls the migration of substrates and growth factors from the inner cheese to the rind (Leclercq-Perlat et al. 2000).

2.5.1.3 Modeling of Microbial Interactions

A preliminary and first attempt to model microbial behavior in food was done a century ago to quantify the heat reduction in food matrices (Beigelow 1921). Interest in predictive modeling was focused in the late 1980s with the first reports of major food-borne outbreaks. Nowadays, predictive modeling has become an important tool in conducting risk assessments (McMeekin et al. 1997). Interaction between one pathogen of interest and background microflora can be studied by different approaches proposed in the literature. Predictive microbiology has made remarkable contributions to food safety risk assessment and risk management (McMeekin et al. 1997). The growth of microorganisms is difficult to model through polynomial equations while considering the role of the food matrix and preservative factors. Microbial modeling systems aim to simplify the ecosystem so that interaction between components could be easily understood (Jessup et al. 2004). Predictive models have been proposed to describe the growth rate and growth boundaries of microorganisms (Augustin et al. 2005). Usually the microbial models are based on monospecific cultures under static environmental condition, while in a food ecosystem the community dynamics are highly complex, reflecting nonsteady-state environmental conditions and microorganisms having different physiological status (McMeekin et al. 1997; Powell et al. 2004). The growth models are valuable tools in assessment of risks and HACCP studies (Hazard Analysis Critical Control Points), e.g., the probability of infection as a function of the pathogen level in a food product (Buchanan and Whiting 1996).

Competitive inhibition of food-borne pathogens was demonstrated for the first time for *Salmonella*, where suppression of growth of all microorganisms occurred when the growth matrix was exhausted (Jameson 1962). When two intestinal organisms, which do not interact with each other, are grown in coculture, the first follows a similar growth pattern to individual culture and remains unaware of its competitor until the population density of one or the other has risen to a level near molar concentration, and then both organisms end rapid multiplication. This effect has also been reported for *L. monocytogenes* (Buchanan and Bagi 1999; Nilsson, Gram, and Huss 1999). The minority population decelerates when the majority or the total population count reaches its maximum (Devlieghere et al. 2001). This phenomenon was called the *Jameson effect* 28 years after its first description (Ross, Dalgaard, and Tienungoon 2000).

The simple predictive growth models, where mixed cultures and/or microbial interaction in foods are only based on the growth curves of the involved species, have realistic application under *in situ* conditions (Augustin et al. 2011). Buchanan and Bagi (1999) showed that the coculture growth of *L. monocytogenes* and *P. fluorescens* varied depending on temperature, acidity, and water activity. This indicates that, considering the complexity of the microbial communities, dynamic modeling would require detailed knowledge of the food, the microbial composition, inoculation levels, etc. Prediction of microbial community dynamics in food ecosystems would be more realistic if it were based on multispecies growth trails. Practically, an experimental program that includes all possible combinations of abiotic and biotic environmental factors is near to impossible because of constraints on time and resources. Moreover, there are practical limits for enumerating pathogens and competitors using current classical or molecular techniques (Powell, Schlosser, and Ebel 2004). Microbial-interaction models usually focus only on reduction of maximum population density, without any consideration on the lag time or growth rate (Buchanan and Bagi 1999).

The Lotka-Volterra competition model provides a basic model for population growth of two interacting species (Brown and Rothery 1993). This growth model is an extended application of a basic logistic model for population growth of a single species limited by the maximum carrying capacity of a particular habitat. The monospecific growth model describes a limited population growth rate that decreases linearly with population density due to intraspecific competition. This is extended to cover the competition between species by incorporating an additional reduction in the population growth rate that is proportional to the population density of other species (Powell, Schlosser, and Ebel 2004). This model was further modified by adding a physiological phase representation before the onset of the exponential growth phase (Dens, Vereecken, and Van Impe 1999). This model has been used for yeast-yeast and yeast-bacteria interaction during cheese ripening (Mounier et al. 2008). The growth of *L. monocytogenes* was wrongly predicted, as microbial interactions were not taken into account by growth boundary models (Augustin et al. 2005). In food products, usually a heterogeneous background of microfloral and microbial interactions is involved that has not been included in most available models. Moreover, most classical models only include intraspecies interaction but not interspecies interaction, as extensive data sets for mixed culture growth are not available (Vereecken, Dens, and Van Impe 2000).

2.5.2 Mechanisms of Inhibition

2.5.2.1 Bacteriocins

A large number of bacteriocins have been reported as active against food-borne pathogenic bacteria such as *Listeria monocytogenes* (Table 2.5). A *B. linens* strain isolated from cheese has been reported to produce a bacteriocin-like compound, linocin M18, active against *Listeria* (Valdès-Stauber and Scherer 1994). Furthermore, two other substances—linecin A and linenscin OC2—have been characterized from two strains of *B. linens* (Kato et al. 1991; Boucabeille et al. 1997). Some *Staphylococcus* spp. isolated from a cheese surface were able to produce bacteriocin-like substances (Gori, Mortensen, and Jespersen 2010). A multibacteriocin-producing strain of *E. faecium* successfully inhibited the growth of *L. monocytogenes* in a cheese ecosystem, but recovery of active bacteriocins from the cheese matrix was not achieved. It was concluded that supplementary use of bacteriocin-producing *E. faecium* could be an option to control *L. monocytogenes* in production lines (Izquierdo et al. 2009). However, bacteriocin-resistant mutants often occur, which makes use of bacteriocin questionable as a primary means of food preservation (Loessner et al. 2003). Nevertheless, there was an added value, in that this strain, *E. faecium* WHE18, could produce several bacteriocins with different structures (Ennahar et al. 2001) having synergistic activity, as shown between enterocin A and enterocin B (Casaus et al. 1997). Moreover, it was shown that survivors of exposure to one of these two bacteriocins do not acquire resistance to the other bacteriocin (Casaus et al. 1997). Thus this could be used as a multibacteriocin hurdle approach (Izquierdo et al. 2009). Nevertheless, resistant mutants could occur in real cheese ripening conditions and compete with the ripening consortia. Indeed, if a resistant mutant is produced at the early stage of ripening, in later stages it could be able to establish in cheese more easily, as in case of "old young smearing." For example, *L. monocytogenes* WSLC 1364R (a pediocin-resistant mutant) at the 29th day was found up to 10^8 cfu/cm^2 in laboratory experiments (Loessner et al. 2003). But practically, in a real cheese production line, establishment of resistant mutants is a rare event, as only a very low concentration of *Listeria* is usually found in the brine. Moreover, it has been reported that the fitness cost of pediocin resistance of *Listeria* can reduce

TABLE 2.5

Examples of Bacteriocins Active against
Listeria monocytogenes

Bacteriocin	References
Acidocin A	(Canzek Majhenic et al. 2003)
Acidocin B	
Cerein 8A	(Bizani et al. 2008)
Curvacin A	(Remiger et al. 1996)
Cytolysin	(De Vuyst et al. 2003)
Divergicin M35	(Tahiri et al. 2004)
Enterocin 31	(De Vuyst et al. 2003)
Enterocin 416K1	(Iseppi et al. 2008)
Enterocin A	(De Vuyst et al. 2003)
Enterocin AS48	
Enterocin B	
Enterocin L50A	
Enterocin L50B	
Enterocin P	
Enterocin 1071A, B	(Balla et al. 2000)
Gassericin A	(Kawai et al. 1998)
Helveticin J	(Trmcic et al. 2008)
Lacticin 3147	(O'Sullivan et al. 2006)
Lacticin 481	(Rodríguez et al. 2000)
Lacticin RM	(Yarmus et al. 2000)
Lactococcin 972	(Martinez et al. 1999)
Lactococcin A	(Rodríguez et al. 2000)
Lactococcin B	(Van Belkum et al. 1992)
Lactococcin G	(Moll et al. 1996)
Linencin A	(Kato et al. 1991)
Linenscin OC2	(Boucabeille et al. 1997)
Linocin M18	(Valdès-Stauber and Scherer 1994)
Nisin NC8	(Rodríguez et al. 2000)
Piscicolin	(Jasniewski et al. 2009)
Plantaricin A	(Maldonado et al. 2004)
Plantaricin S	(Remiger et al. 1996)
Sakacin P	(Remiger et al. 1996)

the maximum growth rate by 44% (Gravesen et al. 2002a). However, the possibility that this may happen cannot be excluded, and resistance to class II bacteriocins is a potential obstacle for their application in food (Gravesen et al. 2002b).

Another important consideration is the *in situ* effectiveness of bacteriocins. For example, linocin M18 produced by *B. linens* has a molecular mass of 31 kDa, which might reduce its diffusion from the surface to the cheese core (Gori et al. 2010). As bacteriocins are secreted in large aggregates (Valdès-Stauber and Scherer 1994), which reduces their ability to diffuse in the matrix, their access to the cheese core is questionable (Gori, Mortensen, and Jespersen 2010). During production of Camembert cheese, a 3.3-log reduction in *Listeria* growth during the second week of ripening has been observed, due to the addition of a nisin-producing strain of *Lactococcus lactis*. But there was an increase in *Listeria* growth in the following weeks, principally on the surface of the cheese (Maisnier-Patin et al. 1992). Inhibition of *Listeria* by piscicolin 126 in milk and Camembert cheese was also observed, but resistance and inactivation of the bacteriocin in cheese, by proteolyic enzymes, occurred during ripening (Wan et al. 1997).

2.5.2.2 Metabolites

Microbial metabolism leads to the production of many compounds. Some of them could be involved in communication or defense (Schulz and Dickschat 2007). Millet et al. (2006) have established that a low concentration of *L. monocytogenes* in cheese was associated with comparatively high contents of D-lactate. But a low correlation was found between L-lactate, D-lactate, or lactose contents and listerial growth. The nondissociated lactic acid could be involved in inhibition, but its concentration was below 8 mM, and alone, it could not explain inhibition of *L. monocytogenes* growth in the presence of microbial consortia. Another inhibitory metabolite is D-3-phenyllactic acid, discovered by Dieuleveux and Guéguen (1998), and produced by *G. candidum*. However, the effectiveness of these metabolites in a cheese matrix is questioned because it depends on several factors, especially pH, that vary with ripening (Dieuleveux and Guéguen 1998). Natural fatty acids, including medium-chain saturated fatty acid (mainly caprylic, capric, and lauric acids) and long-chain unsaturated acids (linoleic and linolenic acids), either in milk or in the conditioning solution utilized for storage of Mozzarella cheese, could inhibit both *L. monocytogenes* growth and its invasion in the intestinal cell line (Petrone et al. 1998).

On the cheese surface, limited substrates are available to support the growth of microorganisms. Several bacterial species develop in favorable pH conditions and have the ability to use the lactate and casamino acids for growth (Mounier et al. 2007). While *Listeria* spp. can only use a limited range of carbon source for growth as glucose, glycerol, fructose, and mannose, no growth occurs on lactate or casamino acids (Pine et al. 1989; Premaratne, Lin, and Johnson 1991; Tsai and Hodgson 2003; Lungu, Ricke, and Johnson 2009). With the exception of a few strains, *Listeria monocytogenes* cannot ferment lactose and galactose (Mira-Gutierrez, Perez de Lara, and Rodriguez-Iglesias 1990). It was also demonstrated that *L. monocytogenes* could use alternative carbon sources like N-acetylglucosamine and N-acetylmuramic acid, which are major components of bacterial cell walls (Premaratne, Lin, and Johnson 1991; Barreteau et al. 2008). Mannan glycopeptides with mannose from yeast cell walls can also be metabolized by *Listeria* spp. (Sentandreu and Northcote 1969). This can also stimulate the growth of *L. monocytogenes* in cheese smear. Lactic acid bacteria can metabolize the lactose and galactose into acetate by the Leloir or the T-tagatose pathways (Bertelsen, Andersen, and Tvede 2001) and can also metabolize lactate into acetate and formate, which could be inhibitory to *L. monocytogenes* (Östling and Lindgren 1993; Liu 2003; Saubusse et al. 2007). However, the lactate is consumed by yeasts like *G. candidum*, *D. hansenii* (Leclercq-Perlat et al. 1999; Brennan et al. 2002; Bonaiti et al. 2004), and *Y. lipolytica* (Mansour, Beckerich, and Bonnarme 2008). Moreover, it was demonstrated that acetate could act as an inducer factor for the production of class IIa bacteriocin in *Carnobacterium divergens* (formerly *C. piscicola*) (Nilsson et al. 2002).

Some synergistic effect of multiple inhibitory factors has been reported. Addition of a bacteriocin-producing *Enterococcus* strain to the ripening culture led to a decrease of *L. monocytogenes* growth up to 3 logs (Giraffa, Neviani, and Tarelli 1994). A nonbacteriocinogenic food-grade strain of *S. equorum* producing micrococcin P_1 was able to reduce the listerial growth significantly in an *in situ* cheese-ripening experiment under laboratory conditions. Indeed, a phenomenon of complete inhibition occurred when this strain was used as the sole starter at an early stage of ripening and at low concentration (100 cfu/ml of brine). However, it was not effective when added with other surface-ripening cultures under laboratory conditions, even though *S. equorum* was established dominantly on the surface (Carnio et al. 2000). The biocontrol phenomenon on soft cheese by protective cultures is more complicated than the addition of a single antilisterial strain to a well-established smear-cheese flora (Carnio et al. 2000).

2.5.2.3   Competition for Nutrition and Space

Bacteria have developed mechanisms to communicate and compete with each other for limited environmental resources. This interbacterial communication can be mediated by soluble secreted factors and direct cell-cell contact. This enables the microbes to assess their own population density and the presence of competitors (Slechta and Mulvey 2006). Cross talk and quorum-sensing sorts of interactions between different strains in a food might be involved in production of bacteriocin or other inhibitory substances (Fontaine et al. 2007). An in-depth understanding of the cell-to-cell signaling mechanism of pathogens like *L. monocytogenes* can be used to control the growth in food systems by identifying the compounds that can act as quorum-sensing antagonists (Smith, Fratamico, and Novak 2004). The signal transduction and sensing can be intraspecific or interspecific. There are many examples that bacteria respond to chemical substances produced by other organisms, but no conclusive example is available for communication systems that have specifically evolved for interactions between species (Keller and Surette 2006). It is possible that, for complex microbial communities, a stable association between species does not depend on chemical signals. An example for this interaction is the association between *S. cerevisiae* and *P. putida*, commonly found on grapes. *Pseudomonas putida* can grow but cannot survive because of acid production in monoculture, but in coculture they both survive, as the yeast alters the environment by deacidification (Romano and Kolter 2005). This kind of interaction does not depend on chemical production but, rather, only on alteration in an environment.

The activation of sigma stress factor (σ^B) leads to an increase in stress resistance in order to protect the cell against adverse conditions. So σ^B could serve as an indicator to assess the resistance of strains to preservation methodology. Its knowledge could help in designing processes that could not lead to stress, allowing a better subsequent control of pathogenic or spoilage bacteria.

Inhibition of *Listeria* by nonbacteriocin-producing bacteria in broth medium has been reported (Nilsson, Gram, and Huss 1999; Cornu, Kalmokoff, and Flandrois

2002). Very little work has been done to investigate inhibition mechanisms that involve no apparent inhibitory substance. It has been demonstrated that a contact-dependent communication pathway in *Escherichia coli* can inhibit the growth of other microbes within mixed microbial populations (Aoki et al. 2005). Some studies have demonstrated that lactic acid can inhibit listerial growth without production of bacteriocins in food products (Buchanan and Bagi 1997; Nilsson, Gram, and Huss 1999). A plasmid-cured derivative of *Carnobacterium piscicola* A9b (nonbacteriocin-producing mutant) that did not produce any known extracellular antimicrobial compound showed antilisterial activity, demonstrating the presence of nonbacteriocin-dependent inhibition. In this case, *L. monocytogenes* entered into a stationary phase earlier in cocultures with *C. piscicola* as compared with monoculture, because of nutrition competition (Buchanan and Bagi 1997; Nilsson et al. 2004). In another study, it was demonstrated that the inhibition of *L. monocytogenes* by *C. divergens* (formerly *C. piscicola*) could be related to a combination of glucose deficiency and toxicity of acetate and possibly additional unidentified factors (Nilsson et al. 2005).

Regarding microbial competition for space, a study was focused on the early death of *Kluyeromyces thermotolerans* and *Torulaspora delbrueckii* in mixed culture with *Saccharomyces cerevisiae*. The inhibition was not explained by nutrition depletion or the presence of an antimicrobial agent, but it seemed to be linked with a cell-to-cell contact mechanism at high cell density of *S. cerevisiae*, and the comparative lesser ability of *K. thermotolerans* and *T. delbrueckii* to compete for space.

Gene-transcription and protein-expression studies can help to figure out the mechanisms of survival under adverse conditions. Stress protein expression in *L. monocytogenes* under conditions of low temperature and high osmotic environment was revealed by 2-D gel electrophoresis (Duché, Trémoulet, and Namane 2002; Liu et al. 2002). Membrane proteins were underrepresented in 2-D gel electrophoresis, so the inhibitory mechanism was not explained by this methodology (Ramnath et al. 2003). To evaluate the inhibitory mechanism described previously, *L. monocytogenes* was challenged by the fermentate of *C. piscicola* (nonbacteriocin-producing strain), which increased its lag phase and reduced its growth rate. However, the protein profile of the treated *L. monocytogenes* culture was not different from the control culture (Nilsson et al. 2005).

Use of whole-DNA microarray can be helpful in evaluating the inhibitory mechanism (Hong et al. 2003; Utaida et al. 2003; Hansen et al. 2004). For *L. monocytogenes* exposed to *Carnobacterium*, array data supported that there is competition for the glucose. The increase in transcription of the propanediol (*pdu*) and ethanolamine (*eut*) utilization regulon indicates that *Listeria* was searching for an alternative carbohydrate source. Transcription of the vitamin B_{12} biosynthesis (*cbi*) regulon was also increased. The decrease in expression of the pyrimidine biosynthetic genes, on the other hand, could indicate that *C. piscicola* (renamed as *C. divergens*) does not deplete pyrimidine precursors as efficiently as *L. monocytogenes* (Nilsson et al. 2005). In a microarray experiment conducted with antilisterial complex microbial consortia, Hain et al. (2007) found that the genes involved in the energy supply were up regulated after 4 hours of contact between *Listeria* and the consortium, which indicates that *Listeria* enters a state of starvation. The competition for nutrition is unlikely to cause development of resistance and could represent an effective supplement to biopreservation with other approaches (Nilsson et al. 2005).

2.5.3 BIOACTIVE PEPTIDES

Among fermented foods, dairy products are a major source of bioactive peptides (Fiat et al. 1993). During fermentation, active peptides are produced and released in fermented milks and cheeses. This peptides production is mediated by the action of proteolytic enzymes during cheese fermentation and ripening (I. Ferreira et al. 2006). Bioactive peptides derived from caseins putatively play a role in the nervous, cardiovascular, digestive, and immune systems (Silva and Malcata 2005). The k-casein fragment f103-111 (LSFMAIPPK) has been found to be involved in antithrombotic (prevent blood clotting) activity (Fiat et al. 1993). Blood pressure regulation effects of bioactive peptides are also found in cheese (Addeo et al. 1994; Fitzgerald and Meisel 2000). Bioactive peptides can be used as dietary supplements in functional foods. Regarding their safe use, more attention should be given to stability and pharmacological activities.

2.5.4 POLYSACCHARIDES

2.5.4.1 Exopolysaccharides (EPSs)

Lactic acid bacteria produce polysaccharides as cell-wall components or exocellular polymers. These exocellular polymers can form a cohesive layer or capsule covalently linked to the cell surface, or they can form a slime layer attached noncovalently to the cell surface, or they can be secreted to the environment (Ruas-Madiedo, Salazar, and de los Reyes-Gavilan 2009). In the dairy industry, the slime-forming LAB strains have traditionally been used in the production of fermented milk products, e.g., yogurts. It has been generally acknowledged that the EPSs secreted by LAB play an important role in the rheological behavior and texture of the products (De Vuyst and Degeest 1999). These biopolymers could also act as protective agents against a changing environment or as a bacteriophage attachment and could play a role in biofilm development. EPSs can exert beneficial effects on human health via prebiotic properties (claims are for serum cholesterol lowering, modulation of immunity, and anticarcinogenic activity) (Welman 2009).

2.5.4.2 Homopolysaccharides

Homopolysaccharides are a group of polysaccharides composed of one monosaccharide type. Several species of LAB are able to utilize sucrose as a specific substrate to produce dextrans, mutans, and levans (Sutherland 1972). Dextrans are a large class of extracellularly formed glucans produced by the genera *Lactobacillus*, *Leuconostoc*, and *Streptococcus*, of which *Leuc. mesenteroides* and *Leuc. dextranicum* are the well-known dextran producers. Although each bacterial strain produces a unique glucan, a common structural feature of all dextrans is a high percentage (up to 95%) of á-1,6 linkages, with a smaller proportion of á-1,2, á-1,3, or á-1,4 linkages, resulting in a highly branched molecule. Dextrans are synthesized outside the cell by a dextransucrase, which catalyzes sucrose to produce D-fructose and D-glucose, and transfers the latter to an acceptor to form dextran. Several strains of *Lactobacillus* present in foods are able to produce prebiotic substances as fructooligosaccharides (*L. acidophilus*) and inulin-oligosaccharides (*L. reuteri*) (Ruas-Madiedo, Salazar, and de los Reyes-Gavilan 2009). Therefore, they could have functional application through this production.

2.5.5 MISCELLANEOUS FUNCTIONALITIES

Extensive studies have been done on the microorganisms isolated from cheese and cheese smear. However, little research has been carried out on the microbial communities naturally growing as biofilms on the cheese surface, although it is the first step of the investigation of interactions that result in a specific functionality. Functionalities of microbial communities are important for industrial applications, as most of the industrial microbial processes are based on multispecies activities. For a long time, functions have been thought to be the sum of the whole metabolic potential of the microbes present in a particular habitat.

Functionality may not depend on the diversity but, rather, on the equilibrium between microorganisms (Versalovic and Relman 2006). Cheese color, especially in red smear cheese, was thought to be produced by *B. linens* alone, but the subdominance of this bacterium in many traditional cheeses does not correlate with its supposed color-production ability in the context of cheese ripening (Goerges et al. 2008). Now it is believed that cheese color is due to complex interactions into cheese surface microflora (Galaup et al. 2007). The orange pigment production is not only dependent on the bacterial composition of the cheese surface, such as the presence of yellow-pigmented *A. arilaitensis* and *Microbacterium* spp. and other orange-pigmented bacteria, staphylococci, and micrococci (Mounier et al. 2006b; Galaup et al. 2007), but also depends on the yeasts used for deacidification (Leclercq-Perlat et al. 2004b; Mounier et al. 2008). Flavor production in the cheese is the result of complex metabolic pathways of degradation of sugars, proteins, and lipids (Sieuwerts et al. 2008). The cumulative ability of strains in a given cheese microbial community determines flavor production in cheese. The bacteria–yeasts interactions also enhance the flavor production. For example, S-methylthioacetate production was enhanced when *Kluyveromyces lactis* was cocultivated with *B. linens* (Rattray and Fox 1999). Another functionality is overproduction of free amino acids and vitamins as a major result of cheese ripening.

The presence of functional microorganisms also contributes to food functionality (Gobbetti, Cagno, and De Angelis 2010). The presence of natural endogenous probiotic bacteria in fermented food was first described in Camembert (Coeuret, Guéguen, and Vernoux 2004) and described or supposed for other fermented foods (Silva and Malcata 2005; Stanton et al. 2005; Anukam and Reid 2009). Fermented foods are also known as good carriers for exogenous probiotics (Heller 2001).

2.6 CONCLUDING REMARKS

Complex microbial ecosystems are useful for preparation of many important foods, such as bread, sauerkraut, olives, dry sausage, vinegar, alcoholic beverages, and fermented milk products. These microbial ecosystems are evolutive ecosystems that modify progressively with time and space due to changes in available metabolites and in the physicochemical parameters of their *in situ* environment. To illustrate this statement, soft and semisoft cheeses such as smear surface-ripened cheeses were chosen as examples of evolving complex microbial ecosystems. This chapter discussed the successive effects of microbial communities from milk and curd (acid production by lactic acid bacteria, gel formation, and whey expulsion associated with changes of internal physicochemical parameters) to cheese after salt addition and intervention of ripening microbiota at controlled temperature and moisture. Ripening starts with the growth of yeasts—with dominant species like *Geotrichum candidum*, *Debaryomyces hansenii*, *Kluyveromyces* sp., and *Yarrowia lipolytica*—that metabolize the lactic acid and raise the pH on the cheese surface from about 5.0 to about 6.0. This allows the development of a salt-tolerant, usually very complex and undefined bacterial consortium, including coryneform (e.g., *Arthrobacter*, *Brevibacterium*) and gram-negative bacteria. Interactions between the microbiota and the matrix determine the final typical sensorial characteristics of these cheeses and the apparition of health products, if any.

The balance of this microflora is also crucial for avoiding the growth of any undesirable contaminants on the cheese surface. Indeed, the surface pH, which increases as the result of lactate degradation by yeasts, also makes conditions favorable for growth of some contaminating pathogenic bacteria. In the case of *Listeria monocytogenes*, which is one such potential contaminant, it has been demonstrated that the whole surface flora of red smear cheese can display stronger antagonistic effects against *Listeria* sp. than a single strain producing inhibitory compounds. Thus the balance between different microbial communities on the cheese surface is crucial for biopreservation and food safety. This natural biopreservation potential needs to be understood, and the basis of its effects has to be determined.

Complex metabolic pathways are involved during cheese ripening, and their contribution to enhancement of microbial functionalities is evident. Such functionalities involve interactions between microorganisms—such as commensalism, mutualism, cooperation and amensalism, and competition—and can lead to the improvement of cheese sensorial characteristics and safety by production of substances such as metabolites and bacteriocins. Production of free amino acids, vitamins, and polysaccharides-prebiotics and the presence of natural probiotic bacteria are of major interest from a public health point of view. Bioactive peptides result from proteolytic activities, but the stability of their concentration can be questioned, as fermented foods are evolving ecosystems.

Understanding these complex interactions in fermented foods is at the beginning, but it is of high value from the perspective of industrial applications and human health and well-being.

REFERENCES

Addeo, F., L. Chianese, R. Sacchi, S. S. Musso, P. Ferranti, and A. Malorni. 1994. Characterization of the oligopeptides of Parmigiano-Reggiano cheese soluble in 120 g trichloroacetic acid. *Journal of Dairy Research* 61:365–374.
Addis, E., G. H. Fleet, J. M. Cox, D. Kolak, and T. Leung. 2001. The growth, properties, and interactions of yeasts and bacteria associated with the maturation of Camembert and blue-veined cheeses. *International Journal of Food Microbiology* 69:25–36.

Alonso-Calleja, C., R. Capita, J. Carballo, A. Bernardo, and M. L. Garcia-Lopez. 2002. Changes in the Enterobacteriaceae populations throughout manufacturing and ripening of Valdeteja cheese. *Milchwissenschaft* 57:523–525.

Alves, A., O. Santos, I. Henriques, and A. Correia. 2002. Evaluation of methods for molecular typing and identification of members of the genus *Brevibacterium* and other related species. *FEMS Microbiology Letters* 213:205–211.

Anukam, K. C., and G. Reid. 2009. African traditional fermented foods and probiotics. *Journal of Medicinal Food* 12:1177–1184.

Aoki, S. K., R. Pamma, A. D. Hernday, J. E. Bickham, B. A. Braaten, and D. A. Low. 2005. Contact-dependent inhibition of growth in *Escherichia coli*. *Science* 309:1245–1248.

Auberger, B., J. Lenoir, J. L. Bergere, P. Brindejonc, and V. G. Naudot. 1997. Caractérisation partielle des exopeptidases d'une souche de *Geotrichum candidum*. *Sciences des Aliments* 17:655–670.

Augustin, J.-C., H. Bergis, G. Midelet-Bourdin et al. 2011. Design of challenge testing experiments to assess the variability of *Listeria monocytogenes* growth in foods. *Food Microbiology* 28:746–754.

Augustin, J. C., V. Zuliani, M. Cornu, and L. Guillier. 2005. Growth rate and growth probability of *Listeria monocytogenes* in dairy, meat, and seafood products in suboptimal conditions. *Journal of Applied Microbiology* 99:1019–1042.

Balla, E., L. M. T. Dicks, M. Du Toit, M. J. Van Der Merwe, and W. H. Holzapfel. 2000. Characterization and cloning of the genes encoding Enterocin 1071A and Enterocin 1071B, two antimicrobial peptides produced by *Enterococcus faecalis* BFE 1071. *Applied and Environmental Microbiology* 66:1298–1304.

Barnett, J. A., R. W. Payne, and D. Yarrow. 2000. *Yeasts: Characterisation and identification*. Cambridge, U.K.: Cambridge University Press.

Baroiller, C., and J. L. Schmidt. 1990. Contribution à l'étude de l'origine des levures du fromage de Camembert. *Lait* 70:67–84.

Barreteau, H., A. Kovac, A. Boniface, M. Sova, S. Gobec, and D. Blanot. 2008. Cytoplasmic steps of peptidoglycan biosynthesis. *FEMS Microbiology Reviews* 32:168–207.

Beckers, H. J., P. S. S. Soentoro, and E. H. M. Delgou-van Asch. 1987. The occurrence of *Listeria monocytogenes* in soft cheeses and raw milk and its resistance to heat. *International Journal of Food Microbiology* 4:249–256.

Beigelow, W. D. 1921. Logarithmic nature of thermal-death-time curves. *Journal of Infectious Diseases* 29:528–536.

Beresford, T. P., N. A. Fitzsimons, N. L. Brennan, and T. M. Cogan. 2001. Recent advances in cheese microbiology. *International Dairy Journal* 11:259–274.

Beresford, T. P., and A. Williams. 2004. The microbiology of cheese ripening. In *Cheese chemistry, physics, and microbiology*, ed. F. Fox, P. Mc Sweeny, T. Cogan, and T. Guinee, 287–317. New York: Elsevier Academic Press.

Bertelsen, H., H. Andersen, and M. Tvede. 2001. Fermentation of D-tagatose by human intestinal bacteria and dairy lactic acid bacteria. *Microbial Ecology in Health and Disease* 13:87–95.

Bizani, D., J. A. C. Morrissy, A. P. M. Dominguez, and A. Brandelli. 2008. Inhibition of *Listeria monocytogenes* in dairy products using the bacteriocin-like peptide cerein 8A. *International Journal of Food Microbiology* 121:229–233.

Bockelmann, W., U. Krusch, G. Engel, N. Klijn, G. Smit, and K. Heller. 1997. The microflora of Tilsit cheese, Part 1: Variability of the smear flora. *Nahrung* 41:208–212.

Bockelmann, W., S. Portius, S. Lick, and K. J. G. Heller. 1999. Sporulation of *Penicillium camemberti* in submerged batch culture. *Systematic and Applied Microbiology.* 22:479–485.

Bockelmann, W., K. P. Willems, H. Neve, and K. H. Heller. 2005. Cultures for the ripening of smear cheeses. *International Dairy Journal* 15:719–732.

Bonaiti, C., M. N. Leclercq-Perlat, E. Latrille, and G. Corrieu. 2004. Deacidification by *Debaryomyces hansenii* of smear soft cheeses ripened under controlled conditions: Relative humidity and temperature influences. *Journal of Dairy Science* 87:3976–3988.

Bonnarme, P., C. Lapadatescu, M. Yvon, and H. E. Spinnler. 2001. L-methionine degradation potentialities of cheese-ripening microorganisms. *Journal of Dairy Research* 68:663–674.

Bora, N., M. Vancanneyt, R. Gelsomino, et al. 2007. *Agrococcus casei* sp. nov., isolated from the surfaces of smear-ripened cheeses. *International Journal of Systematic and Evolutionary Microbiology* 57:92–97.

———. 2008. *Mycetocola reblochoni* sp. nov., isolated from the surface microbial flora of reblochon cheese. *International Journal of Systematic and Evolutionary Microbiology* 58:2687–2693.

Boucabeille, C., D. Mengin-Lecreulx, G. Henckes, J. M. Simonet, and J. van Heijenoort. 1997. Antibacterial and hemolytic activities of linenscin OC2, a hydrophobic substance produced by *Brevibacterium linens* OC2. *FEMS Microbiology Letters* 153:295–301.

Boutrou, R., L. Kerriou, and J.-Y. Gassi. 2006. Contribution of *Geotrichum candidum* to the proteolysis of soft cheese. *International Dairy Journal* 16:775–783.

Boyle, S. A., J. J. Rich, P. J. Bottomley, J. K. Cromack, and D. D. Myrold. 2006. Reciprocal transfer effects on denitrifying community composition and activity at forest and meadow sites in the Cascade Mountains of Oregon. *Soil Biology and Biochemistry* 38:870–878.

Breer, C., and K. Schopfer. 1988. *Listeria* and food. *Lancet* 1:1022.

Brennan, N. M., R. Brown, M. Goodfellow, et al. 2001. *Microbacterium gubbeenense* sp. nov., from the surface of a smear-ripened cheese. *International Journal of Systematic and Evolutionary Microbiology* 51:1969–1976.

Brennan, N. M., A. C. Ward, T. P. Beresford, P. F. Fox, M. Goodfellow, and T. M. Cogan. 2002. Biodiversity of the bacterial flora on the surface of a smear cheese. *Applied and Environmental Microbiology* 68:820–830.

Brown, D., and P. Rothery. 1993. *Models in biology: Mathematics, statistics, and computing*. New York: Wiley.

Buchanan, R. L., and L. K. Bagi. 1997. Microbial competition: Effect of culture conditions on the suppression of *Listeria monocytogenes* Scott A by *Carnobacterium piscicola*. *Journal of Food Protection* 60:254–261.

———. 1999. Microbial competition: Effect of *Pseudomonas fluorescens* on the growth of *Listeria monocytogenes*. *Food Microbiology* 16:523–529.

Buchanan, R. L., and R. C. Whiting. 1996. Risk assessment and predictive microbiology. *Journal of Food Protection* Suppl.: 31–36.

Callon, C., F. Duthoit, C. Delbes, et al. 2007. Stability of microbial communities in goat milk during a lactation year: Molecular approaches. *Systematic and Applied Microbiology* 30:547–560.

Canzek Majhenic, A., B. Bogovic Matijasić, and I. Rogelj. 2003. Chromosomal location of the genetic determinants for bacteriocins produced by *Lactobacillus gasseri* K7. *Journal of Dairy Research* 70:199–203.

Carnio, M. C., I. Eppert, and S. Scherer. 1999. Analysis of the bacterial surface ripening flora of German and French smeared cheeses with respect to their anti-listerial potential. *International Journal of Food Microbiology* 47:89–97.

Carnio, M. C., A. Holtzel, M. Rudolf, T. Henle, G. Jung, and S. Scherer. 2000. The macrocyclic peptide antibiotic Micrococcin P1 is secreted by the food-borne bacterium *Staphylococcus equorum* WS 2733 and inhibits *Listeria monocytogenes* on soft cheese. *Applied and Environmental Microbiology* 66:2378–2384.

Casaus, P., T. Nilsen, L. M. Cintas, I. F. Nes, P. E. Hernandez, and H. Holo. 1997. Enterocin B, a new bacteriocin from *Enterococcus faecium* T136 which can act synergistically with enterocin A. *Microbiology* 143:2287–2294.

Chen, H., and D. G. Hoover. 2003. Bacteriocins and their food applications. *Comprehensive Reviews in Food Science and Food Safety* 2:82–100.

Choisy, C., M. Desmazeaud, J. C. Gripon, G. Lamberet, and J. Lenoir. 1997. La biochimie de l'affinage. In *Le fromage*, ed. A. Eck and J. C. Gillis, 86–105. Paris: Lavoisier Tec & Doc.

Coeuret, V., M. Guéguen, and J. P. Vernoux. 2004. *In vitro* screening of potential probiotic activities of selected lactobacilli isolated from unpasteurized milk products for incorporation into soft cheese. *Journal of Dairy Research* 71:451–460.

Cogan, T. M., S. Goerges, R. Gelsomino, et al. 2011. Biodiversity of the surface microbial consortia from Limburger, Reblochon, Livarot, Tilsit, and Gubbeen cheese. In cheese and microbes, *American Society Mic Series*, ed. C. Donnelly. Danvers, MA: ASM Press.

Compant, S., B. Duffy, J. Nowak, C. Clement, and E. A. Barka. 2005. Use of plant growth-promoting bacteria for biocontrol of plant diseases: Principles, mechanisms of action, and future prospects. *Applied and Environmental Microbiology* 71:4951–4959.

Cornu, M., M. Kalmokoff, and J. P. Flandrois. 2002. Modelling the competitive growth of *Listeria monocytogenes* and *Listeria innocua* in enrichment broths. *International Journal of Food Microbiology* 73:261–274.

Corsetti, A., J. Rossi, and M. Gobbetti. 2001. Interactions between yeasts and bacteria in the smear surface-ripened cheeses. *International Journal of Food Microbiology* 69:1–10.

Coton, M., C. Delbès-Paus, F. Irlinger, et al. 2012. Diversity and assessment of potential risk factors of Gram-negative isolates associated with French cheeses. *Food Microbiology*, 29:88–98.

Curtin, A. C., M. Gobbetti, and P. L. McSweeney. 2002. Peptidolytic, esterolytic, and amino acid catabolic activities of selected bacterial strains from the surface of smear cheese. *International Journal of Food Microbiology* 76:231–240.

De Buyser, M.-L., B. Dufour, M. Maire, and V. Lafarge. 2001. Implication of milk and milk products in food-borne diseases in France and in different industrialised countries. *International Journal of Food Microbiology* 67:1–17.

Deetae, P. 2009. Role de *Proteus vulgaris* et autres bactéries d'affinage peu étudiées sur la production des arômes du fromage. PhD thesis, AgroParisTech, France.

Deetae, P., P. Bonnarme, H. E. Spinnler, and S. Helinck. 2007. Production of volatile aroma compounds by bacterial strains isolated from different surface-ripened French cheeses. *Applied and Microbiology and Biotechnology* 76:1161–1171.

Delbès, C., L. Ali-Mandjee, and M. C. Montel. 2007. Monitoring bacterial communities in raw milk and cheese by culture-dependent and -independent 16S rRNA gene-based analyses. *Applied and Environmental Microbiology* 73:1882–1891.

Demarigny, Y., C. Berger, N. Desmasures, M. Guéguen, and H. E. Spinnler. 2000. Flavour sulphides are produced from methionine by two different pathways by *Geotrichum candidum*. *Journal of Dairy Research* 67:371–380.

Dens, E. J., K. M. Vereecken, and J. F. Van Impe. 1999. A prototype model structure for mixed microbial populations in homogeneous food products. *Journal of Theoretical Biology* 201:159–170.

Desmasures, N., F. Bazin, and M. Guéguen. 1997. Microbiological composition of raw milk from selected farms in the Camembert region of Normandy. *Journal of Applied Microbiology* 83:53–58.

Devlieghere, F., A. H. Geeraerd, K. J. Versyck, B. Vandewaetere, J. Van Impe, and J. Debevere. 2001. Growth of *Listeria monocytogenes* in modified atmosphere packed cooked meat products: A predictive model. *Food Microbiology* 18:53–66.

De Vuyst, L., and B. Degeest. 1999. Heteropolysaccharides from lactic acid bacteria. *FEMS Microbiology Reviews* 23:153–177.

De Vuyst, L., M. R. Foulquié Moreno, and H. Revets. 2003. Screening for enterocins and detection of hemolysin and vancomycin resistance in enterococci of different origins. *International Journal of Food Microbiology* 84:299–318.

Dieuleveux, V., and M. Guéguen. 1998. Antimicrobial effects of D-3-phenyllactic acid on *Listeria monocytogenes* in TSB-YE medium, milk, and cheese. *Journal of Food Protection* 61:1281–1285.

Duché, O., F. Trémoulet, and A. Namane. 2002. A proteomic analysis of the salt stress response of *Listeria monocytogenes*. *FEMS Microbiology Letters* 215:183–188.

Duthoit, F., J. J. Godon, and M. C. Montel. 2003. Bacterial community dynamics during production of registered designation of origin Salers cheese as evaluated by 16S rRNA gene single-strand conformation polymorphism analysis. *Applied and Environmental Microbiology* 69:3840–3848.

EFSA. 2009. The Community summary report on trends and sources of zoonoses and zoonotic agents in the European Union in 2007. *EFSA Journal* 223.

———. 2010. The Community summary report on trends and sources of zoonoses, zoonotic agents, and food-borne outbreaks in the European Union in 2008. *EFSA Journal* 8:1496.

El-Baradei, G., A. Delacroix-Buchet, and J.-C. Ogier. 2007. Biodiversity of bacterial ecosystems in traditional Egyptian Domiati cheese. *Applied and Environmental Microbiology* 73:1248–1255.

Eliskases-Lechner, F., and W. Ginzinger. 1995a. The bacterial flora of surface ripened cheese with special regard to coryneforms. *Lait* 75:571–584.

———. 1995b. The yeast flora of surface ripened cheese. *Milchwissenschaft* 50:458–462.

Ennahar, S., Y. Asou, T. Zendo, K. Sonomoto, and A. Ishizaki. 2001. Biochemical and genetic evidence for production of enterocins A and B by *Enterococcus faecium* WHE 81. *International Journal of Food Microbiology* 70:291–301.

Eppert, I., E. Lechner, R. Mayr, and S. Scherer. 1995. Listerien und coliforme Keime in "echten" und "fehldeklarierten" Rohmilchweichkäsen. *Archiv für Lebensmittelhygiene* 46:85–88.

Eppert, I., N. Valdes-Stauber, H. Gotz, M. Busse, and S. Scherer. 1997. Growth reduction of *Listeria* spp. caused by undefined industrial red smear cheese cultures and bacteriocin-producing *Brevibacterium lines* as evaluated *in situ* on soft cheese. *Applied and Environmental Microbiology* 63:4812–4817.

Euzéby, J. P. 2011. List of prokaryotic names with standing in nomenclature. http://www.bacterio.cict.fr.

Farber, J. M., and P. I. Peterkin. 1991. *Listeria monocytogenes*, a food-borne pathogen. *Microbiology and Molecular Biology Reviews* 55:476–511.

Ferreira, A. D., and B. C. Viljoen. 2003. Yeasts as adjunct starters in matured cheddar cheese. *International Journal of Food Microbiology* 86:131–140.

Ferreira, I. M., C. Veiros, O. Pinho, A. C. Veloso, A. M. Peres, and A. Mendonça. 2006. Casein breakdown in Terrincho ovine cheese: Comparison with bovine cheese and with bovine/ovine cheeses. *Journal of Dairy Science* 89:2397–2407.

Feurer, C., F. Irlinger, H. E. Spinnler, P. Glaser, and T. Vallaeys. 2004a. Assessment of the rind microbial diversity in a farmhouse-produced vs. a pasteurized industrially produced soft red-smear cheese using both cultivation and rDNA-based methods. *Journal of Applied Microbiology* 97:546–556.

Feurer, C., T. Vallaeys, G. Corrieu, and F. Irlinger. 2004b. Does smearing inoculum reflect the bacterial composition of the smear at the end of the ripening of a French soft, red-smear cheese? *Journal of Dairy Science* 87:3189–3197.

Fiat, A. M., D. Migliore-Samour, P. Jollès, L. Drouet, C. Bal dit Sollier, and J. Caen. 1993. Biologically active peptides from milk proteins with emphasis on two examples concerning antithrombotic and immunomodulating activities. *Journal of Dairy Science* 76:301–310.

Fitzgerald, R. J., and H. Meisel. 2000. Milk protein-derived peptide inhibitors of angiotensin-I-converting enzyme. *British Journal of Nutrition* 84 (Suppl. 1): S33–S37.

Fleet, G. H. 1999. Microorganisms in food ecosystems. *International Journal of Food Microbiology* 50:101–117.

Fontaine, L., C. Boutry, E. Guedon, et al. 2007. Quorum-sensing regulation of the production of Blp bacteriocins in *Streptococcus thermophilus*. *Journal of Bacteriology* 189:7195–7205.

Fox, P. F., T. M. Cogan, and P. L. H. McSweeney. 2000. *Fundamentals of cheese science.* Gaithersburg, MD: Aspen Publications.

Fox, P. F., and J. M. Wallace. 1997. Formation of flavor compounds in cheese. *Advances in Applied Microbiology* 45:17–85.

Frings, E., C. Holtz, and B. Kunz. 1993. Studies about casein degradation by *Brevibacterium linens*. *Milchwissenschaft* 48:130 133.

Fukami, T., and P. J. Morin. 2003. Productivity-biodiversity relationships depend on the history of community assembly. *Nature* 424:423–426.

Galaup, P., A. Gautier, Y. Piriou, A. de Villeblanche, A. Valla, and L. Dufossé. 2007. First pigment fingerprints from the rind of French PDO red-smear ripened soft cheeses Epoisses, Mont d'Or and Maroilles. *Innovative Food Science & Emerging Technologies* 8:373–378.

Gálvez, A., H. Abriouel, R. L. López, and N. B. Omar. 2007. Bacteriocin-based strategies for food biopreservation. *International Journal of Food Microbiology* 120:51–70.

Gans, J., M. Wolinsky, and J. Dunbar. 2005. Computational improvements reveal great bacterial diversity and high metal toxicity in soil. *Science* 309:1387–1390.

Garcia, P., L. Rodriguez, A. Rodriguez, and B. Martinez. 2010. Food biopreservation: Promising strategies using bacteriocins, bacteriophages, and endolysins. *Trends in Food Science and Technology* 21:373–382.

Gavrish, E. Y., V. I. Krauzova, N. V. Potekhina, et al. 2004. Three new species of Brevibacteria, *Brevibacterium antiquum* sp. nov., *Brevibacterium aurantiacum* sp. nov., and *Brevibacterium permense* sp. nov. *Microbiology* 73:176–183.

Gay, M., and A. Amgar. 2005. Factors moderating *Listeria monocytogenes* growth in raw milk and in soft cheese made from raw milk. *Lait* 85:153–170.

Gelsomino, R., M. Vancanneyt, C. Snauwaert, et al. 2005. *Corynebacterium mooreparkense*, a later heterotypic synonym of *Corynebacterium variabile*. *International Journal of Systematic and Evolutionary Microbiology* 55:1129–1131.

Giraffa, G., E. Neviani, and G. T. Tarelli. 1994. Antilisterial activity by Enterococci in a model predicting the temperature evolution of Taleggio, an Italian soft cheese. *Journal of Dairy Science* 77:1176–1182.

Gobbetti, M., R. D. Cagno, and M. De Angelis. 2010. Functional microorganisms for functional food quality. *Critical Reviews in Food Science and Nutrition* 50:716–727.

Goerges, S., J. Mounier, M. C. Rea, et al. 2008. Commercial ripening starter microorganisms inoculated into cheese milk do not successfully establish themselves in the resident microbial ripening consortia of a South German red smear cheese. *Applied and Environmental Microbiology* 74:2210–2217.

Gori, K., C. Mortensen, and L. Jespersen. 2010. A comparative study of the anti-listerial activity of smear bacteria. *International Dairy Journal* 20:555–559.

Grattepanche, F., S. Miescher-Schwenninger, L. Meile, and C. Lacroix. 2008. Recent developments in cheese cultures with protective and probiotic functionalities. *Dairy Science and Technology* 88:421–444.

Gravesen, A., A. M. Jydegaard Axelsen, J. Mendes da Silva, T. B. Hansen, and S. Knochel. 2002a. Frequency of bacteriocin resistance development and associated fitness costs in *Listeria monocytogenes*. *Applied and Environmental Microbiology* 68:756–764.

Gravesen, A., M. Ramnath, K. B. Rechinger, et al. 2002b. High-level resistance to class IIa bacteriocins is associated with one general mechanism in *Listeria monocytogenes*. *Microbiology* 148:2361–2369.

Gripon, J. C. 1993. Mould-ripened cheeses. In *Cheese: Chemistry, physics, and microbiology*, ed. P. F. Fox, 111–131. London: Chapman & Hall.

Guéguen, M., and J. Lenoir. 1975. Aptitude de l'espèce *Geotrichum candidum* à la production d'enzymes proteolytiques. *Lait* 55:145–162.

Guéguen, M., and J. L. Schmidt. 1992. Les levures et *Geotrichum candidum*. In *Les groupes microbiens d'intérêt laitier*, ed. J. Hermier, J. Lenoir, and F. Weber, 165–219. Paris: CEPIL.

Haakensen, M., C. M. Dobson, J. E. Hill, and B. Ziola. 2009. Reclassification of *Pediococcus dextrinicus* (Coster and White 1964) Back 1978 (Approved Lists 1980) as *Lactobacillus dextrinicus* comb. nov., and amended description of the genus *Lactobacillus*. *International Journal of Systematic and Evolutionary Microbiology*, 59:615–621.

Hain, T., S. S. Chatterjee, R. Ghai, et al. 2007. Pathogenomics of *Listeria* spp. *International Journal of Medical Microbiology* 297:54–557.

Hang, Y. D., and E. E. Woodams. 1992. Purification and characterization of lactate dehydrogenase from *Geotrichum candidum*. *Food Chemistry* 45:15–17.

Hannan, Y., and M. Guéguen. 1985. Activités endopeptidasiques du levain fongique *Geotrichum candidum* en fonction de sa croissance. *Sciences des Aliments* 5:147–152.

Hansen, E. H., M. A. Schembri, P. Klemm, T. Schafer, S. Molin, and L. Gram. 2004. Elucidation of the antibacterial mechanism of the Curvularia haloperoxidase system by DNA microarray profiling. *Applied and Environmental Microbiology* 70:1749–1757.

Heller, I. R. 2001. Functional foods: Regulatory and marketing developments. *Food and Drug Law Journal* 56:197–225.

Hemme, D., C. Bouillane, F. Metro, and M. J. Desmazeaud. 1982. Microbial catabolism of amino acids during cheese ripening. *Sciences des Aliments* 2:113–123.

Hennekinne, J. A., M. Gohier, T. Maire, C. Lapeyre, B. Lombard, and S. Dragacci. 2003. First proficiency testing to evaluate the ability of European Union National Reference Laboratories to detect staphylococcal enterotoxins in milk products. *Journal of AOAC International* 86:332–339.

Henri-Dubernet, S., N. Desmasures, and M. Guéguen. 2004. Culture-dependent and culture-independent methods for molecular analysis of the diversity of lactobacilli in "Camembert de Normandie" cheese *Lait* 84:179–189.

Holzapfel, W. H., R. Geisen, and U. Schillinger. 1995. Biological preservation of foods with reference to protective cultures, bacteriocins, and food-grade enzymes. *International Journal of Food Microbiology* 24:343–362.

Hong, R. W., M. Shchepetov, J. N. Weiser, and P. H. Axelsen. 2003. Transcriptional profile of the *Escherichia coli* response to the antimicrobial insect peptide Cecropin A. *Antimicrobial Agents and Chemotherapy* 47:1–6.

Hoppe-Seyler, T. S., B. Jaeger, W. Bockelmann, W. H. Noordman, A. Geis, and K. J. Heller. 2004. Molecular identification and differentiation of *Staphylococcus* species and strains of cheese origin. *Systematic and Applied Microbiology* 27:211–218.

Huston, M. A. 1994. *Biological diversity: The coexistence of species on changing landscapes*. Cambridge, U.K.: Cambridge University Press.

Imran, M., N. Desmasures, and J. P. Vernoux. 2010a. Role of microbial growth dynamics in antilisterial activity of model microbial communities originating from cheese. In *Proceedings of 110th general meeting of the American Society for Microbiology*, San Diego, CA, 23–27 May 2010.

———. 2010b. From undefined red smear cheese consortia to minimal model communities both exhibiting similar anti-listerial activity on a cheese-like matrix. *Food Microbiology* 27:1095–1103.

Irlinger, F., F. Bimet, J. Delettre, M. Lefevre, and P. A. Grimont. 2005. *Arthrobacter bergerei* sp. nov. and *Arthrobacter arilaitensis* sp. nov., novel coryneform species isolated from the surfaces of cheeses. *International Journal of Systematic and Evolutionary Microbiology* 55:457–462.

Irlinger, F., A. Morvan, N. El Solh, and J. L. Bergere. 1997. Taxonomic characterization of coagulase-negative staphylococci in ripening flora from traditional French cheeses. *Systematic and Applied Microbiology* 20:319–328.

Iseppi, R., F. Pilati, M. Marini, et al. 2008. Anti-listerial activity of a polymeric film coated with hybrid coatings doped with Enterocin 416K1 for use as bioactive food packaging. *International Journal of Food Microbiology* 123:281–287.

Izquierdo, E., E. Marchioni, D. Aoude-Werner, C. Hasselmann, and S. Ennahar. 2009. Smearing of soft cheese with *Enterococcus faecium* WHE 81, a multi-bacteriocin producer, against *Listeria monocytogenes*. *Food Microbiology* 26:16–20.

Jameson, J. E. 1962. A discussion of the dynamics of *Salmonella* enrichment. *The Journal of Hygiene* 60:193–207.

Jasniewski, J., C. Cailliez-Grimal, I. Chevalot, J. B. Millière, and A. M. Revol-Junelles. 2009. Interactions between two carnobacteriocins Cbn BM1 and Cbn B2 from *Carnobacterium maltaromaticum* CP5 on target bacteria and Caco-2 cells. *Food and Chemical Toxicology* 47:893–897.

Jay, J. M., M. J. Loessner, and D. A. Golden. 2005. *Modern Food Microbiology*. 7th ed. New York: Springer Science.

Jessup, C. M., R. Kassen, S. E. Forde, et al. 2004. Big questions, small worlds: Microbial model systems in ecology. *Trends in Ecology and Evolution* 19:189–197.

Jollivet, N., J. Chataud, Y. Vayssier, M. Bensoussan, and J.-M. Belin. 1994. Production of volatile compounds in model milk and cheese media by eight strains of *Geotrichum candidum* Link. *Journal of Dairy Research* 61:241–248.

Kagkli, D. M., R. Tache, T. M. Cogan, C. Hill, S. Casaregola, and P. Bonnarme. 2006. *Kluyveromyces lactis* and *Saccharomyces cerevisiae*, two potent deacidifying and volatile-sulphur-aroma-producing microorganisms of the cheese ecosystem. *Applied Microbiology and Biotechnology* 73:434–442.

Kassen, R., A. Buckling, G. Bell, and P. B. Rainey. 2000. Diversity peaks at intermediate productivity in a laboratory microcosm. *Nature* 406:508–512.

Kato, F., Y. Eguchi, M. Nakano, T. Oshima, and A. Murata. 1991. Purification and characterization of Linecin-A, a bacteriocin of *Brevibacterium linens*. *Agricultural and Biological Chemistry* 48:161–166.

Kawai, Y., T. Saito, H. Kitazawa, and T. Itoh. 1998. Gassericin A: An uncommon cyclic bacteriocin produced by *Lactobacillus gasseri* LA39 linked at N- and C-terminal ends. *Bioscience, biotechnology, and biochemistry* 62:2438–2440.

Keller, L., and M. G. Surette. 2006. Communication in bacteria: An ecological and evolutionary perspective. *Nature Reviews Microbiology* 4:249–258.

Klein, N., A. Zourari, and S. Lortal. 2002. Peptidase activity of four yeast species frequently encountered in dairy products: Comparison with several dairy bacteria. *International Dairy Journal* 12:853–861.

Kollöffel, B., L. Meile, and M. Teuber. 1999. Analysis of brevibacteria on the surface of Gruyère cheese detected by *in situ* hybridization and by colony hybridization. *Letters in Applied Microbiology* 29:317–322.

Konopka, A. 2009. What is microbial community ecology? *International Society for Microbial Ecology Journal* 3:1223–1230.

Kumar, C. G., and S. K. Anand. 1998. Significance of microbial biofilms in food industry: A review. *International Journal of Food Microbiology* 42:9–27.

Lafarge, V., J. C. Ogier, V. Girard, et al. 2004. Raw cow milk bacterial population shifts attributable to refrigeration. *Applied and Environmental Microbiology* 70:5644–5650.

Larpin, S., M. Imran, C. Bonaïti, et al. 2011. Surface microbial consortia from Livarot, a French smear-ripened cheese. *Canadian Journal of Microbiology* 57:651–660.

Larpin, S., C. Mondoloni, S. Goerges, J. P. Vernoux, M. Guéguen, and N. Desmasures. 2006. *Geotrichum candidum* dominates in yeast population dynamics in Livarot, a French red-smear cheese. *FEMS Yeast Research* 6:1243–1253.

Leclercq-Perlat, M. N., G. Corrieu, and H. E. Spinnler. 2004a. Comparison of volatile compounds produced in model cheese medium deacidified by *Debaryomyces hansenii* or *Kluyveromyces marxianus*. *Journal of Dairy Science* 87:1545–1550.

———. 2004b. The color of *Brevibacterium linens* depends on the yeast used for cheese deacidification. *Journal of Dairy Science* 87:1536–1544.

Leclercq-Perlat, M. N., A. Oumer, J. L. Bergere, H. E. Spinnler, and G. Corrieu. 1999. Growth of *Debaryomyces hansenii* on a bacterial surface-ripened soft cheese. *Journal of Dairy Research* 6:271–281.

———. 2000. Behavior of *Brevibacterium linens* and *Debaryomyces hansenii* as ripening flora in controlled production of smear soft cheese from reconstituted milk: Growth and substrate consumption dairy foods. *Journal of Dairy Science* 83:1665–1673.

Lenoir, J. 1984. The surface flora and its role in the ripening of cheese. *Bulletin of the International Dairy Federation* 171:3–20.

Lenoir, J., G. Lamberet, and J. L. Schmidt. 1983. L'élaboration d'un fromage: L'exemple du Camembert. *Pour la science* July: 30–42.

Lenoir, J., G. Lamberet, J. L. Schmidt, and C. Tourneur. 1985. La main d'oeuvre microbienne domine l'affinage des fromages. *Revue Laitiere Francaise* 444:50–64.

Little, C. L., J. R. Rhoades, S. K. Sagoo, et al. 2008. Microbiological quality of retail cheeses made from raw, thermized, or pasteurized milk in the UK. *Food Microbiology* 25:304–312.

Liu, S. Q. 2003. Practical implications of lactate and pyruvate metabolism by lactic acid bacteria in food and beverage fermentations. *International Journal of Food Microbiology* 83:115–131.

Liu, S., J. E. Graham, L. Bigelow, P. D. Morse II, and B. J. Wilkinson. 2002. Identification of *Listeria monocytogenes* genes expressed in response to growth at low temperature. *Applied and Environmental Microbiology* 68:1697–1705.

Loessner, M., S. Guenther, S. Steffan, and S. Scherer. 2003. A pediocin-producing *Lactobacillus plantarum* strain inhibits *Listeria monocytogenes* in a multispecies cheese surface microbial ripening consortium. *Applied and Environmental Microbiology* 69:1854–1857.

Loncarevic, S., M. L. Danielsson-Tham, P. Gerner-Schidt, L. Sahlström, and W. Tham. 1998. Potential sources of human listeriosis in Sweden. *Food Microbiology* 15:65–69.

Loncarevic, S., M. L. Danielsson-Tham, and W. Tham. 1995. Occurrence of *Listeria monocytogenes* in soft and semi-soft cheeses in retail outlets in Sweden. *International Journal of Food Microbiology* 26:245–250.

Lu, Z., F. Breidt, V. Plengvidhya, and H. P. Fleming. 2003. Bacteriophage ecology in commercial sauerkraut fermentations. *Applied Environmental Microbiology* 69:3192–3202.

Lungu, B., S. C. Ricke, and M. G. Johnson. 2009. Growth, survival, proliferation, and pathogenesis of *Listeria monocytogenes* under low oxygen or anaerobic conditions: A review. *Anaerobe* 15:7–17.

Maisnier-Patin, S., N. Deschamps, S. R. Tatini, and J. Richard. 1992. Inhibition of *Listeria monocytogenes* in Camembert cheese made with a nisin-producing starter. *Lait* 72:249–263.

Maldonado, A., R. Jimenez-Diaz, and J. L. Ruiz-Barba. 2004. Induction of plantaricin production in *Lactobacillus plantarum* NC8 after coculture with specific Gram-positive bacteria is mediated by an autoinduction mechanism. *Journal of Bacteriology* 186:1556–1564.

Mallet, A., M. Guéguen, and N. Desmasures. 2010. Etat des lieux de la diversité microbienne quantitative et qualitative de laits crus normands destinés à la transformation fromagère. *8ème Congrès National de la SFM*, Marseille, France.

Mansour, S., J. M. Beckerich, and P. Bonnarme. 2008. Lactate and amino acid catabolism in the cheese-ripening yeast *Yarrowia lipolytica*. *Applied and Environmental Microbiology* 74:6505–6512.

Maoz, A., R. Mayr, and S. Scherer. 2003. Temporal stability and biodiversity of two complex antilisterial cheese-ripening microbial consortia. *Applied and Environmental Microbiology* 69:4012–4018.

Martinez, B., M. Fernandez, J. E. Suarez, and A. Rodriguez. 1999. Synthesis of lactococcin 972, a bacteriocin produced by *Lactococcus lactis* IPLA 972, depends on the expression of a plasmid-encoded bicistronic operon. *Microbiology* 145:3155–3161.

Mayr, R., M. Fricker, A. Maoz, and S. Scherer. 2004. Anti-listerial activity and biodiversity of cheese surface cultures: Influence of the ripening temperature regime. *European Food Research and Technology* 218:242–247.

McMeekin, T. A., J. Brown, K. Krist, et al. 1997. Quantitative microbiology: A basis for food safety. *Emerging Infectious Diseases* 3:541–549.

McSweeney, P. L. H., and M. J. Sousa. 2000. Biochemical pathways for the production of flavour compounds in cheeses during ripening: A review. *Lait* 80:293–324.

Millet, L., M. Saubusse, R. Didienne, L. Tessier, and M. C. Montel. 2006. Control of *Listeria monocytogenes* in raw-milk cheeses. *International Journal of Food Microbiology* 108:105–114.

Mira-Gutierrez, J., C. Perez de Lara, and M. A. Rodriguez-Iglesias. 1990. Identification of species of the genus *Listeria* by fermentation of carbohydrates and enzymatic patterns. *Acta Microbiologica Hungarica* 37:123–129.

Molimard, P., and H. E. Spinnler. 1996. Review: Compounds involved in the flavor of surface mold-ripened cheeses: Origins and properties. *Journal of Dairy Science* 79:169–184.

Moll, G., T. Ubbink-Kok, H. Hildeng-Hauge, et al. 1996. Lactococcin G is a potassium ion-conducting, two-component bacteriocin. *Journal of Bacteriology* 178:600–605.

Monnet, C., A. Bleicher, K. Neuhaus, A.-S. Sarthou, M.-N. Leclercq-Perlat, and F. Irlinger. 2010. Assessment of the anti-listerial activity of microfloras from the surface of smear-ripened cheeses. *Food Microbiology* 27:302–310.

Morales, P., E. Fernández-García, and M. Nuñez. 2003. Caseinolysis in cheese by Enterobacteriaceae strains of dairy origin. *Letters in Applied Microbiology* 37:410–414.

———. 2005a. Production of volatile compounds in cheese by *Pseudomonas fragi* strains of dairy origin. *Journal of Food Protection* 68:1399–1407.

———. 2005b. Volatile compounds produced in cheese by *Pseudomonas* strains of dairy origin belonging to six different species. *Journal of Agricultural and Food Chemistry* 53:6835–6843.

Morin, P. J. 2000. Biodiversity's ups and downs. *Nature* 406:463–464.

Mounier, J., R. Gelsomino, S. Goerges, et al. 2005. Surface microflora of four smear-ripened cheeses. *Applied and Environmental Microbiology* 71:6489–6500.

Mounier, J., S. Goerges, R. Gelsomino, et al. 2006a. Sources of the adventitious microflora of a smear-ripened cheese. *Journal of Applied Microbiology* 101:668–681.

Mounier, J., F. Irlinger, M. N. Leclercq-Perlat, et al. 2006b. Growth and colour development of some surface ripening bacteria with *Debaryomyces hansenii* on aseptic cheese curd. *Journal of Dairy Research* 73:441–448.

Mounier, J., C. Monnet, N. Jacques, A. Antoinette, and F. Irlinger. 2009. Assessment of the microbial diversity at the surface of Livarot cheese using culture-dependent and independent approaches. *International Journal of Food Microbiology* 133:31–37.

Mounier, J., C. Monnet, T. Vallaeys, et al. 2008. Microbial interactions within a cheese microbial community. *Applied and Environmental Microbiology* 74:172–181.

Mounier, J., M. C. Rea, P. M. O'Connor, G. F. Fitzgerald, and T. M. Cogan. 2007. Growth characteristics of *Brevibacterium*, *Corynebacterium*, *Microbacterium*, and *Staphylococcus* spp. isolated from surface-ripened cheese. *Applied and Environmental Microbiology* 73:7732–7739.

Nilsson, L., L. Gram, and H. H. Huss. 1999. Growth control of *Listeria monocytogenes* on cold-smoked salmon using a competitive lactic acid bacteria flora. *Journal of Food Protection* 62:336–342.

Nilsson, L., T. B. Hansen, P. Garrido, et al. 2005. Growth inhibition of *Listeria monocytogenes* by a nonbacteriocinogenic *Carnobacterium piscicola*. *Journal of Applied Microbiology* 98:172–183.

Nilsson, L., M. K. Nielsen, Y. Ng, and L. Gram. 2002. Role of acetate in production of an autoinducible class IIa bacteriocin in *Carnobacterium piscicola* A9b. *Applied and Environmental Microbiology* 68:2251–2260.

Nilsson, L., Y. Ng, J. Christiansen, B. Jørgensen, D. Grótinum, and L. Gram. 2004. The contribution of bacteriocin to inhibition of *Listeria monocytogenes* by *Carnobacterium piscicola* strains in cold-smoked salmon systems. *Journal of Applied Microbiology* 96:133–143.

O'Sullivan, L., E. B. O'Connor, R. P. Ross, and C. Hill. 2006. Evaluation of live-culture-producing lacticin 3147 as a treatment for the control of *Listeria monocytogenes* on the surface of smear-ripened cheese. *Journal of Applied Microbiology* 100:135–143.

Ogier, J. C., V. Lafarge, V. Girard, et al. 2004. Molecular fingerprinting of dairy microbial ecosystems by use of temporal temperature and denaturing gradient gel electrophoresis. *Applied and Environmental Microbiology* 70:5628–5643.

Östling, C. E., and S. E. Lindgren. 1993. Inhibition of enterobacteria and *Listeria* growth by lactic, acetic and formic acids. *Journal of Applied Microbiology* 75:18–24.

Pereira-Meirelles, F. V., M. H. M. Rocha-Leão, and G. L. Sant'Anna. 2000. Lipase location in *Yarrowia lipolytica* cells. *Biotechnology Letters* 22:71–75.

Peschel, A., and H. G. Sahl. 2006. The co-evolution of host cationic antimicrobial peptides and microbial resistance. *Nature Reviews Microbiology* 4:529–536.

Petrone, G., M. P. Conte, C. Longhi, et al. 1998. Natural milk fatty acids affect survival and invasiveness of *Listeria monocytogenes*. *Letters in Applied Microbiology* 27:362–368.

Pine, L., G. B. Malcolm, J. B. Brooks, and M. I. Daneshvar. 1989. Physiological studies on the growth and utilization of sugars by *Listeria* species. *Canadian Journal of Microbiology* 35:245–254.

Pini, P. N., and R. J. Gilbert. 1988. The occurrence in the U.K. of *Listeria* species in raw chickens and soft cheeses. *International Journal of Food Microbiology* 6:317–326.

Potera, C. 1999. Microbiology: Forging a link between biofilms and disease. *Science* 283:1837–1839.

Powell, M., W. Schlosser, and E. Ebel. 2004. Considering the complexity of microbial community dynamics in food safety risk assessment. *International Journal of Food Microbiology* 90:171–179.

Premaratne, R. J., W. J. Lin, and E. A. Johnson. 1991. Development of an improved chemically defined minimal medium for *Listeria monocytogenes*. *Applied and Environmental Microbiology* 57:3046–3048.

Ramnath, M., K. B. Rechinger, L. Jansch, J. W. Hastings, S. Knochel, and A. Gravesen. 2003. Development of a *Listeria monocytogenes* EGDe partial proteome reference map and comparison with the protein profiles of food isolates. *Applied and Environmental Microbiology* 69:3368–3376.

Rattray, F. P., and P. F. Fox. 1999. Aspects of enzymology and biochemical properties of *Brevibacterium linens* relevant to cheese ripening: A review. *Journal of Dairy Science* 82:891–909.

Rea, M. C., S. Goerges, R. Gelsomino, et al. 2007. Stability of the biodiversity of the surface consortia of Gubbeen, a red-smear cheese. *Journal of Dairy Science* 90:2200–2210.

Remiger, A., M. A. Ehrmann, and R. F. Vogel. 1996. Identification of bacteriocin-encoding genes in *Lactobacilli* by polymerase chain reaction (PCR). *Systematic and Applied Microbiology* 19:28–34.

Retureau, É., C. Callon, R. Didienne, and M.-C. Montel. 2010. Is microbial diversity an asset for inhibiting *Listeria monocytogenes* in raw milk cheeses? *Dairy Science and Technology* 90:375–398.

Richard, J. 1984. Evolution de la flore microbienne à la surface des Camemberts fabriqués avec du lait cru. *Lait* 64:496–520.

Robinson, R. K., A. Y. Tamime, and M. Wszolek. 2002. Microbiology of fermented milks. In *Dairy microbiology handbook*, 3rd ed., chap. 8. New York: John Wiley & Sons.

Rodgers, S. 2008. Novel applications of live bacteria in food services: Probiotics and protective cultures. *Trends in Food Science & Technology* 19:188–197.

Rodríguez, E., B. González, P. Gaya, M. Nuñez, and M. Medina. 2000. Diversity of bacteriocins produced by lactic acid bacteria isolated from raw milk. *International Dairy Journal* 10:7–15.

Romano, J. D., and R. Kolter. 2005. *Pseudomonas-Saccharomyces* interactions: Influence of fungal metabolism on bacterial physiology and survival. *Journal of Bacteriology* 187:940–948.

Roostita, R., and G. H. Fleet. 1996. Growth of yeasts in milk and associated changes to milk composition. *International Journal of Food Microbiology* 31:205–219.

Ross, T., P. Dalgaard, and S. Tienungoon. 2000. Predictive modelling of the growth and survival of *Listeria* in fishery products. *International Journal of Food Microbiology* 62:231–245.

Roth, E., S. Miescher Schwenninger, M. Hasler, E. Eugster-Meier, and C. Lacroix. 2010. Population dynamics of two antilisterial cheese surface consortia revealed by temporal temperature gradient gel electrophoresis. *BMC Microbiology* 10:74.

Ruas-Madiedo, P., N. Salazar, and C. G. de los Reyes-Gavilan. 2009. Biosynthesis and chemical composition of exopolysaccharides produced by lactic acid bacteria. In *Bacterial polysaccharides: Current Innovations and Future Trends*, ed. M. Ullrich. Norfolk, U.K.: Caister Academic Press.

Rudolf, M., and S. Scherer. 2001. High incidence of *Listeria monocytogenes* in European red smear cheese. *International Journal of Food Microbiology* 63:91–98.

Ryser, E. T. 1999. Incidence and behaviour of *Listeria monocytogenes* in cheese and other fermented dairy products. In *Listeria, Listeriosis, and Food Safety*, ed. E. T. Ryser and E. H. Marth, 411–503. New York-Basel: Marcel Dekker.

Ryser, E. T., S. Maisnier-Patin, J. J. Gratadoux, and J. Richard. 1994. Isolation and identification of cheese-smear bacteria inhibitory to *Listeria* spp. *International Journal of Food Microbiology* 21:237–246.

Sablé, S., V. Portrait, V. Gautier, F. Letellier, and G. Cottenceau. 1997. Microbiological changes in a soft raw goat's milk cheese during ripening. *Enzyme and Microbial Technologies* 21:212–220.

Saubusse, M., L. Millet, C. Delbes, C. Callon, and M. C. Montel. 2007. Application of single strand conformation polymorphism: PCR method for distinguishing cheese bacterial communities that inhibit *Listeria monocytogenes*. *International Journal of Food Microbiology* 116:126–135.

Schubert, K., W. Ludwig, N. Springer, R. M. Kroppenstedt, J. P. Accolas, and F. Fiedler. 1996. Two coryneform bacteria isolated from the surface of French Gruyère and Beaufort cheeses are new species of the genus *Brachybacterium*: *Brachybacterium alimentarium* sp. nov. and *Brachybacterium tyrofermentans* sp. nov. *International Journal of Systematic Bacteriology* 46:81–87.

Schulz, S., and J. S. Dickschat. 2007. Bacterial volatiles: The smell of small organisms. *Natural Product Reports* 24:814–842.

Sentandreu, R., and D. H. Northcote. 1969. Yeast cell-wall synthesis. *Biochemistry Journal* 115:231–240.

Sieuwerts, S., F. A. M. de Bok, J. Hugenholtz, and J. E. T. van Hylckama Vlieg. 2008. Unraveling microbial interactions in food fermentations: From classical to genomics approaches. *Applied and Environmental Microbiology* 74:4997–5007.

Silva, S. V., and F. X. Malcata. 2005. Caseins as source of bioactive peptides. *International Dairy Journal* 15:1–15.

Slechta, E. S., and M. A. Mulvey. 2006. Contact-dependent inhibition: Bacterial brakes and secret handshakes. *Trends in Microbiology* 14:58–60.

Smacchi, E., P. F. Fox, and M. Gobbetti. 1999. Purification and characterization of two extracellular proteinases from *Arthrobacter nicotianae* 9458. *FEMS Microbiology Letters* 170:327–333.

Smith, J. L., P. M. Fratamico, and J. S. Novak. 2004. Quorum sensing: A primer for food microbiologists. *Journal of Food Protection* 67:1053–1070.

Stackebrandt, E., F. A. Rainey, and N. L. Ward-Rainey. 1997. Proposal for a new hierarchic classification system: *Actinobacteria classis* nov. *International Journal of Systematic bacteriology* 47:479–491.

Stanton, C., R. P. Ross, G. F. Fitzgerald, and D. Van Sinderen. 2005. Fermented functional foods based on probiotics and their biogenic metabolites. *Current Opinion in Biotechnology* 16:198–203.

Stiles, M. E., and W. H. Holzapfel. 1997. Lactic acid bacteria of foods and their current taxonomy. *International Journal of Food Microbiology* 36:1–29.

Sutherland, I. W. 1972. Bacterial exopolysaccharides. *Advances in Microbial Physiology* 8:143–212.

Sztajer, H., W. Wang, H. Lünsdorf, A. Stocker, and R. D. Schmid. 1996. Purification and some properties of a novel microbial lactate oxidase. *Applied Microbiology and Biotechnology* 45:600–606.

Tahiri, I., M. Desbiens, R. Benech, et al. 2004. Purification, characterization, and amino acid sequencing of divergicin M35: A novel class IIa bacteriocin produced by *Carnobacterium divergens* M35. *International Journal of Food Microbiology* 97:123–136.

Terplan, G., R. Schoen, W. Springmeyer, I. Degle, and H. Becker. 1986. Vorkommen, Verhalten und Bedeutung von Listerien in Milch und Milchprodukten. *Archiv für Lebensmittelhygiene* 37:131–137.

Tornadijo, E., J. M. Fresno, J. Carballo, and R. Martin-Sarmiento. 1993. Study of Enterobacteriaceae throughout the manufacturing and ripening of hard goat's cheese. *Journal of Applied Bacteriology* 75:240–246.

Torsvik, V., and L. Øvreås. 2002. Microbial diversity and function in soil: From genes to ecosystems. *Current Opinion in Microbiology* 5:240–245.

Trmcic, A., T. Obermajer, I. Rogelj, and B. Bogovic Matijasic. 2008. Culture-independent detection of lactic acid bacteria bacteriocin genes in two traditional Slovenian raw milk cheeses and their microbial consortia. *Journal of Dairy Science* 91:4535–4541.

Tsai, H. N., and D. A. Hodgson. 2003. Development of a synthetic minimal medium for *Listeria monocytogenes*. *Applied and Environmental Microbiology* 69:6943–6945.

Utaida, S., P. M. Dunman, D. Macapagal, et al. 2003. Genome-wide transcriptional profiling of the response of *Staphylococcus aureus* to cell-wall-active antibiotics reveals a cell-wall-stress stimulon. *Microbiology* 149:2719–2732.

Vadyvaloo, V., S. Arous, A. Gravesen, et al. 2004. Cell-surface alterations in class IIa bacteriocin-resistant *Listeria monocytogenes* strains. *Microbiology* 150:3025–3033.

Valdès-Stauber, N., and S. Scherer. 1994. Isolation and characterization of Linocin M18, a bacteriocin produced by *Brevibacterium linens*. *Applied and Environmental Microbiology* 60:3809–3814.

Valdès-Stauber, N., S. Scherer, and H. Seiler. 1997. Identification of yeasts and coryneform bacteria from the surface microflora of brick cheeses. *International Journal of Food Microbiology* 34:115–129.

Van Belkum, M. J., J. Kok, and G. Venema. 1992. Cloning, sequencing, and expression in *Escherichia coli* of lcnB, a third bacteriocin determinant from the lactococcal bacteriocin plasmid p9B4-6. *Applied and Environmental Microbiology* 58:572–577.

Vereecken, K. M., E. J. Dens, and J. F. Van Impe. 2000. Predictive modeling of mixed microbial populations in food products: Evaluation of two-species models. *Journal of Theoretical Biology* 205:53–72.

Versalovic, J., and D. Relman. 2006. How bacterial communities expand functional repertoires. *PLoS Biology* 4: e430.

Viljoen, B. C. 2001. The interaction between yeasts and bacteria in dairy environments. *International Journal of Food Microbiology* 69:37–44.

Viljoen, B. C., A. R. Khoury, and A. Hattingh. 2003. Seasonal diversity of yeasts associated with white-surface mould-ripened cheeses. *Food Research International* 36:275–283.

Waide, R. B., M. R. Willig, C. F. Steiner, et al. 1999. The relationship between productivity and species richness. *Annual Review of Ecology and Systematics* 30:257–300.

Wan, J., K. Harmark, B. E. Davidson, et al. 1997. Inhibition of *Listeria monocytogenes* by piscicolin 126 in milk and Camembert cheese manufactured with a thermophilic starter. *Journal of Applied Microbiology* 82:273–280.

Weber, F., and J. P. Ramet. 1986. *Cheese making: Science and technology*. New York: Lavoisier.

Welman, A. D. 2009. Exploitation of exopolysaccharides from lactic acid bacteria: Nutritional and functional benefits. In *Bacterial polysaccharides: Current innovations and future trends*, ed. M. Ullrich. Norwich, U.K.: Caister Academic Press.

Yarmus, M., A. Mett, and R. Shapira. 2000. Cloning and expression of the genes involved in the production of and immunity against the bacteriocin lacticin RM. *Biochimica et Biophysica Acta* 1490:279–290.

Young, I. M., J. W. Crawford, N. Nunan, W. Otten, and A. Spiers. 2008. Microbial distribution in soils: Physics and scaling. In *Advances in agronomy*, chap. 4. London: Academic Press.

Zottola, E. A., and L. B. Smith. 1991. Pathogens in cheese. *Food Microbiology* 8:171–182.

3 The Role of Fermentation Reactions in the Generation of Flavor and Aroma of Foods

Javier Carballo

CONTENTS

3.1 INTRODUCTION

A food that is undergoing a fermentation process is a living system that is sheltering
a multitude of coexisting microbial and enzymatic activities. The results of these
activities are mainly catabolic processes, although the anabolic ones and their contribution cannot be undervalued. As a consequence of these activities, an almost lim-
itless quantity of compounds is formed from the food components (carbohydrates,
proteins, and lipids). These compounds have an unequal contribution to the flavor
and aroma of the fermented products, depending on their concentrations and perception thresholds that, in turn, depend on the chemical structures and properties of the
molecules. Moreover, these compounds interact with each other, producing synergic
(reinforcing) and masking phenomena.

Describing all the fermentation reactions of interest in an orderly and system-
atized way is not a simple task due to the interlacing and interdependence of many
of these reactions. In some cases, a specific compound or a group of compounds may
be produced following more than one pathway. Some particular reactions may have
been described in a particular microorganism or group of microorganisms, but these
reactions may not follow the same pathway in other microorganisms.

In this chapter, we attempt to explain—in a way that is as complete and simple as possible—the chemical reactions that take place during the processes of fermentation and that produce the compounds responsible for the flavor in fermented foods.

3.2 FLAVOR-PRODUCING MICROORGANISMS

3.2.1 YEASTS AS PRODUCERS OF FLAVOR COMPOUNDS

Yeasts are well known for their oxidative and fermentative metabolism of sugars, giving rise to CO_2 and H_2O in the first case (oxidation) and to ethanol and CO_2 in the second (fermentation). Alcoholic fermentation is treated in Subsection 3.5.2 of this chapter.

Using sugars and amino acids as substrates, yeasts can also produce a wide variety of compounds contributing to the flavor of fermented foods. Flavor compounds produced by the yeasts can be classified in seven groups:

1. Superior alcohols
2. Organic acids
3. Aldehydes and ketones
4. Esters
5. Sulfur compounds
6. Terpenes
7. Lactones

A number of different works are devoted to the description of the flavor compounds in fermented drinks such as beer (Engan 1981; Nykänen and Suomalainen 1983; Potgieter 2006; Hrivnák et al. 2010), whiskey (Swan et al. 1981; Fitzgerald et al. 2000), and wine (Nykänen and Suomalainen 1983; Mateo et al. 2001).

The major flavor compound classes that are produced by yeast are described in the following seven subsections.

3.2.1.1 Superior Alcohols

Apart from ethanol, yeasts can produce other alcohols during their growth phase. For example, approximately 40 different alcohols have been identified in beers (Engan 1981). The most important ones are: propan-1-ol, 2-methyl-1-propanol, 2-methyl-1-butanol, 3-methyl-1-butanol, and phenylethanol. Most of these alcohols derive from the metabolism of amino acids, which are transformed into alcohols through the Ehrlich's pathway (Derrick and Large 1993; Hazelwood et al. 2008) in a sequence of reactions that includes:

1. Oxidative deamination of the amino acid, or transamination, giving rise to the corresponding α-keto acid
2. Decarboxylation of the α-keto acid
3. Reduction of the product

This sequence of reactions gives rise to an alcohol whose molecule contains an atom of carbon less than the amino acid used as substrate. This way, 2-methyl-1-propanol comes from valine, 2-methyl-1-butanol from isoleucine, 3-methyl-1-butanol from leucine, and phenylethanol from phenylalanine.

A superior alcohol may also be formed from an α-keto acid containing an atom of carbon more than the corresponding alcohol. This α-keto acid generally comes from the pathway of synthesis of an amino acid. The formation of superior alcohols depends on the species and even on the strain of the yeast. The increase of the temperature and the oxygenation of the culture medium stimulate the production of superior alcohols by the yeasts. Addition of amino acids to the culture medium also stimulates the synthesis of these alcohols.

3.2.1.2   Organic Acids

More than 100 organic acids have been described in fermented drinks. There are three possible origins for these acids:

1. The acetic, succinic, citric, malic, and α-ketoglutaric acids come from the tricarboxylic acid cycle.
2. Isobutyric (2-methylpropanoic) and isovaleric (3-methylbutanoic) acids come from the routes of synthesis of the amino acids.
3. Most of the organic acids (above all, the medium- and long-chain fatty acids) are synthesized from the malonyl-CoA (coenzyme A) through the action of enzymes called fatty acid synthases.

The nature and quantity of the organic acids formed depends on the composition of the fermentation medium, the species and the strain of the yeast, temperature, pH, and aeration during fermentation, with this last factor being the most important.

3.2.1.3   Aldehydes and Ketones

More than 200 different compounds belonging to the aldehyde/ketone family have been detected in fermented drinks (Nykänen and Suomalainen 1983). Aldehydes are synthesized in the pathway of synthesis of the alcohols by oxidative decarboxylation of the α-keto acids. Most of them are reduced into alcohols, a part are oxidized into acids, but another little part can be accumulated and excreted to the fermentation medium. The factors that increase the speed of fermentation (such as aeration, high temperature, easily fermentable sugars, etc.) favor an accumulation of aldehydes (Engan 1981). In the last phases of the fermentation, the aldehydes can be reabsorbed and reduced or·oxidized by the yeasts.

The diketones can be synthesized by condensation of α-keto acids with an aldehyde, and later decarboxylation. This is the way of formation of some important diketones, such as 2,3-butanedione (diacetyl), that can later be reduced into 3-hydroxy butanone (acetoin), and 2,3-pentanedione. The yeasts can reduce these diketones to 2,3-dialcohols. The quantities of diketones formed and accumulated are thus the result of the balance between their synthesis and their degradation by reduction.

3.2.1.4   Esters

Esters are commonly produced by the yeasts throughout a reaction—catalyzed by an enzyme (alcohol acyltransferase)—between the acetyl-CoA coming from the metabolism of the fatty acids and the free alcohols. The formation of these compounds has been exhaustively studied and described (Howard and Anderson 1976; Kallel-Mhiri and Miclo 1993). The nature of the esters formed depends on the alcohols available in the culture medium: The most frequent are ethyl acetate, isoamyl acetate, and propyl acetate. Ethyl acetate is the most abundant ester in the fermented drinks.

3.2.1.5 Sulfur Compounds

Nykänen and Suomalainen (1983) described more than 50 different sulfur compounds in fermented drinks. The main sulfur compounds produced by the yeasts are SO_2 and SH_2; these compounds come from the metabolic pathways of synthesis of the amino acid cysteine and methionine from sulfate (Berry and Watson 1987). Yeasts can also produce dimethyl sulfide if the precursors S-methyl methionine and dimethyl sulfoxide are present. The catabolism of methionine to methionol and methanethiol in *Saccharomyces cerevisiae* was exhaustively studied (Perpète et al. 2006). Methionine is firstly transaminated to α-keto-γ-methylthiobutyrate that is later decarboxylated into methional that is finally reduced to methionol. Methanethiol is formed from both methionine and α-keto-γ-methylthiobutyrate by a demethiolase activity. Demethiolase activity is induced by the presence of methionine in the growth medium. Recently, Liu and Crow (2010) have demonstrated the production of a wide variety of sulfur compounds, above all methionol and dimethyl disulfide, by yeasts of dairy origin in dairy media (milk and cream).

3.2.1.6 Terpenes

Some yeast species, and the wine yeasts in particular (Mateo and Di Stefano 1997), can produce flavor-active monoterpenols, such as linalool, α-terpineol, citronellol, nerol, and geraniol, from their corresponding glucosidic forms (much less flavor active than the free forms). This ability is due to the action of a β-glucosidase activity, probably present in the cell wall, that liberates the monoterpenols. β-glucosidase activity from *Saccharomyces* is quite glucose independent, but it is inhibited by about 50% with 5% of ethanol in the fermentation medium. Therefore, its action seems to be restricted to the first stages in the wine-making process.

3.2.1.7 Lactones

Some yeast species of the *Sporidiobolus* genus (e.g., *S. salmonicolor*, *S. ruinenii*, and *S. johnsonii*) are able to synthesize and accumulate 4-decanolide (γ-decalactone), a lactone that has a peachlike odor, using ricinoleic acid (12-hydroxy-9-octadecanoic acid) as the substrate (Dufossé et al. 1998). This lactone is an important flavor compound in some specific fermented fruit, dairy, and meat products. Other lactones such as cis-dodecen-4-olide are also produced by these yeasts. The main pathway for the production of 4-decanolide, and other lactones, in yeasts is through the peroxisomal β-oxidation system, which involves shortening the length of fatty acid precursor chain followed by lactonization of the resulting 4- or 5-hydroxy acids. It was reported that the use of a reducing agent such as dithiothreitol improves the 4-decanolide synthesis by *S. ruinenii* and *S. johnsonii* (X. Wang et al. 2000).

3.2.2 Molds as Producers of Flavor Compounds

Apart from their proteolytic and lipolytic activities that generate, respectively, free fatty acids and amino acids that are further degraded, it has been proved that molds can produce a wide variety of flavor compounds such as alcohols, esters, ketones, and pyrazines (Kaminski, Stawicki, and Wasowicz 1974). Among these compounds, 1-octen-3-ol was reported as the predominant, and this yields a characteristic fungal odor. Compounds, mechanisms, and metabolic routes involved in their synthesis are in most cases the same as previously described for the yeasts. We will pay attention to the methyl ketones, for their quantitative importance, and to the pyrazines, for their very low perception threshold, demonstrating their importance when present even in very low concentrations. The routes of production of these two substances have not yet been addressed in this chapter.

3.2.2.1 Methyl Ketones

In some mold-ripened foods (e.g., cheeses), methyl ketones are by far the most important flavor compounds (Molimard and Spinnler 1996). In these mold-ripened foods, the methyl ketones detected were: all the alkan-2-ones from C_4 to C_{13}, 3-methylpentan-2-one, 4-methylpentan-2-one, and methylhexan-2-one. The heptan-2-one is preponderant in blue-veined cheeses (Gallois and Langlois 1990).

The precursors in the synthesis of methyl ketones are the fatty acids. Methyl ketones are formed in a metabolic pathway that is connected to the β-oxidation pathway. *Penicillium camemberti*, *P. roqueforti*, and *Geotrichum candidum* possess an enzymatic system that permits a diversion from the normal β-oxidation pathway. The free fatty acids are firstly oxidized to β-ketoacyl-coenzyme A. The action of a thiolase yields a β-keto acid that is rapidly decarboxylated by a β-keto-acyl-decarboxylase, giving rise to a methyl ketone with one less carbon than the fatty acid precursor.

Nevertheless, the corresponding fatty acids are probably not the only methyl ketone precursors. In fact, as Molimard and Spinnler (1996) pointed out, concentrations of heptan-2-one and nonan-2-one in some cheeses are very high, higher than expected considering the amount of octanoic and decanoic acids present in milk fat. The study of the oxidation of ^{14}C-labeled palmitic and lauric acids by *Penicillium roqueforti* demonstrated the formation of methyl ketones from long-chain fatty acids by successive β-oxidation cycles (Kinsella and Hwang 1976). Methyl ketones can be reduced to the corresponding secondary alcohols by reductase activities.

3.2.2.2 Pyrazines

Two different pyrazines (2,5-dimethylpyrazine and 2-methoxy-3-isopropylpyrazine) have been identified in mold-ripened cheeses. The 2,5-dimethylpyrazine seems to come from the threonine degradation, and 2-methoxy-3-isopropylpyrazine from L-valine degradation (Seitz 1994).

3.2.3 Lactic Acid Bacteria as Producers of Flavor Compounds

Lactic acid bacteria are a heterogeneous bacterial group characterized by its ability to produce a common metabolite (lactic acid) from sugars. The lactic acid bacteria associated with foods now include species of the genera *Carnobacterium*, *Enterococcus*, *Lactobacillus*, *Lactococcus*, *Leuconostoc*, *Oenococcus*, *Pediococcus*, *Streptococcus*, *Tetragenococcus*, *Vagococcus*, and *Weissella*. Their proteolytic and lipolytic activities are generally recognized as weak, and hardly contribute to the flavor of the fermented products. The main activity of this microbial group is the utilization of the carbohydrates that are fermented, giving rise to lactic acid (see Subsection 3.5.1 in this chapter). Besides producing lactic acid, the lactic acid bacteria can produce aromatic compounds such as acetate, diacetyl, acetoin, and acetaldehyde, very important in the flavor of some dairy products.

3.2.3.1 Acetate, Diacetyl, and Acetoin

Most lactic acid bacteria are able to produce acetate, diacetyl, and acetoin from the citrate present in milk. Citrate concentration in milk is around 1.8–1.9 g/L (Braunschweig and Puhan 1999). Citrate alone cannot be utilized as a growth substrate by the lactic acid bacteria, but in the presence of a fermentable substrate and a nitrogen source, some lactic acid bacteria (*Lactococcus lactis* ssp. *lactis* biovar *diacetylactis*, *Lactobacillus plantarum*, *Lactobacillus casei*, *Leuconostoc lactis*, *Leuconostoc cremoris*, *Streptococcus salivarius* ssp. *thermophilus*, *Enterococcus faecium*) can utilize the milk citrate. Citrate metabolism plays an important role in many food fermentations involving lactic acid bacteria. Since citrate is a highly oxidized substrate, no reducing equivalents are produced during its degradation, resulting in the formation of metabolic end products other than lactic acid. Some of these end products have a very particular aroma and significantly contribute to the flavor of the fermented foods. The metabolic pathway of utilization of citrate is the same in the different genera of lactic acid bacteria. It leads to the formation of acetate, diacetyl, acetoin, and 2,3-butanediol (see Figure 3.1).

In *Lactococcus lactis* ssp. *lactis* biovar *diacetylactis*, the synthesis of diacetyl and acetoin takes place during the exponential phase of growth, and it is maximum when citrate is out. The amount of diacetyl produced is low (5–9 μg/mL) and that of acetoin is much higher (200–500 μg/mL) (Cogan 1981). In *Leuconostoc*, the diacetyl production starts at the beginning of the stationary phase of the growth; its optimum pH value for production is 4.3, and diacetyl is also produced in lesser quantities than acetoin (Cogan 1985). In the mixed cultures associating a flavor-producing species (*Lactococcus lactis* ssp. *lactis* biovar *diacetylactis* or *Leuconostoc* spp.) with a lactic acid–producing species, the production pattern of diacetyl and acetoin depends on the producer strain used. In some cases, citrate is metabolized from the starting of the growth, and in some others, citrate is used after a period of latency.

The acetoin production can take place theoretically following two different routes: (a) from the α-acetolactate through a decarboxylation process and (b) from the diacetyl through a reduction. The availability of acetyl-CoA is generally low, which favors the acetoin production from α-acetolactate. The production of 2,3-butanediol is low. The yield of the synthesis of acetoin from citrate is high, but the activity of the acetoin reductase enzyme is very low in *Lactococcus lactis* ssp. *lactis* biovar *diacetylactis* as well as in *Leuconostoc* spp. Most of the enzymes participating in the metabolism of the citrate are constitutive in *Lactococcus lactis* ssp. *lactis* biovar *diacetylactis* and are present in the absence of citrate in the medium (Cogan 1981).

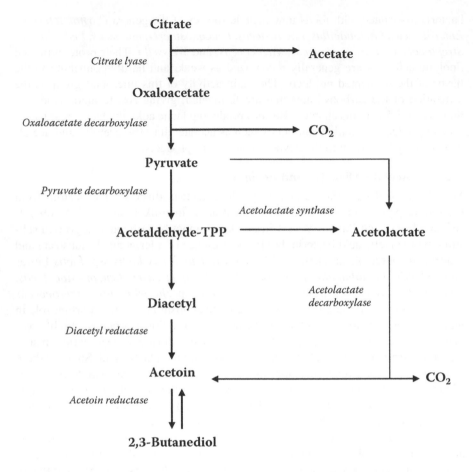

FIGURE 3.1   Citrate metabolism by lactic acid bacteria (Adapted from G. Novel. 1993. *Les bactéries lactiques*. In *Microbiologie industrielle, les micro-organismes d'intérêt industriel*, ed. J. Y. Leveau and M. Bouix, 170–374. Paris: Tec et Doc Lavoisier.)

3.2.3.2   Acetaldehyde

The acetaldehyde is an aroma compound largely produced by the starter cultures of yogurt (*Lactobacillus delbrueckii* ssp. *bulgaricus* and *Streptococcus salivarius* ssp. *thermophilus*) and is the main compound responsible for the flavor of this dairy product. The synthesis of this compound can take place through different metabolic routes:

1. From pyruvate and acetyl-CoA, which come from the metabolism of the sugars (Lees and Jago 1976)
2. From threonine, an amino acid that can be directly transformed into glycine and acetaldehyde by means of the enzyme threonine aldolase (E.C. 2.1.2.1.) (Wilkins et al. 1986)

The latter seems to be an important pathway in the starter cultures used for the production of yogurt (*Lactobacillus delbrueckii* ssp. *bulgaricus* and *Streptococcus salivarius* ssp. *thermophilus*). The threonine aldolase enzyme is also present in *Lactococcus lactis* ssp. *lactis* biovar *diacetylactis* (Lees and Jago 1976) and in several lactobacillus species (*Lb. acidophilus*, *Lb. delbrueckii* ssp. *lactis*, and *Lb. helveticus*) (Hickey, Hillier, and Jago 1983).

The great production and accumulation of acetaldehyde by the species *Lactobacillus delbrueckii* ssp. *bulgaricus* and *Streptococcus salivarius* ssp. *thermophilus* could be explained by the absence of the enzyme ethanol dehydrogenase (E.C. 1.1.1.1) and consequently the inability to produce ethanol from acetaldehyde (Lees and Jago 1976).

3.2.4 Micrococcaceae and Staphylococcaceae as Producers of Flavor Compounds

It is well known that the Gram-positive, catalase-positive cocci belonging to the former Micrococcaceae family (now grouped in two independent families, Micrococcaceae and Staphylococcaceae) actively contribute to the overall flavor of the fermented products (above all, meat products) through their lipase and protease activities. In fact, some specific species such as *Staphylococcus xylosus* and *Staph. carnosus* are widely used in combination with lactic acid bacteria as starter cultures of dry-fermented sausages. Of all the Micrococcaceae and Staphylococcaceae species, *Staphylococcus xylosus* is by far the one most known and studied regarding the ability to produce flavor compounds. Beck, Hansen, and Lauridsen (2002) exhaustively studied the metabolite production by *Staphylococcus xylosus* in a defined synthetic medium (Hussain, Hastings, and White 1991). Seven different groups of metabolites were detected: acids, alcohols, aldehydes, ketones, esters, sulfur compounds, and pyrazines. All these compounds were identified in sausages manufactured using *Staph. xylosus* as starter culture (Johansson et al. 1994). Next we will describe these compounds, as well as their possible synthesis routes. These synthesis pathways have been reported in most cases in other microorganisms distinct from *Staph. xylosus*, but this species could have these same metabolic routes.

3.2.4.1 Acids

Eight different acids were detected: acetic, lactic, propanoic, butanoic, 2-methylpropanoic, 2-methylbutanoic, 3-methylbutanoic, and benzeneacetic. Acetic, lactic, propanoic, and butanoic acids come from the pyruvate obtained from the metabolism of the sugars. The other four acids come from the metabolism of the amino acids: 2-methylpropanoic from valine, 2-methylbutanoic from isoleucine, 3-methylbutanoic from leucine, and benzeneacetic from phenylalanine.

3.2.4.2 Alcohols

Ethanol, 2-propanol, 2-methylpropanol, 3-methyl-3-butene-1-ol, 3-methyl-2-butene-1-ol, 2-methylbutanol, 3-methylbutanol, 2-hydroxy-3-butanone, and 2-phenylethanol were detected. As in the corresponding acids, ethanol and 2-propanol come from the pyruvate. The 2-hydroxy-3-butanone also comes from the pyruvate. The remaining alcohols come from the metabolism of the amino acids, e.g., 2-methylpropanol from valine; 3-methyl-3-butene-1-ol, 3-methyl-2-butene-1-ol, and 3-methylbutanol from leucine; 2-methylbutanol from isoleucine; and 2-phenylethanol from phenylalanine.

3.2.4.3   Aldehydes

Acetaldehyde, 2-methylpropanal, 2-methylbutanal, 3-methylbutanal, benzalde-hyde, and benzeneacetaldehyde were present in the culture medium after incuba-tion. Acetaldehyde comes from the pyruvate (a pathway already mentioned in lactic acid bacteria). The other aldehydes come from the metabolism of the amino acids: 2-methylpropanal from valine, 2-methylbutanal from isoleucine, 3-methylbutanal from leucine, and benzaldehyde and benzeneacetaldehyde from phenylalanine.

The synthesis of the acids, alcohols, and aldehydes coming from the correspond-ing amino acids, although it could also take place following other different routes, it seems that it takes place following the pathway already outlined for the yeasts, in the following steps:

1. Oxidative deamination of the amino acid, giving rise to the corresponding α-keto acid
2. Decarboxylation of the α-keto acid, giving rise to the corresponding aldehyde
3. Reduction of the aldehyde to the corresponding alcohol
4. Oxidation of the aldehyde to the corresponding acid

This sequence of reactions gives rise to an aldehyde, alcohol, or acid whose mol-ecule contains an atom of carbon less than the amino acid used as a precursor. The oxidation of the aldehydes 2-methylpropanal, 2-methylbutanal, and 3-methylbutanal into their corresponding acids 2-methylpropanoic, 2-methylbutanoic, and 3-methyl-butanoic, respectively, was described in the presence of *Staph. xylosus* and its kinet-ics were performed (Beck, Hansen, and Lauridsen 2002).

3.2.4.4   Ketones

Ketones produced by *Staph. xylosus* were acetone, 2-butanone, 2,3-butanodione(diacetyl), 2,3-pentanedione, and acetophenone. Acetone, 2-butanone, and 2,3-butanodione come from pyruvate. For 2,3-pentanedione, L-threonine is a precursor via 2-ketobutyrate, but pyruvate from sugar (e.g., glucose) is the major contributor via activated acetaldehyde (Ott, Germond, and Chaintreau 2000). Acetophenone seems to come from L-phenylalanine, which is firstly deamined into *trans*-cinnamic acid that is later transformed via β-oxidation into benzoic acid, with acetophenone present as a degradation intermediate in this last transformation.

3.2.4.5   Sulfur Compounds

Staphylococcus xylosus produces four sulfur compounds: dimethyldisulfide, 3-methyl-thiopropanal, 3-methylthiopropanoic acid, and 2-methyltetrahydrothiophen-3-one. Dimethyldisulfide and 3-methylthiopropanal come from methionine. The 3-methylthi-opropanoic acid could be formed from 3-methylthiopropanal via an oxidation reaction.

3.2.4.6   Esters

Two different esters were described as produced by *Staph. xylosus*: 2-phenylethylac-etate and 3-methyl-1-butylacetate. The 2-phenylethylacetate seems to come from an esterification reaction of 2-phenylethanol (comes from the amino acid phenylalanine) and acetic acid catalyzed by a lipase (Talon, Montel, and Berdagué 1996). In this same way, 3-methyl-1-butylacetate comes from an esterification reaction of 3-methylbutanol (comes from the amino acid leucine) and acetic acid.

3.2.4.7 Pyrazines

The 2,5-dimethylpyrazine was detected in the culture medium. This compound comes from threonine.

3.3 REACTIONS INVOLVED IN FLAVOR AND AROMA PRODUCTION

The flavor and aroma compounds of fermented foods come from the degradative processes undergone by the main components of the raw materials, proteins, lipids, and carbohydrates. Proteolytic, lipolytic, and glycolytic processes are, therefore, the main sources of the flavor and aroma-active compounds of fermented foods, although in some cases the oxidative processes, and the decomposition of the products of oxidation of certain molecules, can have an outstanding role in the production of flavor and aroma.

3.3.1 PROTEOLYSIS AND VOLATILE COMPOUNDS ARISING FROM AMINO ACIDS

Protein breakdown processes are catalyzed by hydrolytic enzymes that break peptide bonds in the protein molecules. These enzymes, which can be autochthonous from the raw materials (e.g., milk proteases, meat cathepsins, etc.) or of microbial origin (from the pollutant flora or from the added starter cultures), can be initially divided into two different groups: peptidases and proteinases. The peptidases (exopeptidases) hydrolyze the terminal protein or peptide bonds, releasing amino acids or dipeptides; they can be (a) aminopeptidases that catalyze the hydrolysis of the terminal peptide bond at the amino end and (b) carboxypeptidases that catalyze the hydrolysis of the terminal peptide bond at the carboxyl end. The proteinases (endopeptidases) hydrolyze the peptide bonds within the peptide chain, without hydrolyzing the terminal peptide bonds. Further division is possible using other criteria, e.g., the presence of a given amino acid residue in the active site, the presence of metal elements in the molecule, etc. (Belitz, Grosch, and Schieberle 2009a).

As a result of the action of these enzymes, polypeptides (the term *oligopeptides* is used for peptides with 10 or fewer amino acid residues) and free amino acids are freed. Most peptides are neutral in taste, but some others have particularly well-known flavors. The sweet taste of aspartic acid dipeptide esters has been well known for a long time. Some other peptides are bitter in taste. Bitter peptides have been described in cheese (Agboola, Chen, and Zhao 2004; Topçu and Saldamli 2007)—as influenced by the type of rennet used in the manufacture (Agboola, Chen, and Zhao 2004)—in some other fermented foods from soy and fish, and in sake beverages (Maehashi and Huang 2009). These peptides have been extensively studied with the aim of improving the sensory properties of fermented foods and beverages in which these peptides are generated. Maehashi and Huang (2009) reviewed the current knowledge regarding the structural features of bitter peptides and bitter taste receptors. The taste intensity of the peptides does not appear to be dependent on its amino acid sequence, but on the hydrophobicity of the side chains.

Several acidic peptides isolated from fish protein hydrolysates have a flavor-potentiating activity (Noguchi et al. 1975). These peptides (di-, tri-, tetra-, penta-, and hexapeptides) show in their molecules high molar ratios of glutamic acid and have a flavor activity qualitatively resembling that of monosodium glutamate. An octapeptide (Lys-Gly-Asp-Glu-Glu-Ser-Leu-Ala) present in papain-treated beef has an attractive taste qualified as "delicious" (Yamasaki and Maekawa 1978; Nakata et al. 1995). The sensory properties of this peptide have been examined, and its recognition thresholds at different pH values and its synergisms with salt and monosodium glutamate were determined (K. Wang, Maga, and Bechtel 1996).

Regarding the free amino acids, some of them have particular tastes. Table 3.1 (adapted from Belitz, Grosch, and Schieberle 2009a) shows the particular taste, quality, and intensity of the L- and D-isomers of the different amino acids. The taste quality is clearly determined by the particular configuration of the molecule; sweet taste is mainly a character of the D-isomers, while the bitter quality is most frequent among the L-forms. Amino acids with a cyclic lateral chain (1-aminocycloalkane-1-carboxylic acids) are sweet and bitter.

With respect to the flavor intensity, Table 3.1 also shows the threshold values of the different amino acids. The recognition threshold value is the lowest concentration necessary to recognize the compound in a reliable way when assessed by a taste panel. According to data in Table 3.1, L-tryptophan and L-tyrosine are the most bitter amino acids, and D-tryptophan is the sweetest one. When comparing the intensity of the flavor of the amino acids with that of caffeine and sucrose (used as references), it can be seen that there are seven different amino acids (D-asparagine, D-histidine, D-leucine, D-methionine, D-phenylalanine, D-tryptophan, and D-tyrosine) that are sweeter than the sucrose. In any case, when valuing the possible intervention of the free amino acids in the flavor of fermented foods, it is necessary to keep in mind that the L-form is the form presented by the common amino acids that we find in the great majority of the proteins.

L-Glutamic acid has a very characteristic and important role. In high concentrations it has a meat-broth-like flavor, while in low concentrations it enhances the particular flavor of specific foods. L-Glutamic acid is the major flavor-enhancing component of foods, and it is largely used as a flavor enhancer and salt substitute. L-Methionine has a sulfur-like flavor. Free amino acids are further degraded following several different pathways. The amino acid catabolism, carried out principally by enzymatic activities of the microorganisms present, is the real and quantitatively important source of flavor compounds from proteins in fermented foods. The possible catabolism pathways of the amino acids are illustrated in Figure 3.2 and are described in the following subsections.

3.3.1.1   Oxidative Deamination

Through oxidative deamination, the amino acid is converted into the corresponding α-keto acid by the removal of the amine functional group as ammonia, and this amine functional group is replaced by a ketone group. The ammonia can remain in the medium as a flavor compound. The α-keto acids are primarily decarboxylated with the intervention of a decarboxylase (EC 4.1.1.-), giving rise to CO_2 and

TABLE 3.1

Taste of the Different Amino Acids in Aqueous Solution at pH 6–7

Amino acid	L-isomer		D-isomer	
	Quality	Intensity [a]	Quality	Intensity [a]
Alanine	sweet	12–18	Sweet	12–18
Arginine	bitter		Neutral	
Asparagine	neutral		Sweet	3–6
Aspartic acid	neutral		Neutral	
Cystine	neutral		Neutral	
Glutamine	neutral		Sweet	
Glutamic acid	meat broth		Neutral	
Glycine [b]	sweet	25–35		
Histidine	bitter	45–50	Sweet	2–4
Isoleucine	bitter	10–12	Sweet	8–12
Leucine	bitter	11–13	Sweet	2–5
Lysine	sweet bitter	80–90	Sweet	
Methionine	sulfurous		sulfurous Sweet	4–7
Phenylalanine	bitter	5–7	Sweet	1–3
Proline	sweet bitter	25–40 25–27	Neutral	
Serine	sweet	25–35	Sweet	30–40
Threonine	sweet	35–45	Sweet	40–50
Tryptophan	bitter	4–6	Sweet	0.2–0.4
Tyrosine	bitter	4–6	Sweet	1–3

Source: Adapted from Belitz, Grosch, and Schieberle (2009a).

[a] Recognition threshold value (mmol/L).

[b] Not optically active.

an aldehyde containing an atom of carbon less than the corresponding keto acid. This process is also known as the Strecker pathway. The resulting aldehyde can be later oxidized into its corresponding acid or reduced into its corresponding alcohol. Oxidative deamination of amino acids is catalyzed by oxidoreductase enzymes. Following this pathway, several alcohols, aldehydes, and acids with notable influence in the flavor can be formed principally from glutamic and aspartic acids and from valine, leucine, phenylalanine, and methionine. For specific examples, see Section 3.2, which discusses the flavor-producing microorganisms.

3.3.1.2 Transamination

The transamination reaction implies the transfer of an amine group from one molecule to another. It is catalyzed by enzymes named transaminases or aminotransferases

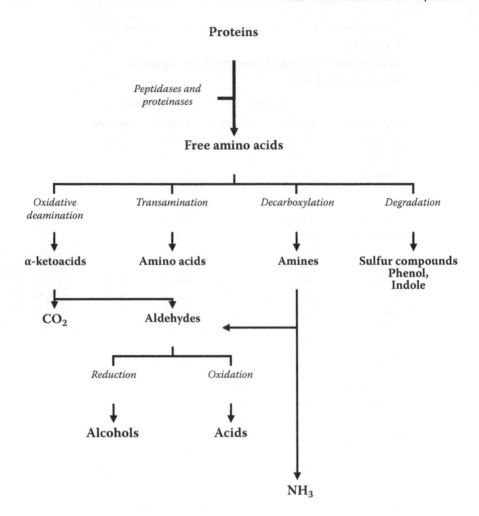

FIGURE 3.2 Routes of formation of flavor compounds from proteins (Adapted from P. Molimard and H. E. Spinnler. 1996. Review: Compounds involved in the flavor of surface mold-ripened cheeses: Origins and properties. *Journal of Dairy Science* 79:169–184.)

(EC 2.6.1.-). The transamination reaction results in the exchange of an amine group on one acid with a ketone group on another acid. The amino acid that gives the group amino is converted into its corresponding α-keto acid, and inversely the α-keto acid is converted into its corresponding amino acid. Through transamination reactions, new amino acids and α-keto acids are formed that can later follow the different catabolic routes that we are discussing for these products.

3.3.1.3   Decarboxylation

Decarboxylation releases carbon dioxide (CO_2) from the amino acid molecule, generating amines containing one atom of carbon less than the corresponding amino acid. This reaction needs the presence of pyridoxal phosphate as coenzyme. Most of the microorganisms of the bacterial flora of the raw materials (e.g., lactic acid bacteria, Enterobacteriaceae, Micrococcaceae, etc.) have decarboxylase activities that intervene in the amino acid decarboxylation reactions (Beutling 1996), and a low oxygen pressure favors these reactions. Numerous volatile amines have particular fruity, alcoholic, or varnish aromatic notes. However, some others have fishy, putty, or several other unpleasant smells. Amines are not the final step in this catabolic route. They can undergo an oxidative deamination, generating the corresponding aldehyde that can be later oxidized or reduced to its respective acid or alcohol.

3.3.1.4 Degradation Reactions

Amino acids can undergo the cleavage of various bonds in their structure, giving rise to different flavor-active compounds. For example, sulfur compounds are formed from the sulfur amino acids. Sulfur compounds come fundamentally from methionine via the cleavage of the bond between carbon and sulfur atoms by a methionine demethiolase (EC 4.4.1.11). This is the route of formation of some important flavor compounds such as the methanethiol (Ferchichi et al. 1985). The degradation of the side chain of tyrosine and tryptophan molecules by the tyrosine phenol-lyase (EC 4.1.99.2) and tryptophan indol-lyase (EC 4.1.99.1) enzymes lead to the formation of phenol and indole, respectively. These two flavor compounds have importance for several reasons in some fermented foods.

Some degradation reactions lead to the formation of short-chain fatty acids (C_1–C_4) or branched-chain fatty acids (C_4 and C_5) from some amino acids by the action of specific microorganisms. For example, mono methyl branched-chain fatty acids are synthesized from the branched-chain amino acids leucine, isoleucine, and valine in a well-known sequence of reactions. In the first step, amino acids are transformed into their corresponding branched-chain α-keto acids by the intervention of a branched-chain aminotransferase (EC 2.6.1.42). Next, the corresponding branched-chain α-keto acids are transformed into 3-methylbutanoyl-CoA, 2-methylbutanoyl-CoA, and isobutyryl-CoA, respectively, by the intervention of a branched-chain α-keto acid dehydrogenase complex that needs thiamine diphosphate, FAD, NAD, lipoate, and coenzyme A as cofactors. Finally, a fatty acid synthase catalyzes the conversion of 3-methylbutanoyl-CoA, 2-methylbutanoyl-CoA, and 2-methylpropanoyl-CoA into 3-methylbutanoic, 2-methylbutanoic, and 2-methylpropanoic acids.

3.3.2 LIPOLYSIS AND VOLATILE COMPOUNDS ARISING FROM FATTY ACIDS

Lipids in foods are a heterogeneous group of compounds grouped by its solubility characteristics rather than for a common chemical structure. They are generally soluble in organic solvents, but with scarce or null solubility in water. For details on their classification and properties, see Nawar (1996). Most lipids have fatty acids in their molecules, and they are denominated acyl lipids. In these acyl lipids, the fatty acids are generally present as esters; only in a very minor group of lipid compounds are they in amide form. Figure 3.3 shows in a concise way the routes of formation of flavor compounds from lipids.

Lipid breakdown in foods during fermentation processes takes place by enzymatic hydrolysis. Hydrolases degrading acyl lipids are present both in raw materials (milk,

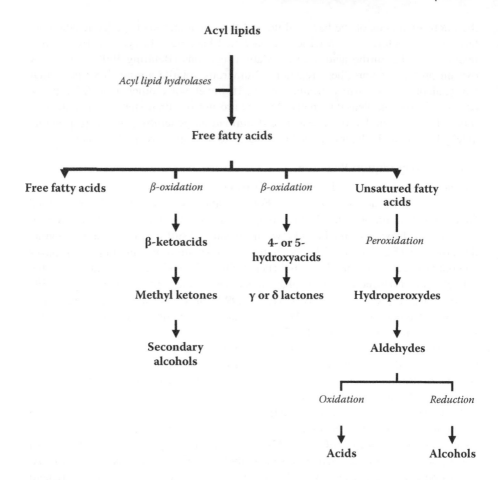

FIGURE 3.3 Routes of formation of flavor compounds from lipids (Adapted from P. Molimard and H. E. Spinnler. 1996. Review: Compounds involved in the flavor of surface mold-ripened cheeses: Origins and properties. *Journal of Dairy Science* 79:169–184.)

meat, fish, vegetables, etc.) and in microorganisms. They belong to the carboxyl-ester hydrolase group of enzymes and can be initially classified in triacylglycerol hydrolases and polar-lipid hydrolases (Belitz et al. 2009b). Triacylglycerol hydrolases (lipases) hydrolyze only acyl-lipids in emulsion. They are active on a water/lipid interface. Lipase activities are present is several foods such as milk, meat, different species of fish, cereals, oilseeds, fruits, and vegetables. Many microorganisms, above all molds, yeasts, Micrococcaceae, Staphylococcaceae, and aerobic Gram-negative rods, release lipases into their culture media. The polar-lipid hydrolases specifically hydrolyze the polar lipids (amphiphilic molecules) and are divided into phospholipases, lysophospholipases, and glycolipid hydrolases, depending on their substrate. These enzymes are also broadly distributed in animal tissues and in bacteria.

Acyl lipid hydrolysis releases free fatty acids. Most of the free fatty acids in foods, having between 4 and 22 atoms of carbon, come from the hydrolysis of the acyl lipids. A lower proportion of free fatty acids, having in general between 2 and 6 atoms of carbon, comes from the degradation of carbohydrates or free amino acids. Free fatty acids, depending on their chain size, and consequently on their volatility, can have a direct role in the flavor of fermented foods. Long-chain fatty acids (>12 carbon atoms) do not play an important role in the flavor due to their high perception thresholds. However, short- and medium-chain fatty acids (C_2 to C_{12}) have much lower perception thresholds and are easily perceived by the smell and taste senses. These acids have characteristic notes. Acetic (C_2) and propionic (C_3) acids have a typical vinegar smell. Butyric (C_4) acid has a rancid, cheesy smell. Isobutyric (2-methylpropanoic) and isovaleric (3-methylbutanoic) acids have mild, sweet, or fruity smells. Hexanoic (C_6) acid has a pungent, blue-cheese flavor note. Octanoic (C_8), 4-methyloctanoic, and 4-ethyloctanoic acids have goaty flavor notes. Decanoic (C_{10}) acid, finally, has a characteristic rancid flavor. Apart from their direct contribution to the flavor, free fatty acids can be further degraded (see Figure 3.3) by several different routes, giving rise to various important flavor compounds.

The free fatty acids can be transformed into methyl ketones. As has already been indicated when considering the importance of the molds as flavor-producing agents, some microorganisms possess an enzymatic system that permits a diversion from the normal β-oxidation pathway. The free fatty acids are first oxidized to β-ketoacyl-coenzyme A. The action of a thiolase (EC 2.3.1.16) yields a β-keto acid that is rapidly decarboxylated by a β-keto-acyl-decarboxylase, giving rise to a methyl ketone with one less carbon than the fatty acid precursor. The importance of the methyl ketones as flavor compounds is addressed in Subsection 3.2.2.1 The methyl ketones can later be reduced into their corresponding secondary alcohols; for example, heptan-2-ol from heptan-2-one and nonan-2-ol from nonan-2-one are two important secondary alcohols in the flavor of some mold-ripened cheeses.

Fatty acids may be catabolized by some microorganisms (yeasts) through the peroxisomal β-oxidation system (Endrizzi et al. 1996), giving rise to the corresponding 4- or 5-hydroxy acids. Hydroxylated fatty acids can also be generated from unsaturated fatty acids by the action of microbial hydratases (Hou 1994) or lipoxygenases (Kato et al. 1996). Hydroxylated fatty acids are direct precursors of lactones, a very important group of flavor compounds. The transformation of hydroxylated acids into lactones (the closing of the ring) can be carried out by microorganisms (Waché et al. 2003) or by the pH action.

Unsaturated fatty acids can undergo peroxidation processes. Peroxidation can be an autoxidation process or a process catalyzed by an enzyme lipoxygenase. The primary products of autoxidation are the monohydroperoxides. The lipoxygenase activity (EC 1.13.11.12) is broadly distributed in fruits and vegetables (Baysal and Demirdöven 2007). It catalyzes the oxidation of several unsaturated fatty acids to their corresponding monohydroperoxides. These hydroperoxides have the same structure as those coming from autoxidation.

The primary products of fatty acid peroxidation, the monohydroperoxides, are odorless and tasteless. However, these compounds can undergo decomposition processes to produce a multitude of volatile compounds with particular flavors. Lipid monohydroperoxides can react again with oxygen to form secondary products such as epoxyhydroperoxides, ketohydroperoxides, dihydroperoxides, cyclic peroxides, and bicyclic endoperoxides (Frankel 1984). These secondary products can in turn undergo decomposition like monohydroperoxides to form volatile breakdown products.

These volatile breakdown products are mainly odor-active carbonyl compounds (aldehydes and ketones). Those formed by autoxidation of the oleic, linoleic, and linolenic fatty acids are well known and have been quantified and well characterized regarding their flavor quality and their odor thresholds both in oil and in water (Belitz, Grosch, and Schieberle 2009b). These different compounds have generally distinct flavor notes: fruity, oily, fatty, tallowy, nutty, cucumber, frying odor, etc.

In mushrooms and plants, fatty acid hydroperoxides can be enzymatically degraded. In these organisms, four different enzymes metabolizing the hydroperoxides are active: hydroperoxide lyase (HPL), hydroperoxide isomerase, allene oxide synthase, and allene oxide cyclase. Specifically, the hydroperoxide lyase reaction can produce—from the hydroperoxides formed by lipoxygenase catalysis of linoleic and linolenic fatty acids—some important flavor compounds such as aldehydes (hexanal, (Z) 3-hexenal, (Z-Z)-3,6-nonadienal) and alcohols (1-octen-3-ol). The mechanism of the cleavage of hydroperoxides by lyases seems to be a β-cleavage (Wurzenberger and Grosch 1986).

3.3.3 GLYCOLYSIS

Glycolytic processes (the reactions of degradation of the carbohydrates) are the third main group of catabolic reactions involved in the flavor and aroma development in fermented foods. Carbohydrates first received this name because of the initial belief that all these compounds were hydrates of carbon. More recently, several compounds that do not fulfill this requirement—but do show common reactions with the carbohydrates—are also considered to belong to this group. Carbohydrates are usually classified into mono-, oligo-, and polysaccharides according to the structure of their molecules. Monosaccharides are polyhydroxyaldehydes or polyhydroxyketones. Oligosaccharides are polymers formed by fewer than 10 monosaccharide units, while polysaccharides are high-molecular-weight polymers containing more than 10 monosaccharide units. A wide variety of carbohydrates are present in foods of animal and vegetal origin, but only a reduced number of them are quantitatively important and/or significantly degraded during food fermentations.

These carbohydrates are glucose and fructose among the monosaccharides; lactose, maltose, and sucrose among the disaccharides; and glycogen and starch among the polysaccharides. D-Glucose or dextrose is widespread in animal and plant tissues. D-Fructose is abundantly present in foods of vegetal origin (principally fruits) and in honey. Lactose (B-D-galactopyranosyl-(1→4)-d-glucopyranose) is the characteristic and the only quantitatively important carbohydrate in milk. Maltose (α-D-glucopyranosyl-(1→4)-D-glucopyranose) is present in foods of vegetal origin and in honey. Sucrose (β-D-fructofuranosyl-α-D-glucopyranoside) is widely spread in vegetables. Glycogen is the storage carbohydrate in animals, and it is present in meat and fish.

Starch is the storage carbohydrate in vegetables. It is a mixture of two different glucans, amylose and amylopectin. Amylose is a linear polymer of residues α-D-glucopyranosyl linked by 1→4 bonds. Amylopectin is a branched polymer formed by a principal linear chain of residues α-D-glucopyranosyl linked by 1→4 bonds, with lateral chains bound in the 6-carbon of the glucose residues of the principal chain.

Starch is widely distributed in several plant organs, mainly in cereal grains, roots, and tubers.

Monosaccharides and disaccharides have a sweet taste (see data on this property in Table 3.2) and directly contribute to the flavor of the foods in which they are present. Moreover, when carbohydrates are catabolized, they give rise to different taste and aroma compounds that have an important role in some fermented foods. Oligo- and polysaccharides should be initially hydrolyzed into their constitutive monosaccharides before further degradation. Enzymes degrading oligo- and polysaccharides are diverse and widely distributed in foods and microorganisms.

Lactose is specifically hydrolyzed by an enzyme hydrolase named lactase or β-D-galactosidase (EC 3.2.1.23), giving rise to the constitutive monosaccharides glucose and galactose. β-D-Galactosidase activity is widely distributed in yeasts, bacteria (mainly lactic acid bacteria), and fruits. Sucrose is hydrolyzed to glucose and fructose by the enzyme β-fructofuranosidase (EC 3.2.1.26), also called invertase. This enzymatic activity is largely present in yeasts and vegetal tissues.

Several hydrolase enzymes are able to cleave polysaccharides. Starch, and in some cases glycogen, are hydrolyzed by enzymes called amylases. Four different amylases have been described: α-amylase, β-amylase, glucoamylase, and pullulanase. The α-amylase (EC 3.2.1.1) hydrolyzes starch, glycogen, and other 1,4-α-glucans. This enzyme randomly cleaves the 1,4-α-D-glucosidic bonds between adjacent glucose residues placed inside the linear molecule (endoglucanase). First, dextrins, polymers of low molecular weight containing six to seven glucose units, are formed. These dextrins are further degraded, giving rise to α-maltose as final product. The enzyme activity descends rapidly with the decrease of the degree of polymerization of the substrate. The α-amylase activity is widespread in cereal grains, yeasts, bacteria, and molds. α-Amylases of meat origin have also been described (Abbiss 1978; Skrede 1983).

TABLE 3.2
Recognition and Detection Threshold Values in Water and Relative Sweetness of Sugars

Sugar	Recognition Threshold*	Detection Threshold*	Relative Sweetness [a]
Fructose	0.052	0.02	114
Galactose			63
Glucose	0.090	0.065	69
Lactose	0.116	0.072	39
Maltose	0.080	0.038	46
Saccharose	0.024	0.011	100

Source: Adapted from Belitz, Grosch, and Schieberle (2009c).
[a] Relative sweetness to saccharose in a 10% aqueous solution.
*(mol/L)

The β-amylase (1,4-α-D-glucan maltohydrolase) (EC 3.2.1.2), also called gly-cogenase, is an exoglucanase that catalyzes the hydrolysis of 1,4-α-D-glucosidic bonds in polysaccharides, freeing successively maltose units starting from the non-reducing end. It acts on starch, glycogen, and related polysaccharides, producing β-maltose as final product. Because the amylopectin is a branched polymer, it is not completely degraded by this enzyme. The hydrolysis reaction stops upon arriving at the branch points. The β-amylase activity is present in cereal grains, yeasts, bacteria, and molds.

The glucoamylase (glucan-1,4-α-D-glucosidase) (EC 3.2.1.3) is an exoglucanase that successively frees b-D-glucose units starting from the nonreducing end of the polymer chains. In the amylopectin molecule, the α-1,6-bonds are also cleaved, but at a lower speed than the α-1,4-bonds. This enzyme is also present in cereal grains and several microorganisms, as are the other amylases. The pullulanase (α-dextrin endo-1,6-α-glucosidase) (EC 3.2.1.41), also called pullulan-6-glucanohydrolase, is a debranching enzyme that specifically hydrolyzes the 1,6-α-glucosidic bonds (the branch points) in branched polysaccharides (e.g., amylopectin, glycogen, and pul-lulan), forming linear oligosaccharide fragments. This enzyme, like the other amy-lases, is widely distributed in vegetables and microorganisms.

Most of the monosaccharides are further metabolized by microorganisms, fol-lowing the glycolysis (Embden-Meyerhoff-Parnas pathway) route (Figure 3.4), to pyruvate. Glucose and fructose, after phosphorylation, can enter as intermediary compounds in the glycolysis. Galactose should be first phosphorylated to galactose-L-phosphate by a galactokinase enzyme (EC 2.7.1.6). Galactose-L-phosphate is later converted into glucose-L-phosphate by the intervention of two different enzymes: galactose-L-phosphate uridyltransferase (EC 2.7.7.12) and UDP-D-galactose-4-epimerase (EC 5.1.3.2.). Pyruvate is the source of a wide variety of flavor compounds in fermented foods, as has already been discussed in Section 3.2. Pyruvate can be alternatively transformed into ethanol or lactic acid in, respectively, the alcoholic and lactic fermentations, two typical fermentation reactions that, due to their importance, are treated separately in Section 3.5. In some lactic acid bacteria (*Streptococcus salivarius* ssp. *thermophilus*), a part of the pyruvate not transformed in lactate can be converted in acetaldehyde following the route shown in Figure 3.1 via the inter-vention of a pyruvate decarboxylase enzyme. This acetaldehyde can be further trans-formed into diacetyl, acetoin, and 2,3-butanediol.

3.4   AROMA COMPOUNDS

Aroma compounds originated during fermentation are a group of molecules diverse in their structures and origins, coming from the degradation of the main compo-nents of foods (proteins, lipids, and carbohydrates) through the action of autoch-thonous enzymes and microorganisms. Although the degrading processes are the main source of these substances, some synthetic reactions can also be the origin of them. In the previous sections of this chapter, the microorganisms and reactions originating these compounds have been discussed. In this section, we will describe the main aroma compound groups, their origin, and the most important molecules within each group.

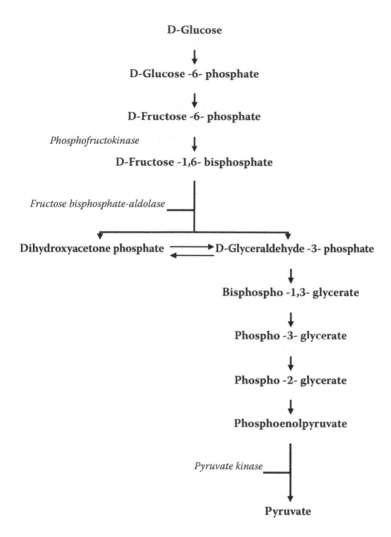

FIGURE 3.4 Glycolysis (Embden-Meyerhoff-Parnas pathway)

3.4.1 ACIDS

Organic acids are one of the most important aroma compound groups in fermented foods. As noted previously, the origin of these compounds is diverse. Most of the fatty acids come from the hydrolysis of lipids, both neutral and polar lipids, by action of the hydrolases. Short-chain linear or branched fatty acids come from the metabolism of the amino acids or carbohydrates, as noted previously. Lactic and acetic acids are the main products of specific fermentations that will be discussed in Section 3.5 of this chapter.

Some organic acids (e.g., lactic, acetic, and succinic) are minor products of the alcoholic fermentation by *Saccharomyces cerevisiae* strains (Ramon-Portugal et al. 1999). In this case, lactic and acetic acids come from the pyruvate originated in the glycolysis. Succinic acid can come from both the pyruvate and the fumarate previously originated from malic acid via the action of a fumarase enzyme (EC 4.2.1.2).

As noted previously, long-chain fatty acids (more than 12 carbon atoms) hardly contribute to the aroma of fermented foods because of their high threshold perception values. In contrast, short- and medium-chain fatty acids have lower threshold perception values, and their contribution is prominent. Threshold perception values and flavor notes of the short-chain fatty acids have been reviewed by Molimard and Spinnler (1996). The characteristics of the sour taste of some common organic acids have also been reviewed and discussed (Da Conceicao Neta, Johanningsmeier, and McFeeters 2007).

3.4.2 ALCOHOLS

Primary and secondary alcohols are a very important group of aroma compounds in fermented foods. Threshold perception values and flavor notes of the most important alcohols have been reviewed (Renger, van Hateren, and Luyben 1992; Molimard and Spinnler 1996). Alcohols can be synthesized following several different metabolic routes:

1. Carbohydrate metabolism: Ethanol can come from the alcoholic fermentation of glucose or lactose, or from glucose via the pentose phosphate pathway.
2. Amino acid metabolism: Most alcohols generated during fermentation come from the amino acid catabolism. The Ehrlich's pathway, described in Section 3.2, is the route for their production. Some specific amino acid degradation processes are responsible for the formation of particular alcohols, as described in Section 3.3.
3. Methyl ketone reduction: Methyl ketones formed from the fatty acid metabolism can be reduced into their corresponding secondary alcohols.
4. Long-chain unsaturated fatty acids degradation: In some foods, such as mold-ripened cheeses, linoleic and linolenic fatty acids are the source of different eight-carbon secondary alcohols (Molimard and Spinnler 1996). The lipoxygenase and the hydroperoxide lyase enzymes of molds seem to be the main enzymes catalyzing these routes (C. Chen and Wu 1984).

3.4.3 AMINES, AMIDES, AND OTHER NITROGEN COMPOUNDS

Amines in fermented foods come from the catabolic reactions of decarboxylation of the amino acids. Numerous amines have been described in fermented foods, above all in protein foods that undergo appreciable proteolysis during their fermentation/ripening. Amines can be volatile or nonvolatile. Volatile amines reported in fermented foods are methylamine, ethylamine, propylamine, butylamine, methylpropylamine, amylamine, hexylamine, ethanolamine, dimethylamine, diethylamine, dipropylamine, and dibutylamine. The nonvolatile amines tyramine, tryptamine, histamine, putrescine, and cadaverine have been reported. The aroma of volatile amines has been described as alcoholic, fruity, or varnishy. Trimethylamine has a typical fishy flavor. According to some authors (Dougan and Howard 1975), amines could collaborate in the ammoniacal odor of some fish sauces. Tertiary amines usually have much lower detection threshold values than the primary and secondary ones (Molimard and Spinnler 1996).

Among the amides, L-glutamine, the amide of the L-glutamic amino acid, was reported in small amounts in some fermented foods such as yogurt and miso. L-Glutamine is synthesized from L-glutamic by the enzyme glutamine synthase. L-Asparagine was reported in bread dough, but its quantities decrease as a consequence of the yeast fermentation. Anyway, these amides, at least in their L-forms, have a neutral taste. More research is needed to elucidate the role of the amide in the flavor of fermented foods.

Ammonia, coming from amino acid deamination, is responsible for the ammoniacal taste and odor of some ripened protein foods such as cheese.

3.4.4 CARBONYL COMPOUNDS

Carbonyl compounds formed during the fermentation reactions include aldehydes and ketones. Aldehydes are usually formed as products of the amino acid catabolism by two different routes:

1. Amino acid deamination or transamination, giving rise to the corresponding α-keto acid that is later decarboxylated into an aldehyde
2. Strecker degradation of the amino acids, forming initially an imine intermediate that is later hydrolyzed to CO_2, NH_2, and the corresponding aldehyde

Some aldehydes, such as acetaldehyde, can be produced from glucose by lactic acid bacteria, as discussed previously. Acetaldehyde (ethanal) can also be produced by yeasts when, in the alcoholic fermentation (see Subsection 3.5.2), alcoholic dehydrogenase activity is less than pyruvate decarboxylase activity. Other aldehydes can come from the metabolism of the unsaturated fatty acids via peroxidation reactions (Figure 3.3). In foods undergoing a fermentation process, aldehydes are usually transitory compounds that, after their formation, are immediately transformed into their corresponding acids (oxidation) or alcohols (reduction).

The main aldehydes in fermented foods are: ethanal, 2-methyl-propanal, 2-methyl-butanal, 3-methyl-butanal, hexanal, heptanal, octanal, nonanal, decanal, dodecanal, and benzaldehyde. The smell of acetaldehyde resembles green apples. Hexanal has a green note, of immature fruit. Octanal, nonanal, decanal, and dodecanal have a fruity or floral aroma, resembling orange in the case of octanal. Decanal has a scent of orange peel. Dodecanal (lauryl aldehyde) smells like violets.

The metabolic pathway of biosynthesis of the methyl ketones has already been described in Subsection 3.2.2. Diacetyl (2,3-butanedione) and acetoin (3-hydroxy-2-butanone) come from pyruvate originated from the citrate or from the glucose metabolism.

Flavor notes of the ketones are diverse. Propan-2-one and butan-2-one have a typical acetone note. Ketones from 5 to 13 atoms of carbon have usually fruity or musty notes, except for heptan-2-one, which has a characteristic blue-cheese flavor. Acetoin and diacetyl are the main compounds responsible for a buttery flavor. The perception thresholds of the main aldehydes and ketones in different solvents and foods have been reviewed and reported by Molimard and Spinnler (1996).

3.4.5 ESTERS

Esters come from esterification reactions occurring between short- to medium-chain fatty acids and alcohols (mainly ethanol) derived from carbohydrate fermentation or from amino acid catabolism. These reactions are catalyzed by enzymes, typically carboxylesterases (EC 3.1.1.1) and arylesterases (EC 3.1.1.2) present in microorganisms, mainly yeasts. Ester formation has been an object of special attention in fermented beverages, in which they play an important aromatic role (Shindo, Murakami, and Koshino 1992). The most common esters are those formed from acetyl-CoA and an alcohol, mainly ethanol, by the action of alcohol-acyltransferases (Yoshioka and Hashimoto 1984). The most habitual notes describing these compounds are melon, pear, banana, apricot, pineapple, floral, rose, honey, and wine. Some of these esters have very low perception thresholds in water, such as ethylhexanoate (1 ppb) or isoamylacetate (2 ppb). Esters having a low carbon number have perception thresholds approximately 10 times lower than those of their corresponding alcohols. Threshold perception values of the most important esters have been reviewed (Renger, van Hateren, and Luyben 1992).

3.4.6 LACTONES

Lactones are cyclic esters formed by condensation of the acid and alcohol functions forming part of the same molecule. Lactones are important flavor compounds in some fermented fruit, dairy, and meat products. They have been described to have, above all, fruity notes (peach, pear, apricot, and coconut). The role of these compounds in the flavor has been reviewed and the perception threshold values have been reported by Dufossé, Latrasse, and Spinnler (1994). Pathways of synthesis of the lactones have already been described in the previous sections of this chapter.

The most common and important lactones in fermented foods are γ-butyrolactone, δ-octalactone, γ-octalactone, δ-decalactone, γ-decalactone, δ-dodecalactone, and γ-dodecalactone. The γ-butyrolactone (4-butanolide) has a creamy, oily, fatty, or caramel odor and a milky, creamy taste, with fruity peachlike after-notes. It contributes to the aroma and flavor of some products such as shoyu (Yokotsuka 1985) and has been described in wines during fermentation (Zea et al. 1995). The flavor of the δ-octalactone (5-octanolide) has been described as animal, goaty in some cases, and as coconut or wine in some other. The flavor of the γ-octalactone (4-octanolide) was described as fruity or coconut.

The δ- and γ-decalactones and the δ- and γ-dodecalactones have fruity flavor notes (peach, apricot, coconut, pear, plum, etc.), but there is also a buttery note in the δ- and γ-dodecalactones. These four lactones have been identified among the flavor compounds of the Camembert and blue-vein cheeses. In general, δ-lactones have higher detection thresholds than γ-lactones, and the values increase as the length of the chain increases.

3.4.7   Pyrazines

Pyrazines are heterocyclic, nitrogen-containing compounds that significantly contribute to the flavor of many fermented foods. They are responsible for different flavors, depending on the nature of the alkyl substituents. Usually they are considered responsible for nutty, roasty, or toasty flavors, but they are also described as contributors of green, pealike, or bell pepper notes in raw vegetables. In the past, evidences that they can be formed by microorganisms were obtained.

The two most important pyrazines in fermented foods are 2,5-dimethylpyrazine and tetramethylpyrazine. The 2,5-dimethylpyrazine has been reported in Camembert cheese (Dumont, Roger, and Adda 1976); it imparts a toasted hazelnut note (Masuda and Mihara 1988). The odor of the tetramethylpyrazine has been described as pungent, walnut, and green. Tetramethylpyrazine was described in Chinese black vinegar (He et al. 2004), and its content was found to increase during the storage and aging process. The 2,5-dimethylpyrazine and tetramethylpyrazine are major contributors to the flavor of fermented cocoa-bean and soybean products such as soy sauce, natto, and miso (Maga 1992).

The first report of production of a pyrazine by microorganisms was made by Kosuge and Kamiya (1962), who isolated crystalline tetramethylpyrazine in cultures of *Bacillus subtilis*. Another microorganism such as *Corynebacterium glutamicum* has been revealed as a good producer of tetramethylpyrazine (Janssens et al. 1992).

The pathways of production of these two pyrazines were definitively reported by Larroche, Besson, and Gros (1999). Results obtained from these authors in a *Bacillus subtilis* strain clearly showed that threonine is the precursor of 2,5-dimethylpyrazine, with aminoacetone being an intermediary compound in this route, while tetramethylpyrazine is synthesized from acetoin and ammonia. This acetoin seems to proceed from the pyruvate coming from the metabolism of sugars (glycolysis). Recently, J. Chen et al. (2010) positively correlated the contents of tetramethylpyrazine and acetoin in vinegars.

3.4.8   Sulfur Compounds

As it was pointed out in previous sections of this chapter, sulfur compounds come from the synthesis and degradation routes of the sulfur amino acids, mainly methionine. Most sulfur compounds come from methionine degradation via the cleavage of the bond between carbon and sulfur atoms by a demethiolase activity. In this way, compounds such as methionol or methanethiol are formed. Methanethiol is later transformed in other sulfur compounds such as dimethyldisulfide and dimethyltrisulfide. Besides *Saccharomyces cerevisiae* (Perpète et al. 2006), some other microorganisms such as *Penicillium camemberti*, *Geotrichum candidum*, and *Brevibacterium linens* (Molimard and Spinnler 1996) are able to produce methanethiol from methionine.

The flavor of the sulfur compounds has been described as garlic, cabbage, cooked cabbage, cauliflower, and cooked cauliflower (Siek et al. 1969; Cuer et al. 1979; Karahadian, Josephson, and Lindsay 1985). These compounds are very volatile, and their threshold-of-perception values are very low (Molimard and Spinnler 1996).

3.5 TYPICAL REACTIONS IN SOME FERMENTED FOODS

Some specific metabolic reactions are typical in food fermentations and merit special attention for their abundance, for their importance, or for the special and unique characteristics that their products confer to the foods in which they take place.

3.5.1 LACTIC ACID FERMENTATION

Lactic acid fermentation is the main fermentation reaction in most dairy, meat, and vegetable fermented foods. It is carried out in anaerobic conditions by microorganisms belonging to the lactic aid bacteria group. This fermentation can be homolactic or heterolactic (see metabolic routes in Figure 3.5). When fermentation is homolactic, the followed route is the Embden-Meyerhof-Parnas pathway, and lactic acid is the only final product. When fermentation is heterolactic, the followed route is the pentose phosphate pathway (phosphogluconate pathway), and the final products are lactate, ethanol, and eventually acetate. On the basis of the type of lactic acid fermentation carried out, the lactic acid bacteria species are usually divided into

1. Obligatory homofermentative
2. Obligatory heterofermentative
3. Facultatively heterofermentative

The obligatory heterofermentative species are unable to degrade the fructose-1,6-bisphospate into dihydroxyacetone-phosphate and glyceraldehyde-3-phosphate. The enzyme responsible for the catalysis of this transformation, the fructose-1,6-bisphosphate aldolase (EC 4.1.2.13), would be absent or repressed (Moat 1985).

In the facultatively heterofermentative species, the type of fermentation followed depends on the abundance of fermentable carbohydrates in the medium. These microorganisms are homofermentative when carbohydrates (lactose or glucose) are abundant in the culture medium, and they become heterofermentative when these carbohydrates are in limiting concentration. The choice of the cells between the two mechanisms of fermentation would depend on the fructose-1,6-bisphospate concentration. The fructose-1,6-bisphospate activates the lactate dehydrogenase activity in the lactic acid bacteria, and it could inhibit the NADP-dependent glucose-6-phosphate dehydrogenase enzyme. Therefore, at low carbohydrate concentration and low fructose-1,6-bisphosphate concentration, the heterofermentative pathway could work in these homofermentative species.

In the heterofermentative lactic acid bacteria, the ethanol produced come through the pentose phosphate pathway from the reduction of the acetyl-phosphate. This route also lapses with the production of CO_2 in non-negligible quantities. Contrary

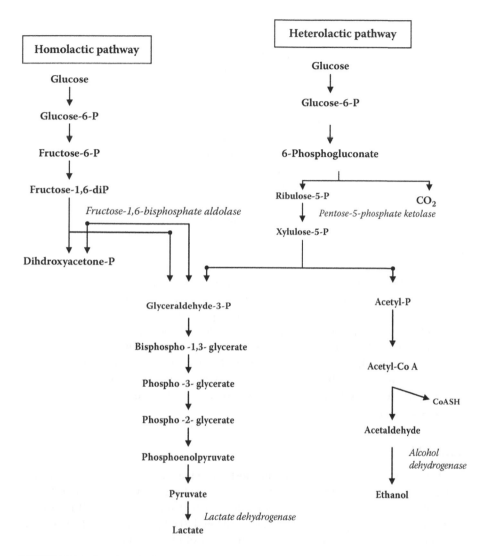

FIGURE 3.5 Lactic acid fermentation (homolactic and heterolactic pathways)

to what happens in plants or in animals, lactic acid bacteria can produce not only D(+)-lactic acid, but also L(-)-lactic acid, or the two isomers. It depends on the type of lactate dehydrogenase (LDH) present: L(+)-LDH, D(-)-LDH, or both. Lactic acid in lactic fermented foods has an important and very varied role, serving as a pH regulator, preservative, flavoring, and flavor enhancer.

3.5.2 Alcoholic Fermentation

Alcoholic fermentation is the characteristic fermentation reaction in alcoholic drinks and bread dough. The pathway of the alcoholic fermentation is shown in Figure 3.6.

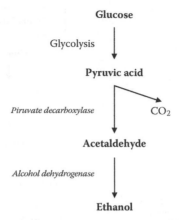

FIGURE 3.6   Alcoholic fermentation pathway

Yeasts first degrade the hexoses (glucose, fructose) via glycolysis to pyruvic acid. Pyruvate is later decarboxylated by a pyruvate decarboxylase enzyme (EC 4.1.1.1) to acetaldehyde, also producing CO_2, which is responsible for the volume increase of the bread dough during fermentation. Finally, acetaldehyde is transformed into ethanol by an alcohol dehydrogenase enzyme (EC 1.1.1.1). Ethanol plays a fundamental role in the flavor of the alcoholic drinks. Apart from its characteristic smell and taste, the ethanol enhances the fruity aroma. Part of the explanation for this enhancement is a physical effect of ethanol on the vapor pressure of the other volatiles (Williams and Rosser 1981).

3.5.3   Malolactic Fermentation

Malolactic fermentation takes place in most red wines, and in some white wines, after the alcoholic fermentation. Malic acid is transformed into lactic acid by the action of specific lactic acid bacteria species, which also produce CO_2. It is usually assumed that for each gram of malic acid fermented, 0.67 g of lactic acid and 0.37 g of carbon dioxide are produced. The stoichiometry of the reaction is

$$COOH\text{-}CHOH\text{-}CH_2\text{-}COOH \rightarrow CH_3\text{-}CHOH\text{-}COOH + CO_2$$

The species that carry out this transformation, called malolactic fermentation bacteria, belong to the *Lactobacillus*, *Leuconostoc*, *Pediococcus*, and *Oenococcus* genera, with *Oenococcus oeni* being the most desirable species. As a consequence of this fermentation, wines become rounder, softer, more equilibrate, and less abrasive to the palate through the transformation of a harsher dicarboxylic acid (malic) into a softer monocarboxylic acid (lactic).

The activity of the malolactic fermentation bacteria is not entirely selective on the malic acid. They also ferment the citric acid and the glucose. The metabolism of the citric acid gives rise to acetic acid, lactic acid, and carbon dioxide. As a consequence of the malolactic fermentation, the titratable acidity of the wine is reduced by 1–4.6 g/L. The pH increases between 0.1 and 0.45 units, and the volatile acidity (acetic acid content) increases by 0.05–0.2 g/L. Regarding the effect of the malolactic fermentation on the aroma and taste of wines, there are discrepancies in the literature. Some authors consider this effect nonexistent (Davis et al. 1985b), while some others have pointed out flavor differences in wines that undergo this fermentation (McDaniel et al. 1987). The production by the malolactic fermentation bacteria of some flavor compounds such as acetoin, diacetyl, 2-butanol, 2,3-butanediol, 1,3-propanediol, etc., would be responsible for these differences. Aspects related to the malolactic fermentation have been extensively reviewed (Davis et al. 1985a; Liu 2002).

3.5.4 ACETIC ACID FERMENTATION

The acetic acid fermentation is the basis of the manufacture of vinegar. It is an oxidation of ethanol to acetic acid via acetaldehyde carried out by Gram-negative bacteria belonging to the *Acetobacter*, *Frateuria*, and *Gluconobacter* genera, with *Acetobacter aceti* being the most efficient and commonly used bacteria for this transformation. The stoichiometry of this reaction is as follows:

$$CH_3\text{-}CH_2OH + O_2 \rightarrow CH_3\text{-}COOH + H_2O$$

The pathway is shown in Figure 3.7. Ethanol is transformed into acetaldehyde by the action of an alcohol dehydrogenase (EC 1.1.1.1), and the acetaldehyde is further oxidized into acetyl-coenzyme A (CoA) by an aldehyde dehydrogenase (EC 1.2.1.3). Acetyl-CoA is later transformed into acetyl-phosphate by a phosphotransacetylase enzyme (EC 2.3.1.8). Finally, the acetyl-phosphate is dephosphorylated into acetate by an acetate kinase (EC 2.7.2.12). This simple transformation has been the object

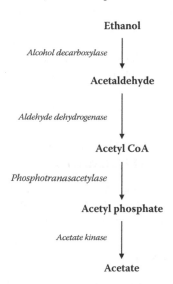

Ethanol

Alcohol decarboxylase

Acetaldehyde

Aldehyde dehydrogenase

Acetyl CoA

Phosphotranasacetylase

Acetyl phosphate

Acetate kinase

Acetate

FIGURE 3.7 Acetic acid fermentation pathway

of multiple improvements in recent years, consisting in the development of more efficient genetically modified acetic acid bacteria (Fukaya, Park, and Toda 1992) and immobilization of the producing cells (Krisch and Szajáni 1996).

3.5.5 GLYCEROPYRUVIC FERMENTATION

Glyceropyruvic fermentation is responsible for the formation of glycerol in wines. Glycerol is the third component in the wine in quantitative terms, after water and ethanol. Glyceropyruvic fermentation usually takes place at the beginning of the alcoholic fermentation, during the consumption of the first 50 g sugar/L. In the beginning of the alcoholic fermentation, yeasts initially grow in the presence of oxygen, and their pyruvate decarboxylase and alcoholic dehydrogenase activities are still weakly expressed. In this situation, the formation of ethanol is very limited: The reoxidation of the NADH is not carried out by the formation of ethanol, but by the transformation of the dihydroxyacetone phosphate into glycerol-3-phosphate. Glycerol-3-phosphate is later transformed into glycerol. Pyruvate, in the absence of enough pyruvate dehydrogenase activity, is transformed into secondary products such as α-ketoglutaric and succinic acids, 2,3-butanediol, diacetyl, etc. The pathway of the glyceropyruvic fermentation is shown in Figure 3.8.

The glyceropyruvic fermentation also takes place in the presence of sufficient quantities of sulfite (high sulfite environment). In such an environment, acetaldehyde combines with sulfite. Acetaldehyde combined with sulfite cannot be reduced into ethanol, and dihydroxyacetone-1-phosphate becomes the final electron acceptor. Dihydroxyacetone-1-phosphate is reduced to glycerol-3-phosphate, which is later dephosphorylated into glycerol.

The production of glycerol is influenced by several factors that include the yeast strain as well as temperature and pH values. Glycerol production is promoted by high pH values in musts, and it is accompanied by the production of other flavor compounds, as noted previously. Glycerol has a sweet taste. However, the amount of glycerol necessary to produce a detectable increase in the sweetness of the wines is around 5 g/L. Glycerol is responsible for taste notes such as oiliness, persistence, and mellowness in mouth. However, glycerol is not involved in modification of the aroma in white wines (Lubbers, Verret, and Voilley 2001).

3.6 FACTORS THAT AFFECT THE REACTIONS OF FLAVOR COMPOUND FORMATION

The biochemical reactions of formation of flavor compounds are mainly enzymatic-catalyzed reactions. In such reactions, the rate is influenced by the concentration of substrate, temperature, characteristics of the medium, and presence of inhibitors.

3.6.1 CONCENTRATION OF SUBSTRATE

For simple (not catalyzed) chemical reactions with a single substrate, the rate of product formation shows a linear relationship with the substrate concentration. However,

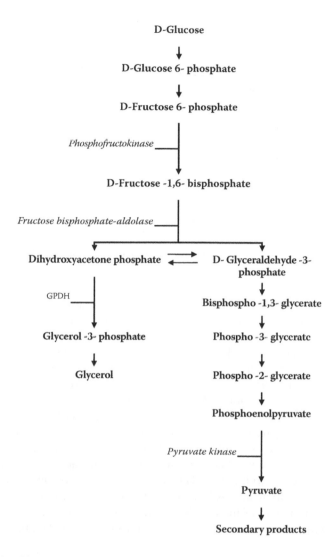

FIGURE 3.8   Glyceropyruvic fermentation pathway

in the enzyme-catalyzed reactions, the relationship between the rate of reaction and the substrate concentration is usually hyperbolic (see Figure 3.9).

At low substrate concentration, the rate of reaction increases as the concentration of the substrate increases. In these conditions, the catalytic site of the enzyme is empty and ready to bind substrate, and the rate of product formation is only limited by the concentration of substrate available. As the concentration of substrate increases, the catalytic site of the enzyme becomes saturated with the substrate. As soon as the catalytic site of the enzyme is vacant, more substrate is ready to bind. In such conditions, the rate of product formation depends only on the activity of the enzyme itself, and the addition of an excess of substrate does not affect the rate of reaction in a significant way.

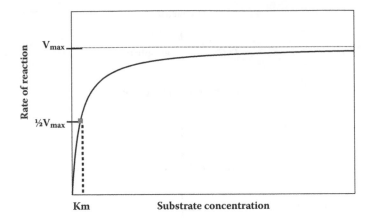

FIGURE 3.9   Effect of substrate concentration on the rate of reaction for monosubstrate enzyme-catalyzed reactions

The rate of reaction achieved when the enzyme is saturated with the substrate is the maximum rate of reaction (Vmax.). The concentration of substrate at which the rate of the reaction is half of the maximum rate is usually denominated Km (Michaelis-Menten constant). The value of Km gives an idea of the affinity of the enzyme for a specific substrate. The higher the value of Km, the lower is the affinity of the enzyme for the substrate. The comparison between the Km value and the physiological concentration of the substrate will give us an idea of whether or not the availability of substrate will affect the rate of product formation. When the Km value is low compared to the physiological concentration of the substrate, the enzyme is saturated with the substrate, and the enzyme activity is independent of the substrate concentration. Therefore, the reaction will lapse to constant speed. Conversely, an enzyme with a high Km value in relation to the physiological concentration of the substrate acts as if it were not saturated with substrate. Therefore, its activity—and the rate of the reaction—will vary with the concentration of the substrate, and the rate of product formation will strongly depend on the availability of substrate.

3.6.2   TEMPERATURE

The effect of temperature on the rate of chemical reactions is well known. In general, the speed of reaction increases with temperature. It is assumed that an increase of 10°C in temperature doubles the reaction speed until an optimal temperature value— the temperature at which the speed of reaction is maximal—is achieved. The optimal temperature value is different for different enzymes. For temperatures above the optimal, the increase in reaction speed due to the increase in temperature is counteracted by the loss of activity of the enzyme due to a protein-denaturation process. This enzymatic denaturation is clearly perceptible above 45°C. Microbial enzymes involved in fermentation reactions often have optimal temperatures between 30°C and 45°C.

3.6.3   CHARACTERISTICS OF THE MEDIUM

The medium of fermentation affects the rate of enzymatic reactions via its pH value and the presence of cofactors. The enzymes, constructed from amino acids, include ionizable chemical groups such as carboxyl, amino, thiol, imidazole, etc. Depending on the pH value of the medium, these groups can have positive, negative, or neutral charge. The protein conformation depends on the electrical charge and, therefore, on the pH value. Thus, for each enzyme, there is an optimal pH value at which its conformation is optimal for its catalytic activity. Enzymes are, in general, very sensitive to pH variations. The optimal pH value varies greatly among the enzymes involved in food fermentation. However, most of them have an optimal pH value near neutrality. The catalytic function of some enzymes depends on specific cofactors that collaborate in the catalysis. These cofactors can be inorganic ions (Ca^{+2}, Fe^{+2}, Mg^{+2}, Mn^{+2}, Zn^{+2}, etc.) or organic molecules (coenzymes).

3.6.4   PRESENCE OF INHIBITORS

The catalytic action of an enzyme can be inhibited by certain molecules. These inhibitors can temporarily occupy the active site of the enzyme due to structural similarity with the substrate (competitive inhibitors), or they can alter the space conformation of the enzyme, impeding its binding to the substrate (noncompetitive inhibitors). Product inhibition can also occur in some catabolic reactions. In such cases, the product of the reaction binds to the enzyme and inhibits its activity. In other cases, feedback inhibition can occur. In feedback inhibition, the end product of a pathway inhibits the first enzyme of the route when its concentration achieves a certain value. In this case of inhibition, there is a second binding site in the enzyme molecule where the inhibitor binds. However, note that the structure of the inhibitor is not necessarily similar to that of the substrate.

3.7   CONCLUDING REMARKS

The flavor compounds of fermented foods are generated as products of the catabolic reactions of proteins, lipids, and carbohydrates. (This does not imply that the contribution of the anabolic reactions is unimportant.) Most of the catabolic reactions are enzyme catalyzed. Amino acids and fatty acids are further degraded, giving rise to a wide variety of compounds that show different flavor notes and perception-threshold values according to their molecular characteristics. In the carbohydrate catabolism, the pyruvate is the key product from which the most relevant flavor compounds are formed.

REFERENCES

Abbiss, J. S. 1978. *Enzyme-mediated changes in carbohydrate in the British fresh sausage.* Ph.D. thesis, University of Bath, U.K.

Agboola, S., S. Chen, and J. Zhao. 2004. Formation of bitter peptides during ripening of ovine milk cheese made with different coagulants. *Lait* 84:567–578.

Baysal, T., and A. Demirdöven. 2007. Lipoxygenase in fruits and vegetables: A review. *Enzyme and Microbial Technology* 40:491–496.

Beck, H. C., A. M. Hansen, and F. R. Lauridsen. 2002. Metabolic production and kinetics of branched-chain aldehyde oxidation in *Staphylococcus xylosus*. *Enzyme and Microbial Technology* 31:94–101.

Belitz, H. D., W. Grosch, and P. Schieberle. 2009a. Amino acids, peptides, proteins. In *Food Chemistry*, 8–92. Berlin: Springer-Verlag.

———. 2009b. Lipids. In *Food Chemistry*, 158–245. Berlin: Springer-Verlag.

———. 2009c. Carbohydrates. In *Food Chemistry*, 248–339. Berlin: Springer-Verlag.

Berry, D. R, and D. C. Watson. 1987. Production of organoleptic compounds. In *Yeast Biotechnology*, ed. D. R. Berry, I. Russel, and G. G. Stewart, 345–364. London: Allen & Unwin.

Beutling, D. 1996. *Biogene Amine in der Ernaehrung*, 59–67. Berlin: Springer-Verlag.

Braunschweig, M., and Z. Puhan. 1999. Correlation between κ-casein variants and citrate content in milk quantified by capillary electrophoresis. *International Dairy Journal* 9:709–713.

Chen, C. C., and C. M. Wu. 1984. Studies on the enzymic reduction of 1-octen-3-one in mushroom (*Agaricus bisporus*). *Journal of Agricultural and Food Chemistry* 32:1342–1344.

Chen, J. C., Q. H. Chen, Q. Guo, S. Ruan, H. Ruan, G. Q. He, and Q. Gu. 2010. Simultaneous determination of acetoin and tetramethylpyrazine in traditional vinegars by HPLC method. *Food Chemistry* 122:1247–1252.

Cogan, T. M. 1981. Constitutive nature of the enzymes of citrate metabolism is *Streptococcus lactis* subsp. *diacetylactis*. *Journal of Dairy Research* 48:489–495.

———. 1985. The *Leuconostocs*: milk products. In *Bacterial starters cultures for foods*, ed. S.E. Gilliland, Vol. 3, 25–40. Boca Raton: CRC Press.

Cuer, A., G. Dauphin, A. Kergomard, S. Roger, J. P. Dumont, and J. Adda. 1979. Flavour properties of some sulfur compounds isolated from cheeses. *Lebensmittel Wissenschaft und Technologie* 12:258–261.

Da Conceicao Neta, E. R., S. Johanningsmeier, and R. F. McFeeters. 2007. The chemistry and physiology of sour taste: A review. *Journal of Food Science* 72:33–38.

Davis, C. R., D. Wibowo, R. Eschenbruch, T. H. Lee, and G. H. Fleet. 1985a. Practical implications of malolactic fermentation: A review. *American Journal of Enology and Viticulture* 36:290–301.

———. 1985b. Practical implications of malolactic fermentation in wine. *Journal of Applied Bacteriology* 63:513–521.

Derrick, S., and P. J. Large. 1993. Activities of the enzymes of the Ehrlich pathway and formation of branched-chain alcohols in *Saccharomyces cerevisiae* and *Candida utilis* grown in continuous culture on valine or ammonium as sole nitrogen source. *Journal of General Microbiology* 139:2783–2792.

Dougan, J., and G. E. Howard. 1975. Some flavouring constituents of fermented fish sauces. *Journal of the Science of Food and Agriculture* 26:887–894.

Dufossé, L., G. Feron, G. Mauvais, P. Bonnarme, A. Durand, and H. E. Spinnler. 1998. Production of γ-decalactone and 4-hydroxy-decanoic acid in the genus *Sporidiobolus*. *Journal of Fermentation and Bioengineering* 86: 169–173.

Dufossé, L., A. Latrasse, and H. E. Spinnler. 1994. Importance des lactones dans les âromes alimentaires: Structures, distribution, propriétés sensorielles et biosynthèse. *Sciences des Aliments* 14:17–50.

Dumont, J. P., S. Roger, and J. Adda. 1976. L'ârome du Camembert: Autres composés mineurs mis en evidence. *Lait* 56:595–599.

Endrizzi, A., Y. Pagot, A. Le Clainche, J. M. Nicaud, and J. M. Belin. 1996. Production of lactones and peroxisomal beta-oxidation in yeasts. *Critical Reviews in Biotechnology* 16:301–329.

Engan, S. 1981. Beer composition: Volatile substances. In *Brewing sciences*, ed. J. R. A. Pollock, Vol. 2, 93–165. London: Academic Press.

Ferchichi, M., D. Hemme, M. Nardi, and N. Pamboukdjian. 1985. Production of methanethiol from methionine by *Brevibacterium linens* CNRZ 918. *Journal of General Microbiology* 131:715–723.

Fitzgerald, G., K. J. James, K. MacNamara, and M. A. Stack. 2000. Characterisation of whiskeys using solid-phase microextraction with gas chromatography-mass spectrometry. *Journal of Chromatography A* 896:351–359.

Fukaya, M., Y. S. Park, and K. Toda. 1992. A review: Improvement of acetic acid fermentation by molecular breeding and process development. *Journal of Applied Bacteriology* 73:447–454.

Frankel, E. N. 1984. Lipid oxidation: Mechanisms, products, and biological significance. *Journal of the American Oil Chemists' Society* 61:1908–1917.

Gallois, A., and D. Langlois. 1990. New results in the volatile odorous compounds of French cheeses. *Lait* 70:89–106.

Hazelwood, L. A., J. M. Daran, A. J. van Maris, J. T. Pronk, and J. R. Dickinson. 2008. The Ehrlich pathway for fusel alcohol production: A century of research on *Saccharomyces cerevisiae* metabolism. *Applied and Environmental Microbiology* 74:2259–2266.

He, Z. Y., Z. H. Ao, J. Wu, G. Q. Li, and W. Y. Tao. 2004. Study of the mensuration of tetramethylpyrazine in Zhenjiang vinegar and its generant mechanism. *China Condiment* 2:36–39.

Hickey, M. W., A. J. Hillier, and G. R. Jago. 1983. Enzymatic activities associated with lactobacilli in dairy products. *Australian Journal of Dairy Technology* 38:154–157.

Hou, C. T. 1994. Production of hydroxy fatty acids from unsaturated fatty acids by *Flavobacterium* sp. DS5 hydratase, a C-10 positional- and *cis* unsaturation-specific enzyme. *Journal of the American Oil Chemists' Society* 72:1265–1270.

Howard, D., and R. G. Anderson. 1976. Cell-free synthesis of ethyl acetate by extracts from *Saccharomyces cerevisiae*. *Journal of the Institute of Brewing* 82:70–71.

Hrivnák, J., D. Smogrovicová, J. Lakatosová, and P. Nádaský. 2010. Technical note: Analysis of beer aroma compounds by solid-phase microcolumn extraction. *Journal of the Institute of Brewing* 116:167–169.

Hussain, M., J. G. M. Hastings, and P. J. White. 1991. A chemically defined medium for slime production by coagulase-negative *Staphylococci*. *Journal of Medical Microbiology* 34:143–147.

Janssens, L., H. L. De Pooter, N. M. Schamp, and E. J. Vandamme. 1992. Review: Production of flavours by microorganisms. *Process Biochemistry* 27:195–215.

Johansson, G., J. L. Berdagué, M. Larsson, N. Tran, and E. Borch. 1994. Lipolysis, proteolysis and formation of volatile compounds during ripening of a fermented sausage with *Pediococcus pentosaceus* and *Staphylococcus xylosus* as starter cultures. *Meat Science* 38:203–218.

Kallel-Mhiri, H., and A. Miclo. 1993. Mechanism of ethyl acetate synthesis by *Kluyveromyces fragilis*. *FEMS Microbiology Letters* 111:207–212.

Kaminski, E., S. Stawicki, and E. Wasowicz. 1974. Volatile flavor compounds produced by molds of *Aspergillus*, *Penicillium*, and *Fungi imperfecti*. *Applied Microbiology* 27:1001–1004.

Karahadian, C., D. B. Josephson, and R. C. Lindsay. 1985. Contribution of *Penicillium* sp. to the flavors of Brie and Camembert cheese. *Journal of Dairy Science* 68:1865–1877.

Kato, T., T. Watanabe, T. Hirukawa, N. Tomita, and T. Namai. 1996. Production of hydroxy unsaturated fatty acids using crude lipoxygenase obtained from infected rice plants. *Bulletin of the Chemical Society of Japan* 69:1663–1666.

Kinsella, J. E., and D. H. Hwang. 1976. Biosynthesis of flavors by *Penicillium roqueforti*. *Biotechnology and Bioengineering* 18:927–938.

Kosuge, T., and H. Kamiya. 1962. Discovery of a pyrazine in a natural product: Tetramethylpyrazine from cultures of a strain of *Bacillus subtilis*. *Nature* 193:776.

Krisch, J., and B. Szajáni. 1996. Effects of immobilization on biomass production and acetic acid fermentation of *Acetobacter aceti* as a function of temperature and pH. *Biotechnology Letters* 18:393–396.

Larroche, C., I. Besson, and J. B. Gros. 1999. High pyrazine production by *Bacillus subtilis* in solid substrate. *Process Biochemistry* 34:667–674.

Lees, G. J., and G. R. Jago. 1976. Acetaldehyde: An intermediate in the formation of ethanol from glucose by lactic acid bacteria. *Journal of Dairy Research* 43:63–73.

Liu, S. Q. 2002. Malolactic fermentation in wine: Beyond deacidification. *Journal of Applied Microbiology* 92:589–601.

Liu, S. Q., and V. L. Crow. 2010. Production of dairy-based, natural sulfur flavor concentrate by yeast fermentation. *Food Biotechnology* 24:62–77.

Lubbers, S., C. Verret, and A. Voilley. 2001. The effect of glycerol on the perceived aroma of a model wine and a white wine. *Lebensmittel Wissenschaft und Technologie* 34:262–265.

Maehashi, K., and L. Huang. 2009. Bitter peptides and bitter taste receptors. *Cellular and Molecular Life Sciences* 66:1661–1671.

Maga, J. A. 1992. Pyrazine update. *Food Reviews International* 8:479–558.

Masuda, H., and S. Mihara. 1988. Olfactive properties of alkyl-pyrazines and 3-substitued 2-alkyl-pyrazines. *Journal of Agricultural and Food Chemistry* 36:584–587.

Mateo, J. J., and R. Di Stefano. 1997. Description of the β-glucosidase activity of wine yeasts. *Food Microbiology* 14:583–591.

Mateo, J. J., M. Jiménez, A. Pastor, and T. Huerta. 2001. Yeast starter cultures affecting wine fermentation and volatiles. *Food Research International* 34:307–314.

McDaniel, M., L. A. Henderson, B. T. Watson Jr., and D. Heatherbell. 1987. Sensory panel training and screening for descriptive analysis of the aroma of Pinot Noir wine fermented by several strains of malolactic bacteria. *Journal of Sensory Studies* 2:149–167.

Moat, A. G. 1985. Biology of the lactic, acetic, and propionic acid bacteria. In *Biology of industrial microorganisms*, ed. A. L. Demain and N. A. Salomon, 143–188. Menlo Park, CA: Benjamin/Cummings Publishing.

Molimard, P., and H. E. Spinnler. 1996. Review: Compounds involved in the flavor of surface mold-ripened cheeses: Origins and properties. *Journal of Dairy Science* 79:169–184.

Nakata, T., M. Takahashi, M. Nakatani, R. Kuramitsu, and M. Tamura. 1995. Role of basic and acidic fragments in delicious peptides (Lys-Gly-Asp-Glu-Glu-Ser-Leu-Ala) and the taste behaviour of sodium and potassium salts in acidic oligopeptides. *Bioscience, Biotechnology and Biochemistry* 59:689–693.

Nawar, W. W. 1996. Lipids. In *Food chemistry*, 3rd ed., ed. O. R. Fennema, 225–319. New York: Marcel Dekker.

Noguchi, M., S. Arai, M. Yamashita, H. Kata, and M. Fujimaki. 1975. Isolation and identification of acidic oligopeptides occurring in a flavor potentiating fraction from a fish protein hydrolysate. *Journal of Agricultural and Food Chemistry* 23:49–53.

Novel, G. 1993. Les bactéries lactiques. In *Microbiologie industrielle, les micro-organismes d'intérêt industriel*, ed. J. Y. Leveau and M. Bouix, 170–374. Paris: Tec et Doc Lavoisier.

Nykänen, L., and H. Suomalainen. 1983. *Aroma of beer, wine, and distilled alcoholic beverages*. Dordrecht, Netherlands: Reidel Publishing.

Ott, A., J. E. Germond, and A. Chaintreau. 2000. Vicinal diketone formation in Yoghurt: [13]C precursors and effect of branched-chain amino acids. *Journal of Agricultural and Food Chemistry* 48:724–731.

Perpète, P., O. Duthoit, S. De Maeyer, L. Imray, A. I. Lawton, K. E. Stravropoulos, V. W. Gitonga, M. J. Hewlings, and J. R. Dickinson. 2006. Methionine catabolism in *Saccharomyces cerevisiae*. *FEMS Yeast Research* 6:48–56.

Potgieter, N. 2006. *Analysis of beer aroma using purge-and-trap sampling and gas chromatography*. Magister Scientiae Report. University of Pretoria, South Africa.

Ramon-Portugal, F., I. Seiller, P. Taillandier, J. L. Faravel, F. Nepveu, and P. Strehaiano. 1999. Kinetics of production and consumption of organic acids during alcoholic fermentation by *Saccharomyces cerevisiae*. *Food Technology and Biotechnology* 37:235–240.

Renger, R. S., S. H. van Hateren, and K. Ch. A. M. Luyben. 1992. The formation of esters and higher alcohols during brewery fermentation: The effect of carbon dioxide pressure. *Journal of the Institute of Brewing* 98:509–513.

Seitz, E. W. 1994. Fermentation production of pyrazines and terpenoids for flavors and fragrances. In *Bioprocesses production of flavour, fragrance, and color ingredients*, ed. A. Gabelman, 95–134. New York: Wiley.

Shindo, S., J. Murakami, and S. Koshino. 1992. Control of acetate ester formation during alcohol fermentation with immobilized yeasts. *Journal of Fermentation and Bioengineering* 73:370–374.

Siek, T. J., I. A. Albin, L. A. Sather, and R. C. Lindsay. 1969. Taste thresholds of butter volatiles in deodorized butteroil medium. *Journal of Food Science* 34:265–267.

Skrede, G. 1983. Enzymic starch-degrading ability of meat and blood plasma in products after processing. *Food Chemistry* 12:15–24.

Swan, J. S., D. Howie, S. M. Burtles, A. A. Williams, and M. J. Lewis. 1981. Sensory and instrumental studies of Scotch whisky flavour. In *The quality of food and beverages: Chemistry and Technology*, ed. G. Charalambous and G. E. Inglett, 201–225. New York: Academic Press.

Talon, R., M. C. Montel, and J. L. Berdagué. 1996. Production of flavor esters by lipases of *Staphylococcus warneri* and *Staphylococcus xylosus*. *Enzyme and Microbial Technology* 19:620–622.

Topçu, A., and I. Saldamli. 2007. Determination of peptides caused bitterness in Turkish white cheese and Kasar cheese. *Journal of Food Technology* 5:131–134.

Waché, Y., M. Aguedo, J. M. Nicaud, and J. M. Belin. 2003. Catabolism of hydroxyacids and biotechnological production of lactones by *Yarrowia lipolytica*. *Applied Microbiology and Biotechnology* 61:393–404.

Wang, K., J. A. Maga, and P. J. Bechtel. 1996. Taste properties and synergisms of beefy meaty peptide. *Journal of Food Science* 61:837–839.

Wang, X. D., G. Mauvais, R. Cachon, C. Diviès, and G. Feron. 2000. Addition of reducing agent dithiothreitol improves 4-decanolide synthesis by the genus *Sporidiobolus*. *Journal of Bioscience and Bioengineering* 90:338–340.

Wilkins, D. W., R. H. Schmidt, R. B. Shireman, K. L. Smith, and L. B. Kennedy. 1986. Evaluating acetaldehyde synthesis from L-[14C(U)] threonine by *Streptococcus thermophilus* and *Lactobacillus bulgaricus*. *Journal of Dairy Science* 69:1219–1224.

Williams, A. A., and P. R. Rosser. 1981. Aroma enhancing effects of ethanol. *Chemical Senses* 6:149–153.

Wurzenberger, M., and W. Grosch. 1986. Enzymic oxidation of linoleic acid to 1,Z-5-octadien-3-ol, Z-2,Z-5-octadien-1-ol and 10-oxo-E-8-decenoic acid by a protein fraction from mushrooms (*Psalliota bispora*). *Lipids* 21:261–266.

Yamasaki, Y., and K. Maekawa. 1978. A peptide with delicious taste. *Agricultural and Biological Chemistry* 42:1761–1765.

Yokotsuka, T. 1985. Fermented protein foods in the Orient with emphasis on shoyu and misu in Japan. In *Microbiology of Fermented Foods*, ed. B. J. B. Wood, Vol. 1, 197–248. Amsterdam: Elsevier.

Yoshioka, K., and N. Hashimoto. 1984. Ester formation by brewers' yeasts during fermentation. *Agricultural and Biological Chemistry* 48:333–340.

Zea, L., J. Moreno, J. M. Ortega, J. C. Mauricio, and M. Medina. 1995. Comparative study of the γ-butyrolactone and pantolactone contents in cells and musts during vinification of three *Saccharomyces cerevisiae* races. *Biotechnology Letters* 17:1351–1356.

4 Effect of Fermentation Reactions on Rheological Properties of Foods

Robert Z. Iwański, Marek Wianecki,
Izabela Dmytrów, and Krzysztof Kryża

CONTENTS

4.1 INTRODUCTION

Fermentation processes have been used by humankind since the dawn of history, when it was discovered that simple actions can induce substantial sensory changes in prepared meals. Initially, these changes were uncontrolled and occurred spontaneously with a high risk of failure. With the beginning of mass-scale production, the problems of product quality and shelf life have emerged. These concerned mainly such products as beer and wine, where apart from developing a new product with altered taste and odor, the combined effect of preservation and change of consistency, structure, and shape was achieved. Whitaker (1978) describes changes in texture as one of the fundamental objectives of fermentation. The fermentative transformations evoked mainly by bacteria, fungi, and molds involve profound changes in the final product in comparison to the raw material. These changes affect not only the organoleptic characteristics, but also the fundamental rheological properties of a product.

Fermentative changes proceed mainly in raw food materials containing fermentable sugars (monosaccharides, disaccharides) and polysaccharides (starch after prior enzymatic hydrolysis) as well as milk sugar (lactose). The quality and rate of these changes are determined by the share of saccharides in the total weight and the presence of nutrients essential for the growth of fermentation microorganisms, i.e., proteins (nitrogen). The progress of fermentation depends also on its nature (spontaneous fermentation, or evoked deliberately and intentionally) as well as on the applied and existing microflorae.

The rheology of fermentative changes is also determined by appropriate treatment of raw material, before and during fermentative changes (mixing, aeration, homogenization, enzymatic modification, modification of fermentation atmosphere, emulsification). The fermentative changes directly affect the basic measurable rheological attributes of the product, including: hardness, consistency, adhesiveness, and viscosity, among which the hardness and viscosity are described most often. According to Bourne (1982a), the measurement of rheological properties merges three basic methods of measurements: fundamental principles, empirical methods, and imitation. Therefore, the appropriate fundamental measurement should always be supported by sensory evaluation performed by a properly trained panel.

4.2 CHANGES IN CARBOHYDRATES AND RELATED RHEOLOGICAL CHANGES

Some microorganisms, including lactic acid bacteria, are able to produce polysaccharides called exopolysaccharides (EPS), which act as thickening agents. EPS of microbiological origin are extracellular polysaccharides, which may be both bound to the surface of cells and secreted to the extracellular environment in the form of suspension or mucilage (De Vuyst and Degeest 1999). They are regarded as safe (GRAS—generally recognized as safe) and play a beneficial role in the formation of rheological properties, texture stability, and mouthfeel of fermented milk products, such as yogurt (Cerning 1995; Wellman and Maddox 2003).

It has been shown that the structure of EPS depends on the composition of the environment, the source of carbon and nitrogen in particular; on the conditions of bacterial growth, i.e., pH, temperature, oxygen concentration; and on the strain of lactic acid bacteria (LAB) (Ruas-Madiedo, Alting, and Zoon 2005). For example, structural analysis of EPS produced by *L. delbrueckii* subsp. *Bulgaricus NCFB 2772* showed that EPS contained repeating units of glucose and galactose (in the ratio of 1:2.4) when the growth was taking place on fructose. If the bacteria were grown on a mixture of fructose and glucose, the repeating units would consist of glucose, galactose, and rhamnose in the ratio of 1:7:0.8. The dependence of EPS structure on the source of carbon was not noted in the case of the growth of, e.g., *Lactobacillus sake*. At lower temperatures, the production of EPS by LAB increases. For example, the amount of EPS produced by mesophilic LAB is about 50% greater when the bacteria growth takes place at 25°C instead of 30°C (Cerning 1995).

The chemical structure of EPS produced by LAB is controversial. It is believed, still, that exopolysaccharides are built from repeating branched units containing α- and β-links. Despite differences in the structure, the molecules of d-galactose, d-glucose, and l-rhamnose monomers, which occur in various proportions, may be distinguished. The apparent molecular weight of EPS produced by LAB ranges from $1.0 \cdot 10^4$ to $26.5 \cdot 10^6$ Da and depends on the source of nitrogen in the growth medium of LAB (Degeest, Vaningelgem, and De Vuyst 2001; Vaningelgem et al. 2004; Ruas-Madiedo, Alting, and Zoon 2005; Lin and Chien 2007). It is one of the factors determining functional properties of EPS. However, the physical and the rheological properties of polysaccharides in solutions strictly depend on their spatial structure or conformation, as well as on their capability to form intermolecular associations (De Vuyst and Degeest 1999). For instance, EPS produced by LAB can be divided into two groups. The first group comprises homopolysaccharides, including α-d-glucans, β-d-glucans, fructans, and polygalactans. The second group is constituted by heteropolysaccharides, which play an important role in the development of rheological properties, texture, and oral sensation (De Vuyst and Degeest 1999). Heteropolysaccharides consist of repeating units of di- to heptasaccharides, which may also contain nonsugar molecules (Ruas-Madiedo, Hugenholtz, and Zoon 2002).

At the optimal conditions of growth, the LAB may produce 0.15–0.60 g/L of EPS (Cerning 1995; Degeest, Vaningelgem, and De Vuyst 2001). However, in the environment with high saccharose concentration, i.e., up to 100 g/L, some LAB can produce more than 25 g/L of EPS (van Geel-Schutten et al. 1998). The disadvantage is a low and unstable production of EPS by thermophilic LAB, but attempts are being made to increase the efficiency of EPS production through the application of two-stage fermentation (Degeest, Vaningelgem, and De Vuyst 2001). The efficiency of EPS production up to 1000 mg/L may be reached when a single LAB starter culture is applied (Duboc and Mollet 2001).

EPS exhibit thickening properties, thus affecting the rheology and texture of fermented products to a greater extent than other thickeners. Biothickeners applied to foods, such as yogurts, other fermented milk products, milk desserts, soups, or dressings, have to exhibit thixotropic and pseudoplastic properties. Those properties significantly decrease the viscosity during shaking, mixing, or pouring, yet the viscosity returns to initial values once shearing stress does not occur. The functional properties of biothickeners shall be stable in a wide range of pH, ionic strength, and temperature (De Vuyst and Degeest 1999).

Although it is evident that EPS produced by various LAB differ in their molecular weight, composition, charge, spatial structure, rigidity, presence of nonsugar components, and the capability to interact with proteins—which indicates no strict correlation between the concentration of EPS and apparent viscosity—some regularities can still be noticed. The EPS of low molecular weight ($200–5.8 \cdot 10^3$) in no way affect viscosity (Cerning 1995). In order to achieve high viscosity, the molecular weight of EPS should be high and the chain relatively rigid. The polysaccharide chains with β (1,4) bonds are more rigid than those with α (1,4) or β (1,3) bonds. Also of importance is the charge of sugars. Electrically neutral EPS increase the viscosity of the system without changing the flexibility; in turn, EPS with a negative charge affect the flexibility without any impact on the viscosity (Jolly et al. 2002). The benefits of EPS application result from their ability to bind water, their interactions with proteins, and their capability to enhance viscosity. For example, a favorable texture of yogurts without syneresis may be achieved at EPS concentration below 1%. Unfortunately, one type of EPS does not usually affect all textural properties; thus LAB starter cultures producing various types of exopolysaccharides should be applied to this end (Jolly et al. 2002).

EPS form structures with homogenous proteins, which increases the viscosity and decreases the permeability of stirred yogurts. Thus, the viscosity of the product resulting from fermentation of specific cultures producing EPS is responsible for adhesiveness, whereas firmness and elasticity are connected with the protein matrix rather than with EPS formation. The studies of Shihata and Shah (2002) have revealed, however, that improvement of firmness is associated with the linking of mucogenic strains to the protein matrix by EPS. These authors believe that it is difficult to establish a good correlation between the amount of polysaccharides produced and the viscosity of yogurts. The difficulties are posed by the changes of the three-dimensional configuration of polymers or their interactions with milk proteins, casein in particular.

The studies of Girard and Schaffer-Leguart (2007a) indicate the existence of distinct phases between the protein network and EPS in sour milk. The authors explain that phenomenon by the considerable viscosity of EPS, which restricts its diffusion to milk. A key problem is knowledge about the physical properties of various EPS in solutions and their interactions with various components of a food product. Most of EPS can be described as polymers in the shape of a random coil with a variable tertiary structure.

In mixtures with proteins, two types of phenomena may appear. At low concentrations of EPS, firm solutions are formed; thus EPS and proteins are cosoluble. With an increasing concentration of biopolymers, the system becomes unstable depending on the type of interactions (Kleerebezem et al. 1999). The interactions are frequently ionic and occur between negatively charged EPS molecules and proteins with a positive charge (Girard and Schaffer-Leguart 2008). For example, the microstructure of yogurt contains a matrix of aggregated molecules of casein with embedded fat particles. There are niches filled with whey and bacteria cells in the gel. An envelope of EPS attaches to bacteria chains (Duboc and Mollet 2001). EPS play an important role as a natural biothickener, which improves the rheological properties of fermented products, and as a stabilizer that binds water and limits syneresis. Since during fermentation both gel forming and EPS biosynthesis occur, a highly cross-linked structure is formed. The rheology of products is affected by the viscosity associated with such features as liquidity, mucosity, flexibility linked with shelf life, and the gumminess of fermented milk products.

The texture of a product is developed by the presence of biothickeners in the aqueous phase (whey), the presence of protein gel consisting mainly of casein, the interactions between proteins and polysaccharides, the presence of bacteria cells and EPS fibrils attached to them, and the amount of bound hydrating water. For example, the viscosity of a product with neutral EPS has been reported to be approximately 10 times higher than the viscosity of the control sample obtained from the strains that do not produce EPS (Duboc and Mollet 2001). The structure of the casein gel being formed depends also on temperature. At lower temperatures, the gel exhibits poorer permeability and capability for spatial rearrangement (Ruas-Madiedo, Alting, and Zoon 2005). Since polysaccharides produced by various LAB differ substantially in structure, charge, spatial shape, rigidity, and ability to interact with proteins, so far there is no clear correlation between EPS concentration and product viscosity. EPS are tasteless, but fermented products are characterized by a higher viscosity, which increases the time of their residence in the mouth and thus extends duration of their contact with taste receptors, which finally results in higher ratings scored for their palatability. The higher ratings are due to improved conditions for the release of volatile odor compounds of yogurt (Duboc and Mollet 2001). For instance, a research study by Piermaria, de la Canal, and Abraham (2008) has revealed that kefiran (EPS-produced by kefir grains) affects viscosity to a lesser extent than other polysaccharides used as food additives, i.e., locust bean gum or guar gum. At concentrations of up to 0.001 g/L, kefiran solutions were reported to behave as Newtonian liquids, whereas above that concentration they exhibited pseudoplastic characteristics. These properties were confirmed by the studies of Rimada and Abraham (2006), who concluded that kefiran might be a very good thickener in the production technology of fermented milk drinks. Kefiran exhibits also gel-forming capability during freezing, cold storage, and thawing.

Purwandari, Shah, and Vasiljevic (2007) studied the effect of EPS with different structure—capsule and capsule-chain—on the rheology of yogurts. They found that yogurt containing EPS in the capsule-chain form was more rigid but exhibited enhanced syneresis than the yogurt with chain-structured EPS. Also, only a weak correlation was observed between the content of EPS and the rheological properties of yogurt. These authors pointed to the higher usefulness of the chain-structured EPS for the production of yogurts.

It has also been revealed (Girard and Schaffer-Leguart 2007b) that an increasing content of polysaccharide in milk leads to the formation of a dense casein network, which is only partly able to reconstruct a homogenous cross-linked gel structure after shearing (stirring). It was also found that the application of EPS with a low electrical charge and a low molecular weight, e.g., dextran, leads to the best reconstitution of the milk gel structure after shearing. These authors revealed also that EPS with a high molecular weight, a high negative charge, and containing sulfate groups were the most effective in reducing the duration of gel formation and increasing gel firmness.

Dlamini et al. (2009) demonstrated the usefulness of EPS produced by *Klebsiella oxytoca* to stabilize yogurt. The addition of EPS (0.03%) diminished syneresis by 6–13 points in comparison to the control sample and at the same time increased the strength of gel expressed as an increase in its hardness and gumminess. For example, an increase in EPS content produced by *Lb. delbrueckii* subsp. *bulgaricus* resulted in an increase in the apparent viscosity of yogurt and improved its ability to reconstitute its structure after stirring. The effect evoked by the increase in EPS content is not always beneficial. For instance, EPS produced by *S. thermophilus* increases susceptibility to syneresis (Laws and Marshall 2001).

For example, in order to obtain proper density and texture of yogurt and to prevent syneresis, a number of additives are used, i.e., skimmed milk powder, whey powder, modified starches, and hydrocolloids. However, those additives increase the cost of yogurt production and lower its natural taste and aroma. The usage of starter yogurt cultures that can produce heteropolysaccharides may prove to be a promising alternative. An important feature of starter cultures is rapid souring of milk, protection from microorganisms, texture formation, and health benefits (Ruas-Madiedo, Hugenholtz, and Zoon 2002). Consumers demand products with low sugar and fat contents, but also with a low content of food additives as well as with a low price. These arguments prove that EPS may be a viable alternative to other structure- and texture-forming additives. They also exhibit health-promoting properties: Due to longer gastric residence, they enhance colonization of probiotic bacteria, exhibit anticancer properties, and lower blood cholesterol level (Duboc and Mollet 2001; Wellman and Maddox 2003). Although the usage of EPS produced by LAB is widespread in the dairy industry, the research on EPS production and their impact on dough and finally bread quality is sparse (Ketabi et al. 2008).

Yeasts also exhibit the ability to produce EPS. The study of Pavlova et al. (2004) suggests the possibility of using the psychrophilic yeast *Sporobolomyces salmonicolor* for biosynthesis of exopolysaccharides. These authors achieved substantial effectiveness of EPS production at a level of 5.18 g/L. They also suggested the application of EPS as a thickening or gelling substance. LAB of bakery starters are able to produce glucans (reuteran, dextran, mutan) and fructans (levan, inulin), which found industrial application to improve dough rheology and bread texture. In addition, LAB strains producing fructans are capable of producing prebiotics such as fructooligosaccharides and heterooligosaccharides (erlose and arabsucrose) from saccharose (Gänzle, Vermeulen, and Vogel 2007).

The exopolysaccharides also develop rheological properties of bread produced using natural sourdough. The beneficial effect was exhibited by, for example, dextrans with poorly branched structure and with a high molecular weight ($2 \cdot 10^6 - 4 \cdot 10^6$ Da) (Katina et al. 2009). The studies of these authors revealed that dextran produced *in situ* by *Weissella confusa* increased viscosity of sourdough and also the volume of wheat bread up to »4.4 mL/g. For comparison, the volume of yeast wheat bread accounts for 3.5–4.6 mL/g. The bread enriched with dextran was characterized by delicate crumb and prolonged shelf life. The authors concluded that dextran delayed starch crystallization, which resulted in inhibition of bread staling. *Lactobacillus mesenteroides* produces about 25% of dextran of bakery starter, which clearly improves moisture of the crumb. Furthermore, the crumb structure is characterized by the fibrous structure and higher porosity (Decock and Capelle 2005). The addition of dextran at a level of 5 g/kg of flour improved viscoelastic properties of wheat dough and bread volume. The dextran operated more effectively than reuteran or levan. It was also revealed that EPS produced *in situ* functioned more effectively in comparison to extrinsic exopolysaccharides. Dextrans with a high molecular weight and a linear structure more effectively improve bread volume in comparison to dextrans with a high molecular weight but branched structure (Lacaze, Wick, and Cappelle 2007). According to U.S. Patent 2983613 (Bohn 1961), bread with dextrans added was characterized by 20% higher volume than loaves without the addition of dextrans. In this case, dextrans with a molecular weight from $2 \cdot 10^6$ to $4 \cdot 10^6$ Da were applied. The presence of dextrans in sourdough increases water absorption of dough, which improves bread freshness. Dough stability and ability to retain gases also increase as a result of structure formation performed by dextrans and their interactions with a gluten network. The crumb structure forms larger pores, characteristic of bread with a longer fermentation time. The oral impressions are better due to the sensation of moisture associated with the hydrophilic properties of dextrans (Lacaze, Wick, and Cappelle 2007).

EPS produced by lactobacilli beneficially affect one or more of the following dough or bread properties:

- Water absorption of dough
- Rheology of dough and machine-made processing of dough (stirring, cutting, shaping)
- Dough stability during cold storage
- Loaf volume
- Bread staling

Additionally, LAB bacteria used in the bakery improve odor, texture, and shelf life of bread (De Vuyst et al. 2001; Tieking and Gänzle 2005; Arendt, Ryan, and Bello 2007). Given the previous discussion, the authors concluded that EPS produced by LAB may be a viable alternative to other additives used in bakery. The structure of EPS determines their usage, but the final effect is affected also by flour quality, recipe and technological parameters of dough formation, and then baking. The study of Ketabi et al. (2008) on the application of EPS to improve rheological properties of dough showed little effect of the polysaccharides added at a level of 2.5% on dough stability. The authors also reported on a tangible increase in dough hardness at polysaccharides content above 0.25%.

4.3 CHANGES IN PROTEINS AND RELATED RHEOLOGICAL CHANGES

Chemical modifications are commonly used to alter certain chemical and biochemical functions of active centers of molecules, but they also enable transformations of physical properties of proteins (Feeney 1977). The change of physical properties may also modify biochemical functions. In practice, physical properties of proteins are modified in order to observe selected effects of these changes on the biochemical properties of proteins. Modifications that change the charges of amino acids constituting a polypeptide chain usually cause complete transformation of protein properties. The most obvious change is the one associated with the isoelectric point of protein, but the changes in electrical charges may also affect protein conformation and therefore its overall functionality. The application of lactic fermentation to modify the rheological properties of proteins is most common in the manufacture of fermented milk products, since the main process occurring during their production process is agglomeration of milk proteins into a three-dimensional network structure (Haque, Richardson, and Morris 2001).

The gel formation by milk proteins is a critical step in the manufacture of cheese, fermented milk, and many other dairy products. Milk proteins are a heterogeneous group of compounds that differ in composition and properties. Milk proteins may be divided into a casein fraction and whey proteins, based on their behavior under the influence of changes in hydrogen ion concentration. At pH 4.6, casein precipitates, whereas whey proteins remain in the solution. In the case of cheese, a major protein undergoing coagulation is casein, whereas in yogurt and other products made from milk subjected to heating, both casein and whey proteins are subject to gelatinization (van Vliet, Lakemond, and Visschers 2004).

Casein is a phosphoprotein; in its basic composition—besides C, H, O, and S— it additionally contains phosphorus. Using electrophoretic methods, 20 fractions of casein were determined to differ in phosphorus content, amino acid composition, molecular weight, saccharide content, and properties. Casein is a heterogeneous protein occurring in the form of spherical, highly porous aggregates called micelles. It consists of five basic fractions: a_{s1}-, a_{s2}-, β-, γ-, and κ-, the latter being almost exclusively located on the surface of micelles. Its C-terminal fragment is strongly hydrophilic and ensures the stability of micelles due to steric repulsion. In milk aqueous phase, the micelles form a colloidal solution (sol). They contain about 94% protein and 6% low-molecular-weight compounds, such as calcium, phosphates, magnesium, and citrates, in dry matter. The spatial structure of casein micelles is stabilized by various types of bonds. In the recent literature, two models of casein micelle structure are well documented (Jaworski and Kuncewicz 2008). The first one assumes that each micelle comprises 100–1000 submicelles, and each submicelle consists of 12–22 molecules of individual casein fractions, whose hydrophobic fragments are directed to the interior of the submicelles, where they are linked by hydrophobic bonds. The hydrophilic fragments containing, among others, phosphate groups esterified with serine, are located on the surface of submicelles, which allows coupling the submicelles into a micellar structure involving calcium ions and colloidal calcium phosphate (CCP). Calcium ions adhere to dissociated free carboxyl groups and phosphate residues of polypeptide chains and form calcium bridges:

$$\text{protein}^- - Ca^{2+} - \text{protein}^-$$

$$\text{protein}^- - Ca^{2+} - HPO_4^{2-} - {}^-\text{protein}$$

Citrates are also involved in the formation of those bonds:

$$\text{protein}^- - Ca^{2+} - H - \text{citrate} - Ca^{2+} - HPO_4^{2-} - Ca^{2+} - {}^-\text{protein}$$

Submicelles differ in the composition of casein fractions. Some of them are composed mainly of a_s- and β-casein fractions, while the others are composed of a_s- and κ-casein. Submicelles comprising a_s- and β-caseins locate themselves inside micelles, and submicelles containing molecules of κ-casein constitute the outer layer of a casein micelle. C-terminal fragments of the polypeptide chain of κ-casein—also containing a highly hydrophilic saccharide chain composed of galactose, N-acetylgalactosamine, and N-acetylneuraminic (sialic) acid—are directed into the aqueous phase, thus forming "hair" sticking out from the micelles (the so-called hairy layer). Those hairs, which ensure colloidal stabilization of the micelles, are called a glycomacropeptide part of κ-casein (GMP) (de Kruif et al. 1995). Due to its polyelectrolytic and polyampholytic properties, and by its electrostatic and spherical interactions, the GMP prevents further attachment of submicelles to created casein micelles and determines casein-micelle repulsion in milk at normal pH, which prevents flocculation (Jaworski and Kuncewicz 2008). In fresh milk (pH about 6.6), casein micelles are negatively charged due to the preponderance of dissociated acidic groups over the base ones. This determines the formation of hydrating layers that coat the micelles as a result of water molecules binding. The hydration layers with like electrical charges repel each other, thus stabilizing the casein colloidal solution. The second model of casein assumes that casein micelles are composed of elastic casein molecules arranged as tangled strands, forming a ball with a structure similar to that of gel stabilized by CCP microgranules and κ-casein molecules (Holt 1992; Lucey and Singh 1997). κ-Casein molecules form the outer layer of micelles, and their C-terminal fragments are arranged on the outside of the micelles in the form of "hair."

In recent years, it has been shown that the casein molecules are not globular proteins or fibrillar ones, but they have other, elastic spatial structure with a configuration resembling an a-helix. That structure has been defined as rheomorphic. The destabilization of protein solutions may occur under the influence of a variety of factors, such as ionic power, temperature, pressure, density, and pH change. Milk may be soured by bacterial cultures that hydrolyze lactose to lactic acid by addition of acids, i.e., lactic acid or HCl, or upon the application of gluconic acid lactone (GDL), when GDL hydrolysis to gluconic acid lowers pH. During fermentation, the natural pH of milk lowers from 6.6 to less than 4.6 as a result of lactic acid production from lactose. In consequence, casein forms gel.

4.3.1 Gelation of Casein

Considering the mechanism of gelation as a result of lactic fermentation, the process of milk souring should be followed thoroughly. According to Laligant et al. (2003), it may be divided into two stages. The first one is related to the production of lactic acid, which is dissociated to lactate and protons. Then, these protons are bound by acids with higher pKa, such as citric and phosphoric acids. When the pH value of the aqueous phase decreases, CCP dissolves in it because its solubility increases. This phenomenon depends on the pH of the aqueous phase, but it does not require the diffusion of protons and acids inside the molecules. At the second stage, more or less superimposed with the first one, protons and the above-mentioned acids are shifting to reach the phosphoserine groups of micelles and dissociate the bound calcium. The production of lactic acid by bacteria is confined to a few small areas, because bacteria in milk grow in macrocolonies, and milk during fermentation consists of areas with and without such colonies, meaning that diffusion of acids is essential.

The number of macrocolonies is probably similar to that at inoculation (10^6–10^7 cfu/mL). The aqueous phase surrounding colonies is characterized by very high pH (Laligant et al. 2003). As a result, the first layer of casein around those void areas (Kalab 1979; Kalab, Allan-Wojtas, and Phipps-Todd 1983; Bottazzi and Bianchi 1986) borders on the area with quite low pH, at which CCP is already dissolved and casein is gelled. This gel layer at the interface between the already-created three-dimensional network and the aqueous phase is an additional barrier for the diffusion of protons to the inside of casein and is responsible for increased gel heterogeneity.

Probably, despite its high porosity, diffusion of acids in a colloid is limited (Mariette and Marchal 1996). The first phase of souring is critical for the networking and aggregation process, which seems to be more complicated than previously thought. The only thing it does not require is the pH value necessary for the κ-casein to lose its hydrophilic properties (de Kruif 1997, Matia-Merino and Singh 2007).

Initially, it was thought that submicelles were held together in micelles by calcium bridges, but it is now known that CCP does not play a decisive role in maintaining the integrity of micelles (Holt 1992). In turn, hydrophobic and hydrogen bonds are of decisive significance, though the role of CCP is important as well, for its removal causes disruption of micelle structure (Horne 1998). During souring, the dissolution of CCP takes place, especially at pH £ 6, which evokes an accompanying increase in electrostatic repulsion between exposed phosphoserine groups. The rate and range of CCP solubility affect the texture of fermented milk products. Along with CCP loss, casein molecules lose their internal rigidity (they become more elastic). As the pH decreases, CCP is dissolving, and at pH 5.5 casein micelles are swelling. Below pH 5, the firmness of gel increases significantly and reaches the maximum at pH close to the isoelectric point. The firmness of gel is increasing with time.

The rearrangement of casein molecules is a dynamic process and occurs before, during, and after gel formation, and is evoked by a change in the type and strength of casein interactions occurring with decreasing pH. The binding of casein molecules takes place during and after gel formation (Lucey 2002). Casein molecules, which form a gel at pH » 4.8, do not contain CCP and are of a different nature than the native casein micelles. In raw milk, gelation occurs at low pH. Under such conditions, the whole CCP is already dissolved before the three-dimensional matrix is formed; thus the gel is not subjected to internal disruption and reveals the maximal ratio of viscosity to elasticity (maximum loss tangent). In heated milk, gelation begins at high pH (e.g., 5.3), and as a result part of CCP dissolves when the micelles are already part of the three-dimensional network. This causes the internal weakening of gel molecules, and the matrix becomes more fluid (at least at a critical time of gel formation).

Lucey et al. (1997a) suggest that gels obtained from nonheated milk can undergo extensive molecular rearrangement during gel formation, which leads to the formation of compact clusters of aggregated casein molecules that combine to form a gel. The molecules constituting those clusters take part in cross-linking with difficulty. Therefore, gels from the nonheated milk are characterized by a lower modulus of elasticity. Casein gels are usually classified as particle gels, although currently they are observed to differ in important aspects from gels formed from hard particles (Mellema et al. 2002; Horne 2003). The properties of gel are determined by the structure of the particles. After gel formation, they are coupling into homogenous strands, and the original particles are no longer visible (Walstra and van Vliet 1986). The internal structure of the original casein molecules strongly determines the properties of the strands. These strands behave like a concentrated gel with rigid polymeric molecules cross-linked by physical bonds. Despite the fact that, during gel formation and immediately after, the casein gels can behave like particle gels, quite soon they exhibit properties ascribed to both particle gels and polymers. The image obtained using an electron microscope confirms that acid gels consist of a coarse particle-network of casein molecules combined into clusters, chains, and strands (Kalab, Allan-Wojtas, and Phipps-Todd 1983). This network is porous and has void spaces in which the aqueous phase is trapped. The diameter of pores may range from 1 to 30 μm, and is higher in gels obtained at higher gelation temperature and from milk with low protein content. The water trapped therein is less mobile than in milk; therefore, interactions between water and proteins are limited (Laligant et al. 2003).

4.3.2 GELATION OF WHEY PROTEINS

The gelation of whey proteins is a result of physical (electrostatic and hydrophobic) and chemical interactions between protein molecules. The destabilization of the native tertiary structure of the proteins enhances their reactivity to the level that finally causes formation of a stable network and gelation. The destabilization can also be caused by hydrostatic pressure, addition of chemical substances, pH change, heating, cooling, and partial enzymatic hydrolysis. Each of those factors causes partial or complete unfolding of the whey protein structure, thus resulting in aggregation and eventually gelation.

The kinetics of gelation and gel properties depend substantially on the presence of casein and the method of gelation (Schorsch et al. 2001). Upon heating of whey proteins and further addition to a suspension of casein micelles, acid coagulation occurs faster and leads to a more coarse microstructure in comparison to gel obtained from simultaneous heating of whey proteins and casein (Guyomarch et al. 2003; Vastbinder et al. 2003). Aggregates of whey proteins can bind and modify the surface of casein micelles through their destabilization at higher temperature, whereas dispersed aggregates of whey proteins are precipitated around their isoelectric point.

The heating of milk preceding souring affects to a substantial extent the rheological properties of the gel formed (van Vliet and Keetels 1995; Surel and Famelart 2003; Vastbinder, Rollema, and de Kruif 2003), since it leads to the binding of whey proteins to casein micelles (Mottar et al. 1989; Law 1996; Lucey et al. 1997; Corredig and Dalgleish 1999; Anema and Li 2003). If gelation takes place in unheated milk, only casein participates in gel formation. An opposite phenomenon occurs in heated milk, which is related to the denaturation of whey proteins. In the case of acid coagulation of milk, the heating leads to the formation of much more rigid gels undergoing lesser syneresis (Lucey 2001; Lucey 2002). Temperature in the range of 70°C–100°C hardly influences the milk casein fraction; however, heating unfolds the major whey proteins, i.e., a-lactalbumin and β-lactoglobulin. In the case of β-lactoglobulin, reactive thiol groups (–SH groups), which can form disulfide bonds with casein micelles by exchange reaction of thiol groups–disulfide bridges with κ-casein (Jang and Swaisgood 1990), are being exposed (Mulvihill and Donovan 1987). It is believed that the first stage of this process is of a physical nature (not covalent), while further on the disulfide bonds are being formed (Lakemond and van Vliet 2008). a-Lactalbumin does not possess free –SH groups but has four disulfide bridges; therefore, it may participate in the exchange reactions of thiol groups–disulfide bridges only in the presence of β-lactoglobulin (Mulvihill and Donovan 1987). However, not all whey proteins get attached to casein micelles during heating; indeed, a considerable portion remains in serum as aggregates.

The extent of whey protein denaturation is a function of temperature and is also affected by pH. Many authors agree that a pH change between 6 and 7 contributes to the different modes of aggregation (Singh et al. 1996). Milk heating (pasteurization, sterilization) preceding the souring process is the main factor determining the sensory properties of acidic gels, with dry-matter content in the range of 10% to 20% being the second factor determining those properties (Pereira et al. 2003). The protein network of most fermented products is produced from a casein molecule; hence, the increase in its content leads to immediate enhancement of firmness and viscosity. If milk is heated (which is essential to denature whey proteins and combine them with casein micelles), an increasing content of whey proteins and a diminishing ratio of casein to whey proteins enhance gel cohesiveness (Lucey 2004). In turn, the excess of whey proteins leads to the formation of a granular (fibrous) texture (Lucey and Singh 1997).

The incubation temperature plays a major role in texture development of fermented milk products. In the case of yogurts, a lower incubation temperature (e.g., 40°C instead of 45°C) slightly prolongs incubation time and thus allows the formation of a strengthened and more viscous gel with less susceptibility to whey drip. The lowering of incubation temperature allows the production of fermented milk with a low dry-matter content and/or stabilizers that is characterized by desirable cohesiveness. In the case of cottage cheese production, the low temperatures (e.g., 20°C) enable the formation of a soft and elastic gel in comparison to a temperature of, e.g., 26°C. It is probably caused by a very slow aggregation and gelation process. An increase in gel strength is observed along with a decrease in pH from 5.1 to 4.6. If pH values are very low (e.g., 4.0), gel strength decreases again because of excessive repulsion of molecules with the same charge.

Gel formation also depends on souring temperature. Souring at low temperatures significantly decreases pH in which the gel is formed (pH of gelation), in contrast to souring at room temperature (de Kruif and Roefs 1996). At lower temperatures there are fewer hydrophobic interactions, which enables molecules to aggregate by the higher number of bonds between two molecules and casein serum, and therefore there is less migration of molecules during gel formation. The low values of G¢ (storage modulus) noted for gels formed at higher temperatures may be due to extensive rearrangements during gel formation, since fewer bonds are formed between molecules. The extensive rearrangements trigger the formation of dense clusters of aggregated particles that, in turn, aggregate to form gel. Many particles constituting those clusters do not participate in networking and, as a consequence, a weak gel is formed (Lucey et al. 1997b). An alternative method of gel formation is the heating of cold-soured milk (van Vliet and Keetels 1995; de Kruif and Roefs 1996).

Souring is commonly applied to extend the shelf life of many types of food products (e.g., pickles, cheddar cheese, salami) as well as to improve their aroma (Barbut 2005). Products of lipolysis and proteolysis, i.e., peptides, amino acids, carbonyls, and odor compounds, contribute to the development of a characteristic aroma and texture of, among others, meat products (Diaz et al. 1997). It is common knowledge that amino acids released by starter bacteria and intracellular proteases play an essential role during ripening. They directly affect the primary taste of slightly dried sausages and indirectly contribute to the development of their typical flavor.

According to data in the literature, utilization of LAB improves the consistency of cured meat batches and reduces the phenomenon of "wild fermentation." It was also reflected in the reduction/elimination of *Escherichia coli* 0157:H7 count in the product (Pond et al. 2001). It is important that the lactic fermentation of meat products be slow, because a rapid decrease in pH causes extensive protein denaturation, which makes the product unacceptable. Usage of traditional LAB is popular due to their unique impact on product flavor and bacteriocins production. The lactic fermentation applied during the production of beef chuck meat causes slow production of lactic acid, which contributes to protein cross-linking before the scalding process and an increase in G¢ value, in comparison to the nonsoured control sample. Oliver and Brock (1997) reported that fermentation weakened ongoing interactions in cracker dough. The rearrangement of proteins in a continuous system of gluten (pasted gluten) was in fact achieved by simultaneous shearing and stirring.

Lactic fermentation is also applied to modify functional properties of surimi proteins (e.g., from meat of *Hypophthalmichthys molitrix*). The texture of surimi is determined by interactions or bonds that occur between myofibrillar proteins after dissolving in salt (Careche, Alvarez, and Tejada 1995). Xu et al. (2010) studied molecular interactions of proteins involved in the formation of a gel network from fermented minced carp, inoculated with *Pediococcus pentosaceus*, and its gelation in respect of their rheological properties and microstructure. The results of rheological analyses confirmed that the work of penetration was increasing along with time. After 24 h, the fermented fish meat showed excellent textural properties. In the course of fermentation, the strength of gel was increasing dramatically, which corresponded with the increasing solubility of the samples in solutions breaking the hydrophobic interactions and disulfide bonds (NaCl and urea, separately and together at different concentrations), and was accompanied by an increase in the content (%) of insoluble proteins. The increasing gel strength is mainly due to covalent and hydrophobic bonds.

4.4 CHANGES IN LIPIDS AND RELATED RHEOLOGICAL CHANGES

The fermentation processes utilize the normal level of fat storage and fat conversion into fatty acids, which serve as feed to biological deposits, constituting systems composed of microorganisms and selected enzymes, microelements, and mineral salts. In turn, the souring stage of this process involves fermentation of simple molecules of organic fats to fatty acids (acetic, propionic, and butyric).

4.4.1 CHANGES IN FATTY ACID COMPOSITION

Under the influence of ethanolic and thermal stress, the composition of fatty acids in biological membranes may change significantly. It is mainly reflected in the increasing degree of saturation of fatty acids and membrane sterols. In this way, a microorganism reduces excessive fluidity of membranes and prevents their breakdown (Steels, Learmonth, and Watson 1994). In the case of *S. cerevisiae,* the increase in saturation of lipid membranes is accompanied by enhanced protein synthesis (Carratu et al. 1996). The main role of trehalose is to protect the stability of hydrated chemical compounds and to seal membranes, which prevents a loss of electrolytes and soluble cell components. Trehalose also protects lipids against oxidation. This sugar plays an important role not only in maintaining the proper structure of biological membranes, but it also protects proteins by stabilizing hydrogen bonds during temperature changes (Grajek and Szymanowska 2008).

Most stresses connected with ethanolic fermentation lead to an increase in permeability of cytoplasmic membrane and result in changes of the composition of membrane lipids and proteins (Bishof et al. 1995). A result of these changes is disturbance in membrane functioning. One of the major functions of biological membranes is to control the transport of nutrients into the cell and the output of the metabolites produced. The damage to the structure of phospholipid bilayer membranes and partial denaturation of membrane proteins is particularly visible in a prolonged thermal shock. It results in a loss, at least partial, of control over exchange of mass and electrolyte leakage. A good indicator of changes within the cytoplasmic membrane of yeast may be modifications in the structure of its lipid fraction (Simonin, Beney, and Gervais et al. 2007).

The length of a hydrocarbon chain of fatty acids present in triacylglycerols, the degree of their unsaturation, their positional distribution and molecular configuration (*cis*-, *trans*-isomers), as well as the polymorphic state of fat—all together these affect the physical properties of food, consistency in particular (e.g., viscosity, melting characteristics, crystalline state, contractility, spreadability, texture, dispersion) (Drozdowski 2002). Fat affects the taste of a product; it takes on the role of a carrier of flavor substances and imparts specific mechanical properties, including spreadability. The product with reduced fat content is harder and exhibits a more cohesive structure. In the case of dough obtained with yeast, demonstrating a spongy structure, use is made of dissolved and cooled fats that also serve in a texture-forming role (Biller 2005). The functionality of lipids in fermented food products depends on the qualitative and quantitative composition of the lipid fraction, including the quality of the fatty acids and the activity of the lipolytic enzymes (Ghosin and Bhattacharyya 1997; Zaręba 2009). During storage, the lipid fraction of those products is subjected not only to consistency changes, but also to oxidative changes. Among the products of microbial degradation of lipids in food, including fermented food, free fatty acids, aldehydes, and ketones should be named among the important ones (Oberman, Piątkowicz, and Żakowska et al. 1997).

A number of fatty acids were identified in model soy yogurts: (C_{12}) lauric, (C_{16}) palmitic, ($C_{16:1}$ n-9) palmitoleic, (C_{17}) margaric, (C_{19}) nonadecylic, ($C_{18:1}$ n-9) oleic, ($C_{18:1}$ n-11) vaccenic, ($C_{18:2}$ n-9, n-12) linoleic, (C_{20}) arachidic, and ($C_{18:3}$ n-9, n-12, n-15) linolenic (Ghosin and Bhattacharyya 1997; Zaręba 2009). Analyses of the fatty acid composition of these products confirmed the substantial contribution of the acids with chain length of $C_{16:0}$–$C_{22:0}$. The predominant fatty acids of soy yogurts were linoleic ($C_{18:2}$ n-9, n-12), oleic ($C_{18:1}$ n-9), palmitic ($C_{16:0}$), and linolenic ($C_{18:3}$ n-9, n-12, n-15) acids. The acids with a lower number of carbons in the chain were not detected, and during the extraction process of soy seeds, the initial content of constituents decreased about 10 times, which could affect the loss of short-chain fatty acids (Zaręba 2009). The low content or absence of short-chain fatty acids may also be due to the lipolytic activity of *Lactobacillus* and *Streptococcus* bacteria. Lactic fermentation bacteria (especially these of *Lactococcus* and *Lactobacillus* genera) are considered as weak lipolytic bacteria in comparison to bacteria of such genera as *Pseudomonas, Acinetobacter,* and *Flavobacterium.* Their considerable lipolytic activity is most often observed during rennet cheese ripening. They are probably responsible for the release of a substantial amount of short-chain free fatty acids that influence the sensory properties of a food product (Zaręba, Obiedziński, and Ziarno 2008; Zaręba et al. 2008). At least 20% of all psychrotrophic bacteria can cause decomposition of proteins and lipolytic rancidity. More than 70% of isolated bacteria classified as *P. fluorescens* exhibit lipolytic and proteolytic activity.

In ripening and mold cheeses, fatty acids (also the saturated ones) are at P_1 and P_3 position; therefore, they are more difficult to be absorbed by a human body. In nonfermented products, it looks different in terms of bioavailability. Both native lipases and those of microbiological origin may occur in fermented food products. Lipase-containing preparations are used in dairy technology to accelerate the ripening of cheese. Lipids hydrolysis during cheese ripening, including mold cheeses, contributes to the development of the proper taste and odor of cheeses, which makes it a desirable stage in their production technology (Pijanowski et al. 1997).

Lipases and esterases of lactic fermentation bacteria are intracellular enzymes; thus, a long ripening time and the lysis of cells enable their higher lipolytic activity during long-term ripening, which takes place during the production of ripening cheese and is not observed in fermented beverages (Lortal and Chapot-Chartier 2005). Cheese ripening involves decomposition of proteins and lipids, which leads to the development of a desirable aroma and taste bouquet as well as consistency (Babuchowski 2003). Hydrolytic changes proceeding in dairy products are referred to as rancidity and affect short-chain fatty acids (especially butyric), and these are responsible for the formation of unfavorable taste and odor. In turn, long-chain fatty acids contribute to a soapy taste.

The content of free fatty acids is a measure of the freshness and quality of fermented milk products (Drozdowski 1996). Rennet ripening cheeses are characterized by a considerable shelf life. Ripening cheeses should be stored at a temperature of 0°C–2°C, which inhibits the ripening process and microbial spoilage of cheese. At lower temperatures, the ripening process of decomposing proteins and lipids is still taking place, contributing to changes in odor, taste, and texture. Most often, the changes manifest themselves as the darkening of cheese inside and the appearance of fat on its skin (Palich 2006).

Lactic bacteria exhibit the activity of esterases and lipases that are able to hydrolyze a number of fatty acid esters: tri-, di-, and monoacylglycerols. Those enzymes are the main lipolytic factors in ripening cheeses. Bacteria of the genus *Lactobacillus* (*Lb. helveticus, Lb. delbrueckii* subsp. *bulgaricus, Lb. delbrueckii* subsp. *Lactis,* and *Lb. acidophilus*) exhibit the esterolytic activity against substrates with a chain longer than $C_{5:0}$, with the two latter species being the most active. None of these species hydrolyzes esters of o- and p-nitrophenol or fatty acids with an even number of carbon atoms from 6 to 14. The maximum lipolytic activity of enzymes achieved as a result of lysis of *Lb. casei* subsp. *casei* LLG (during the Late logarithmic growth phase) cells was observed at pH 7.2 and a temperature of 37°C. In turn, bacteria of *Lb. Fermentum,* constituting a starter culture for Parmesan cheese production, contain esterase bound to the cell surface and specific toward a substrate with $C_{4:0}$ chain length, which may hydrolyze esters of β-naphthol and fatty acids with 2–10 carbon atoms.

The molecular weight of intracellular lipase of *Lb. plantarum* strain isolated from cheddar cheese is $65 \cdot 10^3$ Da, and this enzyme exhibits its optimum activity at pH 7.5 and a temperature of 35°C. It is stable at a temperature of up to 65°C, while heating at 75°C for 2 min causes its irreversible inactivation. This enzyme exhibits the highest activity against tributyrin, and lower activity toward trilaurnate and tripalmitate, and none against glycerol trioleate. The profiles of *sn*-2-monoacylglycerols after hydrolysis revealed that *sn*-2-monoacylglycerols with fatty acids with a chain length of $C_{14:0}$–$C_{18:1}$ were produced, and those that contained acids shorter than $C_{14:0}$ were hydrolyzed (Collins, McSweeney, and Wilkinson 2003b).

Streptococcus thermophilus produces three intracellular esterases, two of which were identified as esterase I and esterase II. Their molecular weights are about $34 \cdot 10^3$ and $60 \cdot 10^3$ Da, respectively. Esterase I hydrolyzes esters of p-nitrophenol and fatty acids with a chain length of two to eight carbon atoms, and esterase II hydrolyzes similar esters, but contains fatty acids with a chain length of two to six carbon atoms. Esterase I is capable of hydrolyzing di- and monoacylglycerols containing fatty acids up to $C_{14:0}$ (Collins, McSweeney, and Wilkinson 2003a). *Lactic streptococci* show a higher lipolytic activity than the lactic bacilli. The scientific reports indicate a successive reduction in the number of viable cells of lactic acid bacteria under the influence of cold storage, and thus a reduction in their lysis and release of intracellular lipolytic enzymes (Donkor et al. 2007). The lipolytic enzymes, called lipases, are used to improve the sensory characteristics of cheeses, including taste and odor. Their role during the ripening of certain types of cheese is very important (Gruyère, cheddar, Emmentaler, Brie, Camembert, and Roquefort-type cheeses) because, due to the properly advanced lipolysis of milk fat, specific products with favorable flavor and consistency properties are produced (Bednarski 1996). This may additionally explain the presence of only the acids with a chain length above C_{16} in the fatty acid profile of cheeses.

The influence of storage time on the fatty acid content was also demonstrated in selected yogurts, including a decrease reported in contents of linoleic and oleic acids, which is a prerequisite to a change in an active acidity (pH). Moreover, the observed ratio of bacilli to streptococci, predominantly *Streptococcus*, may have increased the enzyme activity in the product (Zaręba 2009). As explained by scientific research, the assimilation of unsaturated fatty acids by lactic bacteria cells results in a change in the fatty acid profile of the bacterial cell membrane. According to data in the literature, $C_{18:1}$ acid added to the culture medium can be converted to the cyclopropanoic acid and embedded in the cell membrane, which may explain the diminished fatty acids content in the samples with relatively high pH values and thus with the smallest factor slowing down the activity of the bacterial cells (Johnsson et al. 1995; Itoh et al. 2001; Kankaampaa et al. 2004). Changes in the fatty acid content—or absence of these changes—can be explained by changes in pH. It is also claimed that the change in medium pH may have a direct impact on the biochemical activity of microflorae (Zaręba 2009).

A study by Zaręba (2009) did not demonstrate any statistically significant effect of culture type on the content of the analyzed fatty acids, which may give rise to the inference that the soy yogurts are a good source of unsaturated fatty acid, the content of which is stable during fermentation and storage. Furthermore, the addition of glucose to soy milk does not cause a significant change in fatty acid profile of the fermented soybean milk, and mostly showed an impact on the acidity of the product and indirectly on the enzymatic activity of microflorae.

The synthesis of fat by microorganisms is determined by genetic factors and environmental conditions. Microorganisms are characterized by a vast number of different genotypes. Microbial fat production is closely linked to the biosynthesis of cell mass. Because fat is not excreted outside the cell, it can constitute the cellular structures or provide a storage material for the cell. Microorganisms capable of fat synthesis include strains of *Rhodotorula* and *Lipomyces* yeast as well as *Penicillium* and *Mucor* molds (Bednarski 1996; Pijanowski et al. 1997).

The fermentation of ripe menhaden by *Candida lipolytica* strain reduces the content of fat in the fish by 16% during 72 h and at the same time increases the protein content by 13%. *Geotrichum candidum* reduces the fat content in the meat of young menhaden by 33% and increases that of protein by 20%. Model studies have shown that the production of lipase in a medium containing inorganic nitrogen is four times higher than in a medium with organic nitrogen. The use of labeled triglycerides has revealed that 35% of fatty acids and 43% of glycerol are embedded into the cell of a microorganism (Zaleski 1978).

The growth of undesirable bacteria is prevented by the production of lactic bacteria metabolites such as short-chain fatty acids. During long-term ripening of sausages, in addition to carbohydrate changes in cured meat products, significant changes occur in lipids. A pH value of around 5.3 enables the transformation of a solution of myofibrillar proteins into gel, and the resulting matrix immobilizes the particles of meat and fat. This enables reaching the binding, which denotes the internal cohesion of the particles and adhesion between them, and thus easier slicing. Raw fats, like meat, are infected with a diverse population of microflorae. Particularly dangerous are the microorganisms that trigger their hydrolysis. As a result of the action of lipases, free fatty acids are released. Lipase is produced by the bacilli, which cause the generation of dyes and meat glowing, e.g., *Pseudomonas fluorescens* and *P. pyocyanea*, *Chromobacterium prodigiosum*, *Sarcina lutea*, and *Bacillus subtilis*. Some of these bacteria, for example, *Chromobacterium* and *Pseudomonas*, both exert proteolytic properties and cause the putrefaction of proteins (e.g., fibers of meat remaining at fat and its connective tissue). The same bacilli of *Pseudomonas* and *Chromobacterium* and streptococci produce oxidases (lipoxidases). Changes caused by these enzymes result in spoilage referred to as rancidity of fats (Blottière et al. 2003).

4.4.2 OXIDATIVE CHANGES IN LIPIDS

In a study by Dmytrów et al. (2007), it was found that di- and triene fatty acids acid in acid curd cheese, packed in a barrier foil with a thickness of 40 μm (PE/PA) and 80 μm (PE/EVOH/PA), were transformed into conjugated diene (240 nm) and triene (278 nm) structures during three-week storage at 5°C, whereas the formation of peroxides and secondary oxidation products proceeded to a lesser extent. The availability of oxygen during the production of cottage cheese, and the other factors accelerating the oxidation reactions (e.g., metals), can affect the dynamics of the course of lipid autoxidation. Another research study by Dmytrów, Kryża, and Dmytrów (2007) revealed the formation of peroxides and secondary products in the cottage cheese with a starter culture FLDAN and CHN-11 during five-week storage at 5°C, with their number being higher in the cheese with the CHN-11 starter. Neither starter culture differentiated changes in the content of conjugated diene and triene fatty acids. The transformation of polyunsaturated fatty acids may occur during the oxidative changes, and as a result peroxides and secondary oxidation products may undergo further changes and form conjugated polyenes.

Positional isomerization of polyunsaturated fatty acids may lead to the formation of systems of conjugated double bonds. Resulting from the oxidation of lipids, radicals and the other compounds of a different nature can interact not only among themselves, but also with other nonlipid food constituents, causing permanent sensory and textural changes. The oxidation products also form many permanent bonds with proteins. For instance, aldehydes may react with the primary amine group. These bonds are sparingly digestible, and the essential amino acids bound in them lose their biological activity (Drozdowski 1996).

In the case of fermented meat products, the microbial species used in starter cultures may differ not only between one another but also exhibit different properties within the same species. They have different abilities to ferment sugars and to produce catalase, nitrate reductase, and peroxides. Moreover, they display different tolerance to changes in temperature and water activity, and are more or less antagonistic to other strains (Babuchowski 2003). The oxidative processes may be stimulated by the presence of water. The increase in the water content in the reaction medium enhances the mobility and solubility of heavy metal ions and the swelling properties of proteins. This creates the opportunity for radicals, formed during lipid oxidation, to react with proteins. A consequence of this type of reaction is hardening and a change in consistency (Karel 1980; Karel 1985). As far as the water activity is reducing to the value corresponding to the monomolecular layer, the oxidation reaction is gradually extinguishing in the aqueous phase, and reactions proceeding in the lipid phase are minimized (Świtka 1996).

During processing and storage, food proteins and lipids are oxidized by a singlet oxygen, peroxide anion, or hydroxyl radical, which are produced in the environment as affected by light, cations, and enzymes. The radicals initiate the formation of cross-linking bonds between molecules of proteins as well as between proteins and lipids. Oxidized lipids can also break the peptide chains of proteins, probably by the formation of peroxide groups on the alpha carbon with the participation of active oxygen. The action of lipid hydroperoxides most rapidly damages these amino acid residues, which easily form radicals. The proteins also react with products of lipid hydroperoxide decomposition, mainly with aldehydes and epoxy compounds. The disruption of other labile amino acids as a result of their oxidation and cross-linking by covalent bonds significantly reduces the biological value of proteins (Sikorski 1996). The application of transglutaminase evokes proteins cross-linking, thus modifying their functional properties and affecting the texture of protein substances. Kryża and Dmytrów (2010) found that the enzyme transglutaminase applied to the curd cheese, stored for three weeks under refrigerating conditions, limited the oxidation of the lipid fraction to peroxides. In addition, it actively inhibited the formation of conjugated diene (240 nm) and triene (278 nm) fatty acids, whereas the degree of lipids extractability was observed to increase during storage, which may be due to changes in the availability of lipid particles from a cottage cheese matrix.

Fat metabolism also occurs in beer yeast. The yeast cell membranes contain fatty substances combined with protein and phosphorus. During fermentation, yeasts are increasing their weight four to six times, which necessitates the building of new cells and producing adequate amounts of lipids, which in turn requires adequate amounts of oxygen. In order to propagate, the yeast need oxygen. Its deficiency causes depletion of yeast in lipids and fatty acids, inhibition of cell growth, prolongation of fermentation, and an increase in the number of dead cells. The minimal amount of oxygen in the wort accounts for 8–10 mg. Oxygen is consumed by yeast fermentation in the first phase completely and has no adverse effects on beer quality (Dylkowski 1993; Pazera and Rzemieniuk 1998; Kunze 1999; Michael, Lewis, and Young 2001). Rancidity is the most common cause of spoilage, including lipids degradation, hydrolysis, and oxidation. Hydrolysis occurs under the influence of lipases present in the raw material (native lipases or those produced by *Pseudomonas, Acinetobacter, Aspergillus* bacteria), and is accelerated by the presence of water, microorganisms, temperature (the higher the temperature, the higher the rate of hydrolysis), and access of light (UV rays accelerate hydrolysis). Oxidation of the resultant fatty acids is induced by their own lipoxidases and peroxidases or those produced by microorganisms (they are highly heat resistant, especially lipoxygenases).

Acetic bacteria belong to the *Acetobacter* genus, are gram-negative, and are typically aerobes and mesophiles. Their development depends also on the optimal pH of the medium, in this case from 4 to 6.5. Proteolytic cleavage of amino acids leads to deamination, which in turn leads to the formation of fatty acids. The decomposition of lipids is also associated with bacteria; lipids are especially strongly broken down by psychrophilic bacteria of *Pseudomonas* genus, in particular. In addition, molds cause various defects in dairy products; they attack fats, which leads to rancidity and color changes (*Geotrichum candidum* mold). The functional constituents produced during fermentation include, among others: CLA (conjugated linoleic acid); fatty acids (Sip and Grajek 2004); and bacterial enzymes initiating fat hydrolysis, thereby increasing the bioavailability (Blottière et al. 2003). In the interaction with food components, their addition to food as supplements does not always initiate the formation of a food product with health-promoting properties (Steinka 2009).

In addition to bacteria, molds are claimed to be detrimental microflorae to lipids. Hydrolytic rancidity of fat occurs in fats containing the water and protein that are essential for the bacteria that produce lipases and proteases. In turn, oxidative rancidity dominates (a) in fats whose dehydration favors the oxidation of fats and (b) in fats stored in the frozen state. This is due to higher thermoresistance of lipoxidases than lipases and a higher activity of lipoxidase compared to lipase at lower temperatures. Peroxides, aldehydes, and ketones with shorter hydrocarbon chains or volatile carbonyl compounds are formed during oxidation of fatty acids. The appropriate adjustment of the composition of probiotics and prebiotics may influence the quantitative and qualitative composition of short-chain fatty acids produced in the large intestine during fermentation. Short-chain fatty acids or their salts, especially butyrates, regulate the cell growth *in vitro* and *in vivo*. Butyrates are the main source of energy for colonic epithelial cells. In addition, the short-chain fatty acids directly affect genes regulating cell growth (Blottière et al. 2003).

4.5 MIXED CHANGES INVOLVING LIPIDS, PROTEINS, AND CARBOHYDRATES

From the viewpoint of food-product rheology, the vital activity of saprophytic microorganisms in food products, including the lipid fraction, leads to structural and textural changes, especially due to hydrolytic processes (Kołożyn-Krajewska 2003). In the case of fermented meat products, the aim of fermentation is the souring of meat or meat batter, which prevents the growth of unfavorable or pathogenic microflorae and prolongs its shelf life (Babuchowski 2003). Lactic acid is formed as a result of the probiotic souring of meat and lowers its pH, which favors the gelling of soluble proteins and imparts stable and cohesive texture to meat (sausage), which makes sausage slicing easier. *Lactobacillus* bacteria growing in meat batter consume fat nutrients, and their metabolic products enable the growth of other microorganisms. The level of acidification will depend mainly on the quantity and type of saccharides added, which can be hydrolyzed by lactic bacteria. The characteristic feature of *Pediococcus* bacteria is their capability to produce, apart from lactic acid, other metabolites such as acetic, formic, propionic, and pyruvic acids. Furthermore, bacteria of the genera *Staphylococcus* and *Micrococcus* are able to produce catalase, an enzyme decomposing hydrogen peroxide. The removal of peroxides from meat is important because oxidation of meat constituents leads not only to color changes, but also to fat rancidity. At the same time, *Staphylococcus* bacteria producing nitrate reductase cause faster reduction of nitrate (v) to nitrate (III), which can influence meat color and reduce lipid oxidation (Babuchowski 2003). Meat batter is an emulsion system in which hydrolytic degradation in proteins will affect the change of emulsification with the lipid fraction and the physical stability of the emulsion, and in consequence will affect the strength, hardness, texture, and other important rheological properties of gelling.

The formulations containing *Debaryomyces* yeast and *Streptomyces* bacteria are additionally used for some special types of raw cured meat products. Due to enzymatic conversion of fats and proteins, these microorganisms contribute to the formation of a specific aroma typical of dried, but not smoked, sausages. Also, mold fungi *Penicillium nalgiovensis* and *Penicillium candidum* are applied in the manufacture of cured meat products, for they demonstrate both lipolytic and proteolytic activity (Babuchowski 2003).

In the case of bakery products, the formation of a starch-lipids complex is an important issue. There is a correlation between the ability of monoacylglycerol to form a complex with starch and its ability to reduce staling, the so-called increase in crumb hardness, for it has been demonstrated that 1% addition of monopalmitoyl glycerol or monostearoyl glycerol prolongs the shelf life of bread by as many as two days, whereas bread with 1% addition of monooleyl glycerol may get staled with the same rate. Moreover, it is believed that amylose is better complexing with saturated fatty acids (chain length of 12–20 molecules of carbon) than with *trans*-unsaturated C_{18} and *cis*-unsaturated C_{18} fatty acids (Stauffer 2001).

4.6   CHANGES OF MICROSTRUCTURE AND RHEOLOGICAL PROPERTIES OF FERMENTED FOODS

Food microflorae are highly diversified and are dependent on the type of food and the expected character of a food product (crispy, soft, flowing, and smooth). Hence a division can be made between microflorae appropriate for a selected type of fermentation and undesirable spoilage microflorae. Both these types of microflora affect, to a substantial extent, the rheological profile of food products. From the rheological point of view, food microflora is responsible for changes in the sensory sensation perceived during consumption and for the ease of food ingestion. This involves such microbiological processes as food raising, imparting the desired extensibility, and preserving the relaxation capability of a food product. From the microbiological standpoint, food rheology is determined not only by the microorganisms themselves, but also by the apt preparation of microflora for the process (including strain storage, standard inoculation, proliferation, and propagation). These processes shall, additionally, proceed under strictly specified technological conditions (temperature, pressure, pH, NaCl concentration).

Also of significance is process duration, which strongly determines the proper microflorae of the process. A number of bacterial, yeast, fungal, and mold strains have been claimed responsible for imparting appropriate texturometric characteristics to food products. Investigations are continuously underway in order to introduce new, more sophisticated, and specialized strains even within a genotype. A number of single strains are crossed with each other to achieve desirable characteristics. Very often, use is made of the so-called microbiological cocktails, i.e., mixtures of many strains within one species or a few species. All these procedures affect, directly or indirectly, the end product. From the technological point of view, the objective of microorganisms is to trigger a specified sensory effect, appropriate porosity in particular, which is finally reflected in product hardness. The most appropriate microflorae that modify porosity to the greatest extent are yeasts (*Saccharomyces cerevisiae, Saccharomyces ivanovii*) and lactic acid bacteria (LAB). From a rheological point of view, the heterofermentative aspects of lactic acid cultures are also of great significance.

In bakery technology, an ideal combination of advantages of these two types of microorganisms is a bakery starter. As emphasized by Arendt, Ryan, and Bello (2007), a bakery starter is a structure known from the ancient times, and extensively described owing to its properties affecting food product quality improvement. The microflorae of lactic acid and yeast cooperate on the basis of prevalence and utilization of resources from the medium. As a result, this enables the creation of products with less porosity but with a longer shelf life. Also worthy of notice is the problem of bacteriocins produced by lactic acid bacteria. They significantly affect the medium, thus contributing to diminished porosity by reducing fermentation capability. Another consequence of their activity is a product with greater hardness and a considerably stronger aroma profile.

Gänzle and Vogel (2003) investigated the effect of a low-molecular-weight antibiotic, namely reutericyclin produced by *Lactobacillus reuteri*, on the behavior of bakery starter microflorae. They demonstrated a high susceptibility of the *Lactobacillus sanfranciscensis* strain, responsible for the production of carbon dioxide in bakery starter before the addition of baker's yeast, to the presence of the antibiotic, which might exert a direct impact on the final texture of bread. In turn, Hammes and Gänzle (1998) reported on the beneficial effect of reduced pH of baker's starter medium by lactic acid bacteria, which resulted in improved water binding and reduced retention of fermentation gases from flour. Kawamura and Yonezawa (1982) claim that—from the rheological point of view—the fermentation processes are natural processes taking part in flour structure development. The best example in this case is a decrease in the viscosity of a solution of proteins constituting gluten. In addition, they emphasize that the protein structures are the major factor affecting the structure of both wheat and rye breads.

Weegels, Hamer, and Schofield (1996) claim that the gluten-constituting proteins determine the rheology of fermenting wheat dough. In turn, Di Cagno et al. (2002) have applied empirical models to investigate the rheology of wheat dough. They noticed a decrease of resistance to extension and increased softness. They speculated that the fermenting baker's starter affected rheological changes of a gluten network of wheat dough, which in turn determined the capability for retention of fermentation gases in the dough. According to Clarke et al. (2004) and Schober, Dockery, and Arendt (2003), acidification of medium through lactic fermentation influences the gluten network of dough. At pH below 4.0, the positively charged molecules exert a positive effect on protein solubility and, as a consequence, prevent the formation of new bonds, which facilitates the elasticity of the gluten network. Reduction of disulfide bridges causes dissolution of the gluten network and enables its greater penetration through proteolytic enzymes (Thiele, Ganzle, and Vogel 2002). This finally affects the taste of a food product and changes its rheological properties, which has a direct impact on the product's texture.

As reported by Corsetti et al. (1998), the process of medium pH reduction affects the macropolymer of gliadin, which is the main determinant of aroma and texture of wheat bread. One of the most popular tests used to examine rheological properties of fermented doughs, both those made of wheat and rye flour, is the farinographic evaluation of wheat flours according to ICC or AACC Standards (ICC 2007; Iwański et al. 2006), amylographic assessment of rye flours, and a universal texture profile analysis (TPA) test (General Food Corporation Technical Center) (Iwański et al. 2008; Szczepanik, Ptak, and Iwański 2010). According to Bourne (1982b), devices produced by the General Food Corporation Technical Center imitate the process of chewing in tests of double compression, and the number of textural parameters is correlated with the sensory assessment. The first peak in the double-compression test denotes hardness expressed in the force necessary to deform the sample at a given measuring section. While the first peak is being plotted, a significant fracture is likely to occur, which is referred to as fracturability. In the case of food products with poor elasticity, a peak may appear below the starting line of the measurement, the so-called null. This peak is referred to as adhesiveness. The ratio of surface area of the first to the second peak is referred to as cohesiveness, and peak size—the so-called negative force—as adhesiveness. Tenacity is represented by the distance between the end of the first peak and the beginning of the second peak. In addition, the double-compression test enables determining the so-called gumminess (Figure 4.1).

Under specified conditions, an appropriately prepared dose of lactic fermentation bacteria is likely to evoke rheological changes in, for example, dough or bread. These changes are determined by the dose of microorganisms; composition (the mutual ratio of different technological cultures of bacteria); species ratio of, for example, yeast and lactic acid bacteria; origin and type of flour used for bread making (wheat, rye, triticale); and the quality of flour. (Flours produced from sprouted grain or those exceeding the period of technological usability display similarity to samples deliberately subjected to enzymatic treatment, which may indicate the negative effect of the enzymatic apparatus of flour on the resultant rheological characteristics of the finished product.)

In one study, analyses were conducted for rheological properties of wheat-rye bread made with a baker's starter modified with preparations of lactic acid bacteria (added in the process of dough making) in a suspension of UHT (ultra-high temperature) milk at a dose of 5% of dough weight. As reported by Gobbetti et al. (2005), lactic acid bacteria are predominating microorganisms in the sourdough. Their presence is of decisive significance to the rheology, aroma, and nutritional properties of rye bread made with baker's starter. According to this author, the baker's starter is a complex ecosystem with unpredictable interactions between various groups of

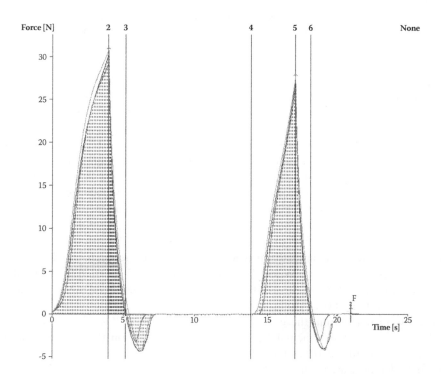

FIGURE 4.1   An exemplary series of TPA plots for wheat bread

microorganisms (Gobbetti et al. 1996). In the case of instrumental analysis, it is therefore difficult to state explicitly which strain is responsible for particular texture parameters. The first stage of the experiment involved the control baking of wheat-rye bread with the addition of preparations containing control strains. The application of the strains in a 5% dose was observed to increase both the baking loss and the total baking loss. The increase was, however, statistically insignificant and accounted for 1–3 degrees. Similar results were obtained for other bread varieties. As compared to the control, these samples were characterized by a higher acidity, with the increase being significant and accounting for 1.5–3 degrees. These changes could have been due to the addition of different bacterial cultures and to the method of their propagation.

For comparison, analyses were also carried out to determine acidity for 10% and 15% content of the starter in UHT milk. Results achieved were similar, and the addition of ACIDOLAC® preparation was found to exert a positive effect on bread volume. The samples with the 5% addition of ACIDOLAC preparation in respect of dough weight were characterized by the highest volume. The study also demonstrated that the addition of strain cultures in the form of commercial starters applied at a dose of 5% in respect to dough weight had a positive effect on bread porosity. The lowest porosity was reported for the control bread. In turn, the highest porosity as compared to the control sample was reported for the samples with the addition of a mixture of TRILAC® and Ferment Mesophile Aromatique Type B® strains. The preparations were also observed to effect an increase in moisture content. In the case of adhesiveness, analyses demonstrated the lowest value of this parameter in bread with the addition of the TRILAC strain. In contrast, the highest adhesiveness was shown for the samples with the addition of the ACIDOLAC strain. These samples were also characterized by the highest tenacity values. Further on, the highest cohesiveness was reported for the samples with the addition of the ACIDOLAC preparation, whereas the lowest cohesiveness was reported for those with the addition of the TRILAC strain. In the case of gumminess, the highest values of this parameter were determined for the samples with the addition of the TRILAC preparation, and the lowest ones for the samples with the addition of ACIDOLAC. In turn, the highest chewiness was noted in the case of the samples with the addition of the TRILAC preparation, and the lowest one was in the case of the samples with the addition of the ACIDOLAC strain. In analyses of elasticity, the samples with the addition of ACIDOLAC preparation exhibited the highest value of that parameter.

The rheological properties of food products are additionally influenced by the method of crop cultivation (conventional crops or ecological fertilization systems, i.e., those fertilized organically instead of applying the recommended nitrogen dose) (Iwański et al. 2009; Wianecki et al. 2009). Therefore, analyses were also conducted for the effect of the TRILAC preparation on the bread made of wheat, wheat-rye, and rye flours from conventional and ecological cultivation systems. Breads were made from these flours with the addition of TRILAC preparation and analyzed for the principal baking properties, based on the control baking and detailed texturometry of their technological usability. The analysis of results obtained demonstrated that the addition of the TRILAC preparation stabilized the basic parameters of the quality assessment of bread as well as dough behavior during processing. The addition of lactic acid bacteria was found to improve dough stability in the farinographic test, especially in the samples made of wheat-rye flours. It additionally leveled out the moisture content of the samples, which oscillated around 40%. In the case of bread porosity, stagnation of results may also be stated, particularly in the assessment according to Dallman scale values, (Horubałowa and Haber 1989) where most of the results fluctuated between 65 and 100 points, which may indicate small, evenly distributed pores of the bread crumb. This observation was confirmed by hardness analysis, which revealed results similar to those noted for the wheat and wheat-rye bread. The rye bread was significantly harder than the other products examined, yet its results differed within a given group of flour and fertilization system (organic or conventional). The significance of differences between the rye bread and the other breads analyzed is natural and results from the specific character of the amylase-carbohydrate and protein complex (Iwański 2009).

4.7   CONCLUDING REMARKS

Fermentation processes are able to give completely new structures, sensory characteristics, and rheological properties to products. The scale of these processes is determined mainly by fermentation conditions and the microbiological purity of the strain used. Another main point is the raw material and its basic chemical composition. A type of fat, particularly the hydrocarbon chain length of the fatty acid, determines the rheological and sensory properties as well as the storage life of fermented food products. Fermentation is not inherent to the nature of the protein, but changes of the environmental pH significantly affect the metabolism of proteins and the ultimate texture of the product, as the proteins form a matrix of most of the food products. From another point of view, EPS (exopolysaccharides) produced by LAB (lactic acid bacteria) have a positive effect on water absorption, rheology, and the stability of the fermenting dough, affecting final porosity and improving resistance to retrogradation.

REFERENCES

Anema, S. G., and Y. Li. 2003. Effect of pH on the association of denatured whey proteins with casein micelles in heated reconstituted skim milk. *J. Agric. Food Chem.* 51:1640–1646.

Arendt, E. K., L. A. M. Ryan, and F. D. Bello. 2007. Impact of sourdough on the texture of bread. *Food Microbiology* 24:165–174.

Babuchowski, A. 2003. Technologie fermentacyjne [Fermentation technology]. In *Biotechnologia żywności* [Biotechnology of food], ed. W. Bednarski and A. Reps, 373–384. Warsaw, Poland: WNT.

Barbut, S. 2005. Effects of chemical acidification and microbial fermentation on the rheological properties of meat products. *Meat Science* 71:397–401.

Bednarski, W. 1996. *Ogólna technologia żywności* [Food technology], 291–296. Olsztyn, Poland: Wyd. ART.

Biller, E. 2005. *Technologia żywności: Wybrane zagadnienia* [Food technology: Selected issues], 196. Warsaw, Poland: Wyd. SGGW.

Bishof, J. C., J. Padanilam, W. H. Holmes, R. M. Ezzell, R. C. Lee, R. G. Tompkins, M. L. Yarmush, and M. Toner. 1995. Dynamics of cell membrane permeability changes at supraphysiological temperatures. *Biophys. J.* 68:2608–2614.

Blottière, H. M., B. Buecher, J. P. Galmiche, and Ch. Cherbut. 2003. Molecular analysis of the effect of short-chain fatty acids on intestinal cell proliferation. *Proc. Nutr. Soc.* 62:101–106.

Bohn, R. T. 1961. Addition of dextran to bread doughs. U.S. Patent 2,983,613.

Bottazzi, V., and F. Bianchi. 1986. Types of microcolonies of lactic acid bacteria: Formation of void spaces and polysaccharides in yoghurt. *Sci. Tec. Latt. Casaria* 37:297–315.

Bourne, C. M. 1982a. Principles of objective texture measurement. *Food Texture and Viscosity* 3:48–49.

———. 1982b. Principles of objective texture measurement. *Food Texture and Viscosity* 3:114–117.

Careche, M., C. Alvarez, and M. Tejada. 1995. Suwari and kamaboko sardine gels: Effect of heat treatment on solubility of networks. *Journal of Agriculture and Food Chemistry* 43 (4): 1002–1010.

Carratu, L., S. L. Franceschelli, C. L. Pardini, G. S. Kobayashi, I. Horvath, L. Vigh, and B. Maresca. 1996. *Proc. Natl. Acad. Sci. USA* 93:3870–3875.

Cerning, J. 1995. Production of exopolysaccharides by lactic acid bacteria and dairy propionibacteria. *Lait* 75:463–472.

Clarke, C. I., T. J. Schober, P. Dockery, K. O'Sullivan, and E. K. Arendt. 2004. Wheat sourdough fermentation: Effect of time and acidification on fundamental rheological properties. *Cereal Chem.* 81:409–417.

Collins, Y. F., P. L. H. McSweeney, and M. G. Wilkinson. 2003a. Lipolysis and free fatty acid catabolism in cheese: A review of current knowledge. *Int. Dairy J.* 13:841–866.

———. 2003b. Evidence of a relationship between autolysis of starter bacteria and lipolysis in cheddar cheese during ripening. *J. Dairy Res.* 70:105–113.

Corredig, M., and D. G. Dalgleish. 1999. Effect of temperature and pH on the interactions of whey proteins with casein micelles in skim milk. *Food Res. Int.* 29:49.

Corsetti, A., M. Gobbetti, F. Balestrieri, F. Paoletti, L. Russi, and J. Rossi. 1998. Sourdough lactic acid bacteria effects on bread firmness and staling. *J. Food Sci.* 63:347–351.

Decock, P., and S. Cappelle. 2005. Bread technology and sourdough technology. *Trends in Food Science and Technology* 16:113–120.

Degeest, B., F. Vaningelgem, and L. De Vuyst. 2001. Microbial physiology, fermentation kinetics, and process engineering of heteropolysaccharide production by lactic acid bacteria. *International Dairy Journal* 11:747–757.

De Kruif, K. G. 1997. Skim milk acidification. *J. Colloids Interface Sci.* 185:19.

De Kruif, K. G., M. A. M. Hoffmann, M. E. van Marle, P. J. J. M. van Mil, S. P. F. M. Roefs, M. Verheul, and N. Zoon. 1995. Gelation of proteins from milk. *Faraday Discuss.* 101:185–200.

De Kruif, K. G., and S. P. F. M. Roefs. 1996. Skim milk acidification at low temperatures: A model for the stability of casein micelles. *Neth. Milk Dairy J.* 50:113.

De Vuyst, L., and B. Degeest. 1999. Heteropolysaccharides from lactic acid bacteria. *FEMS Microbiology Reviews* 23:153–177.

De Vuyst, L., F. De Vin, F. Vaningelgem, and B. Degeest. 2001. Recent developments in the biosynthesis and applications of heteropolysaccharides from lactic acid bacteria. *International Dairy Journal* 11:687–707.

Diaz, O., M. Fernández, G. D. Garcia de Fernando, L. De la Hoz, and J. A. Ordóñez. 1997. Proteolysis in dry fermented sausages: The effect of selected exogenous proteases. *Meat Science* 46:115–128.

Di Cagno, R., M. De Angelis, P. Lavermicocca, M. De Vincenzi, C. Giovannini, M. Faccia, and M. Gobbetti. 2002. Proteolysis by sourdough lactic acid bacteria: Effect on wheat flour protein fractions and gliadin peptides involved in human cereal intolerance. *Appl. Environ. Microbiol.* 68:623–633.

Dlamini, A. M., P. S. Peiris, J. H. Bavor, and K. Kailasapathy. 2009. Rheological character-istics of an exopolysaccharide produced by a strain of *Klebsiella oxytoca*. *Journal of Bioscience and Bioengineering* 3:272–274.

Dmytrów, I., K. Kryża, and K. Dmytrów. 2007. The effect of starter inoculation type on selected qualitative attributes of acid curd cheeses (tvarogs) stored under cooling condi-tions. *EJPAU* 10 (4). http://www.ejpau.media.pl/volume10/issue4/art-04.html.

Dmytrów, I., K. Kryża, K. Dmytrów, and S. Lisiecki. 2007. Wpływ opakowania na wybrane cechy jakościowe sera twarogowego kwasowego przechowywanego w warunkach chłodniczych [The influence of packing on selected qualitative features of acid-curd cheese stored in cooling conditions]. *Żywność Nauka Technologia Jakość* 1 (50): 64–76.

Donkor, O., S. Nilmini, P. Stolic, T. Vasiljevic, and N. Shah. 2007. Survival and activity of selected probiotic organisms in set-type yoghurt during cold storage. *Int. Dairy J.* 17:657–665.

Drozdowski, B. 1996. Lipidy [Lipids]. In *Chemiczne i funkcjonalne właściwości składników żywności* [Chemical and functional properties of food ingredients], ed. Z. E. Sikorski, 208–221. Warsaw, Poland: Wyd. WNT.

————. 2002. Charakterystyka ogólna tłuszczów jadalnych [Characteristics of fats]. In *Chemia Żywności* [Food chemistry], ed. Z. E. Sikorski, 239. Warsaw, Poland: Wyd. WNT.

Duboc, P., and B. Mollet. 2001. Applications of exopolysaccharides in the dairy industry. *International Dairy Journal* 11:759–768.

Dylkowski, W. 1993. *Browarnictwo* [Brewing], 29. Warsaw, Poland: Wydawnictwa Szkolne Pedagogiczne.

Feeney, R. E. 1977. Chemical modification of food proteins. In *Food proteins: Improvement through chemical and enzymatic modification*, ed. R. E. Feeney and J. R. Whitaker, 3–36. Washington, DC: American Chemical Society.

Gänzle, M. G., N. Vermeulen, and R. F. Vogel. 2007. Carbohydrate, peptide, and lipid metabo-lism of lactic acid bacteria in sourdough. *Food Microbiology* 24:128–138.

Gänzle, M. G., and R. F. Vogel. 2003. Contribution of reutericyclin production to the stable per-sistence of *Lactobacillus reuteri* in an industrial sourdough fermentation. *International Journal of Food Microbiology* 80:31–45.

Ghosin, M., and D. Bhattacharyya. 1997. Soy lecithin-monoester interchange reaction by microbial lipase. *JAOCS* 74 (6): 761–763.

Girard, M., and Ch. Schaffer-Leguart. 2007a. Gelation and resistance to shearing of fermented milk: Role of exopolysaccharides. *International Dairy Journal* 17:666–673.

————. 2007b. Gelation of skim milk containing anionic exopolysaccharides and recovery of texture after shearing. *Food Hydrocolloids* 21:1031–1040.

————. 2008. Attractive interactions between selected anionic exopolysaccharides and milk proteins. *Food Hydrocolloids* 22:1424–1434.

Gobbetti, M., M. De Angelis, A. Corsetti, and R. Di Cango. 2005. Biochemistry and physi-ology of sourdough lactic acid bacteria. *Trends in Food Science and Technology* 16:57–69.

Gobbetti, M., E. Smacchi, P. Fox, L. Stepaniak, and A. Corsetti. 1996. The sourdough micro-flora: Cellular localization and characterization of proteolytic enzymes in lactic acid bacteria. *Lebensm.-Wiss.u.Technol.* 26:561–569.

Grajek, W., and D. Szymanowska. 2008. Stresy środowiskowe działające na drożdże *Saccharomyces cerevisiae* w procesie fermentacji etanolowej [Environmental stresses acting on the yeast *Saccharomyces cerevisiae* during ethanol fermentation]. *Biotechnologia* 3 (82): 46–63.

Guyomarch, F., C. Queguiner, A. J. R. Law, D. S. Horne, and D. G. Dalgleish. 2003. Role of the soluble and micelle-bound heat-induced protein aggregates on network formation in acid skim milk gels. *J. Agric. Food Chem.* 51:7743–7750.

Hammes, W. P., and M. G. Gänzle. 1998. Sourdough breads and related products. In *Microbiology of fermented foods*, ed. B. J. B. Woods, vol. 1, 199–216. London: Blackie Academic/Professional.

Haque, A., R. K. Richardson, and E. R. Morris. 2001. Effect of fermentation temperature on the rheology of set and stirred yogurt. *Food Hydrocolloids* 15:593–602.

Holt, C. 1992. Structure and stability of bovine casein micelles. *Advances in Protein Chemistry* 43:63–151.

Horne, D. S. 1998. Casein interactions: Casting light on the black boxes, the structure in dairy products. *International Dairy Journal* 8:171–177.

———. 2003. Casein micelles as hard spheres: Limitations of the model in acidified milk gels. *Colloids Surf. A: Physicochem. Eng. Aspects* 213:255–263.

Horubałowa, A., and T. Haber. 1989. Wypiek laboratoryjny [Baking test]. In *Analiza techniczna w piekarstwie* [Technical analysis in the bakery], 143–156. Warsaw, Poland: WSiP.

ICC Standard Methods. 2007. No. 115/1.

Itoh, Y. H., A. Sugai, I. Uda, and T. Itoh. 2001. The evolution of lipids. *Adv. Space Res.* 28 (4): 719–724.

Iwański, R. 2009. A rheology of bread affected by lactic acid bacteria (LAB). Ph.D. dissertation. West Pomeranian University of Technology, Poland.

Iwański, R., M. Wianecki, J. Cebulska, M. Adamkiewicz, and A. Rumińska. 2006. Influence of the enzymatic modification on the wheat and wheat-rye bread quality. *Folia Univ. Agric. Stetin. Scientia Alimentaria* 251 (5): 23–32.

Iwański, R., M. Wianecki, M. Mierzwa, M. Szymczak, K. Zieliński, P. Stanisz, D. Klinkosz, E. Kłosowska, E. Księżak, and M. Tubacka. 2008. Influence of storage temperature on the texture and structure of bread. *Folia Univ. Agric. Stetin. Agricultura, Alimentaria, Piscaria et Zootechnica* 262 (6): 37–48.

Iwański, R., M. Wianecki, G. Tokarczyk, and S. Stankowski. 2009. The influence of conventional and ecological tillage system method of triticale on bakery value of flour and quality of bread. *Folia Pom. Univ. Tech. Stetin. Agricultura, Alimentaria, Piscaria et Zootechnica* 9:19–31.

Jang, H. D., and H. E. Swaisgood. 1990. Disulfide bond formation between thermally denaturated β-lactoglobulin and κ-casein in casein micelles. *J. Dairy Sci.* 73:900.

Jaworski, J., and A. Kuncewicz. 2008. Właściwości fizykochemiczne mleka [Physicochemical properties of milk]. In *Mleczarstwo*. ed. S. Ziajka, 71–78. Olsztyn, Poland: Wyd. UWM.

Johnsson, T., P. Nikkila, L. Toivonen, H. Rosenqvist, and S. Laakso. 1995. Cellular fatty acid profiles of *Lactobacillus* and *Lactococcus* strains in relation to the oleic acid content of the cultivation medium. *Appl. Environ. Microbiol.* 61 (12): 4497–4499.

Jolly, L., S. J. F. Vincent, P. Duboc, and J.-R. Neeser. 2002. Exploiting exopolysaccharides from lactic acid bacteria. *Antonie van Leeuwenhoek* 82:367–374.

Kalab, M. 1979. Scanning electron microscopy of dairy products: An overview. *Scan. Electr. Microsc.* 3:261–272.

Kalab, M., P. Allan-Wojtas, and B. E. Phipps-Todd. 1983. Development of microstructure in set-style nonfat yoghurt: A review. *Food Microstructure* 2:51–66.

Kankaampaa, P., B. Yang, H. Kallio, E. Isolauri, and S. Salminen. 2004. Effects of polyunsaturated fatty acids in growth medium on lipid composition and on physiochemical surface properties of lactobacilli. *Appl. Environ. Microbiol.* 70 (1): 129–136.

Karel, M. 1980. Teoria procesów suszenia [Theory of drying processes]. In *Nowe metody zagęszczania i suszenia żywności* [New methods of compacting and drying foods], ed. E. Spicer, 53–100. Warsaw, Poland: Wyd. WNT.

———. 1985. Effects of water activity and water content on mobility of food components and their effects on phase transitions in food systems. In *Properties of water in foods*, ed. D. Simatos and J. L. Multon, 153–166. Dordrecht, Netherlands: Martinus Publish.

Katina, K., N. H. Maina, R. Juvonen, L. Flander, L. Johansson, L. Virkki, M. Tenkanen, and A. Laitila. 2009. *In situ* production and analysis of *Weissela confusa* dextran in wheat sourdough. *Food Microbiology* 26:734–743.

Kawamura, Y., and D. Yonezawa. 1982. Wheat flour proteases and their action on gluten proteins in dilute acetic acid. *Agric. Biol. Chem.* 46:767–773.

Ketabi, A., S. Soleimanian-Zad, M. Kadivar, and M. Sheikh-Zeinoddin. 2008. Production of microbial exopolysaccharides in the sourdough and its effects on the rheological properties of dough. *Food Research International* 41:948–951.

Kleerebezem, M., R. van Kranenburg, R. Tuinier, I. C. Boels, P. Zoon, E. Looijesteijn, J. Hugenholtz, and W. M. de Vos. 1999. Exopolysaccharides produced by *Lactococcus lactis*: From genetic engineering to improved rheological properties? *Antonie van Leeuwenhoek* 76:357–365.

Kołożyn-Krajewska, D. 2003. *Higiena produkcji żywności* [Hygiene in food production], 132–133. Warsaw, Poland: Wyd. SGGW.

Kryża, K., and I. Dmytrów. 2010. Wpływ transglutaminazy na zmiany oksydacyjne tłuszczów w serach twarogowych w czasie chłodniczego przechowywania. Badania własne [Effect of transglutaminase on the oxidative changes in fat in cheese curd during refrigerated storage. Own research]. West Pomeranian University of Technology, Szczecin, Poland. http://www.chlodnictwo.zut.edu.pl.

Kunze, W. 1999. *Technologia piwa i słodu* [Technology of beer and malt], 302–308. Warsaw, Poland: Piwochmiel/VLB Berlin.

Lacaze, G., M. Wick, and S. Cappelle. 2007. Emerging fermentation technologies: Development of novel sourdoughs. *Food Microbiology* 24:155–160.

Lakemond, C. M. M., and T. van Vliet. 2008. Rheological properties of acid skim milk gels as affected by the spatial distribution of the structural elements and the interaction forces between them. *Int. Dairy J.* 18:585–593.

Laligant, A., M. H. Famelart, D. Paquet, and G. Brulé. 2003. Fermentation by lactic bacteria at two temperatures of pre-heated reconstituted milk, II: Dynamic approach of the gel construction. *Lait* 83:307–320.

Law, A. J. R. 1996. Effects of heat and acidification on the dissociation of bovine casein micelles. *J. Dairy Res.* 63:35–48.

Laws, A. P., and V. M. Marshall. 2001. The relevance of exopolysaccharides to the rheological properties in milk fermented with ropy strains of lactic acid bacteria. *International Dairy Journal* 11:709–721.

Lin, T. Y., and M.-F. C. Chien. 2007. Exopolysaccharides production as affected by lactic bacteria and fermentation time. *Food Chemistry* 100:1419–1423.

Lortal, S., and M. P. Chapot-Chartier. 2005. Role, mechanisms, and control of lactic acid bacteria lysis in cheese. *Int. Dairy J.* 15:857–871.

Lucey, J. A. 2001. The relationship between rheological parameters and whey separation in milk gels. *Food Hydrocolloids* 15:603–608.

———. 2002. Formation and physical properties of milk protein gels. *J. Dairy Sci.* 85:281–294.

———. 2004. Cultured dairy products: An overview of their gelation and texture properties. *International Journal of Dairy Technology* 57 (2/3): 77–84.

Lucey, J. A., and H. Singh. 1997. Formation and physical properties of acid milk gels: A review. *Food Research International* 30:529–542.

Lucey, J. A., C. T. Teo, P. A. Munro, and H. Singh. 1997a. Rheological properties at small (dynamic) and large (yield) deformations of acid gels made from heated milk. *J. Dairy Res.* 64:591–600.

Lucey, J. A., T. van Vliet, K. Grolle, T. Geurts, and P. Walstra. 1997b. Properties of acid casein gels made by acidification with GDL, 1: Rheological properties. *Int. Dairy J.* 7:381.

Mariette, F., and P. Marchal. 1996. NMR relaxation studies of dairy processes. *J. Magn. Reson. Anal.* 2:290–296.

Matia-Merino, L., and H. Singh. 2007. Acid-induced gelation of milk protein concentrates with added pectin: Effect of casein micelle dissociation. *Food Hydrocolloids* 21:765–775.

Mellema, M., P. Walstra, J. H. J. van Poheusden, and T. van Vliet. 2002. Effects of structural rearrangements on the rheology of rennet-induced casein particle gels. *Adv. Colloid Interface Sci.* 98:25–50.

Michael, J., T. Lewis, and W. Young. 2001. *Piwowarstwo* [Brewing], 144. Warsaw, Poland: Wydawnictwo Naukowe PWN.

Mottar, J., A. Bassier, M. Joniau, and J. Baert. 1989. Effect of heat-induced association of whey proteins and casein micelles on yoghurt texture. *J. Dairy Sci.* 72:2247–2256.

Mulvihill, D. M., and M. Donovan. 1987. Whey proteins and their thermal denaturation: A review. *Irish Journal of Food Science and Technology* 11:43–75.

Oberman, H., A. Piątkowicz, and Z. Żakowska. 1997. Surowce żywnościowe pochodzenia roślinnego jako źródło zagrożeń mikrobiologicznych [Raw foods of plant origin as a source of microbiological hazards]. Materiały konferencji naukowej: Bezpieczeństwo mikrobiologiczne produkcji żywności. Warsaw, Poland: 20.

Oliver, G., and C. J. Brock. 1997. A rheological study of mechanical dough development and long fermentation processes for cream-cracker dough production. *Journal of the Science of Food and Agriculture* 74 (3): 294–300.

Palich, P. 2006. *Podstawy technologii i przechowalnictwa żywności* [Technology of food storage]. *Wyd. Akademia Morska w Gdyni*: 119–120.

Pavlova, K., L. Koleva, M. Kratchanowa, and I. Panchev. 2004. Production and characterization of an exopolysaccharide by yeast. *World Journal of Microbiology and Biotechnology* 20:435–439.

Pazera, T., and T. Rzemieniuk. 1998. *Browarnictwo* [Brewing], 38. Warsaw, Poland: Wydawnictwa Szkolne i Pedagogiczne.

Pereira, R. B., H. Singh, P. A. Munro, and M. S. Luckma. 2003. Sensory and textural characteristic of acid milk gels. *Int. Dairy J.* 13:655–667.

Piermaria, J. A., M. L. de la Canal, and A. G. Abraham. 2008. Gelling properties of kefiran, a food-grade polysaccharide obtained from kefir grain. *Food Hydrocolloids* 22:1520–1527.

Pijanowski, E., M. Dłużewski, A. Dłużewska, and A. Jarczyk. 1997. Procesy biotechnologiczne w technologii żywności [Biotechnological processes in food technology]. In *Ogólna Technologia Żywności* [Food technology], ed. E. Pijanowski et. al., 280–290. Warsaw, Poland: Wyd. WNT.

Pond, T. J., D. S. Wood, I. M. Mumin, S. Barbut, and M. W. Griffiths. 2001. Modeling the survival of *Escherichia coli* O157:H7 in uncooked, semidry, fermented sausage. *Journal of Food Protection* 64:759–766.

Purwandari, U., N. P. Shah, and T. Vasiljevic. 2007. Effects of exopolysaccharide-producing strains of *Streptococcus thermophilus* on technological and rheological properties of set-type yogurt. *International Dairy Journal* 17:1344–1352.

Rimada, P. S., and A. G. Abraham. 2006. Kefiran improves rheological properties of glucono-δ-lactone induced skim milk gels. *International Dairy Journal* 16:33–39.

Ruas-Madiedo, P., A. C. Alting, and P. Zoon. 2005. Effect of exopolysaccharides and proteolytic activity of *Lactococcus lactis* subsp. *cremoris* strains on the viscosity and structure of fermented milks. *International Dairy Journal* 15:155–164.

Ruas-Madiedo, P., J. Hugenholtz, and P. Zoon. 2002. An overview of the functionality of exopolysaccharides produced by lactic acid bacteria. *International Dairy Journal* 12:163–171.

Schober, T. J., P. Dockery, and E. K. Arendt. 2003. Model studies for wheat sourdough systems using gluten, lactate buffer, and sodium chloride. *Eur. Food Res. Technol.* 217:235–243.

Schorsch, C., D. K. Wilkins, M. G. Jones, and I. T. Norton. 2001. Gelation of casein-whey mixtures: Effects of heating whey proteins alone or in the presence of casein micelles. *J. Dairy Res.* 68:471–481.

Shihata, A., and N. P. Shah. 2002. Influence of addition of proteolytic strains of *Lactobacillus delbrueckii* subsp. *bulgaricus* to commercial ABT starter cultures on texture of yoghurt, exopolysaccharide production and survival of bacteria. *International Dairy Journal* 12:765–772.

Sikorski, Z. E. 1996. Białka: Budowa i właściwości [Protein: Structure and properties]. In *Chemiczne i funkcjonalne właściwości składników żywności* [Chemical and functional properties of food ingredients], ed. Z. E. Sikorski, 266–268. Warsaw, Poland: Wyd. WNT.

Simonin, H., L. Beney, and P. J. Gervais. 2007. Cell death induced by mild physical perturbations could be related to transient plasma membrane modifications. *J. Membr. Biol.* 216 (1): 37–47.

Singh, H., M. S. Roberts, P. A. Munro, and C. T. Teo. 1996. Acid-induced dissociation of casein micelles in milk: Effects of heat treatment. *Journal of Dairy Science* 97:1340–1346.

Sip, A., and W. Grajek. 2004. Zastosowanie bakteriocyn i bakteriocynogennych bakterii w przemyśle spożywczym [The use of bacteriocins and bakteriocinogens bacteria in the food industry]. In *Bakterie fermentacji mlekowej* [Lactic acid bacteria], 121. Lodz, Poland: Wyd. politechniki łódzkiej.

Stauffer, C. E. 2001. *Emulgatory* [Emulsifiers], 103–104. Warsaw, Poland: Wyd. WNT.

Steels, E. S., R. P. Learmonth, and K. Watson. 1994. Stress tolerance and membrane lipid unsaturation in *Saccharomyces cerevisiae* grown aerobically and anaerobically. *Microbiology* 140:569–576.

Steinka, I. 2009. Innowacje technologiczne a bezpieczeństwo żywności [Technological innovation and food safety]. *Ann. Acad. Med. Gedan.* 39:123–132.

Surel, O., and M. H. Famelart. 2003. Heat induced gelation of acid milk: Balance between weak and covalent bonds. *J. Dairy Res.* 70:253–256.

Świtka, J. 1996. Woda jako składnik żywności [Water as a food ingredient]. In *Chemiczne i funkcjonalne właściwości składników żywności* [Chemical and functional properties of food ingredients], ed. Z. E. Sikorski, 65. Warsaw, Poland: Wyd. WNT.

Szczepanik, G., K. Ptak, and R. Iwański. 2010. Influence of the cheese addition on physicochemical and textural changes in wheat bread during frozen storage. *EJPAU* 13 (1): #10. http://www.ejpau.media.pl/volume13/issue1/art-10.html.

Thiele, C., M. G. Ganzle, and R. F. Vogel. 2002. Contribution of sourdough lactobacilli, yeast, and cereal enzymes to the generation of amino acids in dough relevant for bread flavour. *Cereal Chem.* 79:45–51.

Tieking, M., and M. G. Gänzle. 2005. Exopolysaccharides from cereal-associated lactobacilli. *Food Science and Technology* 16:79–84.

Van Geel-Schutten, G. H., F. Flesch, B. ten Brink, M. R. Smith, and L. Dijkhuizen. 1998. Screening and characterization of *Lactobacillus* strains producing large amounts of exopolysaccharides. *Appl Microbiol Biotechnol.* 50:697–703.

Vaningelgem, F., R. van der Meulen, M. Zamfir, T. Adriany, A. P. Laws, and L. De Vuyst. (2004). *Streptococcus thermophilus* ST 111 produces a stable high-molecular-mass exopolysaccharide in milk-based medium. *Int. Dairy J.* 14:857–864.

Van Vliet, T., and S. P. F. M. Keetels. 1995. Effect of preheating of milk on the structure of acidified milk gels. *Neth. Milk Dairy J.* 49:27.

Van Vliet, T., C. M. M. Lakemond, and R. W. Visschers. 2004. Rheology and structure of milk protein gels. *Current Opinion in Colloid and Interface Science* 9:298–304.

Vastbinder, A. J., A. C. Alting, R. W. Vischers, and C. G. de Kruif. 2003. Texture of acid milk gels: Formation of disulfide cross-links during acidification. *Int. Dairy J.* 13:29–38.

Vastbinder, A. J., H. S. Rollema, and C. G. de Kruif. 2003. Impaired rennetability of heated milk: Study of enzymatic hydrolysis and gelation kinetics. *J. Dairy Sci.* 86:1548–1555.

Walstra, P., and T. van Vliet. 1986. The physical chemistry of curd making. *Neth. Milk Dairy J.* 40:241–259.

Weegels, P. L., R. J. Hamer, and J. D. Schofield. 1996. Functional properties of wheat glutenin. *J. Cereal Sci.* 23:1–18.

Wellman, A. D., and I. S. Maddox. 2003. Exopolysaccharides from lactic acid bacteria: Perspectives and challenges. *Trends Biotechnol.* 21 (6): 269–274.

Whitaker, J. R. 1978. Biochemical changes occurring during the fermentation of high-protein foods. *Food Technology* 5:175–180.

Wianecki, M., R. Iwański, S. Stankowski, G. Tokarczyk, and K. Felisiak. 2009. The influence of conventional and ecological methods of rye tillage on flour baking value and bread quality. *Folia Pom. Univ. Techn. Stet. Agricultura, Alimentaria, Piscaria et Zootechnica* 274 (12): 75–90.

Xu, Y., W. Xia, F. Yang, and X. Nie. 2010. Protein molecular interactions involved in the gel network formation of fermented silver carp mince inoculated with *Pediococcus pentosaceus*. *Food Chemistry* 120:717–723.

Zaleski, S. J. 1978. Mikrobiologia żywności pochodzenia morskiego [Microbiology of seafood], 335. Gdansk, Poland: Wyd. Morskie.

Zaręba, D. 2009. Profil kwasów tłuszczowych mleka sojowego fermentowanego różnymi szczepami bakterii fermentacji mlekowej [Fatty acid profile of soy milk fermented different strains of lactic acid bacteria]. *Żywność. Nauka. Technologia. Jakość* 6 (67): 59–71.

Zaręba, D., M. Obiedziński, and M. Ziarno. 2008. Porównanie profilu lotnych związków mleka fermentowanego i niefermentowanego przez bakterie jogurtowe i szczepy pro biotyczne [Comparison of volatile compounds profile of fermented milk and unfermented yoghurt bacteria and probiotic strains]. *Żywność. Nauka. Technologia. Jakość* 15 (3): 18–32.

Zaręba, D., M. Ziarno, M. Obiedziński, and A. Bzducha, 2008. Profil lotnych związków modeli mleka niefermentowanego i fermentowanego przez bakterie jogurtowe [Profile of volatile compounds and models of unfermented milk fermented by yogurt bacteria]. *Żywność. Nauka. Technologia. Jakość* 2 (57): 60–73.

5 The Role of Fermentation Reactions in Changing the Color of Foods

Esther Sendra, Maria E. Sayas-Barberá, Juana Fernández-López, and Jose A. Pérez-Alvarez

CONTENTS

5.1 INTRODUCTION

Color is the main aspect that defines a food's quality, and a product may be rejected simply because of its color, even before other properties, such as aroma, texture, and taste, can be evaluated (Pérez-Alvarez, Fernández-López, and Sayas-Barberá 2000a). All foods have a number of visually perceived attributes that contribute to their color and appearance as well as to their overall quality. The appearance can be divided in three different categories: color, cesia, and spatial properties or spatiality (Lozano 2006).

Color is related to the optical power spectral properties of the stimulus detected by observers (Hutchings 2002). Cesia include transparency, translucence, gloss, luster, haze, lightness, opacity, and matt, and is related to the properties of reflecting, transmitting, or diffusing light by foods evaluated by human observation. Spatial properties are divided into two main groups:

1. Modes of appearance in which color is modified depending on the angle of observation related to the light incidence angle, such as metallic, pearlescent, or iridescent materials
2. Modes of appearance related to optical properties of surfaces or objects in which effects of ordered patterns (textures) or finishing characteristics of food (as roughness, polish, etc.)

The color of foods can be studied in two main ways: chemically by analyzing the pigments present or physically by measuring the interaction of light. The last one is supported by objective values that consist of numerous combinations of tristimulus values such as Hunter Lab, CIE XYZ, or CIELAB (Pérez-Alvarez 2006a). The use of CIELAB color space was adopted as an internal standard using as illuminant D_{65} and 10° as a standard observer. L* is a measure of lightness, where 0 equals black and 100 equals white. High, positive values of a* indicate redness, and large, negative values indicate greenness; b* values indicate yellowness to blueness (AMSA 1991). C* is the color saturation, and h_{ab} is the hue. Another method to determine color is using reflectance measurements. This is a good method for examining the amount of pigment and the pigments' chemical state *in situ*. This technique closely relates to what the eye and brain see. With this method, repeated measurements over time can be made on the same sample. In addition, the procedure is rapid and relatively easy. In foods, reflectance measurements are affected by food structure, surface moisture, fat content, and additives and pigment concentrations (Gallego-Restrepo, Pérez-Alvarez, and Ochoa 2010). Food structures are also associated with the light-scattering properties. Several foods colors can be described by their reflectance spectra. The present chapter aims to review the basis of color changes during food fermentation of meat, fish, dairy products, and vegetables (olives).

5.2 COLOR CHANGES IN DRY-CURED MEAT PRODUCTS

5.2.1 FERMENTED MEAT PRODUCTS

The most important meat products in which microorganisms play an important role upon color, among other characteristics, are the dry-cured fermented meat products. They can be classified into whole-muscle dry-cured meat products (dry-cured ham: Serrano ham, Teruel ham, Guijuelo Iberian ham, Parma ham, San Daniele ham, etc.) and dry-cured fermented sausages (DCFS). Most of the dry-cured meat products are traditionally from Mediterranean countries and now are widely distributed around the world (Pérez-Alvarez 2006b). Table 5.1 shows CIELAB color parameters of traditional dry-cured fermented meats.

TABLE 5.1
CIELAB Color Parameters of Several Commercial Dry-Cured Fermented Meat Products

Fermentation temperature	Paprika	Sausage type	Diameter (mm)	Lightness (L*)	Red-green coordinate (a*)	Yellow-blue coordinate (b*)	Chroma (C*)	Hue (h_ab)
Refrigeration (4°C–6°C)	No	Raw Longaniza blanca	20–23	34.9	19.3	13.5	23.5	34.8
	Yes	Raw Longaniza roja		21.5	35.2	36.8	51.0	46.2
Low temperature (10°C–12°C)	No	Salchichon pepper surface covered	80	41.7	20.5	8.7	22.3	22.9
	Yes	Chorizo pepper surface covered		33.9	21.9	16.5	27.5	35.8
	Yes	Red dry Longaniza	20–23	21.6	20.9	25.5	33.0	50.6
	No	White dry Longaniza		34.0	14.7	11.6	18.8	38.1
	Paprika marinated (adobo)	Dry-cured loin		38.0	9.2	13.7	16.5	56.5
	Yes	Iberian acorn-fed dry-cured loin		27.1	12.2	13.525	18.225	48.123
Medium (20°C–22°C)	Yes	Galician smoked chorizo	30–33	23.3	20.0	16.7	26.1	39.4
	Yes	Chorizo Pamplona	50	29.1	24.6	31.6	40.2	51.3
	Yes	Iberian acorn-fed Chorizo		30.8	23.5	24.1	33.7	45.5
	Yes	Chorizo vela		39.0	24.6	20.6	32.2	39.1
	Yes	Majorca Sobrasada IGP	60	36.6	37.6	62.8	73.2	59.1
	Yes	Hot paprika Majorca Sobrasada IGP		30.3	41.3	52.0	66.4	51.5
	No	Salchichon Vic IGP	50	34.5	12.7	11.7	17.370	42.7
	No	Salchichon Vallés		25.9	14.5	11.3	18.4	37.7
	No	Iberian acorn-fed Salchichon		28.0	11.1	11.7	16.2	45.9
	No	Artisanal Salchichon		38.8	17.8	7.7	19.4	23.4

Note: CIELAB color space was measured by a Minolta Spectro Colorimeter CM-2600 (Minolta Camera Co., Osaka, Japan), with illuminant D_{65} and an observer of 10°, SCI mode, 11-mm aperture of the instrument for illumination and 8 mm for measurement. Spectrally pure glass (CR-A51; Minolta Co.) was put between the samples and the equipment.

Dry-cured fermented sausages are intermediate-moisture foods (shelf stable to room temperature), and their color characteristics are a result of the chemical reaction between compounds derived from added nitrite/nitrate and myoglobin (MbFeII), leading to the simultaneous formation of the bright red nitrosylmyoglobin (MbFeIINO) (Pérez-Alvarez 2006a). This red bright color is a most important quality characteristic; if this property is defective, the consumer's choice at the moment of purchase can be negative. Indeed, the lack of what is generally considered as "typical" color entails a negative judgment on the acceptability and mistrust about the safety.

From a macroscopic point of view, the DCFS color is not homogeneous. Such properties depend on different characteristics such as

1. The integrity of the muscle tissues (higher in dry-cured products than in fine-paste products)
2. The physical properties of the meat product (geometry of the muscle and fat fractions and the optical properties of the meat and nonmeat ingredients)
3. Intra- and intermuscular spaces
4. Emulsion formation
5. Air occlusion in the structural matrix (microbiota establish and initiate its metabolic activity)

All these properties directly affect both color and the water transport mechanisms toward the exterior. It should also be remembered that during drying, the movement of water, cells, and tissues is inevitably accompanied by deformations of the product's structure (mainly of the cell membranes) (Pérez-Alvarez and Fernández-López 2009).

The use of starter cultures in industrial DCFS has become well established in recent times as a means to increase processing rates and improve product consistency. The microbiota used in these starter cultures depends on the use of nitrite, nitrate, or both as curing agents (Hugas and Monfort 1997). Thus, the starter cultures used for nitrite-added fermented sausages contain only lactic acid bacteria (LAB) and staphylococci, whereas starters for nitrate-added sausages include lactic acid bacteria, staphylococci, and micrococci (Scannell, Kenneally, and Arendt 2004). All of these microorganisms possess the nitrate reductase activity.

In traditional (spontaneous fermentation) and industrial processing, the indigenous microorganism or added starter culture mainly metabolize the carbohydrate, lowering the sausage pH value (Perez-Alvarez et al. 1999). As a consequence of this, the muscle protein coagulates, resulting in lightness of color and a sliceable, firm, and cohesive product. One of the advantages to the use of starter cultures is that the microbiota reduces the variability between commercial batches. It must also be taken into account that starter cultures play several roles: inhibition of pathogenic and spoilage bacteria; enhancement of flavor and taste; reduction of nitrite and/or nitrate content; and enhancement of color properties (Hugas and Monfort 1997).

From a chemical point of view, the bright red is caused by nitrosylmyoglobin (MbFeIINO), in which an axial ligand nitric oxide (NO) is coordinated to the central FeII in the myoglobin heme group (Møller and Skibsted 2002). The color development depends on the reduction of nitrite into NO. Alternatively, the NO radical is a possible intermediate in the bacterial reduction of nitrite to dinitrogen (Watmough et al. 1999) and, thus, the nitrogen dissimilation of potential meat starter cultures in the meat matrix is important. The utilization of nitrate/nitrite as an alternative electron acceptor in the respiratory chain by staphylococci is a possible pathway for NO generation. The reduction of nitrate into nitrite by *Staphylococcus carnosus* is suggested to be due to a membrane-bound type of nitrate reductase involved in respiratory energy conservation (Neubauer and Götz 1996). According with Gøtterup and coworkers (2007), nitrate and nitrite reductase activities of meat starter cultures are traditionally proposed to have great significance in the color development of fermented cured meat products, since reduction of the curing agents into NO is a critical reaction step in the generation of MbFeIINO.

Several studies were made related to the use of specific starter cultures for fermented meat products. Thus Hugas and coworkers (1993) reported 254 strains of lactobacilli isolated from Spanish spontaneously DCFS. These authors mention that *Lactobacillus sake* was the most isolated (55%), followed by *Lactobacillus curvatus* (26%), *Lactobacillus bavaricus* (11%), and *Lactobacillus plantarum* (8%). Now the strains most used in these types of meat products are *L. sake* and *L. curvatus*. In traditional dry Chinese sausages (Rai, Zhang, and Xia 2010), the addition of starter culture *L. casei* subsp. *casei*-1.001 as starter culture was found to be the most effective strain to improve red color.

In general, as pH decreases, caused by microbiota metabolism of carbohydrates, the color of meat becomes lighter, due to reduced water-holding capacity (WHC) and a corresponding increase in the light-scattering properties in the muscle fiber (Warriss 2000). Water plays a predominant role in DCFS color. However, no simple model satisfactorily explains its relation with properties (such as reflection) or the components of color (red-green component or yellow-blue component, etc.), since water has unusual properties, such as a high boiling point and latent heat of evaporation, low density, and high dielectric constant, among others. These properties are responsible for strong molecular interactions, which play an important role in certain properties of the DCFS as a result of the relation/reaction between the water and the different components of these products. Figure 5.1 shows reflectance spectra of several dry-cured fermented sausages. Microbiota metabolic activity can modify the state (free, semibound, and bound) and availability of water, thus modifying the DCFS color (Pérez-Alvarez and Fernández-López 2009).

The composition, curing agents, microbiota type and its metabolic activity, and processing conditions all modify the biochemical and chemical changes related to the color characteristics of these types of meat products.

In coarse-ground dry-cured fermented sausages, fat plays an important role. According to Pérez-Alvarez, Fernández-López, and Sayas-Barberá (2000a), fat can modify the color coordinates, thus increasing the formula fat content, lightness (L*), and yellowness (b*) and reducing redness (a*). Pérez-Alvarez (1996) defined the color coordinates' behavior during processing. Thus, in all dry-cured meat products, L* increased during fermentation processing and decreased during ripening caused by dehydration (there is less water free at the surface). Redness increased by nitrosomyoglobin formation and dehydration, which reduces the dilution effect of water content in the meat product (Pérez-Alvarez et al. 1997), and b* is reduced by changes of the myoglobin states during processing and the

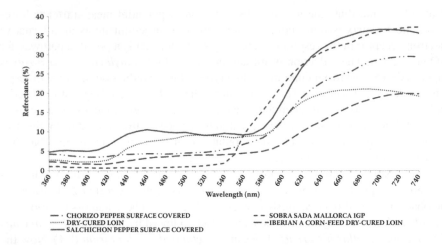

FIGURE 5.1   Reflectance spectra of several dry-cured fermented meat products (whole muscle: dry-cured loins; pepper- or paprika-added sausages: chorizo, *sobrasada, salchichon*). (CIELAB color space was measured by a Minolta Spectro Colorimeter CM-2600 [Minolta Camera Co., Osaka, Japan], with illuminant D_{65} and an observer of 10°, SCI mode, 11-mm aperture of the instrument for illumination and 8 mm for measurement. Spectrally pure glass [CR-A51; Minolta Co.] was put between the samples and the equipment.)

changes that take place upon the DCFS microstructure (Fernández-López et al. 2000). As previously noted, fat plays an import role in DCFS color. Soyer, Erta, and Üzümcüoglu (2005) reported that sausages between 10%–20% fat content, in high ripening temperature (24°C–26°C), have a good color development among other technological characteristics such as rapid decline in pH, water activity, and lactic acid production by LAB. The same authors mentioned that the increases in fat concentration at the same conditions provoke a poor color development.

5.2.1.1   Color Stability

Many factors can affect this type of meat product's color stability. These factors can include pH, temperature, relative humidity, microbiota load and type, lipid oxidation, partial oxygen pressure, light intensity, metmyoglobin-reducing systems, type of muscle, and antioxidant concentration, among others.

Most lactobacilli are capable of forming hydrogen peroxide, and this peroxide leads to the discoloration of the nitroso heme pigment. The metabolic characteristics of several microbial strains contain the catalase activity, and by using those together with the catalase-positive cocci, this type of problem can be avoided (Hugas and Monfort 1997). The worldwide success encountered by meat and meat products packed in a protective atmosphere is a clear proof of the importance of preventing or delaying meat discoloration (Zanardi et al. 1999).

5.2.1.2   Processing Conditions

As a processing parameter, the temperature has a strong effect upon microbiota behavior in DCFS. This parameter also has a strong correlation with fermentation, ripening, and drying conditions. Thus temperature can modify the evolution of pH, water activity, and texture as well as microbiota growth and its metabolism and color.

The characteristic cured color of DCFS is formed early in the processing stage (mixing, resting), but its stability depends on several technical and technological characteristics. Thus sausages with higher casing diameter (>80 mm) requires the use of fermentation conditions of 10°C–12°C (slow fermentation). These temperatures reduce the excessive microbiota metabolic activity. At these temperatures, the lactic acid formation is lower, thereby avoiding the acquisition of a lighter, less reddish color and a strong acid flavor for these sausages. During ripening, Soyer, Erta, and Üzümcüoglu (2005) reported that the formation of nitrosylmyoglobin is fast at high ripening temperature. Thus, the desirable red color of DCFS is better and more homogeneous.

During ripening, the mold surface growth must be taken into account. From a color point of view, molds play two main roles: (a) They use lactic acid as a substrate, thus increasing the DCFS pH. In this way the WHC increases and the optical properties changes (decrease in light-scattering properties), and (b) they prevent oxygen diffusion through the sausage, thereby reducing oxidation reactions that can produce dry-cured color fading (Pérez-Alvarez 2006b).

Technological processing leaves "fingerprints," and the spectrum can be used to quantify the degree to which DCFS color is altered by such processes. However, in addition to reflectance spectra, other aspects have to be considered in the spectrum such as the nature of the surface (smooth, rough) or the particles (nonmeat ingredients) or the molecules dispersed or dissolved in the aqueous phase (salt, phosphates, etc.) (Perez-Alvarez et al. 2011).

5.2.1.3   Paprika-Added Dry-Cured Fermented Sausages

In traditional DCFS, different spices and herbs are added to contribute to the characteristic flavor and color of the end product. In Mediterranean countries, it is common to add paprika (also called *pimentón*) to these types of meat products, the most famous of which are Spanish chorizo and Italian pepperoni. These paprika-added DCFS differ on this spice concentration, among other ingredients, and technological processes (Pérez-Alvarez, Fernández-López, and Sayas-Barberá 2000b). Paprika is added as a flavoring and coloring agent. However, paprika can also be a source of many other substances, such as sugars, nitrates, and metallic ions (Aguirrezabal et al. 1998). Paprika among other spices used in DCFS formulation has been found to accelerate lactic acid production by the lactic acid bacterial starter culture (Nes and Skjelkvåle 1982). Hagen, Næs, and Holck (2000) concluded that the degree of stimulation of lactic acid production was related to the manganese content of the spices. However, other authors (Verluyten, Leroy, and de Vuyst 2004) reported that the stimulatory effect of paprika on lactic acid production is probably due to components other than manganese.

Fernandez-López and coworkers (2002) reported that color characteristics of paprika-added DCFS mainly depend on paprika color properties and not on other ingredients incorporated. Paprika color characteristics (smoked, oil added, etc.) are reflected in these types of meat products. Figure 5.1 shows that the shape of the reflectance spectra of paprika-added DCFS are similar. Paprika generally interfered with meat reflectance spectra, mainly in the range from 500 to 580 nm. At these wavelengths, the typical meat spectra are masked by paprika spectra. Any color modifications observed in paprika-added DCFS are due principally to paprika color changes more than meat color changes.

5.2.1.4 Functional Fermented Meats

Although there is no exact definition of what a functional meat products is, in a simple way, the definition given by Dipplock and coworkers (1999) can be adapted for DCFS. Thus, a functional DCFS must remain as it is and must demonstrate its effects in amounts that can normally be expected to be consumed in the diet as part of the normal food pattern. The challenge to the meat industry is to provide visually appealing functional meat products that retain the traditional characteristics of flavor and color and meet the consumer's demands on health, quality, and price.

Thus color is an important consideration in the development of new functional meat products. However, one thing that must be taken into account when producing a functional DCFS is that chemical, biochemical, physical, and microbiological factors are closely interrelated. Thus, when one factor is modified, color characteristics and stability can be very different from the expected result. This is particularly true in functional meat products, when functional ingredients such as polyphenols, olive oil, probiotics, or dietary fiber, among others, are added to the formula. All of these ingredients can modify the metabolic pattern of several types of microorganism, and thus they can modify the color. Probiotic meat starter cultures seem to be a good option to introduce in DCFS, but this microbiota must not alter the technological and sensory properties of the product, as it can lead to product rejection. Several authors have proposed the use of probiotics in the manufacture of DCFS (Erkkilä and Petäjä 2000), but several hurdles must be solved to achieve success. The most important hurdles are (a) the meat environment, (b) the high content in curing salt, and (c) low pH and water activity (caused during fermentation and drying) (de Vuyst, Falony, and Leroy 2008). Pérez-Alvarez (2008) reported the role of several functional ingredients on the color of these new DCFS meat products.

From a color point of view, food technologists must check the effect of functional ingredients upon lipid oxidation, protein-lipid interaction, modification of indigenous or starter culture microbiota, changes in pH and WHC, etc., to control the influence of these new ingredients.

5.3 COLOR CHANGES IN FERMENTED FISH PRODUCTS

5.3.1 THE COLOR OF FRESH FISH

The color of fish (both fresh and processed fish), particularly its uniformity, is a very important quality parameter for consumers and is used to make purchasing decisions (Breithaupt 2007; Sánchez-Zapata and Pérez-Alvarez 2007). Myoglobin has been known to be a major contributor to the color of muscle, depending on its redox states (purple reduced myoglobin [DeoMb], red oxymyoglobin [OxyMb], and brown metmyoglobin [MetMb]) and concentration (Faustman and Cassens 1990; Y. Liu and Chen 2001; Pérez-Alvarez and Fernández-López 2007). Furthermore, the changes in the myoglobin states have a close relation with lipid oxidation. The lipid oxidation in fish muscle is promoted by the oxidation of myoglobin (Sohn et al. 2005; Thiansilakul, Benjakul, and Richards 2011), and both processes provoke changes in fish muscle. Myoglobin is made up of a single polypeptide chain, globin, and a prosthetic heme group, an iron (II) protoporphyrin-IX complex (Pegg and Shahidi 1997) with a molecular weight of 14–18 kDa; fish myoglobin is generally smaller than the mammalian counterpart (Ueki and Ochiai 2004; Chaijan et al. 2007). The stability of myoglobin varies with species, due to the different amino acid sequences and secondary structures of globin (Pérez-Alvarez and Fernández-López 2007). Chaijan et al. (2007) demonstrated that sardine (*Sardina pilchardus*) myoglobin was prone to oxidation and denaturation at temperature above 40°C and at very acidic or alkaline pH. The stability of myoglobin, which affects color as well as the lipid oxidation of meat, is associated with its autoxidation rate. Tuna has been known to provide one of the most remarkable red muscles, in which myoglobin is found at high concentrations (Chow, Ochiai, and Watabe 2004; Ueki and Ochiai 2004; Sánchez-Zapata et al. 2010).

The color of fish muscle is also related to the muscle retention of other natural pigments, such as different carotenoids in salmon and rainbow trout or melamines in crustacea (Czeczuga 1979; Y. Liu and Chen 2001; Chaijan et al., 2007; Pérez-Alvarez, Fernández-López, and Sayas-Barberá 2009a). The typical red to pink muscle color of rainbow trout is an important criterion of quality, and it is due to astaxanthin (3,3¢-dihydroxy-β,β-carotene-4,4¢-dione), a carotenoid of dietary origin (Bernhard 1990). Czeczuga, Bartel, and Czeczuga-Semeniuk (2002) determined that the fish muscle color depends on the total amount of carotenoids and also on the ratio of red pigments (astaxanthin and canthaxanthin) and yellow pigments (lutein and zeaxanthin). Czeczuga (1979) determined that the muscle of rainbow trout (typically a red to pink color) had a carotenoid content of 0.212 g/g fresh weight, mainly in the form of astaxanthin and canthaxanthin. In contrast, the carotenoid content of sardine muscle (white muscle) was 10.537 μg/g fresh weight (Czeczuga 1979).

All of these differences in myoglobin content, proportion of different myoglobin states, lipid content, development of lipid oxidation, amount and proportion of different carotenoids, etc., determine the color of fish muscle, which can be easily evaluated using the CIELAB color space (Sánchez-Zapata and Pérez-Alvarez 2007). These authors reported that a* and b* coordinates can be used to classify the color of fish muscle: Higher a* and b* values are characteristics of red and yellow fish muscle, and lower a* and b* values are characteristics of white fish muscle. Table 5.2 shows the CIELAB color coordinates—lightness (L*), redness (a*), and yellowness (b*)—of different fish species.

5.3.2 THE COLOR IN FERMENTED FISH PRODUCTS

Fish and processed fish products are among the most important foods in the world. Fish fermentation is an old preservation technique that is used to treat both freshwater and marine fish. This preservation technique improves the sensory and hygienic quality of the end product (Fernández-López et al. 2002). In some countries, most fish fermentation is still conducted as spontaneous processes at household or small-scale levels. Initiation of a spontaneous fermentation process takes a relatively long

TABLE 5.2
CIELAB Color Coordinates of Different Fish Species

	L*	a*	b*
Hake (*Merluccius merluccius*)	48.0	−0.1	−2.6
White tuna (*Sarda sarda*)	31.8	1.3	0.9
Sea bass (*Dicentrarchus labrax*)	40.7	1.5	−1.2
Sardine (*Sardina pilchardus*)	37.2	8.2	2.1
Rainbow trout (*Oncorhynchus mykiss*)	37.2	10.3	9.7
Salmon (*Salmo salar*)	48.3	11.8	13.4
Tuna (*Thunnus thynnus*)	31.6	17.2	5.8

Note: CIELAB color space was measured by a Minolta Spectro Colorimeter CM-2600 (Minolta Camera Co., Osaka, Japan), with illuminant D_{65} and an observer of 10°, SCI mode, 11-mm aperture of the instrument for illumination and 8 mm for measurement. Spectrally pure glass (CR-A51; Minolta Co.) was put between the samples and the equipment.

time, and the quality of the products varies considerably, limiting their acceptability and commercial importance (Twiddy, Cross, and Cooke 1987). A very complex flora develops on those products. For example, Leistner et al. (1994) isolated 58 strains of lactobacilli and carnobacteria from gravid salmon.

The use of starter cultures in food fermentation has been studied widely and introduced into commercial practice to increase processing rates and product consistency. Interest in the use of lactic acid bacteria for fish products has increased in recent years. Lactic acid fermentation could cause rapid acidification of the raw material through the production of organic acids, mainly lactic acid and acetic acid, and also produce a variety of antimicrobial substances that can consequently prevent the growth of most hazardous food microorganisms (Bover-Cid, Izquierdo-Pulido, and Vidal-Carou 2000; Yin, Pan, and Jiang 2002; Hu, Xia, and Ge 2007, 2008; Z. Liu et al. 2010). Fermentation processes for fish can be developed both on whole fish fillets and on minced fish.

5.3.2.1 Noncomminuted Fermented Fish

The traditional fermentation process involves curing fish fillets with sugar and salt (in order to reach the final concentration of 10 and 25 g per kg of fish, respectively) and usually herbs or spices. This mixture is applied evenly by rubbing onto the visceral side of the fillets, which are then placed on a metal rack and stored (usually at 12°C overnight) to allow the diffusion of the ingredients into the flesh. Subsequently, the undissolved ingredients remaining on the surface of the fillets are removed (Morzel, Fitzgerald, and Arendt 1997). Then, the fillets are packed in polyamide-polyethylene bags and incubated at 12°C–35°C for a short (72 h) or long (several years) period of time. The specific fermentation characteristics (temperature and time) depend on the fish species and the country where the fish are processed. For example, in many European countries, fatty fishes such as salmon, herring, mackerel, sardine, white tuna, tuna, etc., are frequently preserved by salting and spontaneous fermentation for a short period (Pérez-Alvarez, Fernández-López, and Sayas-Barberá 2009b). Long fermentation periods are more characteristics in Asia, Philippines, Japan, Korea, etc.

The color in fermented fish products depends on the process characteristics, although it is also influenced by the color of the raw fish fillets (Hatlen, Jobling, and Bejerken 1998). Several authors have reported that these differences could be reduced by the effect of the salting and fermenting process on the raw material color parameters (Sánchez-Zapata et al. 2008; Fuentes et al. 2010). Table 5.3 shows the values of CIELAB color coordinates in salted and fermented fish fillets from several fish species.

Morzel, Fitzgerald, and Arendt (1997) studied the color changes in fermented salmon fillets during fermentation (low temperature [12°C] and short period [72 h]) and reported that the CIELAB color coordinates in salmon fillets decreased during storage overnight at 12°C before increasing throughout the remaining period of the process. The a* and b* coordinates initially decreased by 3–8 and 4–10 units, respectively, indicating a loss of redness and yellowness compared to fresh salmon, which was due to the salting process. During the fermentation and storage period, the fillets regained redness and yellowness compared to day 0. The L* values decreased by 8–10 units during the first phase of processing (storage overnight), indicating that samples acquired a darker shade compared to fresh salmon, again probably due to salting. The fillets increased in brightness to a very similar extent during fermentation at 12°C.

TABLE 5.3
CIELAB Color Coordinates of Salted and Fermented Fillets from Different Fish Species

	L*	a*	b*
White tuna (*Sarda sarda*)	43.5	1.7	6.2
Sardine (*Sardina pilchardus*)	45.5	2.3	6.6
Salmon (*Salmo salar*)	47.3	11.6	12.4
Tuna (*Thunnus thynnus*)	37.8	8.9	5.8

Note: CIELAB color space was measured by a Minolta Spectro Colorimeter CM-2600 (Minolta Camera Co., Osaka, Japan), with illuminant D_{65} and an observer of 10°, SCI mode, 11-mm aperture of the instrument for illumination and 8 mm for measurement. Spectrally pure glass (CR-A51; Minolta Co.) was put between the samples and the equipment.

Mah and Hwang (2009) studied a popular product in Korea: a salted and fermented anchovy (*Engraulis japonicus*). These anchovies, which commonly contain abundant amino acids, are salted (around 20%) and fermented for a long period of time (several years). Salt-ripened anchovy (*Engraulis encrasicolus*) is a traditional fish product commercially produced in several Mediterranean countries (Fernández-López, Sayas-Barberá, and Pérez-Alvarez 2002). In this case, the final product has a characteristic brown-red color that is attributed mainly to the formation of Maillard reaction products (MRPs) (Morales and Jiménez-Pérez 2001). MRPs are formed by the reaction between reducing sugars and amino compounds, and their formation increased significantly as fermentation was prolonged (Peralta et al. 2008).

5.3.2.2   Fermented Minced Fish

Fish paste and fish sauce are the most popular products due to their salty, slightly cheeselike flavor as well as possessing a characteristic appetite-stimulating aroma. Fish paste is obtained through the natural fermentation process of whole/minced fish or shrimp in the presence of 20%–25% salt under ambient conditions. Fermented minced fish, which is an excellent source of protein, is widely consumed throughout Southeast Asia. It is typically composed of freshwater fish species, salt, a carbohydrate source, and spices. Fish fermentation can be conducted spontaneously or induced by the use of a starter culture (Paludan-Müller, Huss, and Gram 1999; Xu et al. 2010). Twiddy, Cross, and Cooke (1987) reported that the use of a minced fish-salt-glucose system with *Lactobacillus plantarum* and *Pediococcus pentosaceus* as starter culture promote a rapid fermentation at 30°C; Yin, Pan, and Jiang (2002) used several LAB, including *L. plantarum*, *Lactococcus lactis*, and *L. helveticus*, as starter culture to ferment mackerel mince at 37°C; Glatman, Drabkin, and Gelman (2000) inoculated *P. pentosaceus*, *L. plantarum*, and *Leuconostoc mesenteroides* into the minced frozen fillet of yellowfin tuna and fermented at 8°C. In all cases, the organoleptic qualities (color included) and digestibility of the product were improved.

On the other hand, fish sauce is a straw yellow to amber clear liquid extracted through the complete hydrolysis of fish/salt mixture for 9–12 months (Crisan and Sands 1975; Lopetcharat et al. 2001; Peralta et al. 2008). In Korea, salted and fermented anchovy sauce is the most widely used additive for unique taste development in various traditional foods (Park, Chang, and Kim 1997; Kim et al. 2004). The typical color of this sauce has been described as "amber," although Lee, Homma, and Aida (1997) reported that this type of sauce became darker with melanoidin produced by the Maillard reaction during storage. These Maillard reaction products have also been responsible for the typical color of the salt-fermented shrimp paste (Kim, Shahidi, and Heu 2003; Peralta et al. 2008).

5.4   COLOR CHANGES IN FERMENTED DAIRY PRODUCTS

For dairy foods, color is the first sensory characteristics perceived by consumers and even tends to modify other perceptions such as flavor and aroma. Fermented dairy foods can be grouped into three main categories: cheese, yogurt and other fermented milks, and butter. In this section we will explore the mechanisms that lead to color changes during the manufacture and storage of such products. These mechanisms are mainly related to: modifications in protein structure (as affected by pH and clotting), dehydration, and microbial metabolism. Color additions will not be considered in this discussion.

Milk whey contains riboflavin, providing a yellow-greenish color, and milk fat contains carotenoids that are responsible for its yellowish color. The white color of milk is mainly due to scattering of visible light by fat globules and casein micelles. Caseins disperse the blue light better than red light, and so skimmed milk has a characteristic bluish color (Walstra and Jenness 1984). Several technological treatments modify milk color. Heat treatment of milk induces an enhancement of white color (of unknown cause), and severe heat treatments enhance browning due to Maillard reactions. It has been hypothesized that low-heat milk whitening may be due either to the presence of denatured whey proteins in the media, thereby increasing the number of dispersed particles, or to the increase in colloidal phosphate in casein micelles (Walstra and Jenness 1984). Milk homogenization results in a whiter milk due to the increased scattering of light by smaller, homogenized fat globules. So, all factors affecting fat globules or casein structure and integrity will modify milk color as well as riboflavin and carotenoid content.

The first factor affecting milk color is milk origin: Cow's milk is yellower than that of other milking ruminants due to the transfer of carotenoids to milk, especially in late spring and summer (Kosikowski and Brown 1969) due to feeding. Cows fed on grass produce more yellow-colored milk than cows fed on hay or concentrates. Genetics of the specific breed also affects carotene content in milk. Sheep and goats do not transfer carotenoids to their milks, and so they are whiter than cow's milk. That color difference among milks has led to interest in whitening cow's milk to manufacture cheeses in which a whitish color is valued by consumers, such as mozzarella and white-brined cheeses. The carotenoids may be bleached by using peroxides (e.g., H_2O_2 or benzoyl peroxide) or masked (e.g., with chlorophyll or titanium oxide) (Fox and McSweeney 1998).

5.4.1 CHEESE

The mechanisms of cheese making have been reviewed by several authors (Walstra and Jenness 1984; Robinson 1995; Fox and McSweeney 1998). In a brief summary, cheese making has mainly five steps: coagulation, acidification, syneresis (expulsion of whey), molding (shaping), and salting. These steps partially overlap, and changes in the conditions of each of them enable the cheese maker to control cheese composition and thus cheese ripening and quality. Milk proteins are mainly caseins (70%–80%) and whey proteins (20%–30%). Caseins are the proteins that form the curd matrix, and consist of four types of phosphoproteins: αs_1, αs_2, β, and κ. Whey proteins are only present in cheese if made from ultrafiltered milk or from milk that has suffered an intense heat treatment, in which case they get absorbed onto the surface of caseins. The phosphorylated residues in caseins enable the binding of calcium ions, and they impart an acidic nature to caseins, with an isoelectric point at around 4.5–5.0 (Robinson 1995). Caseins combined with calcium and calcium phosphate are organized into units containing thousands of casein micelles. The α and β caseins are mainly in the center, whereas κ caseins, which are glycoproteins, are mainly on the surface, thereby protecting α and β caseins from spontaneous aggregation and precipitation.

Milk coagulation is usually achieved through enzymatic action or acidification. Most cheeses are obtained by enzymatic coagulation, which involves the modification of casein micelles by limited proteolysis. Coagulation enzymes hydrolyze κ casein (between Phe105-Met106), thereby leaving the α and β caseins unprotected. In the presence of calcium ions, the casein micelles aggregate to form a curd, entrapping fat globules and other milk constituents. Enzymatic coagulation has two phases: The first one goes from the addition of coagulant enzymes up to about the point where 85% of κ-casein is hydrolyzed; the second phase goes from 85% hydrolysis and up, and ionic calcium plays an essential role.

Color changes induced by enzymatic coagulation are presented in Figure 5.2 (a,b,c). The trials were run in commercial pasteurized milk, and color was assessed at 5-min intervals until coagulation was completed (within 45 min). The main changes took place within the first 20 min, probably coinciding with the first phase of coagulation, and stabilized afterwards. It is interesting to note that partially skimmed and skimmed milk experienced greater color changes than whole milk, and always toward whole-milk values. Overlapped with the action of coagulant enzymes, starter cultures are growing and induce some acidification, which may follow a homofermentative route (more than 90% of the fermented lactose yields lactic acid) or a heterofermentative route (lactic acid is also the main product of lactose hydrolysis, but many other compounds are formed). Such acidification affects coagulant activity, solubilizes colloidal calcium phosphate, promotes syneresis, and influences enzyme activity during ripening.

If milk coagulation is entirely achieved through acidification, it is usually due to lactose fermentation by lactic acid bacteria. When the pH is reduced and approaches the isoelectric point of caseins, they become unstable, and a soft gel is obtained in which k caseins remain intact. Changes in color due to the growth of lactic acid bacteria are discussed in Section 5.4.2 in the context of Figure 5.3. All technological factors affecting water retention in the curd—coagulation mechanisms, cutting and whey removal conditions, cooking, salting and ripening conditions—affect cheese color, mainly through ultrastructural modifications.

Some cheeses are always produced with added colorants (e.g., annatto), which mainly define the product's color. Other cheeses, such as pickled cheese, are whitish in color, as they are stored in brine. But in some other cheeses, microbial metabolism play an important role in developing the characteristic color, such in mozzarella, surface-ripened cheeses, and blue-veined cheeses. Mozzarella cheese is commonly used for baked foods (pizza and others), and so the presence of residual galactose can modify the color of the melted cheese on pizza. Galactose (a monocarbonyl) may be involved in Maillard reactions together with an amine (e.g., amino acid) and develop an undesirable brownish color. The choice of galactose-positive strains of *Lactobacillus delbrueckii* subsp. *bulgaricus* and *Streptococcus thermophilus* could minimize mozzarella browning (Tamine 2000).

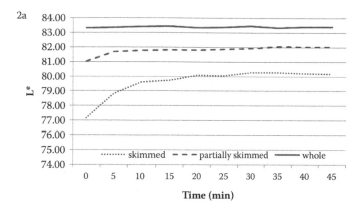

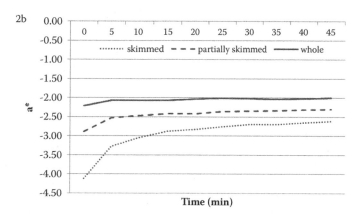

FIGURE 5.2 Assessment of color changes during milk enzymatic milk coagulation. Pasteurized whole, partially skimmed, and skimmed milk added with calcium chloride (56 mg/L) and bovine rennet (activity 1:10,000) and incubated at 30°C until coagulation. (CIELAB color space was measured by a Minolta Colorimeter CR-300 with a liquid accessory CR-A70 [Minolta Camera Co., Osaka, Japan], with illuminant D_{65} and an observer of 10°. Three measures per sample were taken at pH = 6.7.) 5.2(a): L*; 5.2(b): a* (-green, +red); 5.2(c): b* (-blue, +yellow).

Processed cheese needs a special mention, as its tendency to brown during and after processing is undesirable. As well as in mozzarella cheese, galactose content in cheese is the most important factor involved in browning, so the longer the aging of cheese before processing, the lower the browning due to the reduction in galactose content. Rapid cooling of processed cheese also reduces cheese browning (Bley, Johnson, and Olson 1985).

Brevibacterium linens is commonly associated with smear surface-ripened cheeses (El Soda 2000). It produces colonies that are yellow to deep orange-red, due to the formation of carotenoids. The main pigment produced by *B. linens* is

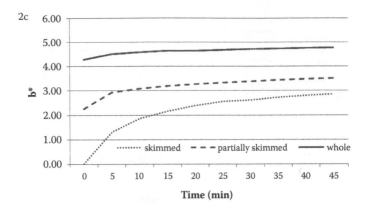

FIGURE 5.2 (CONTINUED) Assessment of color changes during milk enzymatic milk coagulation. Pasteurized whole, partially skimmed, and skimmed milk added with calcium chloride (56 mg/L) and bovine rennet (activity 1:10,000) and incubated at 30°C until coagulation. (CIELAB color space was measured by a Minolta Colorimeter CR-300 with a liquid accessory CR-A70 [Minolta Camera Co., Osaka, Japan], with illuminant D_{65} and an observer of 10°. Three measures per sample were taken at pH = 6.7.) 5.2(a): L*; 5.2(b): a* (-green, +red); 5.2(c): b* (-blue, +yellow).

the aromatic carotenoid 3,3¢-dihydroxy-isorenieratene (φ,φ-carotene-3,3¢-diol). Minor components are the corresponding monohydroxy compound and the hydrocarbon, isorenieratene. The dihydroxyphenyl carotenoid is responsible for the color shift from yellow-orange to pink-purple observed when colonies of *B. linens* are covered with an alkaline solution (Khol, Achenbach, and Reichenbach 1983). Such carotenoids are of several types: nonhydroxylated, monohydroxylated, and dihydroxylated, depending on the strain and kinetic aspects (Guyomarc, Binet, and Dufossé 2000). Yeasts, especially from genera *Debaryomyces*, *Candida*, and *Torulopsis*, are present in surface-ripened cheeses, and these play an important role in the transformation of the cheese surface, mainly increasing the pH of the surface through lactic acid consumption and by the synthesis of vitamins including riboflavin (which is colored), niacin, and pantothenic acid, thereby stimulating the growth of *B. linens* and micrococci. Yeast usually disappears after 1–20 days ripening, giving way to *B. linens* and micrococci (El Soda 2000). Factors affecting yeast growth, such as ripening temperature, also affect cheese surface color; at 14°C, yeast metabolism is faster than at 10°C, and so surface bacteria growth is enhanced (Mounier et al. 2006).

Recently, some other bacteria have been related to smear-ripened cheeses, and several studies have shown that the relevance of *B. linens* has been overestimated; *Brevibacterium aurantiacum*, *Corynebacterium casei*, *Microbacterium gubbeenense*, and *Staphylococcus saprophyticus*, among others, have been isolated from such types of cheese and have proved to provide color in aseptic cheese curd studies (Mounier et al. 2006). Another important color development in cheeses is that of mold-ripened cheeses, which are mainly of two types: those with white mold *Penicillium camemberti* and those of blue-veined cheeses with *Penicillium roqueforti* (Tamine 2000).

5.4.2 Yogurt

The main fermentation reaction in yogurt making is the conversion of lactose to lactic acid, the mechanisms of which have been well described and reviewed in several reference books (Walstra and Jenness 1984; Tamine and Robinson 1985; Fox and McSweeney 1998). During this process, pH declines and reaches the average isoelectric point of caseins, thereby changing milk ultrastructure and modifying milk color. Figure 5.3(a,b,c,d) presents color and pH changes of whole, half-skimmed, and skimmed milk during fermentation by yogurt cultures. The trial was run on commercial UHT milk. Initial b* values are higher in UHT milk (Figure 5.3) when compared to pasteurized milk (Figure 5.2), probably due to Maillard reactions during UHT heat processing. Milk color changes during fermentation are determined by pH

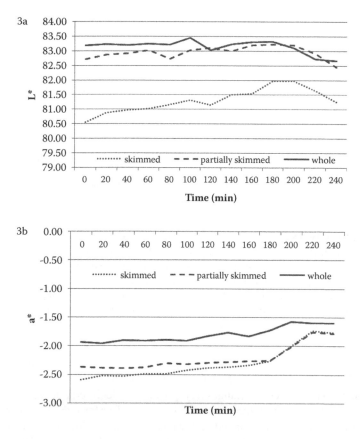

FIGURE 5.3   Assessment of color and pH changes during yogurt fermentation. Ultra-high-treated whole, partially skimmed, and skimmed milk inoculated with *Lactobacillus delbrueckii* subsp. *bulgaricus* and *Streptococcus thermophilus* and incubated at 43°C until pH 4.6. (CIELAB color space was measured by a Minolta Colorimeter CR-300 with a liquid accessory CR-A70 [Minolta Camera Co., Osaka, Japan], with illuminant D_{65} and an observer of 10°. Three measures per sample were taken.) 5.3(a): L*; 5.3(b): a* (-green, +red); 5.3(c): b* (-blue, +yellow); 5.3(d): pH.

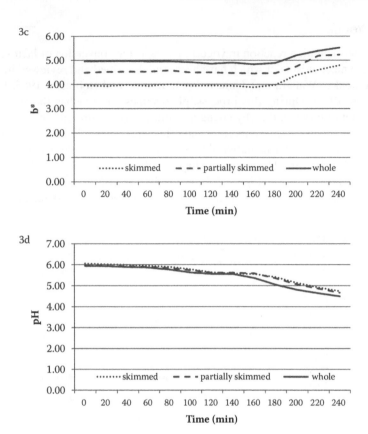

FIGURE 5.3 (CONTINUED) Assessment of color and pH changes during yogurt fermentation. Ultra-high-treated whole, partially skimmed, and skimmed milk inoculated with *Lactobacillus delbrueckii* subsp. *bulgaricus* and *Streptococcus thermophilus* and incubated at 43°C until pH 4.6. (CIELAB color space was measured by a Minolta Colorimeter CR-300 with a liquid accessory CR-A70 [Minolta Camera Co., Osaka, Japan], with illuminant D_{65} and an observer of 10°. Three measures per sample were taken.) 5.3(a): L*; 5.3(b): a* (-green, +red); 5.3(c): b* (-blue, +yellow); 5.3(d): pH.

changes: The first inflexion point is about pH = 5.5, and the second inflexion point occurs at about pH = 5.0 (Figure 5.3). García-Pérez et al. (2005) determined correlation formulas to predict color changes in yogurt as pH declined during fermentation. These authors also reported that yogurt refrigeration after fermentation increased L* (enhanced whiteness), b* (enhanced yellowness), and a* (reduced greenness). Color components were not modified during yogurt refrigerated storage for 28 days.

5.4.3 BUTTER

The color of butter depends mainly on the presence of carotenoids in the milk fat. Butter usually is available as sweet cream butter and cultured cream butter, which is cultured with lactic acid bacteria and so has a lower pH than the sweet butter and also has a characteristic diacetyl flavor. As regards the effect of the fermentation process on butter color, it has been reported that fermentation does not yield significant differences in butter color as measured by a colorimeter, but under sensory analysis, culture butters are detected as yellower and shinier than sweet ones (Jinjarak et al. 2006).

5.5 COLOR CHANGES IN FERMENTED VEGETABLES EXEMPLIFIED IN OLIVES

Traditionally, fermented vegetables have been produced to extend the shelf life of fresh vegetables. The fermentation by lactic acid bacteria may be considered as a simple and valuable biotechnology for maintaining and/or improving the safety, nutritional, sensory, and shelf-life properties of vegetables (Rodriguez et al. 2009). Olives are one of the most important fermented vegetables in the world economy.

The olive fruit is a drupe, from specific varieties of the cultivated olive tree (*Olea europaea sativa*), harvested at the proper stage of ripeness. The olive has a bitter component, a glucoside called oleuropein; a low sugar concentration (2.6%–6.0%); and high oil content (12%–30%), although these values can change with maturity and olive variety (Garrido-Fernández, Férnandez Díaz, and Adams 1997; Arroyo-López et al. 2008). Such characteristics make it a fruit that cannot be consumed directly from the tree. Instead, it has to undergo a series of processes to make it edible, and these differ considerably from region to region and also depend on the variety of the olive (IOOC 2010). The most important industrial preparations are: (a) the green Spanish style (or Sevillian), with about 60% of the production, (b) ripe olives by alkaline oxidation (the so-called Californian style), and (c) naturally black olives (also known as Greek style) (Sánchez Gómez, García García, and Rejano Navarro 2006; Panagou et al. 2008). However, there are many other traditional or industrial table olive preparations according to fermentation conditions (temperature, levels of salt, and type of acid) and raw material (green, turning color, or black olives)

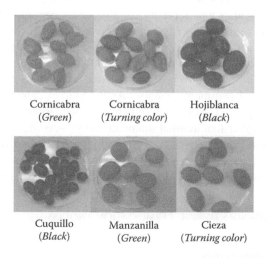

| Cornicabra | Cornicabra | Hojiblanca |
| (*Green*) | (*Turning color*) | (*Black*) |

| Cuquillo | Manzanilla | Cieza |
| (*Black*) | (*Green*) | (*Turning color*) |

FIGURE 5.4 Main commercial preparations of Spanish-style olives

(Sánchez Gómez, García García, and Rejano Navarro 2006; Arroyo-López et al. 2008), as seen in Figure 5.4.

The olives are harvested at the green or black stage, sorted, size graded, and then placed under anaerobic conditions, or in some cases under aerobic conditions, into brine with a salt concentration ranging from 6% to 8% NaCl, and spontaneously fermented for months (Romero et al. 2004a). Osmotic exchange leads to the leaching of substrates such as carbohydrates (glucose mainly, as well as fructose, mannitol, and sucrose) and organic acids (malic, citric, and acetic). Consequently, the aqueous solution (brine) becomes a good medium for the growth of microorganisms. In the processing of green Spanish-style olives, the bitterness is totally removed by hydrolysis with diluted sodium hydroxide solution for a few days. During this period, the hydrosoluble oleuropein and sugars are eliminated and transformed (Piga and Agabbio 2003). Processing of ripe olives (by darkening in an alkaline solution) consists of successive treatment of olives with sodium hydroxide solutions (lyes) over three consecutive days, and during the intervals between lye treatments, the fruits are suspended either in water or dilutes brines through which air is bubble. Finally, ferrous gluconate or a lactate solution is added to fix the color (Brenes-Balbuena, Garcia-Garcia, and A. Garrido-Fernandez 1992).

During fermentation, the use of NaCl and the progressive decrease in pH select lactic microflora, which increase from 1% of the total microbial population in the fresh brine to 80% after a few days. Indigenous lactic acid bacteria (LAB) change spontaneously during natural fermentation, and at the end of the process, *Lactobacillus* species are involved, mainly *Lactobacillus plantarum* (Rodríguez et al. 2009) at a percentage of 39.9% of the total (Sanchez-Gomez, García García, and Rejano Navarro 2006), although other LAB species such as *Lactobacillus pentosus* and *Leuconostoc mesenteroides,* among others, have also been isolated (Rodriguez et al. 2009). During the first fermentation phase, the gram-negative bacilli are the most characteristic microorganisms, their metabolism giving rise to carbon dioxide, hydrogen, acetic acid, lactic acid, ethyl alcohol, etc., as final products. During the second phase, as a consequence of the fall in pH (pH of up to 4.5), a strong development of Lactobacilli begins; they produce only lactic acid as a final product of glucose fermentation. During the third phase, when the pH reaches 4.0 or less, acid formation ceases (Minguez-Mosquera, Garrido-Fernandez, and Gandul-Rojas 1989), and when the sugars are exhausted, the fermentation period can be considered finished and the storage period begins.

After fermentation, olives are kept in the same brine until curing is achieved, before its packing and commercialization. As the market demands, the olives are sorted, graded, and packed. In some commercial presentations, they can be broken or cut along their higher longitudinal diameter and/or seasoned with natural products. The addition of preservatives or a thermal treatment of pasteurization is frequently used in the packing of fermented vegetables to ensure their microbiological stability during storage (Sanchez-Gómez, García García, and Rejano Navarro 2006).

5.5.1 COLOR CHANGES DURING FERMENTATION PROCESS

Color is one of the parameters that most directly reflect the quality of fermented vegetables and, most importantly, the sensory characteristics from a commercial point of view. When olives are placed in brine, they are subjected to physical and chemical changes that modify the fruits and their brine solution. The water-soluble components are withdrawn from the olive flesh due to diffusion of these components into the brine (Poiana and Romero 2006). The sugars diffused in brine are used by microorganisms present in the raw material and the processing environment (Sanchez-Gomez, García García, and Rejano Navarro 2006) and converted to organic acids. Several factors can affect the growth of native microflora: Sodium concentration and pH of brine are the main control parameters acting during fermentation (Garrido-Fernandez, Férnandez Díaz, and Adams 1997; Poiana and Romero 2006); the polyphenols in fruits inhibit LAB growth (Piga and Agabbio 2003); the presence of oleuropein and related compounds possess a specific inhibitory effect (Rodriguez et al. 2009).

In table olives, *phenolic compounds* have a significant influence on color, and they play a crucial role in the sensory and nutritional characteristics as well (Marsilio, Campestre, and Lanza 2001; Othman et al. 2009; Romeo et al. 2009; Rodríguez et al. 2009). Differences between the phenol composition of fresh and processed fruits are the result of reactions occurring during the fermentation period (Romero et al. 2004a); polyphenols undergo chemical transformations and, in general, their concentration in olives diminishes (Romero et al. 2004b). During the fermentation period, an increase in hydroxytyrosol to the detriment of oleuropein was observed (Esti, Cinquanta, and La Notte 1998; Romero et al. 2004a). Hydroxytyrosol is the most important phenolic compound detected in the final products (Romero et al. 2004a). Browning reactions that occur during processing are often caused by polymerization of phenolic compounds (Romeo et al. 2009).

During processing of table olives, there are important changes in phenolic quantity and quality that are generally due to several mechanisms that occur during the fermentation (Othman et al. 2009), as the alkaline treatment has no practical effect on the initial pigment composition (Minguez-Mosqueta et al. 1989). One of the phenomena is an osmotic exchange between fruit and brine (Othman et al. 2009), mainly affecting the soluble sugars, NaCl, and phenolic compounds. During fermentation, the total phenolic content in the olive fruit is reduced by 32%–51% due to diffusion of phenolic compounds into the brine, and it is closely related to the permeability of the olive skin (Poiana and Romero 2006) and the type of fermentation. It has been shown that phenolic reduction is lower in the controlled fermentation due the development of a biofilm that can act as a barrier to phenolic diffusion (Othman et al. 2009). Romeo, Piscopo, and Poiana (2010) concluded that the behavior of polyphenols during the fermentation stage could be dependent on the cuticle barrier effect, and not on the processing variables (salt content and acidification).

With respect to anthocyanins, they also diffuse into the brine, which assumes a color similar to olives, but this reduction does not stop, and this may be related to the fact that simple anthocyanins are either transformed or degraded as soon as they are diffused (Piga et al. 2005). Loss of monomeric anthocyanins may result in irreversible discoloration of the product. In black olives, Piga et al. (2005) observed that the L* values and the chroma of peel of black olives (Semidana and Kalamata cultivars) increased during processing, so the olives became gradually less black and developed a new color due to the fact that monomeric anthocyanins are replaced by other, more stable pigments, although this increase was independent of anaerobic or aerobic fermentation (Romero et al. 2004a).

Other phenomena related to color changes during the fermentation process include changes and transformations in the pigment composition, catalyzed by the acid pH of the medium and the action of chlorophyllase (Minguez-Mosquera, Garrido-Fernandez, and Gandul-Rojas 1989). During fermentation of naturally black olives, the main reaction that takes place is acid hydrolysis of the glucosides and the aglycones. Polymerization of anthocyanin compounds is the cause of the final color development in olives (Romero et al. 2004a; Othman 2009), and these reactions are of great importance for the nutritional and organoleptic properties of the fruits.

According to Minguez-Mosquera, Garrido-Fernandez, and Gandul-Rojas (1989), chlorophylls present in the fresh green fruit disappear completely in the process, giving rise to a mixture of their corresponding derivatives: pheophytins and pheophorbides. This takes place by two different mechanisms: During the first fermentation phase, chlorophyll *a* gives rise to pheophytin *a*, chlorophyllide *a*, and pheophorbide *a* due to enzymatic hydrolysis with the help of a medium with optimum pH and not in a chemical way. At the end of the second period, when the pH of the medium falls to 4.5, chlorophyll *b* content falls, giving rise to pheophytin *b* and a mixture of chlorophyllide *b* and pheophorbide *b* in an important transformation. After seven months, neither chlorophylls nor chlorophyllides appear in the fruit, and the levels of pheophytins and pheophorbides remain practically constant during this time.

The change in color from brown to black during the oxidation process of a ripe olive preparation is due mainly to the polymerization of hydroxytyrosol and caffeic acid under the effect of oxygen when fruits are exposed to air or aerated in aqueous suspension (Brenes-Balbuena, Garcia-Garcia, and Garrido-Fernandez 1992).

Table 5.4 presents values of color parameters for 10 preparations of Spanish table olives. The differences found were due to the total phenol content and to numerous reactions that can take place during fermentation and storage (hydrolysis, oxidation, polymerization, etc.) and, therefore, to the development of polymers. The values of the green olives are similar to other studies on olives (Esti, Cinquanta, and La Notte 1998; Brenes and Garcia-Garcia 2005; Romeo et al. 2009). The green olives showed the highest values of lightness (L*) and yellowness (b*) and lowest values of redness (a*). Lightness values decreased with increasing degree of ripeness. Lightness depended on the variety, the harvest time, and a greater degree of browning (Esti, Cinquanta, and La Notte 1998; Romeo et al. 2009). The differences between a* values in the same varieties (Cornicabra a and b; Cieza A and B) may be due to reactions of browning. Romeo et al. (2009) attributed the increase of a* to reactions of browning and enzyme-catalyzed or chemical oxidization of o-diphenols of olives in quinones and their subsequent transformation into different dark compounds.

Figure 5.5 shows the reflection spectra of different preparations of table olives, illustrating the differences between the spectra of each variety of olives according to surface color. The type of processing and the olive cultivar have a marked influence on the concentration of polyphenols in table olives (Romero et al. 2004a). This, in turn, influences the reflection spectra.

TABLE 5.4
CIELAB Color Parameters for Different Spanish Table Olives

Type of processing [a]	Varieties [b]	L*	a*	b*
Green olives	Cornicabra (a)	58.0	2.0	47.8
Green olives	Cieza (A)	53.1	0.3	47.8
Turning-color olives	Cornicabra	36.8	16.4	5.6
Ripe black olives	Hojiblanca	3.9	3.1	4.0
Green olives	Verdal	61.1	6.4	55.2
Ripe black olives	Cuquillo	10.4	15.8	11.3
Green olives	Manzanilla	58.2	8.5	63.3
Green olives	Cornicabra (b)	46.0	8.2	36.4
Green olives	Cieza (B)	41.5	2.2	35.5
Turning-color olives	Cieza	24.6	17.5	6.65

Note: CIELAB color space was measured by a Minolta Spectro Colorimeter CM-2600 (Minolta Camera Co., Osaka, Japan), with illuminant D_{65} and an observer of 10°, SCI mode, 11-mm aperture of the instrument for illumination and 8 mm for measurement. Each value is the average of nine measurements in the olive surface.

[a] The Spanish "Reglamentación Técnico Sanitaria para la elaboración, circulación y venta de lasa-ceitunas de mesa" (BOE 2001) distinguishes four elaboration types according to surface color: green, turning color, natural black, and ripe olives.

[b] (a), (A), (b), and (B) represent different commercial brands of green olives.

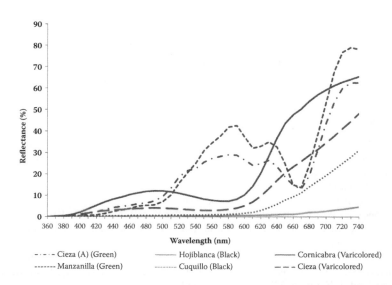

FIGURE 5.5 Reflectance spectra of different table olives (CIELAB color space was measured by a Minolta Spectro Colorimeter CM-2600 [Minolta Camera Co., Osaka, Japan], with illuminant D_{65} and an observer of 10°, SCI mode, 11-mm aperture of the instrument for illumination and 8 mm for measurement.)

5.6 CONCLUDING REMARKS

Food color is a major factor in consumer purchasing decisions, and consequently it is an important issue in food quality assessment. Color changes in fermented foods are caused by several mechanisms, depending on the food type:

1. In fermented meat products, chemical reactions between compounds derived from added nitrite/nitrate and myoglobin are the main factor influencing color, together with meat structural matrix, composition of the product (fat content, addition of coloring spices such as paprika, etc.), microbial metabolism that leads to pH decline and coagulation of muscular proteins, and processing conditions.
2. In fermented fish products, the main factors affecting color are the redox state of myoglobin, lipid oxidation, natural fish pigments, and processing conditions such as salting and the occurrence of Maillard reactions, among others.
3. In fermented dairy products, most color changes are given by milk gelation mechanisms (acidic or combined acidic-enzymatic), which are the result of the structural modification of milk when the pH drops due to microbial growth and milk proteins reach the isoelectric point and form a gel. Specific microbiota may provide distinctive colors to specialty cheeses.
4. In fermented vegetables, exemplified in this chapter by olives, factors such as the cultivar, ripening stage at harvesting, salting, pH changes, phenol composition, or optional technological treatments (such as alkalinization) may lead to changes in pigment composition and thus to changes in color during fermentation.

The direct effect of fermentation on color is minor. Changes in color result mainly from the effect of fermentation on the extractability and pH dependence of added ingredients, e.g., paprika carotenoids, phenols/polyphenols, etc.

REFERENCES

Aguirrezabal, M., J. Mateo, C. Domínguez, and J. M. Zumalacárregui. 1998. Spanish paprika and garlic as sources of compounds of technological interest for the production of dry fermented sausages. *Science des Aliments*. 18:409–414.

AMSA. 1991. Guidelines for meat color evaluation. American Meat Science Association. Chicago: National Live Stock and Meat Board.

Arroyo-López, F. N., A. Querol, J. Bautista-Gallego, and A. Garrido-Férnandez. 2008. Role of yeast in table olive production. *International Journal of Food Microbiology* 128:189–196.

Bernhard, K. 1990. Synthetic astaxanthin: The route of a carotenoid from research to commercialisation. In *Carotenoids: Chemistry and biology*, ed. N. I. Krinsky, M. M. Mathews-Roth, and R. F. Taylor, 337–363. New York: Plenum Press.

Bley M., M. E. Johnson, and N. F. Olson. 1985. Predictive test for the tendency of cheddar cheese to brown after processing. *Journal of Dairy Science* 68:2517–2520.

BOE (Boletín Oficial del Estado). 2001. Real Decreto 1230/2001. Reglamento técnico-sanitario para la elaboración circulación y venta de las aceitunas de mesa. BOE 279, 21/11/2001, 42587–42594.

Bover-Cid, S., M. Izquierdo-Pulido, and M. C. Vidal-Carou. 2000. Mixed starter cultures to control biogenic amine production in dry fermented sausages. *Journal of Food Protection* 63:1556–1562.

Breithaupt, D. E. 2007. Modern application of xanthophylls in animal feeding: A review. *Trends in Food Science and Technology* 18:501–506.

Brenes-Balbuena, M., and P. Garcia-Garcia. 2005. Elaboración de aceitunas denominadas green ripe olives con variedades españolas. *Grasas y Aceites* 56:188–191.

Brenes-Balbuena, M., P. Garcia-Garcia, and A. Garrido-Fernandez. 1992. Phenolic compounds related to the black color formed during the processing of ripe olives. *Journal of Agricultural and Food Chemistry* 40:1192–1196.

Chaijan, M., L. S. Benjaku, W. Visessanguan, and C. Faustman. 2007. Characterization of myoglobin from sardine (*Sardinella gibbosa*) dark muscle. *Food Chemistry* 100:156–164.

Chow, C. J., Y. Ochiai, and S. Watabe. 2004. Effect of frozen temperature on autoxidation and aggregation of bluefin tuna myoglobin in solution. *Journal of Food Biochemistry* 28:123–134.

Crisan, E. V., and A. Sands. 1975. Microflora of four fermented fish sauces. *Applied Microbiology* 29:106–108.

Czeczuga, B. 1979. Carotenoids in *Salmo gairdneri* Rich. and *Salmo trutta morpha fario* L. *Hydrobiology* 64:251–259.

Czeczuga, B., R. Bartel, and E. Czeczuga-Semeniuk. 2002. Carotenoids content in eggs of Atlantic salmon (*Salmo salar* L.) and brown trout (*Salmo trutta* L.) entering Polish rivers for spawning or reared in fresh water. *Acta Ichthyology Piscat* 32:3–21.

De Vuyst, L., G. Falony, and F. Leroy. 2008. Probiotics in fermented sausages. *Meat Science* 80:75–78

Dipplock, A. T., P. J. Agget, M. Ashwell, F. Bornet, E. B. Fern, and M. B. Robertfroid. 1999. Scientific concepts of functional foods in Europe: Consensus document. *British Journal of Nutrition* 81 (Suppl. 1): S1–S27.

El Soda, M. A. 2000. Role of specific groups of bacteria. In *Encyclopedia of food microbiology*, ed. R. K. Robinson, C. A. Batt, and P. D. Patel, 393–397. London: Academic Press.

Erkkilä, S., and E. Petäjä. 2000. Screening of commercial meat starter cultures at low pH and in the presence of bile salts for potential probiotic use. *Meat Science* 55:297–300.

Esti, M., L. Cinquanta, and E. La Notte. 1998. Phenolic compounds in different olive varieties. *Journal of Agricultural and Food Chemistry* 46:32–35.

Faustman, C., and R. G. Cassens. 1990. The biochemical basis for discoloration in fresh meat: A review. *Journal of Muscle Foods* 1:217–243.

Fernandez-Lopez, J., J. A. Perez-Alvarez, E. Sayas-Barbera, and V. Aranda-Catala. 2000. Characterization of the different states of myoglobin in pork using color parameters and reflectance ratios. *Journal of Muscle Foods*. 11 (3): 157–167.

Fernández-López, J., J. A. Pérez-Alvarez, E. Sayas-Barberá, and F. López-Santoveña. 2002. Effect of paprika (*Capsicum annuum*) on color of Spanish-type sausages during the resting stage. *Journal of Food Science* 67 (6): 2410–2414.

Fernández-López, J., E. Sayas-Barberá, and J. A. Pérez-Alvarez. 2002. Las salazones de pescado. In *Fundamentos tecnológicos y nutritivos de la dieta Mediterránea*, ed. J. A. Pérez-Alvarez, J. Fernández-López, and E. Sayas-Barberá, 141–159. Alicante, Spain: Universidad Miguel Hernández.

Fox, P. F., and P. L. H. McSweeney. 1998. Milk lipids. In *Dairy chemistry and biochemistry*, 67–145. London: Blackie Academic and Professional.

Fuentes, A., I. Fernández-Segovia, J. M. Barat, and J. A. Serra. 2010. Physicochemical characterization of some smoked and marinated fish products. *Journal of Food Processing and Preservation* 34:83–103.

Gallego-Restrepo, J. A., J. A. Pérez-Alvarez, and O. A. Ochoa. 2010. Use of reflectance spectra for assessing colour of meat and meat products. *Alimentación Equipos y Tecnología* 250:36–41.

García-Pérez, F. J., Y. Lario, J. Fernández-López, E. Sayas, J. A. Pérez-Alvarez, and E. Sendra. 2005. Effect of orange fiber addition on yogurt color during fermentation and cold storage. *Color Research and Applications* 30:457–463.

Garrido-Fernández, A., M. J. Férnandez Díaz, and R. M. Adams. 1997. *Table olives: Production and processing*. London: Chapman & Hall.

Glatman, L., V. Drabkin, and A. Gelman. 2000. Using lactic acid bacteria for developing novel fish food products. *Journal of the Science of Food and Agriculture* 80:375–380.

Gøtterup, J., K. Olsen, S. Knöchel, K. Tjener, L. H. Stahnke, and J. K. S. Møller. 2007. Relationship between nitrate/nitrite reductase activities in meat associated staphylococci and nitrosylmyoglobin formation in a cured meat model system. *International Journal of Food Microbiology* 120:303–310.

Guyomarc, F., A. Binet, and L. Dufossé. 2000. Production of carotenoids by *B. linens*: Variation among strains, kinetic aspects, and HPLC profiles. *Journal of Industrial Microbiology & Biotechnology* 24:64–70.

Hagen, B. F., H. Næs, and A. L. Holck. 2000. Meat starters have individual requirements for Mn^{+2}. *Meat Science* 55:161–168.

Hatlen, B., M. Jobling, and B. Bejerken. 1998. Relationships between carotenoid concentration and colour fillets of Artic charr, *Salvelinus alpinus* (L.), fed astaxanthin. *Aquatic Research* 29:191–202.

Hu, Y. J., W. S. Xia, and C. R. Ge. 2007. Effect of mixed starter cultures fermentation on the characteristics of silver carp sausages. *World Journal of Microbiology Biotechnology* 23:1021–1031.

———. 2008. Characterization of fermented silver carp sausages inoculated with mixed starter culture. *LWT–Food Science and Technology* 41:730–738.

Hugas, M., M. Garriga, T. Aymerich, and J. M. Monfort. 1993. Biochemical characterization of lactobacilli isolated from dry sausages. *International Journal of Food Microbiology* 18:107–113.

Hugas, M., and J. M. Monfort. 1997. Bacterial starter cultures for meat fermentation. *Food Chemistry* 59 (4): 547–554.

Hutchings, J. B. 2002. The perception and sensory assessment of colour. In *Colour in food: Improving quality*, ed. D. B. MacDougall, 9–32. Cambridge, U.K.: Woodhead Publishing.

IOOC (International Olive Oil Council). 2014. Trade standard applying to table olive. Res-2/91-IV/04. IOOC, Madrid.

Jinjarak, S., A. Olabi, R. Jiménez-Flores, and J. H. Walker. 2006. Sensory, functional, and analytical comparisons of whey butter with other butters. *Journal of Dairy Science* 89:2428–2440.

Khol W., H. B. Achenbach, and H. Reichenbach. 1983. The pigments of *B. linens*: Aromatic carotenoids. *Phytochemistry* 22 (1): 207–210.

Kim, J. H., H. J. Ahn, H. S. Yook, K. S. Kim, M. S. Rhee, G. H. Ryu, and M. W. Byun. 2004. Color, flavour, and sensory characteristics of gamma-irradiated salted and fermented anchovy sauce. *Radiation Physics and Chemistry* 69:179–187.

Kim, J., F. Shahidi, and M. Heu. 2003. Characteristics of salt-fermented sauces from shrimp processing byproducts. *Journal of Agriculture and Food Chemistry* 51:784–792.

Kosikowski, F. V., and D. D. Brown. 1969. Application of titanium dioxide to whiten mozzarella cheese. *Journal of Dairy Science* 52 (7): 968–970.

Lee, Y. S., S. Homma, and K. Aida. 1997. Characterization of melanoidin in soy sauce and fish sauce by electrofocusing and high performance gel permeation chromatography. *Nippon Shokuhin Kogyo Gakkaishi* 34:313–319.

Leistner, J. J., J. C. Millan, H. H. Huss, and L. M. Larson, 1994. Production of histamine and tyramine by lactic and bacteria isolated from vacuum-packed sugar-salted fish. *Journal of Applied Microbiology* 76:417–423.

Liu, Y., and Y. R. Chen. 2001. Analysis of visible reflectance spectra of stored, cooked, and diseased chicken meats. *Meat Science* 58:395–401.

Liu, Z. Y., Z. H. Li, M. L. Zhang, and X. P. Deng. 2010. Effect of fermentation with mixed starter cultures on biogenic amines in bighead carp surimi. *International Journal of Food Science & Technology* 45:930–936.

Lopetcharat, K., Y. J. Choi, J. W. Park, and M. A. Daeschel. 2001. Fish sauce products and manufacturing: A review. *Food Reviews International* 17:65–88.

Lozano, D. 2006. A new approach to appearance characterization. *Color Research and Application* 31:164–167.

Mah, J. H., and H. J. Hwang. 2009. Inhibition of biogenic amine formation in a salted and fermented anchovy by *Staphylococcus xylosus* as a protective culture. *Food Control* 20:796–801.

Marsilio, V., C. Campestre, and B. Lanza. 2001. Phenolic compounds change during California-style ripe olive processing. *Food Chemistry* 74:55–60.

Minguez-Mosquera, M. I., J. Garrido-Fernandez, and B. Gandul-Rojas. 1989. Pigment changes in olives during fermentation and brine storage. *Journal of Agricultural and Food Chemistry* 37:8–11.

Møller, J. K. S., and L. H. Skibsted. 2002. Nitric oxide and myoglobins. *Chemical Reviews* 102:1167–1178.

Morales, F. J., and S. Jiménez-Pérez. 2001. Free radical scavenging capacity of Maillard reaction products as related to color and fluorescence. *Food Chemistry* 72:119–125.

Morzel, M., G. F. Fitzgerald, and E. K. Arendt. 1997. Fermentation of salmon fillets with a variety of lactic acid bacteria. *Food Research International* 30:777–785.

Mounier, J., F. Irlinger, M. N. Leclerq-Perlat, A. S. Sarthou, H. E. Spinnler, G. F. Fitzgerald, and T. M. Cogan. 2006. Growth and colour development of some surface ripening bacteria with *Debaryomyces hansenii* on aseptic cheese curd. *Journal of Dairy Research* 73:441–448.

Nes, I. F., and R. Skjelkvåle. 1982. Effect of natural spices and oleoresins on *Lactobacillus plantarum* in the fermentation of dry sausages. *Journal of Food Science* 47:1618–1625.

Neubauer, H., and F. Götz. 1996. Physiology and interaction of nitrate and nitrite reduction in *Staphylococcus carnosus*. *Journal of Bacteriology* 178:2005–2009.

Othman, N. B., D. Roblain, N. Chammen, P. Thonart, and M. Hamdi. 2009. Antioxidant phenolic compounds loss during the fermentation of Chétoui olives. *Food Chemistry* 116:662–669.

Paludan-Müller, C., H. H. Huss, and L. Gram. 1999. Characterization of lactic acid bacteria isolated from a Thai low-salt fermented fish product and the role of garlic as substrate for fermentation. *International Journal of Food Microbiology* 46:219–229.

Panagou, E. Z., U. Schillinger, C. M. A. P. Franz, and G. J. Nychas. 2008. Microbiological and biochemical profile of cv. Conservolea naturally black olives during controlled fermentation with selected strains of lactic acid bacteria. *Food Microbiology* 25: 328–358.

Park, Y. H., D. S. Chang, and S. T. Kim. 1997. *Seafood processing*, 2nd ed., 771–778. Seoul, Korea: Hyung Seoul Publisher.

Pegg, R. B., and F. Shahidi. 1997. Unraveling the chemical identity of meat pigments. *Critical Reviews in Food Science and Nutrition* 37:561–589.

Peralta, E. M., H. Hatate, D. Kawabe, R. Kuwahara, S. Wakamatsu, T. Yuki, and H. Murata. 2008. Improving antioxidant activity and nutritional components of Philippine salt-fermented shrimp paste through prolonged fermentation. *Food Chemistry* 111:72–77.

Pérez-Alvarez, J. A. 1996. Contribución al estudio objetivo del color de productos cárnicos crudo-curados. Ph.D. thesis. Polytechnical University of Valencia, Spain.

———. 2006a. Color. In *Ciencia y tecnología de carnes*, ed. Y. H. Hui, I. Guerrero, and M. R. Rosmini, 161–198. Mexico City: Limusa Noriega Editores.

———. 2006b. Aspectos tecnológicos de los productos crudo-curados. In *Ciencia y tecnología de carnes*, ed. Y. H. Hui, I. Guerrero, and M. R. Rosmini, 463–492. Mexico City: Limusa Noriega Editores.

———. 2008. Overview of meat products as functional foods. In *Technological strategies for functional meat products development*, ed. J. Fernández-López and J. A. Pérez-Alvarez, 1–18. Kerala, India: Transworld Research Network.

Pérez-Alvarez, J. A., and J. Fernández-López. 2007. Chemical and biochemical aspects of color in muscle foods. In *Handbook of meat, poultry, and seafood quality*, ed. L. M. L. Nollet, 25–44. Ames, IA: Blackwell Publishing.

———. 2009. Colour of meat products from a food engineering perspective. *Alimentacion Equipos y Tecnologia* 247:34–37.

Pérez-Alvarez, J. A., J. Fernández-López, and E. Sayas-Barberá. 2009a. El color del pescado y de los productos derivados de la pesca. In *Tecnología de productos de origen acuático*, ed. I. Guerrero-Legarreta, M. R. Rosmini, and R. E. Armenta-López, 133–158. Mexico City: Limusa.

———. 2009b. Productos pesqueros tradicionales de la Cuenca Mediterránea: Salazones Españoles. In *Tecnología de productos de origen acuático*, ed. I. Guerrero-Legarreta, M. R. Rosmini, and R. E. Armenta-López, 361–378. Mexico City: Limusa.

Pérez-Alvarez, J. A., J. Fernández-López, and M. E. Sayas-Barberá. 2000a. Fundamentos físicos, químicos, ultraestructurales y tecnológicos en el color de la carne. In *Nuevas tendencias en la tecnología e higiene de la industria cárnica*, ed. M. R. Rosmini, J. A. Pérez-Alvarez, and J. Fernández-López, 51–71. Elche, Spain: Universidad Miguel Hernández.

———. 2000b. El pimentón en la industria cárnica. In *Nuevas tendencias en la tecnología e higiene de la industria cárnica*, ed. M. R. Rosmini, J. A. Pérez-Alvarez, and J. Fernández-López, 235–254. Elche, Spain: Universidad Miguel Hernández.

Perez-Alvarez, J. A., E. Sanchez-Rodriguez, J. Fernandez-Lopez, M. A. Gago-Gago, C. Ruiz-Peluffo, M. Rosmini, M. J. Pagan-Moreno, F. Lopez-Santovena, and V. Aranda-Catala. 1997. Chemical and color characteristics of "lomo embuchado" during salting seasoning. *Journal of Muscle Foods* 8 (4): 395–411.

Pérez-Alvarez, J. A., M. E. Sayas-Barbera, J. Fernández-López, and V. Aranda. 1999. Physicochemical characteristics of Spanish type dry-cured sausage. *Food Research International* 32 (9): 599–613.

Pérez-Alvarez, J. A., M. E. Sayas-Barberá, E. Sendra, and J. Fernández-López. 2011. Color measurements on edible animal by-products and muscle based foods. In *Handbook of analysis of edible animal by-products*, ed. L. Nollet and F. Toldrá, 87–104. Boca Raton, FL: Taylor & Francis Group.

Piga, A., and M. Agabbio. 2003. Quality improvement of naturally green table olives by controlling some processing parameters. *Italian Journal of Food Science* 2:259–268.

Piga, A., A. Del Caro, M. Pinna, and M. Agabbio. 2005. Anthocyanin and colour evolution in naturally black table olives during anaerobic processing. *LWT–Food Science and Technology* 38:425–429.

Poiana, M., and F. V. Romero. 2006. Changes in chemical and microbiological parameters of some varieties of Sicily olives during natural fermentation. *Grasas y Aceites* 57:402–408.

Rai, K. P., C. Zhang, and W. S. Xia. 2010. Effects of pure starter cultures on physico-chemical and sensory quality of dry fermented Chinese-style sausage. *Journal of Food Science and Technology* 47 (2): 188–194.

Robinson, R. K. 1995. Cheese and fermented milks: Background to manufacture. In *A colour guide to cheese and fermented milks*, 1–30. London: Chapman & Hall.

Rodriguez, H., J. A. Curiel, J. M. Landete, B. de las Rivas, F. López de Felipe, C. Gómez-Cordovés, J. M. Mancheño, and R. Muñoz. 2009. Food phenolics and lactic bacteria. *International Journal of Food Microbiology* 132:79–90.

Romeo, F. V., S. De Luca, A. Piscopo, E. Perri, and M. Poiana. 2009. Effects of post-fermentation processing on the stabilisation of naturally fermented green table olives (cv. nocellara etnea). *Food Chemistry* 119:873–878.

Romeo, F. V., A. Piscopo, and M. Poiana. 2010. Effect of acidification and salt concentration on two black brined olives from Sicily (cv. moresca and giarraffa). *Grasas y Aceites* 215:251–260.

Romero, C., M. Brenes, P. García, A. García, and A. Garrido, A. 2004a. Polyphenol changes during fermentation of naturally black olives. *Journal of Agricultural and Food Chemistry* 52:1973–1979.

Romero, C., M. Brenes, K. Yousfi, and P. García. 2004b. Effect of cultivar and processing method on the contents of polyphenols in table olives. *Journal of Agricultural and Food Chemistry* 52:479–484.

Sánchez Gómez, A. H., P. García García, and L. Rejano Navarro. 2006. Elaboration of table olives. *Grasas y Aceites* 57:86–94.

Sánchez-Zapata, E., J. Fernández-López, E. Sayas, E. Sendra, C. Navarro, and J. A. Pérez-Alvarez. 2008. Orientative study on the colorimetric characterization of different smoked and salted-dried fish products found in Spanish market. *Óptica Pura y Aplicada* 41:273–279.

Sánchez-Zapata, E., and J. A. Pérez-Alvarez. 2007. El color en distintas especies de pescado: Caracterización objetiva. *Alimentación, Equipos y Tecnología* 219:39–43.

Sánchez-Zapata, E., J. A. Pérez-Alvarez, J. Fernández-López, and J. X. Barber-Valles. 2010. Descriptive study of reflectance spectra of hake (*Merluccius australis*), salmon (*Salmo salar*), and light and dark muscle from tuna (*Thunnus thynnus*). *Journal of Food Quality* 33:391–403.

Scannell, A. G. M., Paul M. Kenneally, and E. K. Arendt. 2004. Contribution of starter cultures to the proteolytic process of a fermented non-dried whole muscle ham product. *International Journal of Food Microbiology* 93:219–230.

Sohn, J. H., Y. Taki, H. Ushio, T. Kohata, I. Shioya, and T. Ohshima. 2005. Lipid oxidations in ordinary and dark muscles of fish: Influences on rancid off-odor development and color darkening of yellowtail flesh during ice storage. *Journal of Food Science* 70:490–496.

Soyer, A., A. H. Erta, and U. Üzümcüoglu. 2005. Effect of processing conditions on the quality of naturally fermented Turkish sausages (sucuks). *Meat Science* 69:135–141.

Tamine, A. Y. 2000. Cheese in the marketplace. In *Encyclopedia of food microbiology*, ed. R. K. Robinson, C. A. Batt, and P. D. Patel, 372–381. London: Academic Press.

Tamine, A. Y., and R. K. Robinson. 1985. *Yogurt: Science and technology.* Oxford, U.K.: Pergamon Press.

Thiansilakul, Y., S. Benjakul, and M. P. Richards. 2011. Isolation, characterisation, and stability of myoglobin from Eastern little tuna (*Euthynnus affinis*) dark muscle. *Food Chemistry* 124:254–261.

Twiddy, D. R., D. J. Cross, and R. D. Cooke. 1987. Parameters involved in the production of lactic acid preserved fish-starchy substrate combinations. *International Journal of Food Science and Technology* 22:115–121.

Ueki, N., and Y. Ochiai. 2004. Primary structure and thermostability of bigeye tuna myoglobin in relation to those of other scombridae fish. *Fisheries Science* 70:875–884.

Verluyten, J., F. Leroy, and L. de Vuyst. 2004. Effects of different spices used in production of fermented sausages on growth of and curvacin A production by *Lactobacillus curvatus* LTH 1174. *Applied and Environmental Microbiology* 70:4807–4813.

Walstra P., and R. Jenness. 1984. Other milk properties. Chapter 11 in *Dairy chemistry and physics*, 164–186. New York: John Wiley and Sons.

Warriss, P. D., 2000. Post-mortem changes in muscles and its conversion to meat. In *Meat science: An introductory text*, 93–105. London: CABI.

Watmough, N. J., G. Butland, M. R. Cheesman, J. W. B. Moir, D. J. Richardson, and S. Spiro. 1999. Nitric oxide in bacteria: Synthesis and consumption. *Biochimica et Biophysica Acta* 1411:456–474.

Xu, Y., W. Xia, F. Yang, J. M. Kim, and X. Nie. 2010. Effect of fermentation temperature on the microbial and physicochemical properties of silver carp sausages inoculated with *Pediococcus pentosaceus*. *Food Chemistry* 118:512–518.

Yin, L. J., C. L. Pan, and S. T. Jiang. 2002. Effect of lactic acid bacterial fermentation on the characteristics of minced mackerel. *Journal of Food Science* 67:786–792.

Zanardi, E., E. Novelli, G. P. Ghiretti, V. Dorigoni, and R. Chizzolini. 1999. Colour stability and vitamin E content of fresh and processed pork. *Food Chemistry* 67:163–171.

6 The Role of Fermentation in Providing Biologically Active Compounds for the Human Organism

Afaf Kamal-Eldin

CONTENTS

6.1 INTRODUCTION

The term *dietary bioactive compounds* (or *biologically active compounds*) has been used in the last decades to describe food constituents with extra nutritional value, usually in connection with aging and degenerative diseases. The classification of a certain compound or group of compounds into this category is based on health benefits arbitrarily observed in humans, animals, or even in cell studies. Bioactive compounds range from low-molecular-weight acids, phenols, steroids, alkaloids, etc., to high-molecular-weight carbohydrates and proteins. Their physiological effects can relate to signaling, cholesterol-lowering, lipid modulation, immunity, hypotensivity, or other beneficial effects (Biesalski et al. 2009). Bioactive compounds are also termed *nutraceuticals*, and foods containing them are referred to as "functional foods," i.e., health-promoting, disease-preventing, or medicinal foods.

Bioactive compounds can be native molecules in the food raw material or can be generated or concentrated during processing, e.g., milling, germination, fermentation, cooking, etc. The bioactivity and nutritional value of foods can also be enhanced by alterations of the food matrix rather than the effective molecules themselves. For example, cooking may loosen the food structure and increase the accessibility/bioavailability of minerals and vitamins or provitamins. Fermentation induces different types of bioactivities caused by the hydrolysis and further metabolism of carbohydrates and proteins, biosynthesis of vitamins, structural changes and release of antioxidants and other compounds with potent bioactivities. Fermented milk products are the fermented foods mostly studied in detail (Hayes et al. 2007; Ebringer, Ferencik, and Krajcovic 2008), but scattered studies have also investigated the effects of fermentation on certain bioactive compounds/bioactivities of cereals, legumes, fruits, and vegetables.

6.2 MICROORGANISMS RESPONSIBLE FOR ALTERATIONS IN BIOLOGICALLY ACTIVE COMPOUNDS

Several strains of *Lactobacillus* are capable of producing or enriching bioactive compounds in fermented foods, e.g., *L. acidophilus*, *L. delbrueckii*, *L. helveticus*, *L. lactis*, *L. plantarum*, *L. rhamnosus*, etc. The well-characterized proteolytic activity of lactic acid bacteria (LAB) is caused by cell wall–bound proteinases and a number of intracellular peptidases, including aminopeptidases, dipeptidases, endopeptidases, and tripeptidases. A number of bioactive peptides, e.g., antioxidant, antimutagenic, immunomodulatory, and antihypertensive angiotensin-1 converting enzyme (ACE, EC 3.4.15.1) inhibitory peptides are released from milk proteins through microbial proteolysis (Gobbetti, Minervini, and Rizzello 2004; Korhonen and Pihlanto 2006). The release of bioactive peptides is highly dependent on the choice of bacterial strain ensuring the balance between product integrity and the right level of proteolytic activity needed to release these active peptides. An example is antihypertensive ACE released from milk caseins. In this case, the concentration of ACE inhibitory peptides relies on the balance between their formation and breakdown. Two well known ACE inhibitory tripeptides, valyl-prolyl-proline (Val-Pro-Pro, VPP) and isoleucyl-prolyl-proline (Ile-Pro-Pro, IPP), were isolated from milk fermented with different strain of LAB including *L. delbrueckii*, *L. lactis*, and *L. helveticus*. Examples of ACE-inhibiting proteins and -producing bacteria are shown in Table 6.1. Immunomodulatory properties are imparted in fermented milks by a number of LAB, *inter alia*, *L. helveticus*, *L. paracasei*, and *L. casei* GG, which stimulate IL-10 production, suppress lymphocyte proliferation, inhibit T cell activation, and/or modulate IL-2 expression (Pessi et al. 2001; Le Blanc et al. 2002; Prioult, Pecquet, and Fliss 2004).

Fungi are also active in the generation of bioactive compounds. For example, the baker's yeast, *Saccharomyces cerevisiae*, has probiotic activity and is widely used as a dietary supplement for protection against enteric pathogens, modification of lymphocyte proliferation, and release of bound phenolic antioxidants in some fermented foods such as wine, sourdough, and cheese. *Saccharomyces boulardii* was

TABLE 6.1

Examples of Angiotensin-1 Converting Enzyme (ACE)–Inhibiting Peptides, Their Precursor Proteins, and Producing Microorganisms

Peptide sequence	Precursor protein	Microorganism	Reference
Ile-Pro-Pro-Val-Pro-Pro	β-CN, κ-CN	*L. helveticus* and *S. cerevisiae*	Nakamura et al. 1995
Lys-Val-Leu-Pro-Val-Pro-Gln	β-CN	*L. helveticus* CP790 protease	Maeno, Yamamoto, and Takano 1996
Tyr-pro-Phe-Pro Ala-Val-pro-Tyr-Pro-Gln-Arg Thr-Met-Pro-Leu-Trp	β-CN, α$_{s1}$-CN	*Lactobacillus* GG + enzymes (pepsin & trypsin)	Rokka et al. 1997
Ser-Lys-Val-Tyr-Pro-Phe-Pro-Gly-Pro-Ile	β-CN	*L. delbrueckii* spp. *bulgaricus*	Ashar and Chand 2004
Ser-Lys-Val-Tyr-Pro	β-CN	*L. lactis* spp. *lactis diacetylactis* and *S. thermophilus*	Ashar and Chand 2004

shown to catalyze the breakdown of dietary phytate, to enhance the biofortification of foods with folate, and to modify signaling pathways associated with proinflammatory responses via a mechanism involving blockage of the activation of nuclear factor–kappa B (NF-κB) and mitogen-activated protein kinase (MAPK) (Fleet 2007; Chen et al. 2006). Fermentation of soybean with other fungi, such as *Aspergillus sojae*, elicit the formation of glyceollins, which are major bioactive compounds and shown to have antifungal and anticancer activities (Kim et al. 2010). Molds, e.g., *Rhizopus* spp., produce a number of lipases and proteases that degrade macronutrients into low-molecular-weight compounds including vitamins and bioactive phenols (Nout and Kiers 2005).

6.3 MICROORGANISM GENERATION OF BIOACTIVE COMPOUNDS

Microorganisms are capable of supporting a large number of chemical reactions catalyzed by the vast number of enzyme activities contained in their cells. The microorganisms use these reactions to degrade nutrients via exothermic catabolic pathways into energy and small molecules that are used in the endothermic anabolic pathways. The types of chemical reactions catalyzed by microorganisms can be grouped into the following main categories:

1. *Oxidation*, including hydroxylation, epoxidation, dehydrogenation, deamination, ring fission, and formation of new C-C bonds
2. *Reduction*, including hydrogenation and reductive removal of substituents
3. *Hydrolysis*, including hydrolysis of esters, amines, amides, ethers, lactones, etc.

4. *Condensation*, including dehydration, acylation, esterification, glycosidation, lactonization, and amination
5. *Isomerization*, including rearrangement of double bonds and oxygen functions and epimerizations

Biotransformations, caused by microorganisms, are either catalyzed by enzymes in the normal biosynthetic pathways of the microorganism or catalyzed by induced cytochrome P-450-dependent monooxygenases specific to the transformation of substrates that are foreign to the microorganisms (xenobiotic transformation). Usually fermentation reactions result from multienzyme activities originating from the mixed culture working cooperatively or sequentially. Microorganisms can be broadly divided into two classes: aerobic organisms that can grow in oxygen atmosphere and carry reactions using oxygen and anaerobic organisms that grow and carry reactions in the absence of oxygen. Fermentation reactions refer to all reactions catalyzed by microorganisms and or enzymes that decompose food stuffs into useful compounds.

6.3.1 Generation of Bioactive Compounds from Carbohydrate Metabolism

Carbohydrate metabolism is the most facile way to obtain energy. Sugars released from carbohydrates are fermented to organic acids such as lactic acid by *Lactobacillaceae* and acetic acid by *Acetobacter* species. Lactic acid bacteria can be divided into homofermenters utilizing hexoses (mainly glucose) to produce lactic acid only and heterofermenters that produce lactic acid, ethanol, and carbon dioxide, the overall reactions being

$$C_6H_{12}O_6 \rightarrow 2\ CH_3\text{-}CH_2OH\text{-}COOH$$

$$C_6H_{12}O_6 \rightarrow CH_3\text{-}CH_2OH\text{-}COOH + C_2H_5OH + CO_2$$

Lactic and other organic acids produced by bacterial fermentation add to the antibacterial properties of the producing bacteria against other harmful microorganisms, thereby inducing probiotic effects. In addition, bacteria are able to produce glutamic acid from sugars and use it for growth (Holden and Holman 1959; Waller and Lichstein 1965).

Microorganisms are also able to metabolize complex carbohydrates including dietary fibers, e.g., cellulose, hemicellulose, pectic substances, and fructo-oligosaccharides. A diverse group of microorganisms including cellulolytic bacteria (e.g., *Clostridium, Bacillus*) and fungi (e.g., *Penicillium, Aspergillus, Trichoderma*) possess active cellulases capable of converting cellulose, b(1→4)-linked glucose polymer, to glucose (Bayer et al. 1998). During fermentation, enzymes are released to degrade complex carbohydrates *inter alia* into short-chain fatty acids (SCFA) and simple sugars. According to Levitt, Gibson, and Christ (1995), acetic acid, propionic acid, butyric acid, carbon dioxide, hydrogen, and water are produced from glucose by colonic bacteria. The SCFA, acetic, propionic, and butyric acids are formed at different proportions depending on the substrates and fermenting culture. SCFA has several beneficial health effects, *inter alia*, modulation of gut microflora and colon health (Smith, Yokoyama, and German 1998; Hijova and Chmelarova 2007), signaling and stimulation of leptin production (Xiong et al. 2004), as well as cardioprotective (Demigne et al. 1995) and anticarcinogenic effects (Scharlau et al. 2009).

Besides low-molecular-weight carbohydrate degradation products, microorganisms are capable of biosynthesizing high-molecular-weight polysaccharides, called exopolysaccharides (EPS), as well as lipopolysaccharides (LPS) (Table 6.2). As an example, the structure of lentinan (Figure 6.1) is (β-D-glucopyranosyl-(1→6)-[β-D-glucopyranosyl-(1→3)-[β-D-glucopyranosyl-(1→6)]-β-D-glucopyranosyl-(1→3)-β-D-glucopyranosyl-(1→3)-β-D-glucopyranosyl-(1→3)]-β-D-glucopyranose).

EPS are produced by a number of food-grade microorganisms, particularly lactic acid bacteria, propionibacteria, and bifidobacteria. The EPS produced by LAB can be divided into two groups,

Homopolysaccharides, including
1. α-D-glucans (α-1,6-linked glucose residues or dextrans) produced by, e.g., *Leuconostoc mesenteroides* subsp. *mesenteroides* and *Leuc. mesenteroides* subsp. *dextranicum*, and alternans (*Leuc. mesenteroides*) and mutans (*Streptococcus mutans* and *Streptococcus sobrinus*), both composed of α-1,3- and α-1,6-linkages
2. β-D-glucans (β-1,3-linked glucose residues with β-1,2-branches) produced by, e.g., *Pediococcus* spp. and *Streptococcus* spp.

TABLE 6.2
Examples of Exopolysaccharides (EPS) Produced by Bacteria

EPS	Description
Acetan (=Xylinan)	An acidic, water-soluble heteropolysaccharide produced by *Acetobacter xylinum* that consists of a cellulose backbone with side chains composed of repeating alternative heptasaccharide branches (containing α-L-rhamnose and α-D-mannose).
Kefiran	A pale yellow polysaccharide gel with medicinal and mouthfeel properties from dairy and grains water kefirs. Kefiran is produced by *Lactobacillus kefiranofaciens* and *L. delbrueckii* subsp. *bulgaricus*.
Lentinan	An anticancer β-glucan with a glycosidic β-1,3:β-1,6 linkage. Lentinan is produced by the shiitake mushroom *Lentinus edodes*.
Levan	A homopolysaccharide biopolymer composed of 2,6-D-fructofuranosyl residues having multiple 1,2-branches that can be used as food or a feed additive. Levan is produced by, e.g., *Alcaligenes viscosus* and *Zymomonas mobilis* and has prebiotic and hypocholesterolemic effects.
Xanthan	A viscous polysaccharide gum derived from the coat of the bacteria *Xanthomonas campestris* by the fermentation of glucose, sucrose, and lactose. It is used as a food additive (thickening agent for use in salad dressings, sauces, ice creams, etc.). It is a very long linear polymer (M.wt. 1–10 million Da) composed of D-glucose, D-mannose, and D-glucuronic acid with side chains containing pyruvic acid (2.5%–5% of the molecule).

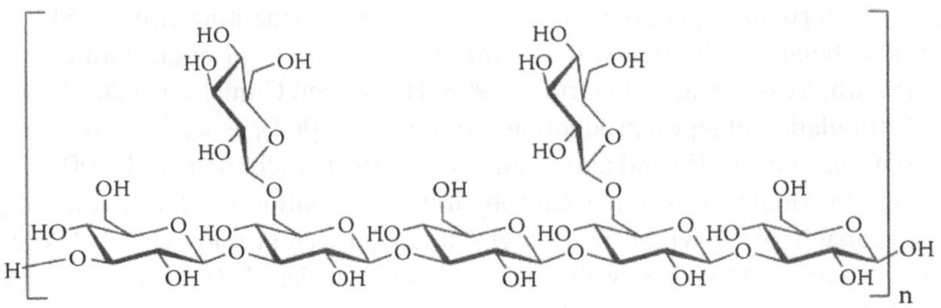

FIGURE 6.1 Lentinan

3. Fructans (β-2,6-linked fructose residues with some β-2,1-branches), example is levan produced by *S. salivarius*
4. Polygalactan having different glycosidic linkages between identical repeating units

and *Heteropolysaccharides* having different glycosidic linkages and variable sugar residues.

EPS contribute to human health by acting as indigestible fiber and for their immunogenic, anticancer, antiulcer, and lipid-lowering properties (de Vuyst and Degeest 1999). These EPS are produced by mesophilic (*Lactococcus lactis* subsp. *lactis*, *L. lactis* subsp. *cremoris*, *Lactobacillus casei*, *Lb. sake*, *Lb. rhamnosus*, etc.) and thermophilic (*Lb. acidophilus*, *Lb. delbrueckii* subsp. *bulgaricus*, *Lb. Helveticus*, and *S. thermophilus*) LAB strains. Heteropolysaccharides play very important roles in the rheology, texture, and mouthfeel of fermented milks.

EPS are used by the food industry in a wide range of applications including thickening agents, emulsifiers, gelling agents, and stabilizers. In addition, a number of beneficial health effects of EPS have been reported, e.g., cholesterol-lowering, antitumor, and immune-modulating effects, e.g., from *Bifidobacterium*. An EPS produced by *Streptococcus macedonicus* Sc136 was found to contain the trisaccharide sequence β-D-GlcpNAc-(1-3)-β-D-Galp(1-4)-β-D-Glcp corresponding to the backbones of lacto-N-tetraose and lacto-N-neotetraose, which are structural elements in several human milk oligosaccharides important for infant nutrition.

6.3.2 GENERATION OF BIOACTIVE COMPOUNDS FROM PROTEIN METABOLISM

While carbohydrates function as important carbon sources, proteins provide both carbon and nitrogen for microbial metabolism. Microorganisms have a number of protease enzymes (hydrolases) including aminopeptidase, carboxypeptidase, and endopeptidase activities able to release amino acids from proteins. Several papers have reviewed the formation of bioactive peptides (including inhibitors of angiotensin-1-converting enzyme, antihypertensive, immunomodulatory, antioxidative, and antimicrobial peptides) from milk proteins via microbial proteolysis (Gobbetti, Minervini, and Rizzello 2004). Mellander (1950) suggested that casein-derived phosphorylated peptides, caseinophosphopeptides (CPPs), enhance vitamin D–independent bone calcification in rachitic infants. Some of the peptides derived from milk casein by the action of the proteolytic enzymes of lactic acid bacteria exhibit significant antihypertensive effects in spontaneously hypertensive rats (Yamamoto 1998; Tsai et al. 2008).

Angiotensin-1 converting enzyme (ACE) is a key enzyme in the blood pressure regulating pathway, the renin-angiotensin system, where it acts by removing a dipeptide from the C-terminal of angiotensin-I to form angiotensin-II. Small peptides, such as Val-Pro-Pro (VPP) and Ile-Pro-Pro (IPP), derived from β-caseins, induce hypotensive effects by acting as ACE inhibitors (Seppo et al. 2003; Huth, Dirienzo, and Miller 2006). These ACE-inhibitory tripeptides form the basis for the antihypertensive effects of the fermented milk functional drinks Calpis® and Evolus® of Japan and Finland, respectively. In addition, the pentapeptide Ile-Ile-Ala-Glu-Lys (IIAGL), derived from β-lactoglobulin, possesses hypocholesterolemic effects (Nagaoka et al. 2001). Recently, milk caseins hydrolyzed with *Bifidobacterium animalis* subsp. *lactis* (Bb12) were found to have hypocholesterolemic effects (Alhaj et al. 2010).

Fermentative bacteria are able to produce a heterologous group of ribosomally synthesized antibacterial peptides known as bacteriocins (De Vuyst and Vandamme 1994). The bacteriocins are cationic peptides having hydrophobic or amphiphilic properties, via which they are able to kill bacterial species that are harmful to the producer bacterium (Table 6.3). Depending on the producer organism and classification criteria, bacteriocins can be classified into several groups (Ennahar et al. 2000; Jack and Jung 2000; Cleveland et al. 2001; McAuliffe, Ross, and C. Hill 2001) in which classes I and II are the most thoroughly studied. Class I, termed *lantibiotics*, constitute a group of small peptides that are characterized by their content of several unusual amino acids (Guder, Wiedeman, and Sahl 2000). The class II bacteriocins are small, unmodified, heat-stable peptides (Nes and Holo 2000). Many bacteriocins are active against food-borne pathogens (Vignolo et al. 1996; De Martinis and Franco 1998; Bredholt, Nesbakken, and Holck 1999). A large number of bacteriocins have been isolated and characterized from lactic acid bacteria, and some have acquired a status as potential antimicrobial agents because of their potential as food preservatives and antagonistic effect against important pathogens. The important ones are

TABLE 6.3
Examples of Bacteriocins Isolated from Different *Lactobacillus* Species

Bacteria	Bacteriocin
L. acidophilus	Acidolin, Acidophilin, Lactacin B, Lactacin F
L. brevis	Lactobrevin, Lactobacillin
L. bulgaricus	Bulgarin
L. helveticus	Lactolin 27, Helveticin J
L. plantarum	Plantaricin SIK-83, Plantaricin A, Lactolin, Plantaricin B
L. reuteri	Reuterin

nisin, diplococcin, acidophilin, bulgarican, helveticins, lactacins, and plantaricins (Nettles and Barefoot 1993). The lantibiotic nisin, which is produced by different *Lactococcus lactis* spp., is the most thoroughly studied bacteriocin to date and the only bacteriocin that is applied as an additive in food worldwide (Delves-Broughton et al. 1996). Besides protein amino acids, microorganisms including bacteria, fungi, and yeasts are able to produce nonprotein amino acids, such as gamma-aminobutyric acid, with beneficial physiological functions including neurotransmission, hypotensive, diuretic, and tranquilizer effects (Li and Cao 2010).

6.3.3 GENERATION OF BIOACTIVE COMPOUNDS FROM LIPID METABOLISM

Several strains of *L. acidophilus* have the ability to produce conjugated linoleic acid (CLA) from linoleic acid by the enzyme linoleate isomerase (EC 5.2.1.5) although with different potential (Macouzet, Robert, and Lee 2010; Ogawa et al. 2005). Due to this action, the levels of CLA are high in cheese and fermented milks. Dietary CLA is claimed to have different beneficial effects including reduction of body fat and protection against cancer and atherosclerosis, but it also impairs insulin sensitivity.

6.3.4 BIOSYNTHESIS OF VITAMINS

Microorganisms are also able to alter the contents and bioavailability of a number of vitamins including vitamins B_1 (thiamin), B_2 (riboflavin), B_5 (pantothenic acid), B6 (pyridoxine, pyridoxamine, and pyridoxal), B_7 (biotin), B_{12} (cyanocobalamin), folate, and K in fermented foods (Vandamme 1992; Ek 2005). For example, fermentation of the Turkish cereal food *tarhana* resulted in significant increases in the levels of vitamins B_2, B_3, B_5, folic acid, and ascorbic acid (Ek 2005).

Thiamin pyrophosphate (vitamin B_1) is an essential cofactor for several important enzymes in the carbohydrate metabolism. The thiamin biosynthesis pathway in bacteria involves separate biosynthesis of thiazole and pyrimidine moieties, which are then coupled to yield thiamin phosphate. In *Escherichia coli* and other anaerobic proteobacteria, the thiazole moiety (hydroxyethylthiazole-P, HET-P) is formed from tyrosine in a reaction catalyzed by ThiH, a radical S-adenosyl methionine (SAM) enzyme; the pyrimidine moiety (hydroxymethylpyrimidine-PP, HMP-PP) is formed by the rearrangement of aminoimidazole ribotide catalyzed by the ThiC gene product, while the ThiE protein catalyzes the formation of thiamin-P by the coupling of HET-P and HMP-PP moieties. In yeast, e.g., *Saccharomyces cerevisiae*, the biosynthesis of these two heterocycles is different from that in bacteria. Here, the imidazole moiety is synthesized from glycine using the distinct gene (THI4), while the pyrimidine moiety of thiamin is synthesized using a distinct gene (THI5), which appears to be synthesized from histidine and pyridoxol-P (Begley et al. 2008). Some of the yeast enzymes involved in thiamin biosynthesis, such as thiamin diphosphokinase and thiazole synthase, and the genes encoding them, have been revealed (Hohmann and Meacock 1998; Rodriguez-Navarro et al. 2002; Kowalska and Kozik 2008). While yeast fermentation was found to increase the level of vitamin B_1 in fermented foods, fermentation with *Lactobacilli* was found to decrease it (Khetarpaul and Chauhan 1989; Iwashima 1989). For example, fermentation of chickpeas and cowpeas in a medium containing *Lactobacillus casei, L. leichmannii, L. plantarum, Pediococcus pentosaceus*, and *P. acidilactici* was found to be associated with decreased levels of B_1 (Zamora and Fields 2006). A similar decrease in B_1 levels in foods fermented with *Lactobacilli* was documented for Sudanese *kisra* (Mahgoub et al. 1999).

Riboflavin (vitamin B_2) is produced by a wide number of microorganisms including lactose-fermenting fungi (e.g., *Ashbya, Candida, Clostridium*, and *Saccharomyces* spp.), bacteria (e.g., *Lactobacillus* spp., *Bacillus subtilis*, and *Ashbya gossypii*), and molds (*Aspergillus* and *Penicillium* spp.) (Kutsal and Özbas 1989). In riboflavin biosynthesis, guanosine triphosphate (GTP) is the committed precursor of supplying the pyrimidine ring, the nitrogen atoms of the pyrazine ring, as well as the ribityl side chain of the vitamin. The GTP enzyme cyclohydrolase II catalyzes the opening of the imidazole ring of GTP, while an enzyme product catalyzes the conversion of 2,5-diamino-6-ribosylamino-4(3H)-pyrimidinone 50-phosphate to 5-amino-6-ribitylamino-2,4(1H,3H)-pyrimidinedione 50-phosphate by two reaction steps involving the hydrolytic cleavage of the position 2 amino group of the heterocyclic ring and the reduction of the ribosyl side chain to the ribityl side chain of the vitamin. The sequence of these reaction steps varies between microorganisms, with the deamination preceding the side-chain reduction in eubacteria and vice versa in yeasts and fungi (Bacher et al. 2000).

Pantothenic acid (vitamin B_5) is a precursor to coenzyme A (CoA) and acyl carrier protein, which are important cofactors for a large number of metabolic processes. Some species of bacteria, e.g., *Escherichia coli, Arthrobacter ureafaciens, Corynebacterium erythrogenes, Brevibacterium ammoniagenes, Corynebacterium glutamicum, Bacillus subtilis*, and also yeasts, such as, e.g., *Debaryomyces castellii*, can produce D-pantothenic acid. In *E. coli* and other bacteria, B_5 is synthesized by the condensation of pantoate, derived from 2-oxoisovalerate (an intermediate in valine biosynthesis) and β-alanine. The yeast *S. cerevisiae* was found to be rate limiting for β-alanine, which is produced via a novel polyamine pathway (White, Gunyuzlu, and Toyn 2001). Since the microorganisms require panthonic acid for growth, fermentation processes may lead to reduction of the vitamin B_5 content as, for example, in wine making.

Vitamin B_6 (pyridoxine, pyridoxamine, and pyridoxal) is produced by microorganisms belonging to the genera *Achromobacter, Bacillus, Flavobacterium, Klebsiella, Pichia, Saccharomyces*, and *Rhizobium* (Pardini and Argoudelis 1968). In *E. coli* and *Sinorhizobium meliloti*, pyridoxol 5¢-phosphate is believed to be synthesized from two precursors, 1-deoxy-D-xylulose 5-phosphate and 4-(phosphohydroxy)-L-threonine by the action of two enzymes, HTP dehydrogenase and PNP synthase encoded by pdxj gene.

Biotin (vitamin B_7 or vitamin H) is a coenzyme that facilitates CO_2 transfer in cellular carboxylation/decarboxylation reactions involved in fatty acid biosynthesis, gluconeogenesis, and amino acid metabolism. Biotin biosynthesis is widespread among microorganisms, with some differences between gram-positive and gram-negative bacteria. The biosynthesis pathway differs between *Bacillus* spp. and *E. coli* in the first step of pimeloyl-CoA synthesis. *Bacillus subtilis* and *B. sphaericus* synthesize pimeloyl-CoA from pimelic acid using pimeloyl-CoA synthase, and *E. coli* synthesize pimeloyl-CoA from L-alanine and/or from acetate via acetyl-CoA, while the conversion of pimeloyl-CoA to biotin is similar in *E. coli* and the *Bacillus* spp.

(Ifuku et al. 1994). Other microorganisms capable of producing biotin include fungi and yeasts, e.g., *Rhizopus delemar, Rhizobium* spp., *Kurthia* spp., *Sphingomonas* spp., *Candida utilis*, and *S. cerevisiae*.

The structure of vitamin B_{12} (cyanocobalamin) is based on a central cobalt atom coordinated by four horizontal pyrrole groups and two axial ligands; the lower ligand is a nucleotide with 5,6-dimethylbenzimidazole as a base, and the upper ligand is either 5¢-deoxyadenosine (adenosylcobalamin) or a methyl group (methylcobalamin) (Watanabe 2007). The biosynthesis of vitamin B_{12} can be divided into three parts:

1. The synthesis of uroporphyrinogen III from either glutamyl-tRNA or glycine and succinyl-CoA
2. The synthesis of the corrin ring that differs between anaerobic organisms, where the insertion of cobalt occurs in an early intermediate, and aerobic organisms, in which the cobalt insertion occurs after corrin ring synthesis
3. The corrin ring adenosylation or the attachment of the amino-propanol arm to the cobalt at the core of the corrin ring

Certain bacteria can synthesize B_{12} and/or related compounds with alternative bases in the lower ligand (e.g., adenine, 2-methyl adenine, pseudovitamin B_{12}). Edible cyanobacteria (viz., *Spirulina platensis, Aphanothece sacrum, Aphanizomenon frosaquae, Nostoc commune*, and *N. flagelliforme*) were reported to contain substantial amounts of pseudovitamin 12 (Watanabe et al. 1999, 2006; Hoffman et al. 2000; Miyamoto et al. 2006; Santos et al. 2007).

Folate can be produced by a number of microorganisms in certain foods, e.g., fermented milks and cereal doughs. With few exceptions (*L. acidophilus, L. plantarum*), lactobacilli are known for their inability to synthesize this vitamin but rather consume it. On the other hand, certain bacteria, e.g., *Lactococcus lactis* and *Streptococcus thermophilus*, have the ability to synthesize folate (LeBlanc et al. 2007). The folate molecule consists of a para-aminobenzoic acid (pABA) group coupled to a pterin moiety, originating from 6-hydroxymethyl-7,8-dihydropterin pyrophosphate (DHPPP). *De novo*, biosynthesis by bacteria produces pABA from the pentose phosphate pathway, while DHPPP is produced from guanosine triphosphate (GTP) in four consecutive steps. Yeast strains belonging to different *Candida* and *Saccharomyces* species, isolated from kefir, were found to produce significant amounts of 5-methyltetrahydrofolate (Patring et al. 2006).

Microorganisms are also involved in the synthesis of fat-soluble vitamins. It is well known that menaquinones (2-methyl-3-multiprenylnaphtho-1,4-quinones, or vitamin K_2) are produced by microorganisms from chorismic acid via isochorismic acid. In addition, bacteria, yeast, and fungi are capable of biosynthesizing of a number of carotenoids (Schmidt-Dannert 2000).

6.3.5 BIOSYNTHESIS OF PHENOLIC ANTIOXIDANTS

Several enzyme activities belonging to microorganisms have the ability to generate or modify the quantity and quality of phenolic antioxidants in foods (Gunatilaka 2006; Okami 1986). Hydroxycinnamic acids, typically *p*-coumaric, ferulic, and caffeic acids, are often esterified with organic acids such as tartaric acid and are present in, for example, grapes and berries in the form of caftaric acid, fertaric acid, and coutaric acid. Microorganisms contain esterases that hydrolyze phenolic acid esters to free aglycones, which will be available for further reactions. For example, p-coumaric and ferulic acid undergo decarboxylation and reduction to 4-ethylphenol and 4-ethylguaiacol in the presence of the yeast *Dekkera bruxellensis*.

Microorganism enzymes (e.g., from *Aspergillus oryzae*, *Rhizopus oligosporus*, and *Bacillus* spp.) increase the antioxidant activities during fermentation of soya flour to miso, tempeh, soy milk, and tofu by converting malonyldaidzin and malonylgenistin into acetyldaidzin and acetylgenistin, and generating new antioxidants such as 3-hydroxyanthranilic acid (HAA) (Esaki et al. 1996). The first step in these reactions is usually the liberation of isoflavone aglycones from their glycosylated derivatives by the action of b-glycosidases on the microorganisms. It is, for example, known that β-glucosidase from *Bacillus subtilis* natto can convert isoflavone glucosides to their aglycones (Ibe et al. 2001). Numerous studies have revealed that the biological effects of isoflavone stem from their aglycones (Hendrich 2002; Kawakami et al. 2005), which are more easily and rapidly absorbed by humans (Izumi et al. 2000).

Other enzyme activities may be involved in phenolic antioxidant release. In addition to a- and β-glucosidase, the food-grade fungus *Rhizopus oligosporus* releases α-amylase, and β-glucuronidase activities that were able to increase the total antioxidant potential of fermented soybeans (McCue and Shetty 2003). Certain bacteria, including *Lactobacillus, Bacillus, Staphylococcus*, and *Klebsiella* use the enzyme tannase (EC 3.1.1.20) to degrade tannins to simple phenols and sugars and/or organic acids for use as energy. Certain Enterobacteriaceae in the natural flora of blueberry are able to increase in total phenolic content and enhance the total antioxidant activity after fermentation of the juice (Martin and Matar 2005). The increase in the antioxidant potential is not only related to the increase in total phenolic content, but also to the release of phenolic aglycones from combined forms (Bhat, Singh, and Sharma 1998).

Kombucha tea (also known as Manchurian mushroom tea, Manchurian fungus tea, or *kwassan*) is a fermented beverage where a mushroom (*Fungus japonicus*) is a source of a symbiotic culture of bacteria and yeasts that produces a sparkling liquid with an apple cider–like taste in a solution of sugar and green or black tea (Dufresne and Farnworth 2000). Among the yeasts identified in *kombucha* are *Schizosaccharomyces pombe, Saccharomycodes ludwigii, Kloeckera apiculata, Saccharomyces cerevisiae, Zygosaccharomyces bailii, Brettanomyces bruxellensis, B. lambicus, B. custersii*, and *Candida* and *Pichia* spp. The main bacteria found in the tea fungus belong to *Acetobacter*, e.g., *A. xylinum, A. xylinoides, A. aceti*, and *A. pasteurianus*. The yeasts convert sucrose into fructose and glucose and produce ethanol, while the acetic acid bacteria convert glucose to gluconic acid and fructose into acetic acid. In a symbiotic association, acetic acid produced by bacteria stimulates the yeast to produce ethanol, which stimulates bacteria to grow and produce acetic acid. Ethanol and acetic acid are able to enhance the extraction of bioactive phenolic compounds and colors from tea. *Kombucha* is believed to have a probiotic effect, i.e., it produces beneficial bacteria able to destroy harmful bacteria. Both ethanol and acetic acid have been reported to have antimicrobial activity against pathogenic bacteria, thereby providing protection against contaminating bacteria. Although no strong scientific evidence exists, *kombucha* tea is regarded by some as an "elixir of life" and is believed to have life-extending properties as well as antitumor, antiaging, vision-improving, and fertility-enhancing treatment.

6.3.6 BIOSYNTHESIS OF ANTIBIOTICS

Bacteria belonging to the genus *Streptomyces* represent the major antibiotics-producing microorganism (Table 6.4). Antibiotic production is a result of secondary metabolic pathways involving simple precursors, including amino acids, short-chain fatty acids, sugars, and nucleic acids. In addition to antibiotics, some streptomycetes produce other medicinal compounds such as the anticancer drugs migrastatin (from *S. platensis*) and bleomycin (from *S. verticillus*), which are antineoplastic (anticancer) drugs. *Penicillium* fungi produce penicillin, which is a group of β-lactam antibiotics derived from and used in the treatment of bacterial infections usually caused by gram-positive bacteria. Penicillin is produced commercially by the fungus *Penicillium chrysogenum*. They include penicillin G (benzylpenicillin), procaine penicillin, benzathine penicillin, and penicillin V (Figure 6.2).

6.4 CONCLUDING REMARKS

Microorganisms, mostly bacteria and fungi, are involved in the production of a wide range of bioactive compounds including polysaccharide polymers, peptides, and low-molecular-weight vitamins, antioxidants, and antibiotics. In addition,

TABLE 6.4
Examples of Antibiotics Produced by Different Species of *Streptomyces*

Antibiotic	Producing bacteria	Antibiotic	Producing bacteria
Chloramphenicol	*S. venezuelae*	Cycloserine	*S. orchidaceus*
Cefoxitin	*S. lactamdurans*	Cyclohexamide	*S. griseus*
Daptomycin	*S. roseosporus*	Erythromycin	*S. erythreus*
Fosfomycin	*S. fradiae*	Kanamycin	*S. kanamyceticus*
Lincomycin	*S. lincolnensis*	Neomycin	*S. fradiae*
Nystatin	*S. noursei*	Puromycin	*S. alboniger*
Rifamycin	*S. mediterranei*	Streptomycin	*S. griseus*
Tetracycline	*S. remosus*	Vancomycin	*S. orientalis*

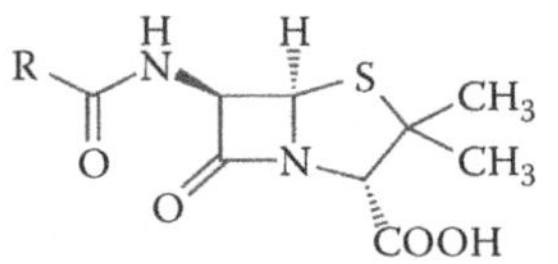

FIGURE 6.2 Penicillins (R variable)

microorganisms are the source of distinctive aromas and flavors that add sensory value to fermented foods. The research in these areas is still limited, and more work is needed to identify the most potent microorganisms, molecular structures of unexplored bioactive compounds, and optimization of yield. This research is vital for the valorization of fermented foods through identification of potent microorganisms and design of starter cultures, identification of potential synergistic effects and symbiotic interactions, and optimization of processing conditions.

REFERENCES

Alhaj, O. A., A. D. Kanekanian, A. C. Peters, and A. S. Tatham. 2010. Hypocholesterolaemic effect of *Bifidobacterium animalis* subsp. *lactis* (Bb12) and trypsin casein hydrolysate. *Food Chem.* 123:430–435.

Ashar, M. N., and R. Chand. 2004. Fermented milk containing ACE-inhibitory peptides reduces blood pressure in middle aged hypertensive subjects. *Milchwissenschaft* 59:363–366.

Bacher, A., S. Eberhardt, M. Fischer, K. Kis, and G. Richter. 2000. Biosynthesis of vitamin B$_2$ (riboflavin). *Annu. Rev. Nutr.* 20:153–67.

Bayer, E. A., H. Chanzy, R. Lamed, and Y. Shoham. 1998. Cellulose, cellulases, and cellulosomes. *Curr. Opin. Struct. Biol.* 8:548–557.

Begley, T. P., A. Chatterjee, J. W. Hanes, A. Hazra, and S. E. Ealick. 2008. Cofactor biosynthesis—Still yielding fascinating new biological chemistry. *Curr. Opinion in Chem. Biol.* 12:118–125.

Bhat, T. K., B. Singh, and O. P. Sharma. 1998. Microbial degradation of tannins—A current perspective. *Biodegradation* 9:343–357.

Biesalski, H. K., L. O. Dragsted, I. Elmadfa, R. Grossklaus, M. Müller, D. Schrenk, P. Walter, and P. Weber. 2009. Bioactive compounds: Definition and assessment of activity. *Nutrition* 25:1202–1205.

Bredholt, S., T. Nesbakken, and A. Holck. 1999. Protective cultures inhibit growth of *Listeria monocytogenes* and *Escherichia coli* O157: H7 in cooked, sliced, vacuum- and gas-packaged meat. *Int. J. Food Microbial.* 53:43–52.

Chen, X., E. G. Kokkotou, N. Mustafa, K., R. Bhaskar, S. Sougioultzis, M. O'Brien, C. Pothoulakis, and C. P. Kelly. 2006. *Saccharomyces boulardii* inhibits ERK1/2 mitogen-activated protein kinase activation both in vitro and in vivo and protects against *Clostridium difficile* toxin A-induced enteritis. *J. Biol. Chem.* 281:24449–24454.

Cleveland, J., T. J. Montvik, I. F. Nes, and M. L. Chikindas. 2001. Bacteriocins: Safe, natural antimicrobials for food preservation. *Int. J. Food Microbiol.* 71:1–20.

Delves-Broughton, J., P. Blackburn, R. J. Evans, and J. Hugenholtz. 1996. Applications of the bacteriocin nisin. *Antonie Van Leeuwenhok* 69:193–202.

De Martinis, E. C. P., and D. G. M. Franco. 1998. Inhibition of *Listeria monocytogenes* in a ponk prod by a *Lactobacillus sakei* strain. *Int. J. Food Microbiol.* 42:119–126.

Demigne, C., C. Morand, M. A. Levrat, C. Besson, C. Moundras, and C. Remesy. 1995. Effect of propionate on fatty acid and cholesterol synthesis and on acetate metabolism in isolated rat hepatocytes. *Br. J. Nutr.* 74:209–219.

De Vuyst, L., and B. Degeest. 1999. Heteropolysaccharides from lactic acid bacteria. *FEMS Microbiol. Revs.* 23:153–177.

De Vuyst, L., and E. J. Vandamme, eds. 1994. *Bacteriocins of lactic acid bacteria: Microbiology, genetics, and applications.* London: Blackie Academics and Professional.

Dufresne, C., and E. Farnworth. 2000. Tea, Kombucha, and health: A review. *Food Research International* 33:409–421.

Ebringer, L., M. Ferencik, and J. Krajcovic. 2008. Beneficial health effects of milk and fermented dairy products—Review. *Folia Microbiol.* 53:378–394.

Ek, R. 2005. The effect of fermentation and drying on the water-soluble vitamin content of tarhana, a traditional Turkish cereal food. *Food Chem.* 90:127–132.

Ennahar, S., T. Sashihara, K. Sonomoto, and A. Ishzaki. 2000. Class IIa bacteriocins: Biosynthesis, structure, and activity. *FEMS Microbiol. Rev.* 24:85–106.

Esaki, H., H. Onozaki, S. Kawakishi, and T. Osawa. 1996. New antioxidant isolated from tempeh. *J. Agric. Food Chem.* 44:696–700.

Fleet, G. H. 2007. Yeasts in foods and beverages: Impact on product quality and safety. *Curr. Opin. Biotechnol.* 18:170–175.

Gobbetti, M., F. Minervini, and C. G. Rizzello. 2004. Angiotensin I–converting-enzyme-inhibitory and antimicrobial bioactive peptides. *International Journal of Dairy Technology* 57:172–188.

Guder, A., I. Wiedeman, and H. G. Sahl. 2000. Posttranslationally modified bacteriocins—The lantibiotics. *Biopolymers* 55:62–73.

Gunatilaka, A. A. L. 2006. Natural products from plant-associated microorganisms: Distribution, structural diversity, bioactivity, and implications of their occurrence. *J. Natural Products* 69:509–526.

Hayes, M., R. P. Ross, G. F. Fitzgerald, and C. Stanton. 2007. Putting microbes to work: Dairy fermentation, cell factories, and bioactive peptides. Part 1: Overview. *Biotechnol. J.* 2:426–434.

Hendrich, S. 2002. Bioavailability of isoflavones. *J. Chromatogr. B Analyt. Technol. Biomed. Life Sci.* 777:203–210.

Hijova, E., and A. Chmelarova. 2007. Short chain fatty acids and colonic health. *Bratisl. Lek. Listy.* 108:354–358.

Hoffman, B., M. Oberhuber, E. Stupperich, H. Bothe, W. Buckel, B. Konrat, and B. Kräutler. 2000. Native corrinoids from *Clostridium cochlearium* are adeninylcobamides: Spectroscopic analysis and identification of pseudovitamin B_{12} and factor A. *J. Bacteriol.* 182:4773–4782.

Hohmann, S., and P. A. Meacock. 1998. Thiamin metabolism and thiamin diphosphate-dependent enzymes in the yeast *Saccharomyces cerevisiae*: Genetic regulation. *Biochim. Biophys. Acta.* 1385:201–219.

Holden, J. T., and J. Holman. 1959. Accumulation of freely extractable glutamic acid by lactic acid bacteria. *J. Biol. Chem.* 234:865–869.

Huth, P. J., D. B. Dirienzo, and G. D. Miller. 2006. Major specific advances with dairy foods in nutrition and health. *J. Dairy Sci.* 89:1207–1221.

Ibe, S., K. Kumada, M. Yoshiba, and T. Onga. 2001. Production of natto which contains a high level of isoflavone aglycones. *Nippon Shofuhin Kagaku Kogaku Kaishi* 48:27–34.

Ifuku, O., H. Miyaoka, N. Koga, J. Kishimoto, S. Haze, Y. Wachi, and M. Kajiwara. 1994. Origin of carbon atoms of biotin: 13C-NMR studies on biotin biosynthesis in *Escherichia coli. Eur. J. Biochem.* 220:585–591.

Iwashima, A. 1989. Microbial synthesis of vitamin B_1 (thiamine). In *Biotechnology of vitamins, pigments, and growth factors*, ed. E. J. Vandamme, 137–148. Essex, U.K.: Elsevier Sci. Publ.

Izumi, T., M. K. Piskula, S. Osawa, A. Obata, K. Tobe, M. Saito, S. Kataoka, Y. Kubota, and M. Kikuchi. 2000. Soy isoflavone aglycones are absorbed faster and in higher amounts than their glucosides in humans. *J. Nutr.* 130:1695–1699.

Jack, R. W., and G. Jung. 2000. Lantibiotics and microcines: Polypeptides with unusual chemical diversity. *Curr. Opinion in Chem. Biol.* 4:310–317.

Kawakami, Y., W. Tsurugasaki, S. Nakamura, and K. Osada. 2005. Comparison of regulative functions between dietary soy isoflavones aglycone and glucoside on lipid metabolism in rats fed cholesterol. *J. Nutr. Biochem.* 16:205–212.

Khetarpaul, N., and B. M. Chauhan. 1989. Effect of fermentation on protein, fat, minerals, and thiamine content of pearl millet. *Plant Foods for Human Nutrition* 39:169–177.

Kim, H. J., H. J. Suh, J. H. Kim, S. Park, Y. C. Joo, and J. S. Kim. 2010. Antioxidant activity of glyceollins derived from soybean elicited with *Aspergillus sojae. J. Agric. Food Chem.* 58:11633–11638.

Korhonen, H., and A. Pihlanto. 2006. Bioactive peptides: Production and functionality. *International Dairy Journal* 16:945–960.

Kowalska, F., and A. Kozik. 2008. The genes and enzymes involved in the biosynthesis of thiamin and thiamin diphosphate in yeasts. *Cellular & Molecular Biology Letters* 13:271–282.

Kutsal, T., and M. T. Özbas. 1989. Microbial production of vitamin B_2 (riboflavin). In *Biotechnology of vitamins, pigments, and growth factors*, ed. E. J. Vandamme, 149–166. Essex, U.K.: Elsevier Sci. Publ.

Le Blanc, J. G., G. S. de Giori, E. J. Smid, J. Hugenholtz, and F. Sesma. 2007. Folate production by lactic acid bacteria and other food-grade microorganisms. In *Communicating current research and educational topics and trends in applied microbiology*, ed. A. Méndez-Vilas, 329–339. Badajoz, Spain: FORMATEX.

Le Blanc, J. G., C. Matar, J. C. Valdez, J. Leblanc, and G. Perdigon. 2002. Immunomodulating effects of peptidic fractions issued from milk fermented with *Lactobacillus helveticus. J. Dairy Sci.* 85:2733–2742.

Levitt, M. D., G. R. Gibson, and S. U. Christ. 1995. Gas metabolism in the large intestine. In *Human colonic bacteria: Role in nutrition, physiology, and pathology*, eds. G. R. Gibson and G. T. Macfarlane, 131–154. Boca Raton, FL: CRC Press.

Li, H., and Y. Cao. 2010. Lactic acid bacterial cell factories for gamma-aminobutyric acid. *Amino Acids* 39:1107–1116.

Macouzet, M., N. Robert, and B. H. Lee. 2010. Genetic and functional aspects of linoleate isomerase in *Lactobacillus acidophilus. Appl. Microbiol. Biotechnol.* 87:1737–1742.

Maeno, M., N. Yamamoto, and T. Takano. 1996. Identification of an antihypertensive peptide from casein hydrolysate produced by a proteinase from *Lactobacillus helveticus* CP790. *J. Dairy Sci.* 79:1316–1321.

Mahgoub, E. O., B. M. Ahmed, M. M. O. Ahmed, and E. A. A. El Agib. 1999. Effect of traditional Sudanese processing of *kisra* bread and *hulu-mur* drink on their thiamine, riboflavin, and mineral contents. *Food Chem.* 67:129–133.

Martin, L. J., and C. Matar. 2005. Increase of the antioxidant capacity of the lowbush blueberry (*Vaccinium angustifolium*) during fermentation by a novel bacterium from the fruit microflora. *J. Sci. Food Agric.* 85:1477–1484.

McAuliffe, O., R. P. Ross, and C. Hill. 2001. Lantibiotics: Structure, biosynthesis, and mode of action. *FEMS Microbiol. Rev.* 25:285–308.

McCue, P., and K. Shetty. 2003. Role of carbohydrate-cleaving enzymes in phenolic antioxidant mobilization from whole soybean fermented with *Rhizopus oligosporus. Food Biotechnology* 17:27–37.

Mellander, O. 1950. The physiological importance of the casein phosphopeptide calcium salts, II: Peroral calcium dosage of infants. *Acta of the Society of Medicine of Uppsala* 55:247–255.

Miyamoto, E., Y. Tanioka, T. Nakao, F. Balra, H. Inui, T. Fujita, F. Watanabe, and Y. Nakano. 2006. Purification and characterization of a corrinoid-compound in an edible cyanobacterium *Aphanizomenon flos-aquae* as a nutritional supplementary food. *J. Agric. Food Chem.* 54:9604–9607.

Nagaoka, S., Y. Futamura, K. Miwa, T. Awano, K. Yamauchi, Y. Kanamaru, K. Tadashi, and T. Kuwata. 2001. Identification of novel hypocholesterolemic peptides derived from bovine milk β-lactoglobulin. *Biochem. Biophys. Res. Commun.* 281:11–17.

Nakamura, Y., N. Yamamoto, K. Sakai, A. Okubo, S. Yamazaki, and T. Takano. 1995. Purification and characterization of angiotensin I: Converting enzyme inhibitors from sour milk. *J. Dairy Sci.* 78:777–783.

Nes, I. F., and H. Holo. 2000. Class II antimicrobial peptides from lactic acid bacteria. *Biopolymers* 55:50–61.

Nettles, C. G., and S. F. Barefoot. 1993. Biochemical and genetic characteristics of bacteriocins of food associated lactic acid bacteria. *J. Food Protection* 56:338–356.

Nout, M. J. R., and J. L. Kiers. 2005. Tempe fermentation, innovation, and functionality: Update into the third millennium. *J. Applied Microbiology* 98:789–905.

Ogawa, A., S. Kishino, A. Ando, S. Sugimoto, K. Mihara, and S. Shimizu. 2005. Production of conjugated fatty acids by lactic acid bacteria. *J. Biosci. Bioeng.* 100:355–364.

Okami, Y. 1986. Marine microorganisms as a source of bioactive agents. *Microbiol. Ecology* 12:65–78.

Pardini, R. S., and C. J. Argoudelis. 1968. Biosynthesis of vitamin B_6 by a yeast mutant. *J. Bacteriol.* 96:672–677.

Patring, J., S. B. Hjortmo, J. A. Jastrebova, U. K. Svensson, T. A. Andlid, and M. Jägerstad. 2006. Characterization and quantification of folates produced by yeast strains isolated from kefir granules. *European Food Research and Technology* 223:633–637.

Pessi, T., E. Isolauri, Y. Sutas, H. Kankaanranta, E. Moilanen, and M. Hurme. 2001. Suppression of T-cell activation by *Lactobacillus rhamnosus* GG-degraded bovine casein. *Int. Immunopharmacol.* 1:211–218.

Prioult, G., S. Pecquet, and I. Fliss. 2004. Stimulation of interleukin-10 production by acidic beta-lactoglobulin-derived peptides hydrolyzed with *Lactobacillus paracasei* NCC2461 peptidases. *Clin. Diagn. Lab. Immunol.* 11:266–271.

Rodriguez-Navarro, S., B. Liorente, M. T. Rodriquez-Manzaneque, A. Ramne, G. Uber, D. Marchesan, B. Dujon, E. Herrero, P. Sunnerhagen, and J. E. Pérez-Ortín. 2002. Functional analysis of yeast gene families involved in metabolism of vitamins B_1 and B_6. *Yeast* 19:1261–1276.

Rokka, T., E. L. Syvaoja, J. Tuominen, and H. Korhonen. 1997. Release of bioactive peptides by enzymatic proteolysis of *Lactobacillus* GG fermented UHT milk. *Milchwissenschaft* 52:659–678.

Santos, F., J. L. Vera, P. Lamosa, P. de Lamosa, F. de Valdez, W. M. de Vos, H. Santos, F. Sesma, and J. Hugenholtz. 2007. Pseudovitamin B_{12} is the corrinoid produced by *Lactobacillus reuteri* CRL 1098 under anaerobic conditions. *FENS Lett.* 581:4865–4870.

Scharlau, D., A. Borowicki, N. Habermann, T. Hofmann, S. Klenow, C. Miene, U. Munjal, K. Stein, and M. Glei. 2009. Mechanisms of primary cancer prevention by butyrate and other products formed during gut flora-mediated fermentation of dietary fibre. *Mutat Res.* 682:39–53.

Schmidt-Dannert, C. 2000. Engineering novel carotenoids in microorganisms. *Curr. Opin. Biotechnol.* 11:255–261.

Seppo, L., T. Jauhiainen, T. Poussa, and R. Korpela. 2003. A fermented milk high in bioactive peptides has a blood pressure–lowering effect in hypertensive subjects. *Am. J. Clin. Nutr.* 77:326–330.

Smith, J. G., W. H. Yokoyama, and J. B. German. 1998. Butyric acid from the diet: Actions at the level of gene expression. *Crit. Rev. Food Sci. Nutr.* 38:259–297.

Tsai, J.-S., T.-J. Chen, B. S. Pan, S.-D. Gong, and M.-Y. Chung. 2008. Antihypertensive effect of bioactive peptides produced by protease-facilitated lactic acid fermentation of milk. *Food Chem.* 106:552–558.

Vandamme, E. J. 1992. Production of vitamins, coenzymes, and related biochemicals by biotechnological processes. *J. Chem. Technol. Biotechnol.* 53:313–327.

Vignolo, G., S. Fadda, M. N. DeKairuz, A. A. P. De Ruiz Holgdo, and G. Olivier. 1996. Control of *Listeria monocytogenes* in ground beef by Lactocin 705, a bacteriocin produced by *Lactobacillus casei* CRL 705. *Int. J. Food Microbiol.* 27:397–402.

Waller, J. R., and H. C. Lichstein. 1965. Biotin transport and accumulation by cells of *Lactobacillus plantarum*. I. General properties of the system. *J. Bacteriol.* 90:843–852.

Watanabe, F. 2007. Vitamin B_{12} sources and bioavailability. *Exp. Biol. Med.* 232:1266–1274.

Watanabe, F., H. Katsura, S. Takenaka, T. Fujita, K. Abe, Y. Tamura, T. Nakatsuka, and Y. Nakano. 1999. Pseudovitamin B_{12} is the predominant cobamide of an algal health food, *Spirulina* tablets. *J. Agric. Food Chem.* 47:4736–4741.

Watanabe, F., E. Miyamoto, T. Fujita, Y. Tanioka, and Y. Nakano. 2006. Characterization of a corrinoid compound in the edible (blue-green) alga, Suizenji-nori. *Biosci. Biotechnol. Biochem.* 70:3066–3068.

White, W. H., P. L. Gunyuzlu, and J. H. Toyn. 2001. Saccharomyces cerevisiae is capable of de novo pantothenic acid biosynthesis involving a novel pathway of β-alanine production from spermin. *J. Biol. Chem.* 276:10794–10800.

Xiong, N., K. Miyamoto, M. A. Shibata, T. Valasek, R. M. Motoike, M. Kedzierski, and M. Yanagisawa. 2004. Short-chain fatty acids stimulate leptin production in adipocytes through the G protein–coupled receptor GPR41. *PNAS* 101:1045–1050.

Yamamoto, N. 1998. Antihypertensive peptides derived from food proteins. *Peptide Sci.* 43:129–134.

Zamora, A. F., and M. L. Fields. 2006. Nutritive quality of fermented cowpeas (*Vigna sinensis*) and chickpeas (*Cicer arietinum*). *J. Food Sci.* 44:234–236.

7 The Role of Fermentation in the Elimination of Harmful Components Present in Food Raw Materials

Aly Savadogo

CONTENTS

7.1 INTRODUCTION

Food fermentation has been practiced for millennia with the result that there is a tremendous variety of flavor and taste-enriched fermented foods derived from milk, meat, and plants. Fermented foods were defined by Campbell-Platt (1987) as those foods that have been subjected to the action of microorganisms or enzymes so that desirable biochemical changes cause significant modification to the food. Food fermentation has shown to have not only preservative effects and the capability of modifying the physicochemical properties of various foods, but also to improve the nutritional and functional quality of foods (Knorr 1998). Fermentation reactions are known to improve the nutritional value of foods (Paredes-López and Harry 1988). Fermentation can enhance the nutritional value of a food product through increased vitamin levels and improved digestibility. Many fruit and vegetable products contain toxins and antinutritional compounds that can be removed or detoxified by the action of microorganisms during fermentation. Fermentation may also lead to the detoxification and destruction of undesirable factors present in raw foods such as phytates, tannins, and polyphenols (Sharma and Kapoor 1996).

Lactic acid bacteria (LAB), generally considered as "food grade" organisms, show special promise for selection and implementation as protective cultures. Fermentation is a process dependent on the biological activity of microorganisms for production of a range of metabolites that can suppress the growth and survival of undesirable microflora in foodstuffs (Klaenhammer 1993; Soomro, Masud, and Anwaar 2002).

The single most important development permitting the formation of civilization was the ability to produce and store large quantities of food. It is beneficial to be able to store as much food as possible in order to minimize the amount of time spent gathering that food. Food fermentation has great economic value, and it has been accepted that these products contribute in improving human health. Several traditional fermented products have been documented in different African countries and include nonalcoholic beverages, alcoholic beverages, breads, pancakes, porridges, cheeses, and milks (Haggblade and Holzapfel 1989; Ashenafi and Busse 1989; Gorbach 1990, 2000; Dirar 1993; Steinkraus 1996; Savadogo et al. 2004).

According to Steinkraus (1996), traditional fermentation of foods (cereal, milk, fruits) serves several functions:

1. Enrichment of the diet through development of a diversity of flavors, aromas, and textures in food substrates
2. Preservation of substantial amounts of food through lactic acid, alcoholic, acetic acid, and alkaline fermentations
3. Enrichment of food substrates biologically with essential amino acids, essential fatty acids, vitamins, and such foods that are more appetizing and easily digestible

Fermentation improves the digestibility of the ingredients for human consumption and enhances the keeping quality and shelf life. This chapter aims to summarize the important role of fermentation in the elimination of harmful components present in food raw materials. The biochemical and microbiological properties of the fermentation process in eliminating harmful components is discussed in this chapter.

7.2   EXAMPLES OF HARMFUL COMPONENTS IN FOOD RAW MATERIALS AND THEIR EFFECTS

7.2.1   PATHOGENIC MICROORGANISMS IN MILK

Raw milk quality depends on the health of the milk-producing animal and the animal feed. *Staphylococcus aureus* and *E. coli* are the most common pathogens causing both clinical and subclinical mastitis. Mastitis can be caused when the microflora associated with this condition enter the udder, usually through the duct at the teat tip. *Staphylococcus aureus* particularly is able to colonize the duct itself, and other typical microflora include *Streptomyces agalactiae, Streptococcus uberis*, and *E. coli*. Pathogenic microorganisms generally occur in milk as *Staphylococcus aureus, Staphylococcus intermedius, Mycobacterium tuberculosis, Mycobacterium bovis, Brucella melitensis, Brucella abortus, Bacillus cereus, Listeria monocytogenes, Yersinia enterocolitica, Salmonella typhimurium, Escherichia coli* O157:H7, and *Campylobacter jejuni* (Mantis 1985; Skovgaard 1990; Teuber 1992). The microorganisms that cause milk spoilage include coliforms, yeast, and mold. *Pseudomonas fluorescens* and *Pseudomonas fragi* can produce proteolytic and lipolytic extracellular enzymes that are heat stable and capable of causing spoilage.

7.2.2 LINAMARIN

Linamarin is a cyanogenic glucoside found in the leaves and roots of plants such as cassava. Cassava is a widely grown root crop that accumulates two cyanogenic glucosides: linamarin and lotaustralin. Linamarin produces hydrogen cyanide (HCN), a toxic compound that can be hazardous to the consumer. Toxicity caused by free cyanide (CN⁻) has been reported. According to Jensen and Abdel-Ghaffar (1979), microorganisms are able to detoxify cyanide by splitting the CN⁻ anion into carbon and nitrogen (Figure 7.1).

7.2.3 MYCOTOXINS

Mycotoxins include a diverse range of molecules that are harmful to animals and humans. Mycotoxins are toxic secondary metabolites produced by fungi. Their contamination depends on geographical, production, and storage conditions. More than a hundred species of filamentous fungi are known to produce mycotoxins and to cause toxic responses under naturally occurring conditions. Mycotoxins have been reported to be carcinogenic, tremorogenic, hemorrhagic, teratogenic, and dermatitic to a wide range of organisms and to cause hepatic carcinoma in humans (Refai 1988; van Egmond 1989; Wary 1981). Figure 7.2 shows the structures of some mycotoxins.

The best way of controlling mycotoxin contamination is by prevention, and this can be accomplished by reducing fungal infection in growing crops through the adoption of suitable cultural practices, by rapid drying, or by the use of suitable preservatives (Sinha 1993; Smith and Moss 1985). Mycotoxins can be eliminated or detoxified by physical, chemical, or biological techniques. Many chemicals including numerous acids, alkalis, aldehydes, oxidizing agents, and several gases have been tested for their ability to degrade or inactivate aflatoxin and many other mycotoxins (Samarajeewa et al. 1990; Smith and Moss 1985; Thanaboripat, Im-erb, and Ruangrattanametee 2002). Cinnamon, cinnamon oil, clove, and clove oil have been demonstrated to have strong antimycotic properties (Bullerman, Lieu, and Seier 1977; Bullerman, Schroeder, and Park 1984). All four substances inhibited growth and aflatoxin production. The essential oils of lemon and orange have been shown to be inhibitory to *Aspergillus niger* and *Aspergillus flavus* and to suppress aflatoxin

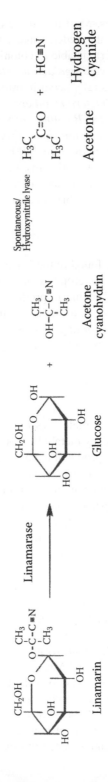

FIGURE 7.1 Linamarin hydrolyzed by linamarase and release of hydrogen cyanide

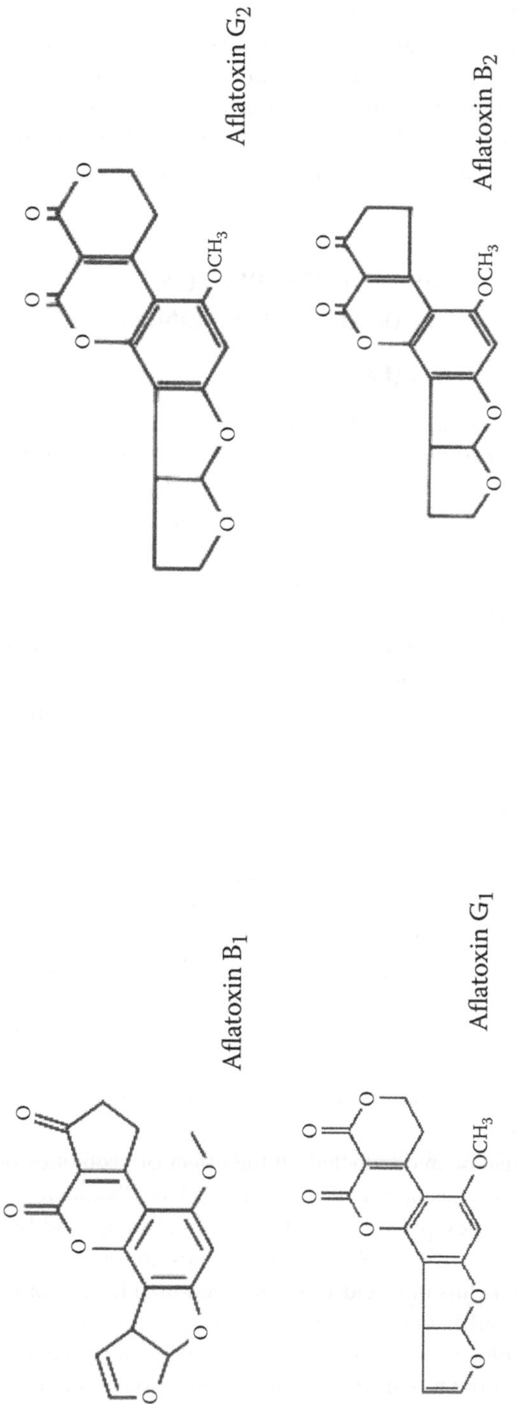

FIGURE 7.2 Structures of some mycotoxins (Aflatoxin B₁, Aflatoxin G₂, Aflatoxin G₁, Aflatoxin B₂)

formation (Subba, Soumithri, and Rao 1967; Alderman and Marth 1976). Various investigators have reported that a number of microorganisms affect the production of aflatoxin in a competitive environment. A mixture of *Lactobacillus* species has been reported to reduce mold growth and inhibit aflatoxin production by *A. flavus* subsp. *parasiticus* (Gourama and Bullerman 1995). *Rhizopus oligosporus*, a fungus used in the preparation of tempeh, was reported to inhibit the growth of *A. flavus* and *A. parasiticus*, thereby inhibiting the formation of aflatoxin (Ko 1978; Thanaboripat et al. 1996).

7.3 MICROORGANISMS RESPONSIBLE FOR ELIMINATION OF HARMFUL COMPONENTS

7.3.1 LACTIC ACID BACTERIA (LAB)

Lactic acid fermentation is carried out by many bacteria, most often by lactic acid bacteria such as *Lactobacillus acidophilus*, *Streptococcus thermophilus*, *Lactobacillus plantarum*, *Lactococcus lactis*, *Leuconostoc mesenteroides*, etc. Foods produced by lactic acid fermentation include yogurt, cheese, and pickles. LAB classification was initiated in 1919 by Orla-Jensen. Lactic acid bacteria comprise a diverse group of gram-positive, nonspore-forming, nonmotile rod and coccus-shaped, catalase-lacking organisms. LAB are chemoorganotrophic and only grow in complex media; carbohydrates and alcohols are used as the energy source to form lactic acid. LAB are able to degrade hexoses to lactate (by homofermentative pathway) or lactate and additional products such as acetate, ethanol, carbon dioxide, formate, or succinate (by heterofermentative pathway).

LAB were referred to as probiotics in the scientific literature by Lilley and Stillwell (1965). Fuller (1989) described probiotics as a "live microbial feed supplement beneficial to the host by improving the microbial balance within its body." According to Schrezenmeir and De Vrese (2001), probiotics are viable microbial food supplements that beneficially influence the health of the host. *Lactobacillus* and *Bifidobacterium* spp. reduce lactose intolerance, alleviate some diarrhea, lower blood cholesterol, increase immune response, and prevent cancer (Marteau and Rambaud 1993, 1996; Gilliland 1996; Salminen et al. 1998). LAB can be effective in preventing gastrointestinal disorders and in the recovery from diarrhea of miscellaneous causes (Marteau et al. 2001). Many bacteria of different taxonomic branches and residing in various habitats produce antimicrobial substances that are active against other bacteria.

Different works have been conducted on the effect of probiotics on the incidence and duration of various types of diarrhea (Isolauri 1996; Bhatnagar et al. 1998). LAB are of interest because they promote good health in animals and humans (Gorbach 1990, 2000). Among the positive effects of probiotics are growth promotion of farm animals (Havenaar and Huis in't Veld 1992), protection of hosts from intestinal infections (Fuller 1989; O'Sullivan et al. 1992), alleviation of lactose intolerance (Fuller 1989; Sawada et al. 1990), relief of constipation (Oyetayo and Oyetayo 2005), anticarcinogenic effects (O'Sullivan et al. 1992; Saïkali et al. 2004), anticholesteremic effects (Tamine 2002), nutrient synthesis and bioavailability (Oyetayo and Oyetayo 2005), prevention of genital and urinary tract infection (McLean and Rosenstein 2000), and immunostimulatory effects (Perdigon and Alvarez 1992; Malin et al. 1997).

Organic acids are important products of lactic acid bacteria fermentation. The concentration and types of organics acids produced during the fermentation process depend on the species of organisms involved. The antimicrobial effect of organic acids lies in the reduction of pH as well as the undissociated form of the molecules (Gould 1991; Podolak et al. 1996). Acids are generally recognized to exert their antimicrobial effect by interfering with the maintenance of cell membrane potential, inhibiting active transport, reducing intracellular pH, and inhibiting a variety of metabolic function (Doores 1993). Organic acids have a very broad action mode and inhibit both gram-positive and gram-negative bacteria as well as yeast and molds (Blom and Mortvedt 1991).

7.3.2 YEASTS AND OTHER MICROORGANISMS

Alcohol fermentation is carried out by many bacteria, e.g., *Clostridium sporogenes*, *Zymomonas mobilis*, *Leuconostoc mesenteroides*, *Streptococcus lactis*, and some yeasts such as *Saccharomyces cerevisiae*. However, yeast fermentations may also involve yeastlike molds such as *Amylomyces rouxii* and moldlike yeasts such as *Endomycopsis* and sometimes bacteria such as *Zymomonas mobilis*.

7.4 MECHANISMS ASSOCIATED WITH ELIMINATION OF HARMFUL COMPONENTS

7.4.1 PRESERVATION OF FOOD AGAINST HARMFUL MICROORGANISMS

Fermentation of foods is one of the oldest forms of biopreservation practiced by mankind. Fermentation involves the breaking down of complex organic substances into simpler ones. The fermentation process involves the oxidation of carbohydrates to generate a range of products, principally organic acids, alcohol, and carbon dioxide (Ray and Daeschel 1992). Fermentations involving the production of ethanol provide foods and beverages that are generally safe (Steinkraus 1979). These include wines, beers, Indonesian *tape ketan/tape ketela*, Chinese *lao-chao*, South African *kaffir*/sorghum beer, and Mexican *pulque*.

Protection of food from spoilage and pathogenic microorganism by LAB is accomplished through the production of organic acids, hydrogen peroxide, and diacetyl (Messens and De Vuyst 2002) as well as antifungal compounds such as fatty acids (Corsetti et al. 1998) or phenyllactic acid (Lavermicocca et al. 2000) and/or bacteriocins (De Vuyst and Vandamme 1994). Lactic acid bacteria play an important role in food fermentation, as the products obtained are characterized by hygienic safety, enhanced storability and stability, and attractive sensory properties. Lactic acid inhibits many pathogenic bacteria, and the undissociated form of the acid is considered to be the active component (Robinson and Samona 1992). The amount of undissociated lactic acid depends on both the concentration of lactic acid and the pH. Kingamkono et al. (1996) added several enteropathogens to cereal gruels prepared from sorghum and inoculated with a lactic acid starter culture. Upon fermentation, *Campylobacter* strains were not detectable after 6 h, while *Salmonella*, *Shigella*, and *Staphylococcus* strains were not detectable after 12 h. After 16 h, no viable *Bacillus* strains were found, and ETEC (enterotoxigenic *E. coli*) strains were completely inhibited after 24 h. On the other hand, in the gruels prepared without the lactic acid starter culture, all enteropathogens increased in number during incubation at 32°C except for *Campylobacter* strains, which decreased after 12 h.

Application of LAB in malting technology accelerates the fermentation and imparts beneficial properties to the malting, including lowering of viscosity and decreasing the β-glucan content of the wort. The use of LAB in malting reduces fungal contamination and leads to a higher quality malt due to the natural variation of microflora of barley (Boivin and Malanda 1997a, 1997b; Haikara, Uljas, and Suurnäkki 1993). The use of LAB has also led to significant effects in the mashing and brewing processes. β-Glucans, which are complex carbohydrates composed of mixed linkage (1-3),(1-4)-β-D-glucose polymers, are known to cause a number of problems during the brewing process. These polysaccharides have been linked with slow lautering, poor beer filtration, and formation of haze during storage of packaged beer.

Lactic acid bacteria are applied in brewing for their ability to improve mash and wort properties while resulting in better beer. Certain lactic acid bacteria (*Lactobacillus acidophilus*, *Lactobacillus fermentum*, *Lactobacillus helveticus*, *Lactobacillus plantarum*, *Lactococcus lactis* ssp. *lactis*) are able to produce antimicrobial compounds that inhibit the growth of harmful gram-negative and -positive bacteria. LAB bacteriocins are able to inhibit a wide spectrum of beer-spoilage organisms. Certain bacteriocins are also capable of inhibiting gram-negative beer-spoiling organisms such as *Acetobacter pasteurianus* and *Gluconobacter oxydans*.

One important attribute of LAB is their ability to produce a range of antimicrobial peptides and proteins, which are collectively referred to as bacteriocins. The bacteriocin called *nisin* (produced by lactic streptococci of the serological group N) was assessed to be safe for food use by the Joint Food and Agriculture Organization/World Health Organization (FAO/WHO) Expert Committee on Food Additives in 1969. The FAO/WHO Codex Committee on milk and milk products accepted nisin as a food additive for processed cheese at a concentration of 12.5 mg pure nisin per kilogram product. This biopreservative was also added to the European food additive list where it was assigned the number E234 (EEC 1983).

The most well known of the lantibiotics is nisin, which was first described in 1928 (Rogers and Whittier 1928), although its structure was not elucidated until 1971 (Figure 7.3). Nisin has been widely used as a natural and safe food preservative in the dairy industry for almost 30 years (Hurst 1981; Delves-Broughton et al. 1996), and it is chiefly the success of nisin that has stimulated the interest in other lantibiotics. Nisin is a natural antimicrobial agent with activity against a wide variety of undesirable food-borne (pathogenic) bacteria. It is a peptide that is produced by the food-grade dairy starter bacterium *Lactococcus lactis*. Nisin is used as a preservative in heat-processed and low-pH foods.

A flexible hinge region in the central part of the type A lantibiotics is indispensable for the pore-forming process, and although this portion of nisin may also contribute to the interaction with lipid, its primary function seems to be to provide the

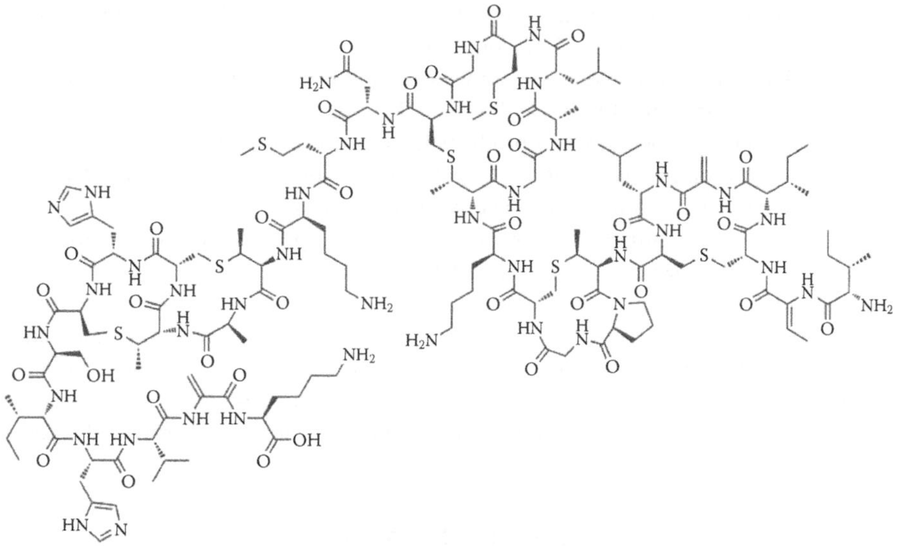

FIGURE 7.3 Nisin structure

peptide with enough conformational freedom to adopt a conducting position in the membrane. Variations in the carboxyl-terminal region of nisin had relatively minor influence on the binding to lipid.

7.4.2 DETOXIFICATION OF CASSAVA

The steeping (fermentation) and the action of the heat play important roles in the treatment of cassava and contribute to the elimination of linamarin, the very toxic substance present in some varieties of cassava. Collard and Levi (1959) noted that detoxification takes place in two stages. First, within 24 hours, *Corynebacterium manioc* decomposes the starch to produce the organic acids that decrease the pH and provokes the hydrolysis of the linamarin, which gives the acidic sparkling *cyanhydrique*. The production of the organic acids stimulate the growth of the mushroom *Geotrichum candida* that produces aldehydes and the esters responsible for the characteristic flavor of *gari*. *Gari* can be transformed into a thick dough with hot water and can be eaten with a soup of vegetables and meat, or it can be mixed in cold water, with or without sugar and milk, and eaten with wads of beans or with coconut. Meuser and Smolnik (1979) have proposed a process to mechanize the production of *gari*.

Since the root of peeled cassava contains about 61% water and a soluble toxic cyanogenetic glucoside (linamarin), the first stage of the process of detoxification consists in removing at least a part of the water that will drag the toxin with it. The major part of the vestigial glucosides (and of water) is eliminated, then, by any form of heating. In some cases, an initial stage consists of letting the root soak for a certain number of days during which the microflora of cassava provokes a fermentation that completes the activity of the endogenous linamarase, which efficiently completes the deterioration. Maduagwu (1981) noted that the inhibition of the activity of the linamarase by the 1,5-gluconolactone (potential inhibitor of the activity of the β-glucosidase) provoked an important reduction of the deterioration of the linamarin (about 35% in 24 hours and 65% in 72 hours).

These numerous processes and products have evolved with the accumulation of experience that took into account the food as well as social, economic, and cultural objectives. Among the important objective was the desire to conserve perishable commodities and to eliminate the toxicity bound to the presence of cyanogenic glucosides from the cassava staple. Giraud, Champailler, and Raimbault (1994) compared the effect of the inoculation of the cassava pulp by *Lactobacillus plantarum* A6 (amylolytic) and *Lactobacillus plantarum* (nonamylolytic), and their results indicated that the linamarin disappeared completely in all cases in less than 5 hours. Djoulde Darman et al. (2003) showed that for fermentation of cassava, the use of a starter composed of *Lactobacillus plantarum* and *Rhizopus oryzae* provided a good output elimination of linamarin.

7.5 SAFETY OF FERMENTED FOODS AS A RESULT OF ELIMINATION OF HARMFUL COMPONENTS

In traditional fermented food preparations, microbes are used to prepare and preserve food products (Achi 1992). Fermentation of food has many advantages such as improvement of nutritional value and protection against bacterial pathogens (Gadaga, Nyanga, and Mutukumira 2004). According to Adams (1990), fermentation can contribute to improved food stability (meat, fish, milk, vegetables, fruits), safety, nutritive value, and acceptability. LAB are added to salad mixtures and olives to prevent the growth of spoilage organisms. Certain LAB are used in biocontrol of aflatoxigenics. Karunaratne, Wezenberg, and Bullerman (1990) demonstrated the suppression of aflatoxin production by *Aspergillus flavus* in the presence of *Lactobacillus acidophilus*, *Lactobacillus bulgaricus*, and *Lactobacillus plantarum*.

Foods prepared under unhygienic conditions and frequently contaminated with pathogenic organisms play a major role in child mortality through a combination of diarrhea diseases, nutrient malabsorption, and malnutrition. *Salmonella, Campylobacter, Shigella, Vibrio, Yersinia,* and *Escherichia* are the most common organisms associated with bacterial diarrhea diseases. Enterotoxigenic bacteria including *Pseudomonas, Enterobacter, Klebsiella, Serratia, Proteus, Providencia, Aeromonas, Achromobacter,* and *Flavobacterium* have been reported by Nout, Rombouts, and Hautvast (1989). Adams (1990) suggested that LAB are inhibitory to many other microorganisms when they are cultured together, and this fact is the basis of the extended shelf life and improved microbiological safety of lactic fermented foods. Fermentation by LAB is one of the oldest forms of biopreservation practiced by mankind. One important attribute of LAB is their ability to produce antimicrobial compounds such as bacteriocins. Fermentation is a process dependent on the biological activity of microorganisms for production of a range of metabolites that can suppress the growth and survival of undesirable microflora in foodstuffs (Klaenhammer 1993; Soomro, Masud, and Anwaar 2002).

Vegetable foods and vegetable/fish/shrimp mixtures are preserved around the world by lactic acid fermentation (Steinkraus 1983). The classic lactic acid vegetable fermentation is sauerkraut (Pederson 1979). Fermentation of maize can increase total soluble solids and nonprotein nitrogen and slightly increase protein content (Yousif and El Tinay 2000). Lactic acid fermentation usually improves the nutritional value and digestibility of cereals. Cereals are limited in essential amino acids such as threonine, lysine, and tryptophan, thus making their protein quality poorer compared with animal meats and milk (Chavan and Kadam 1989). Their protein digestibility is also lower than that of animal meats, due partially to the presence of phytic acid, tannins, and polyphenols that bind to protein, thus making them indigestible (Oyewole 1997). Lactic acid fermentation of different cereals, such as maize, sorghum, and finger millet, has been found to effectively reduce the amount of phytic acid and tannins and improve protein availability (Chavan, Chavan, and Kadam 1988; Lorri and Svanberg 1993). Increased amounts of riboflavin, thiamine, niacin, and lysine due to the action of LAB in fermented blends of cereals have also been reported (Hamad and Fields 1979; Sanni, Onilude, and Ibidabpo 1999). Khetarpaul and Chauhan (1990) reported improved availability of minerals in pearl millet fermented with pure cultures of lactobacilli and yeasts.

The beneficial role of the probiotics is to assist in the breakdown of food while also manufacturing vitamins essential to the body as well as breaking down and destroying some toxic chemicals that may have been ingested with the food. Oral supplement of diet with viable *Lactobacillus acidophilus* of human origin, which is bile resistant, led to a significant decline of three different fecal bacterial enzymes (Goldin and Gorbach 1984). LAB including *Lactobacillus, Leuconostoc, Lactococcus, Pediococcus*, and *Bifidobacterium* are found throughout the gastrointestinal tract.

7.6   CONCLUDING REMARKS

Fermentation is able to protect food by elimination of harmful components. LAB fermentation has long been held in special favor in the production of safe and nutritious foods that may also elicit positive effects on health and well-being. There is increasing evidence the live LAB in these products do indeed have health benefits. Clinical studies have shown that select members of the LAB provide resistance to enteric pathogens, stimulate the immune system, and help maintain a balanced gastrointestinal microflora. Some advantages of traditional fermentation are that it is labor intensive, integrated into village life, and is a familiar technology; it utilizes locally produced raw materials; it is inexpensive, and the added value of fermented foods provides barter potential; and the subtle variations resulting from fermentation processes add interest and tradition to the foods available for local consumers. From this perspective, research leading to new fermentation technologies should be sensitive to social and economic factors in developing countries.

REFERENCES

Achi, O. K. 1992. Microorganisms associated with natural fermentation of *Prosopis africana* seeds for the production of okpiye. *Plant Foods for Human Nutrition* 42:297–304.

Adams, M. R. 1990. Topical aspects of fermented foods. *Trends in Food Science & Technology* 1:141–144.

Alderman, G. G., and E. H. Marth. 1976. Inhibition of growth and aflatoxin production of *Aspergillus parasiticus* by citrus oils. *Z. Lebensm-Unters-Forsch* 160:355–358.

Ashenafi, M., and M. Busse. 1989. Inhibitory effect of *Lactobacillus plantarum* on *Salmonella infantis*, *Enterobacter aerogenes*, and *Escherichia coli* during tempeh fermentation. *Journal of Food Protection* 52:169–172.

Bhatnagar, S., K. D. Singh, S. Sazawal, S. K. Saxena, and M. K. Bhan. 1998. Efficacy of milk versus yoghurt offered as part of a mixed diet in acute noncholera diarrhoea among malnourished children. *J. Pediatr.* 132:999–1003.

Blom, H., and C. Mortvedt. 1991. Anti-microbial substances produced by food-associated micro-organisms. *Biochem. Soc. Trans.* 19:694–698.

Boivin, P., and M. Malanda. 1997a. Improvement of malt quality and safety by adding starter culture during the malting process. *Tech. Q. Master Brew. Assoc. Am.* 34:96–101.

———. 1997b. Pitching of a fermenting agent during malting: A new process to improve the sanitary and biophysical quality of malt. In *Proceedings of the European Brewery Convention Congress, Maastricht*, 117–126. Oxford: IRL Press.

Bullerman, L. B., F. Y. Lieu, and S. A. Seier. 1977. Inhibition of growth and aflatoxin production by cinnamon and clove oils, cinnamic aldehyde and eugenol. *J. Food Sci.* 42:1107–1108.

Bullerman, L. B., L. L. Schroeder, and K. Y. Park. 1984. Formation and control of mycotoxins in food. *J. Food Protec.* 47:637–646.

Campbell-Platt, G. 1987. *Fermented foods of the world—A dictionary and guide*. London: Butterworth.

Chavan, U. D., J. K. Chavan, and S. S. Kadam. 1988. Effect of fermentation on soluble proteins and in vitro protein digestibility of sorghum, green gram, and sorghum green gram blends. *J. Food Sci.* 53:1574–1575.

Chavan, J. K., and S. S. Kadam. 1989. Nutritional improvement of cereals by fermentation. *Crit. Rev. Food Sci. Nutr.* 28:349–400.

Collard, P., and S. Levi. 1959. A two-stage fermentation of cassava. *Nature* 183:620–621.

Corsetti, A., M. Gobebetti, J. Rossi, and P. Damiani. 1998. Antimould activity of sourdough lactic acid bacteria: Identification of a mixture of organic acids produced by *Lactobacillus sanfrancisco* CB1. *Appl. Microbiol. Biotechnol.* 50:253–256.

Delves-Broughton, J., P. Blackburn, R. J. Evans, and J. Hugenholtz. 1996. Applications of the bacteriocin, nisin. *Antonie van Leeuwenhoek* 69:193–202.

De Vuyst, L., and E. J. Vandamme. 1994. *Bacteriocins of lactic acid bacteria: Microbiology, genetics, and applications*. London: Blackie Academic and Professional.

Dirar, H. A. 1993. *The indigenous fermented foods of the Sudan: A study in African food and nutrition*. Wallingford, Oxon, U.K.: CAB International.

Djoulde Darman, R., F. X. Etoa, J. Essia Ngang, and C. M. F. Mbofung. 2003. Fermentation du manioc cyanogene par une culture mixte de *Lactobacillus plantarum* et *Rhizopus oryzae*. *Microb. Hyg. Ali.* 15:9–13.

Doores, S. 1993. Organic acids. In *Antimicrobials in Foods*, ed. P. M. Davidson and A. L. Branen, 95–136. New York: Marcel Dekker.

EEC. 1983. EEC Commission Directive 83/463/EEC.

Fuller, R. 1989. Probiotics in man and animals. *J. Appl. Bacteriol.* 66:365–378.

Gadaga, T. H., L. K. Nyanga, and A. N. Mutukumira. 2004. The occurrence, growth, and control of pathogens in African fermented foods. *African J. Food Agriculture, Nutrition and Development* 4 (1): 20–23.

Gilliland, S. E. 1996. Special additional cultures. In *Dairy starter culture*, ed. T. M. Cogan and J. P. Accolas, 25–46. New York: VCH Publishers.

Giraud, E., A. Champailler, and M. Raimbault. 1994. Degradation of raw starch by a wild amylolytic strain of *Lactobacillus plantarum*. *Appl. Environ. Microbiol.* 60:4319–4323.

Goldin, B. R., and S. L. Gorbach. 1984. The effect of oral administration on *Lactobacillus* and antibiotics on intestinal bacterial activity and chemical induction of large bowel tumors. *Developments in Industrial Microbiology* 25:139–150.

Gorbach, S. L. 1990. Lactic acid bacteria and human health. *Annals of Medicine* 22:37–41.

———. 2000. Probiotics and gastrointestinal health. *American Journal of Gastroenterology* 95 (Suppl. 1): S2–S4.

Gould, G. W. 1991. Antimicrobial compound. In *Biotechnology and food ingredients*, ed. I. Goldberg and R. Williams, 461–483. New York: Van Nostrand Reinhold.

Gourama, H., and L. B. Bullerman. 1995. Inhibition of growth and aflatoxin production of *Aspergillus flavus* by *Lactobacillus* species. *J. Food Prot.* 58:1249–1256.

Haggblade, S., and W. H. Holzapfel. 1989. Industrialisation of Africa's indigenous beer brewing. In *Industrialisation of indigenous fermented foods*, ed. K. H. Steinkraus, 191–283. New York: Marcel Dekker.

Haikara, A., H. Uljas, and A. Suurnäkki. 1993. Lactic starter cultures in malting: A novel solution to gushing problems. In *Proceedings of the European Brewery Convention Congress, Oslo*, 163–172. Oxford, U.K.: IRL Press.

Hamad, A. M., and M. L. Fields. 1979. Evaluation of the protein quality and available lysine of germinated and fermented cereals. *J. Food Sci.* 44:456–459.

Havenaar, R., and J. Huis in't Veld. 1992. Probiotics: A general view. In: *The lactic acid bacteria in health and disease*, ed. B. J. Wood, 209–224. London: Elsevier Applied Science.

Hurst, A. 1981. Nisin. *Advances in Applied Microbiology* 27:85–123.

Isolauri, E. 1996. Studies on *Lactobacillus* GG in food hypersensitivity disorders. *Nutrition Today* 31:28S–31S.

Jensen, H. L., and A. S. Abdel-Ghaffar. 1979. Cyanuric acid as nitrogen sources for microorganisms. *Arch. Microbiol.* 67:1–5.

Karunaratne, A., E. Wezenberg, and L. B. Bullerman. 1990. Inhibition of mold growth and aflatoxin production by *Lactobacillus* species. *Journal of Food Protection* 53:230–236.

Khetarpaul, N., and B. M. Chauhan. 1990. Improvement in HCl-. Extractability of minerals from pearl millet by natural fermentation. *Food Chemistry* 37:69–75.

Kingamkono, R., E. Sjoegren, U. Svanberg, and B. Kaijser. 1996. Inhibition of different strains of enteropathogens in a lactic-fermenting cereal gruel. *World Journal of Microbiology & Biotechnology* 11:299–303.

Klaenhammer, T. R. 1993. Genetics of bacteriocins produced by lactic acid bacteria. *FEMS Microbiology Review* 12:39–85.

Knorr, D. 1998. Technology aspects related to microorganisms in functional foods. *Trends Food Sci. Technol.* 9:295–306.

Ko, S. D. 1978. Self-protection of fermented foods against aflatoxins. In *Proceedings of the 4th International Congress on Food Science and Technology*, 244–253.

Lavermicocca, P., F. Valerio, A. Evidente, S. Lazzaroni, A. Corsetti, and M. Gobetti. 2000. Purification and characterization of novel antifungal compounds from the sourdough *Lactobacillus plantarum* strain 2IB. *Appl. Environ. Microbiol.* 66:4084–4090.

Lilley, D. M., and R. H. Stillwell. 1965. Probiotics: Growth promoting factors produced by microorganisms. *Sci.* 147:747–748.

Lorri, W., and U. Svanberg. 1993. Lactic-fermented cereal gruels with improved in vitro protein digestibility. *Int. J. Food Sci. Nutr.* 44:29–36.

Maduagwu, E. N. 1981. Differential effects on cyanogenic glycosides in fermenting cassava root pulp by B-glucosidase and microbial activities. In *Proceedings of a National Conference on the Green Revolution in Nigeria*. Port Harcourt, University of Science and Technology, Port Harcourt, Nigeria.

Malin, M., P. Verronen, and H. Korhonen. 1997. Dietary therapy with Lactobacillus GG, bovine colostrums or bovine immune colostrums in patients with juvenile chronic arthritis: mallett evaluation of effect of gut defense mechanisms. *Inflammopharmacology* 5: 219–236.

Mantis, A. I. 1985. Hygiene problems of goat's and sheep's milk and their products. Report of International Seminar on Production and Profitability of Dairy Products. Athens, Greece.

Marteau, P. R., M. De Vrese, C. J. Cellier, and J. Schrezenmeir. 2001. Protection from gastrointestinal diseases with the use of probiotics. *Am. J. Clin. Nutr.* 73:4305–4365.

Marteau, P., and J. C. Rambaud. 1996. Therapeutic applications of probiotics in humans in: leeds AR, Rowland IR (eds) Gut flora and Health. Past, present and future, London: *The royal soc. of medicine press Ltd* : 47–56.

Marteau, P., and J.-C. Rambaud. 1993. Potential of using lactic acid bacteria for therapy and immunomodulation in man. *FEMS Microbiology Reviews* 12:207–220.

McLean, N. W., and I. J. Rosenstein. 2000. Characterisation and selection of a *Lactobacillus* species to re-colonise the vagina of women with recurrent bacterial vaginosis. *Journal of Medical Microbiology* 49:543–552.

Messens, W., and L. De Vuyst. 2002. Inhibitory substances produced by *lactobacilli* isolated from sourdoughs—A review. *Int. J. Food Microb.* 72:31–43.

Meuser, F., and H. D. Smolnik. 1979. Processing of cassava to gari and other foodstuffs. *Starch/Stärke* 32 (4): 116–122.

Nout, M. J. R., F. M. Rombouts, and G. J. Hautvast. 1989. Accelerated natural lactic fermentation of infant food formulations. *Food and Nutrition Bulletin* 11:65–73.

Orla-Jensen, S. 1919. *The lactic acid bacteria.* Fred Host and Son, Copenhagen.

O'Sullivan, M. G., G. Thornton, G. C. O'Sullivan, and J. K. Collins. 1992. Probiotic bacteria: Myth or reality. *Trends in Food Science and Technology* 3:309–314.

Oyetayo, V. O., and F. L. Oyetayo. 2005. Potential of probiotics as biotherapeutic agents targeting the innate immune system. *African Journal of Biotechnology* 4:123–127.

Oyewole, O. B. 1997. Lactic fermented foods in Africa and their benefits. *Food Control* 8:289–297.

Paredes-López, O., and G. I. Harry. 1988. Food biotechnology review: Traditional solid-state fermentations of plant raw materials—Application, nutritional significance, and future prospects. *CRC Critical Reviews in Food Science and Nutrition* 27:159–187.

Pederson, C. S. 1979. *Microbiology of Food Fermentations,* 2nd ed. Westport, CT: AVI Publishing.

Perdigon, G., and S. Alvarez. 1992. Probiotics and the immune state. In *Probiotics: The scientific basis,* ed. R. Fuller, 146–180. London: Chapman and Hall.

Podolak, R. K., J. F. Zayas, C. L. Kastner, and D. Y. C. Fung. 1996. Inhibition of *Listeria monocytogenes* and *Escherichia coli* O157:H7 on beef by application of organic acids. *Journal of Food Protection* 59:370–373.

Ray, B., and M. Daeschel. 1992. *Food biopreservatives of microbial origin.* Boca Raton, FL: CRC Press.

Refai, M. K. 1988. Aflatoxins and aflatoxicosis. *J. Egypt. Vet. Med. Assoc.* 48:1–19.

Robinson, R. K., and A. Samona. 1992. Health aspects of "bifidus" products: A review. *International Journal of Food Sciences and Nutrition* 43:175–180.

Rogers, L. A., and E. O. Whittier. 1928. Limiting factors in lactic fermentation. *Journal of Bacteriology* 16:211–219.

Saïkali, J., V. Picard, M. Freitas, and P. Holt. 2004. Fermented milks, probiotic cultures, and colon cancer. *Nutrition and Cancer* 49:14–24.

Salminen, S., M. Deighton, and S. Gorbach. 1993. Lactic acid bacteria in health and disease. In *Lactic acid bacteria,* ed. S. Salminen and A. von Wright, 237–294. New York: Marcel Dekker.

Samarajeewa, U., A. C. Sen, M. D. Cohen, and C. I. Wei. 1990. Detoxification of aflatoxins in foods and feeds by physical and chemical methods. *J. Food Prot.* 53:489–501.

Sanni, A. I., A. A. Onilude, and O. T. Ibidabpo. 1999. Biochemical composition of infant weaning food fabricated from fermented blends of cereals and soybean. *Food Chem.* 65:35–39.

Savadogo, A., C. A. T. Ouattara, P. W. Savadogo, N. Barro, A. S. Ouattara, and A. S. Traore. 2004. Microorganisms involved in Fulani fermented milk in Burkina Faso. *Pakistan Journal of Nutrition* 3:134–139.

Sawada, H., M. Furushiro, K. Hiral, M. Motoike, T. Watanabe, and T. Yokokura. 1990. Purification and characterization of an antihypertensive compound from *Lactobacillus casei*. *Agricultural and Biological Chemistry* 54:3211–3219.

Schrezenmeir, J., and M. De Vrese. 2001. Probiotics, prebiotics, and synbiotics: Approaching a definition. *Am. J. Clinical Nutr.* 73:361S–364S.

Sharma, A., and A. C. Kapoor. 1996. Level of antinutritional factors in pearl millet as affected by processing treatments and various types of fermentation. *Plant Foods Hum. Nutr.* 49 (24): 12–52.

Sinha, K. K. 1993. Mycotoxins. *ASEAN Food J.* 8:87–93.

Skovgaard, N. 1990. Facts and trends in microbial contamination of dairy products. *Bull. of the IDF* 250:31–33.

Smith, J. E., and M. O. Moss. 1985. *Mycotoxins: Formation, analysis, and significance.* New York: John Wiley and Sons.

Soomro, A. H., T. Masud, and K. Anwaar. 2002. Role of lactic acid bacteria (LAB) in food preservation and human health: A review. *Pakistan J. Nutr.* 1:20–24.

Steinkraus, K. H. 1979. Nutritionally significant indigenous fermented foods involving an alcoholic fermentation. In *Fermented food beverages in nutrition,* ed. C. F. Gastineau, W. J. Darby, and T. B. Turner, 36–50. New York: Academic Press.

———. 1983. Lactic acid fermentation in the production of foods from vegetables, cereals, and legumes. *Antonie van Leeuwenhoek.* 49:337–348.

———. 1996. *Handbook of indigenous fermented foods.* New York: Marcel Dekker.

Subba, M. S., T. C. Soumithri, and R. S. Rao. 1967. Antimicrobial action of citrus oils. *J. Food Sci.* 32:225–227.

Tamine, A. Y. 2002. Fermented milks: A historical food with modern applications—A review. *European Journal of Clinical Nutrition* 56:S2–S15.

Teuber, M. 1992. Microbiological problems facing the dairy industry. *Bull. of the IDF* 276:6–9.

Thanaboripat, D., A. Im-erb, and V. Ruangrattanametee. 2002. Effect of Ling Zhi mushroom on aflatoxin production of *Aspergillus parasiticus.* In *Biological control and bio-technology,* ed. Yang Qian, 22–30. Harbin, China: Heilongjiang Science and Technology Press.

Thanaboripat, D., P. Roisoongnoen, S. Nuchnong, and M. Chantrapanthakul. 1996. Inhibition of aflatoxin production during tempeh preparation. *Srinakarinwirot Sci. J.* 12:8–15.

Van Egmond, H. P. 1989. *Mycotoxins in dairy products.* London and New York: Elsevier Applied Science.

Wary, B. B. 1981. Aflatoxin, hepatitis, B-virus, and hepatocellular carcinoma. *New England J. Med.* 305:833–843.

Yousif, Nabila E., and Abdullahi H. El Tinay. 2000. Effect of fermentation on protein fractions in vitro protein digestibility of maize. *Food Chemistry* 70:181–184.

8 Fortification Involving Products Derived from Fermentation Processes

Peter Berry Ottaway and Sam Jennings

CONTENTS

8.1 INTRODUCTION

A properly designed and effectively implemented food fortification program can have a significant effect on public health, particularly in those countries where there are inherent nutritional deficiencies.

Over the past two decades, many food fortification programs have been introduced across the world, with the main fortified micronutrients being vitamin A and iron. Staple foods being fortified include wheat flour, maize flour, cow's milk, sugar, margarine, and cooking fat. Other fortified foods used as part of a national nutrition program include a chocolate drink and soy sauce.

This chapter looks at the fortification of fermented foods and considers the various technical issues surrounding the production and distribution of such foods.

8.2 HISTORY OF FOOD FORTIFICATION

Food fortification with both macro- and micronutrients has been carried out for centuries, often before the scientific rationale became available. There are a number of examples of early food fortification from different parts of the globe. In

Central Europe during the Middle Ages, mothers were known to push iron nails into apples, leave them there for a while, and then feed the apples to listless or ailing daughters. There are reports as early as 1824 of the indigenous Indian population of Columbia, South America, treating goiter with a specific source of salt that was later found to have a high iodine content; in the making of traditional tortillas in Mexico, the corn was first soaked in lime water and a pinch of ground limestone was added to the tortilla itself; we now know this was to provide an intake of calcium (Scala 1985).

One of the first recorded suggestions for the fortification of food was that of Boussingault, a noted French chemist who worked in South America in the early part of his career and reported on the high incidence of goiter. In 1831, he advocated the iodization of salt for the reduction of goiter. The first official addition of an iodide to domestic salt was started in Switzerland in 1900, and the practice has continued in a number of countries to the present day, as it has been estimated that iodine deficiency affects about two billion people (about 30% of the world's population).

Other foods where micronutrient additions have been officially sanctioned or legally required are margarine, sugar, and flour. Apparently, only six years after vitamin A (retinol) had been identified, a leading doctor in London wrote to the chairman of Unilever to the effect that, if the company wanted its margarine to resemble butter, they would have to add the "new-fangled vitamins." A vitamin A deficiency was reported among children in Denmark in 1917, which was apparently found to be due to Denmark exporting a large proportion of its butter production and replacing it in the home market with margarine. As butter was a major source of vitamin A in the Danish diet, the fortification of margarine with fish liver oils was introduced. The legal requirement for the addition of vitamins A and D to margarine continues in many countries (IFST 1989).

In Canada in the early 1900s, a high incidence of beriberi and blindness was found in segments of the population in Newfoundland and Labrador. This led to the Canadian government introducing the mandatory addition of calcium (as bone meal), iron, and B vitamins to flour and vitamin A to margarine. In 1965 the regulations were amended to include the addition of vitamin D to fluid milk (CPHA 2010).

During the Second World War, the United Kingdom introduced legislation making it compulsory to add vitamin B_1 (thiamine), niacin, and an approved source of iron to all flour used for cooking or baking. Some flours were also required to have added calcium carbonate. These requirements are still in force over 60 years later.

In the United States, there have been a number of official food fortification programs, starting with the iodization of salt in the 1920s, the fortification of milk with vitamin D in the 1930s, the enrichment of flour and bread in the 1940s, and the widespread addition of calcium to a large number of products that began in the 1980s.

By 2010 there were 60 countries that were fortifying foods for public health reasons, mainly with vitamin A, iron, or folic acid.

8.3 DEFINITIONS

Over the years there have been a number of definitions of food fortification. One of the more recent ones is "the public health policy of adding micronutrients to foodstuffs to ensure that minimum dietary requirements are met."

The World Health Organization/Food and Agricultural Organization (WHO/FAO) Guidelines on Food Fortification with Micronutrients has the more comprehensive definition of "Fortification is the practice of deliberately increasing the content of an essential micronutrient, i.e., vitamins and minerals (including trace elements) in a food, so as to improve the nutritional quality of the food supply and provide a public health benefit with minimal risk to health." The guidelines also subdivide fortification into mass or universal fortification, which is "the addition of micronutrients to foods commonly consumed by the general public" (e.g., cereals, condiments, and milk), and targeted fortification, which is the fortification of foods designed for specific population subgroups, such as complementary weaning foods for infants.

Another term that is used is *market-driven fortification*, which refers to the situation where a food manufacturer takes the initiative to add one or more micronutrients to processed foods, usually within regulatory limits, to increase the sales and profitability. The practice of fortifying breakfast cereals with vitamins and minerals would fall into this category.

Over the years, a number of terms relating to the addition of nutrients to foods have been used, often indiscriminately. Some agreement on specific definitions was reached within Codex Alimentarius, and this was published in 1987 in the "General Principles for the Addition of Essential Nutrients to Foods" (WHO/FAO Codex Alimentarius 1987).

In addition to food fortification, the following terms were defined:

- **Restoration** is self-evident, involving the replacement, in full or in part, of losses incurred in processing (e.g., loss of iron and B vitamins in milling of cereals to low extraction rates and loss of vitamin C in the preparation of instant potato). It should be noted that, up to now, restoration has been limited. For example, white bread has thiamine, niacin, and iron added back, although other vitamins such as pantothenate, folate, pyridoxine, and tocopherol are also partly lost in the milling process. Similarly, fruit juices might have vitamin C added back, but juices also supply folate and thiamine. In some countries, calcium and riboflavin are also added to bread. This, however, is fortification, since wheat contains only small amounts of these two nutrients.
- **Enrichment** involves increasing the level of nutrients present to make the food a "richer" source. *Enrichment* has frequently been interchanged with *restoration* and *fortification* as a general term. However, for labeling purposes, the use of the term *enriched* is controlled.
- **Standardization** is sometimes used to mean additions to compensate for natural or seasonal variations in nutrient content.
- **Substitution** is the nutrient addition to a substitute product to the levels found in the food that it is designed to resemble in appearance, texture, flavor, and odor and that it is intended to replace partially or completely. An example is the addition of vitamins A and D to fat replacers.
- **Supplementation** is the supply of nutrients (normally micronutrients) singly or in combination in a quantified dose form. Supplements can take a variety of forms such as tablets, capsules, pastilles, measured amounts of liquid, or small sachets of powder.

The term *nutrification* has been occasionally used in the past to mean the addition of nutrients to formulated or fabricated foods that are marketed mainly as meal replacements.

8.4   IMPORTANCE AND NEED OF FORTIFICATION

As already discussed, food fortification and restoration has played an important role for over a century in helping to achieve specific health policies and ensuring the nutritional health of populations across the world. In addition to the enrichment of salt with iodine and the fortification of margarine and bread mentioned previously, there have been programs to combat rickets with vitamin D–fortified milk; the addition of thiamine (vitamin B_1), niacin, and folic acid to cereals to reduce the incidence of beriberi and pellagra, and more recently, the incidence of neural tube syndrome in pregnant women in both the developed and developing countries; and the addition of an iron source to cereals to help reduce anemia, particularly in women and young children.

The fortification of food has the great advantage that it can often be accomplished within the context of an indigenous diet and requires little change to the consumer's dietary behavior and food habits. In most cases, effective fortification can be achieved with little or no effect on the organoleptic properties of the food due to the very small amounts of the added micronutrients. An additional advantage is that the fortification can be achieved for very little extra cost.

In developing countries, where there is the greatest need for micronutrient fortification, the vitamins and minerals can be added to the foods or condiments most regularly consumed by a significant proportion of the population at risk. In many cases this is likely to be the staple cereal such as rice, wheat, or maize flour, but there have been reports where condiments such as iron-fortified fish sauces and soy sauces have been successfully used in programs aimed at reducing iron deficiency, mainly in Asia.

Food fortification can have a positive effect on the health of a population at a very low cost. A report published by UNICEF (2009) based on an evaluation of 129 countries gives a vivid picture of the consequences of micronutrient deficiencies in those countries:

- Vitamin A deficiency is compromising the immune systems of approximately 190 million of the developing world's preschool age children and 19 million of pregnant women.
- An estimated one-third (195 million) of children under five years old in the developing world are stunted, reflecting chronic nutritional deficiency, aggravated by illness.
- Iodine deficiency is estimated to have lowered the intellectual capacity of almost all of the nations reviewed by an average of 13.5%.
- Iron deficiency, and its related anemia, affects about 25% of the world's population, most of them children of preschool age, and women.

These examples are not exhaustive but are indicative of the scale of the problem faced by a large proportion of the world's population. To put it into perspective, a substantial proportion of the global population is at risk of vitamin A, iron, or iodine deficiency, in addition to other macro- and micronutrient deficiencies such as protein, folic acid, and zinc.

In May 2002, the General Assembly of the United Nations (UN) agreed that the key micronutrient deficiencies should be one of the global development goals to be achieved in the first decade of the new millennium. Specifically, the UN targets were for the virtual elimination of iodine deficiency by 2005; the elimination of vitamin A deficiency by 2010; and a reduction of at least 30% of iron-deficiency anemia, also by 2010 (UN General Assembly 2002).

In 2009, UNICEF found that 125 countries were implementing and reporting on salt iodization programs, an increase of 39% since 2002, and around 72% of all households in developing countries were consuming adequately iodized salt. However, despite this progress, about 41 million newborns a year still suffer the consequences associated with iodine deficiency. A WHO report (2009) indicated that vitamin A deficiency is still a clinical problem in 45 countries and a subclinical problem in 122 countries. As of June 2010, 60 countries worldwide had legislation or decrees that mandate fortification of one or more types of flour with either iron or folic acid, and 30% of the world's wheat flour that is produced in large roller mills is fortified either mandatorily or voluntarily (FFI 2010). However, an earlier WHO report (2008) estimated that, globally, anemia was still affecting around 1.62 billion people (just under 25% of the population).

The mandatory fortification programs needed to achieve these ambitious UN 2002 objectives require a political will and commitment at a national level. The scale and severity of micronutrient deficiency is not yet fully appreciated by the politicians, public, or press in most nations of the world. The UNICEF report states that "the goals will not be achieved, and the impact of vitamin and mineral deficiency will not be substantially reduced without a more ambitious, visionary, and systematic commitment to putting known solutions into effect on the same scale as the known problems." It also points out that progress is only likely to be achieved by dynamic alliances involving governments, the private sector, health and nutrition professionals, academics and researchers, civil society, and international agencies.

8.5   PRINCIPLES OF FORTIFICATION

As the object of food fortification is to supply essential nutrients that cannot be obtained from the diet, it is important that the fortification be applied to foods that commonly form an integral part of the diet consumed by the target population.

For a fortified food to be successful, the fortifying ingredients must not be obtrusive and must not affect the organoleptic properties of the food, particularly the color, taste, and texture. In addition, the inclusion of the fortifying ingredients should not significantly affect the accepted stability and shelf life of the product. In essence, the fortification should not in any way dissuade the target consumers from continuing to retain the food in their diet.

Unfortunately, a number of the macro- and micronutrients for which there is a nutritional need, and which should be given priority in a fortification program, are the ones that can affect the stability or organoleptic properties of the food. The addition of such nutrients to food requires the careful selection of the sources of the nutrients, i.e., those that are the least reactive and the most stable in the carrier food matrix.

The economic aspects of fortification also need consideration, not only the increased costs of ingredients and processing, which are normally minimal, but also the costs required to maintain the stability of the product, such as a change of packaging to increase the barrier properties.

The organoleptic properties that need to be given serious consideration during the early stages of the development of a fortified food are:

- Color
- Texture
- Taste
- Odor

Each of these parameters has to be acceptable to the target consumer, and they have to be integrated, as the subsequent color and texture can influence the perception of taste.

The stability of a food can be affected by a number of factors, the main ones being moisture, oxygen, light, and temperature. In addition, the combination of the components of a food can have a significant effect on its stability by inducing reactions during the storage of the food. These effects can be caused by:

- pH
- Oxidizing and reducing agents
- Presence of metallic ions (e.g., iron and copper)
- Components and additives in the food, such as sulfur dioxide
- Reactions between nutrients, such as vitamin-vitamin interactions
- Combinations of the above

In the context of stability, fermented foods can present a significant challenge to micronutrient fortification in that, in general, they tend to have a high moisture content and a low pH. In addition, the process of fermentation produces a number of changes to the components of the food, some of which will continue over a period of time, particularly if viable organisms are present.

Fermentation has been described as the "slow decomposition process of organic substances induced by microorganisms or by complex nitrogenous substances (enzymes) of plant or animal origin" (Walker 1988). It can also be described as biochemical changes brought about by the anaerobic or partially anaerobic oxidation of carbohydrates by either microorganisms or enzymes. The fermentation of food is widespread across the world and is used as a form of food preservation in many developing countries, where there are problems with food distribution and storage. Knowledge of traditional fermentation for indigenous products has been handed down from parent to child through the ages, which can produce localized and potentially significant variations in the composition of the food.

In terms of a fortification program, it is important that there be an agreed standardization of the processing method and the food. The changes to the food components during fermentation can be both advantageous or disadvantageous, depending on the conditions applied. Properly controlled fermentation can be a relatively efficient, low-energy process of food preservation, which can increase the shelf life of the food and decrease the need for low-temperature storage.

For effective fortification of the fermented food, the fermentation needs to be controlled to ensure that all critical parameters are maintained within agreed limits and that the final product is as consistent as technically achievable. This is as important for the stability and shelf life of the food as it is for the stability of the fortifying ingredients.

8.6   FORTIFYING FERMENTED PRODUCTS

Of recent time, a number of fermented foods world wide have either been fortified or investigated for possible fortification. While fermented dairy and cereal products dominate the list, other fermented plant products such as yam flour, soy meals and vegetable (sauerkraut), fish and meat products, and fermented sauces have all been fortified with either macro- or micronutrients.

Although in certain cases the fortification is carried out as a marketing strategy due to perceived or anticipated consumer demand, in many instances the fortification of fermented foods has been studied with a view to finding a simple means of improving the nutritional intake of the local population. Thus, in the Middle East and in Western Asia, studies have concentrated on the fortification of fermented dairy products such as local yogurts, cheeses, or fermented milks, while in many African countries, fortification of fermented maize, sorghum, millet, or yam flour meals or fortification of local beers have been considered as being more beneficial to the target population. In the eastern and southeastern Asian countries, greater interest has been paid to the fortification of fermented fish products and condiments.

8.6.1   Fermented Dairy Products

The products that have been the subject of considerable research as a carrier for fortificants are the yogurts and drinking yogurts, where studies have been carried out on micronutrients such as vitamin D and calcium, fiber, probiotics, fruit juices, and high omega-3 oils (Table 8.1). Yogurts of various compositions are commonly consumed in a large number of countries and are therefore suitable candidates for targeted fortification programs, such as the provision of omega-3 fatty acid DHA (docosahexaenoic acid) in drinking yogurts aimed at school children or nursing or pregnant women in some countries, or the plant sterol/stanol-containing yogurts for consumers striving to lower their cholesterol levels. In India, the yogurt known as *dahi* (made from cow, goat, or buffalo milk) has been considered for fortification projects as a potential means of combating chronic and noncommunicable diseases.

Mineral fortification of yogurts has been undertaken in scientific studies and extended to the commercial market. Investigations of the effect of mineral fortification on yogurt quality and stability have found that it is possible to fortify yogurts with various minerals without adversely affecting the product characteristics or

TABLE 8.1
Examples of Fermented Dairy Products and Related Nutrients That Are Currently in Use or Have Been Studied for Possible Use for Fortification

Food	Fortified with
Yogurt: generally made from cow, goat, or buffalo milk	Minerals, e.g., calcium, iron, iodine, zinc, magnesium, manganese, molybdenum, chromium, and selenium
	Vitamins, e.g., vitamins A, C, D, and K_2
	Dietary fiber
	Probiotic organisms
	Phytosterols/stanols
	Fruit juice
Fermented milks/drinking yogurts	Calcium
	B vitamins
	DHA
	Probiotic organisms
	Phytosterols
	Fruit and vegetable juices (for their antioxidant content)
	Lutein
Cheddar cheese	Omega-3 fatty acids
	Vitamin D
	Calcium
	Phytosterols
Shrikhand: fermented and sweetened milk product from India, usually made from buffalo milk	Fruit pulp
	Probiotic organisms
Kurut: fermented and dried dairy product from Pakistan, usually made from cow or buffalo milk	Protein (soy protein isolate)

consumer acceptability. For example, manganese or magnesium could be added to yogurt at levels ranging from 25% to 90% of the U.S. recommended dietary allowance (RDA) with some small changes in the product quality, such as color and viscosity, but these changes were not detected during the sensory evaluation (Achanta and Aryana 2007; Cueva and Aryana 2008). However, the particular salts used for fortification, the stage at which the fortificant is added, and the amount of fortificant can have an impact on the product quality. For instance, for fortification with calcium, it was found that calcium gluconate had a greater adverse effect on the sensory quality of a yogurt than calcium lactate (Pirkul, Temiz, and Erdem 1997), while calcium carbonate is not recommended for yogurts that are fermented in the package, as the salt's presence raises the pH higher than the pH 3 or 4 needed for fermentation and, in so doing, releases carbon dioxide, which leads to swelling of the container. When Hekmat and McMahon (1997) investigated the fortification of yogurt with iron at levels ranging up to 40 mg iron/kg yogurt, they concluded that the use of a protein-chelated iron, rather than ferric chloride, might help to minimize the production of any oxidized flavors. However, although the trained sensory panelists were able to detect these oxidized flavors, no differences in flavor were detected during the consumer sensory testing.

Fortification of yogurt and other fermented milk products with vitamins raises another issue in addition to that of the sensory acceptability and stability of the product, and that is the stability of the vitamins during manufacture and storage. No vitamin is completely stable in foods, and stability ranges from the relatively stable, such as niacin, to the relatively unstable, such as vitamin B_{12} (Berry Ottaway 2008). The factors affecting the stability of vitamins vary from vitamin to vitamin and include pH, heat, moisture, oxygen, and light. In addition, the form in which the vitamin is added, such as a salt or ester, can have an effect on its stability in the fermented dairy product, as can the species and strain of the bacterial culture used (Papastoyiannidis et al. 2006).

When yogurts produced under commercial manufacturing conditions were fortified with vitamins A and C, it was found that the preferred forms were a water-dispersible form of vitamin A palmitate and ascorbic acid. Combination of vitamins A and C provides a technological challenge, as vitamin A is a fat-soluble vitamin, while vitamin C is water soluble. To help overcome this incompatibility, manufacturers have developed a dispersible form of vitamin A, whereby it is encapsulated in very small beadlets to facilitate a more homogeneous dispersion in a liquid medium. Ascorbic acid was used, as it is more soluble and appeared to have a better retention than its salts. Using the preferred forms, the yogurts were found not to be significantly altered in terms of their pH, consistency, or sensory characteristics. However, there was an up to 40% loss of vitamin A and 75% loss of vitamin C over the 6-week test period (Ilic and Ashoor 1988). It has been found in previous work that the stability of vitamin C can be seriously affected by metallic ions, particularly zinc, copper, and iron, present in processed water and ingredients. Significant improvement in vitamin C stability can be achieved by eliminating or greatly reducing the levels of these metallic ions (Berry Ottaway 2008). Conversely, fortification of yogurt with vitamin D at levels of 2.5 µg and 6.25 µg found no loss of the vitamin during processing or the 42-day shelf life, and the sensory characteristics were unchanged (Hanson and Metzger 2010). As with vitamin A, a water-dispersible form of vitamin D is preferred.

An investigation of the fortification of fermented milk with six B-group vitamins found that the vitamin level throughout fermentation and storage of the products was significantly affected by the species and strain of the culture. Fortification with the B-group vitamins had no apparent impact on the composition or sensory properties of the fermented milk products (Papastoyiannidis et al. 2006). However, there has been considerable experience of "yeasty off-flavors" from the use of the B-group vitamins, particularly thiamine, at higher levels, for example, above RDA levels. The thiamine molecule is sensitive to sulfites and can be destroyed by the presence of very low levels of sulfite in ingredients such as fruit juices and purées used to flavor the products (Berry Ottaway 2008).

When evaluating the shelf life of fermented dairy products fortified with vitamins, serious consideration must be given to the maximum time in distribution and retail. In contrast to many vitamin-fortified foods, fermented milks have a relatively short shelf life, generally only up to around 42 days.

Fiber, found in the cell walls of cereals, fruits, and vegetables, is another fortificant that has been investigated for use in yogurt for its purported health benefits for the general population as well as for those suffering from various physiological disorders. The addition of dietary fiber may cause changes in the texture and flavor of the final product, and thus selection of an appropriate fiber at the correct level is important. The addition of fiber to yogurt has been studied using numerous sources, including wheat bran; oat fiber; fibers from fruits such as apple, date, and orange; and fiber from vegetables such as asparagus.

Oat fiber (1.32%) was found to have no significant effect on fermentation time, pH progression, or orotic acid consumption by the starter bacteria during fermentation, but consumption of hippuric acid appeared to be increased. Levels of acetic acid and propionic acid were significantly higher after eight hours of fermentation and remained high (though not significantly so) after 28 days of storage. Fortification with oat fiber appeared to improve the body and texture of unsweetened yogurt but reduced the overall flavor quality, while the effects on yogurt sweetened with varying levels of sucrose or fructose appeared to depend on the type and quantity of sweetener that was used (Fernández-Garcia, McGregor, and S. Taylor 1998).

Wheat bran, both natural and toasted, added alongside calcium tended to improve the consistency, increase pH, and decrease syneresis with increasing levels of the fiber (at 1.5%, 3.0%, and 4.5% by weight), although natural bran had a greater effect on the consistency than did toasted bran (Aportela-Palacios, Sosa-Morales, and Velez-Ruiz 2005). However, these effects were reversed during storage.

Although fortification with commercially produced apple fiber was found to modify some rheological properties of the yogurt, and also affected the color, the fortified yogurt was considered to be acceptable by a consumer panel who were untrained in sensory perception (Staffolo et al. 2004). When date fiber was investigated, it was found that fortification of yogurt with up to 3% date fiber had no significant effect on the rheological properties and overall acceptance ratings, while greater amounts of the fiber had an adverse effect on the physical and sensory properties (Hashim, Khalil, and Afifi 2009).

Asparagus fiber, resulting as a by-product from the asparagus processing industry, when added to yogurt was found to increase the consistency but had no effect on the viscoelastic behavior. However, the fiber diminished clarity and imparted a yellow-green color to the yogurt (Sanz et al. 2008). When considering fiber fortification of fermented dairy products, consideration must be given to the amount of fiber that can be incorporated and its relevance to dietary requirements in the intended market.

Yogurts, drinking yogurts, and other fermented milk products, such as the locally produced *shrikhand* in India, have been fortified with probiotic organisms both in scientific studies and commercially due to the growing interest in the perceived health benefits to much of the population. The FAO/WHO definition of probiotics is that they are "live microorganisms which when administered in adequate amounts confer a health benefit on the host" (FAO/WHO 2002), while the International Scientific Association for Probiotics and Prebiotics further clarified this definition by adding the requirements that (a) a probiotic must be alive when administered; (b) the probiotic must have undergone controlled evaluation to document health benefits in the target host; (c) the probiotic must be a taxonomically defined microbe or combination of microbes (genus, species, and strain level); and (d) the probiotic must be safe for its intended use (ISAPP 2009).

Milk is an excellent medium in which to carry or generate live and active cultured probiotic organisms due to its composition and combination of nutrients, and products such as yogurt, drinking yogurt, and other fermented milk products are a popular basis for fortification. The viability and stability of probiotic organisms in fermented dairy products has been a technological challenge for industrial producers, as factors such as high temperature, low pH, water activity, and oxygen content may all adversely impact the organisms' survival (Khurana and Kanawjia 2007). Numerous methods have been investigated to enhance the survival of probiotic organisms during manufacture and storage of fermented dairy products, including the addition of lactulose, inulin, high amylase corn starch powder, or beta-glucan (Tabatabaie and Mortazavi 2008; Donkor et al. 2007; Vasiljevic, Kealy, and Mishra 2007); the addition of cysteine or ascorbic acid (Bari et al. 2009; Dave and Shah 1997a, 1997b, 1998); concentration of the milk solids content or addition of whey powder or whey protein concentrate (Yeganehzad, Mazaheri-Tehrani, and Shahidi 2007; Antunes, Cazetto, and Bolini 2005; Dave and Shah 1998); addition of dried fruit extracts (Herzallah 2005); adjustment of acidity levels (Donkor et al. 2006); concentration of starter culture and storage time (Bari et al. 2009); and microencapsulation of the probiotic organisms (Anal and Singh 2007; Krasaekoopt, Bhandari, and Deeth 2006).

A further issue that has to be considered is the probiotic organisms' effects on the fermented milk product in terms of its consumer acceptability. It has been found that the choice of probiotic culture can have an impact on the sensory properties of fermented milk products (Papastoyiannidis et al. 2006), and therefore the selection of the appropriate probiotic organism for the product is important in terms of both technological issues and consumer acceptance. However, one of the essential requirements of a probiotic fortified fermented milk is that a high level of viability of the probiotic organisms is retained for the intended shelf life of the product.

Fats and oils high in omega-3 polyunsaturated fatty acids (PUFAs) have become popular fortificants over the past decade, and fermented dairy products containing omega-3 fatty acids such as alpha-linolenic acid (ALA), eicosapentaenoic acid (EPA), and docosahexaenoic acid (DHA) in fats and oils are being investigated and marketed worldwide. For example, targeted drinking yogurts containing various high omega-3 oils have been produced, aimed at schoolchildren in countries such as Japan and at pregnant and nursing mothers in Russia. High omega-3 oils have also been considered for the fortification of cheese. ALA is commonly sourced from flaxseed oil, whilst EPA and DHA have been generally sourced from fish oils.

However, there are a few problems that can be encountered when substituting milk fat with oils rich in omega-3 fatty acids, including possible adverse effects on the product consistency (such as whey separation) and unpleasant aroma and flavor attributes (Khurana and Kanawjia 2007; Martín-Diana et al. 2004; Barrantes et al. 1996), although these could potentially be overcome by the addition of whey protein concentrate to the product (Martín-Diana et al. 2004). When fish oils containing EPA and/or DHA are added to some fermented dairy products, they can cause oxidative instability, which may lead to potential oxidized, rancid, and fishy off-flavor formation, though the extent of the off-flavor formation can depend on the fortification level used (Martini et al. 2009). Fish-oil fortified yogurt, though, appears to have a higher oxidative stability than other dairy products, which may be due to antioxidant peptides released during the fermentation by lactic acid bacteria and/or to the lower oxygen content of yogurt, which thus reduce the oxidative stress on the fish oil fortificant (Farvin et al. 2010).

For a number of years, yogurts have been fortified with phytosterols and phytostanols, which are added for their proven cholesterol-lowering properties. Following a general consensus on the efficacy of the phytosterols, a number of food companies market yogurts containing plant sterols in many countries, and there are now numerous competing brands. The effectiveness of plant sterols, stanols, and their esters has been extensively documented, and studies show that a regular consumption of 1 to 3 g of plant sterols a day lowers LDL cholesterol by 6% to 11% (Katan et al. 2003). However, there is evidence that the cholesterol-lowering effects may differ with the food type and matrix (Clifton et al. 2004; European Food Safety Authority 2009).

Other fortificants that have been explored for fermented dairy products include carotenoids, such as lutein and antioxidant substances derived from various fruits and vegetables, in fermented milk beverages (Granado-Lorencio et al. 2010; Salem, Gafour, and Eassawy 2006) and protein from soy beans in indigenously produced foods in areas where protein deficiency is a nutritional concern (Shakeel et al. 2009).

8.6.2 FERMENTED PLANT PRODUCTS

Fermented plant products, whether cereal, grain, fruit, or vegetable based, are consumed throughout the world and are of particular importance in developing countries. Cereals are an especially valuable substrate for fermented food products in Africa, Asia, and the Indian subcontinent, often forming the basis of the main part of the diet and making a major contribution to the carbohydrate and other macronutrient intakes. However, other fermented plant products have also been considered for fortification purposes (Table 8.2).

Fermented cereal porridges made from maize, sorghum, or millet form an essential staple food product in many African countries, particularly as weaning foods for infants from around 9 months old (Osungbaro 2009; Kohajdová and Karovičová 2007; Parveen and Hafiz 2003). However, the nutritional value in terms of protein content is not high for the fermented cereal porridge, and thus, given the importance of these porridges in the diet of the indigenous populations, and the shortage and high cost of animal protein, considerable research has been undertaken as to how to improve the nutritional quality of the product. Most of the research has focused on increasing the protein content, although targeted iodine fortification has also been considered.

Fortificants that have been used to increase the protein content and quality of fermented cereal porridge include whole soybeans, soy flour, bambara groundnut, okra seed meal, and fish flour, and both large-scale commercial and household/small-scale fortification processes have been considered. Soy-fortified fermented maize dough samples have been shown to be a good source of protein, fat, and minerals, and a fermented cereal porridge containing 10% soy flour was developed by the Federal Institute of Industrial research in Nigeria (Osungbaro 2009). Fortification of

TABLE 8.2
Examples of Fermented Plant Products and Related Nutrients That Are Currently in Use or Have Been Studied for Possible Use for Fortification

Food	Fortified with
Fermented cereals	Protein (soybean/soybean flour, bambara nut, okra seed, cowpea, and fish)
	Iodine
Fermented root and tuber crops	Protein (soybean flour and soy-melon blend)
African sorghum beer	Thiamine
Sauerkraut	Folate

fermented maize cereal with soy flour at replacement levels of 10% and 20% showed an improvement in protein content from a base level of 10.3% up to around 13% and 18%, respectively, and a greater concentration of most of the essential amino acids (Plahar, Nti, and Annan 1997). However, soy flour fortification greatly reduced all viscosity characteristics of the fermented cereal, with the 10% soy flour reducing peak viscosity by around 32% and the 20% level reducing it by approximately 75%. In comparison, peak viscosity following fortification with whole boiled soybeans was similar to the unfortified fermented maize, particularly at the 10% level. There were no significant effects on the flavor of the cooked porridge with fortification with either soy flour or boiled whole soybeans, though a slight but significant color difference was noted in the fortified fermented maize dough with both forms of fortificant (Plahar, Nti, and N. T. Annan 1997). Soy flour fortification has also been investigated for fortifying fermented sorghum porridge and was found to almost double the protein content (Baker, El Tinay, and Yagoub 2010), while the sensory characteristics appeared to remain acceptable (Adelekan and Oyewole 2010).

Bambara groundnut (*Vigna subterranean* L.) is another fortificant that has been considered for fortifying fermented cereal meals, as its seeds were found to consist of 19% protein on average (Mbata, Ikenebomeh, and Ezeibe 2009). Fortification with 10% and 20% raw or boiled bambara-nut raised the protein content of the fermented maize dough from a base of 10.1% up to over 12% and 16%, respectively, with boiled bambara nut providing a greater protein content. Although peak viscosity was reduced in the dough fortified with boiled bambara nut, particularly at 20% fortification, the fortification with raw bambara nut had minimal effect. All the fortified dough samples displayed a consistent gelling tendency and desirable starch stability, but only the samples fortified with boiled bambara nut fell within the limits considered acceptable for a weaning food. The samples fortified with raw bambara nut were found to be suitable for adult consumption. Bambara nut fortification did not appear to have any detrimental effects on the organoleptic qualities of the maize dough (Mbata, Ikenebomeh, and Alaneme 2009; Mbata, Ikenebomeh, and Ezeibe 2009b; Mbata, Ikenebomeh, and Ahonkhai 2006).

Another protein-rich fortificant investigated for use in fermented maize porridge is okra seed meal. Okra (*Abelmoschus esculentus* L.) seeds consist of approximately 24% crude proteins (Ndangui et al. 2010), and fortification of fermented maize porridge at a 20% level with either roasted okra seed meal or defatted okra seed meal raised the protein content of the porridge from around 10.5% to 16%–18% (Aminigo and Akingbala 2004), which was a similar level to that achieved with soy flour. However, the fortified fermented maize porridge was not considered as acceptable as the market sample of fresh unfortified porridge, although the difference in acceptability when the fortified porridge was compared to the laboratory-prepared unfortified control was not significant. Fortification with okra seed meal had an adverse effect on the texture of the fermented maize, attributed to the lower viscosity of the fortified cereal, and also on its aroma and color.

A protein-rich flour, both fermented and nonfermented, was made from deskinned and gutted Nile tilapia fish (*Oreochromis niloticus*) and used to produce fortified fermented maize dough at ratios of maize flour to tilapia flour ranging from 6.25:1 to 9.30:1. The protein content of the dough increased from around 7.63% to between 14% and 16%. The addition of tilapia flour also increased the water and oil absorption capacity, increased the gelling ability, and decreased the viscosity of the fermented mixture (Fasasi, Adeyemi, and Fagbenro 2007). Although sensory characteristics were not given in this particular investigation, the use of Java tilapia (*Oreochromis mossambicus*) flour as a protein source in dried crackers revealed that the greater the amount of fish flour used, the greater the adverse effect on the color, taste, and smell of the fortified food (Lelana, Purnomosari, and Husni 2003).

In addition to fortifying with protein, fermented cereal has also been considered as a possible means of iodine fortification in an iodine-deficient region of Africa. The fermented maize meal was fortified with 10% w/w codfish (*Gadus virens*) powder, thereby raising the iodine content from 0.039 mg/kg to 2.13 mg/kg. This fortified maize meal was further prepared by mixing with water, cooking, and the addition of sugar or salt as applicable to make the intervention porridge or dumplings. The amount of iodine provided by the intervention diet (one serving of porridge and one serving of dumplings daily) was 460 µg per person per day. The intervention diet was successful in improving the iodine status of goiterous individuals, and it was concluded that fortification of the fermented maize was an alternative means of supplementing the iodine intake of specific populations where the use of iodized salt is less efficient. However, no information was provided as to the effect of the cod powder on the physical and organoleptic properties of the fermented maize meal (Maage et al. 2008).

Root and tuber crops are important staples in many tropical and subtropical countries, and products made from fermented cassava and yam flours often form a significant part of the daily diet. However, the food products produced from these flours are of low protein quality and thus have been investigated for fortification with readily affordable, accessible, and good-quality protein sources. Cofermentation of cassava and soybean has been examined with soy flour levels of 10%, 20%, and 30%. It was found that levels of 10% and 20% did not significantly affect the physicochemical properties of the product, such as pH, starch content, and pasting characteristics, but fortification with 30% soy flour had significant adverse effects on both the physicochemical and, thus, the organoleptic properties of the fermented product. It was concluded that fortification with 20% soy flour would be a feasible option to improve protein quality while maintaining acceptable quality of the food (Afoakwa et al.

2010). When fermented yam flour was fortified with 10%, 20%, 30%, and 40% soy flour, it was again found that fortification with up to 20% soy flour increased the protein content without causing significant differences in color, taste, flavor, or overall acceptability. Although fortification with 30% and 40% soy flour greatly increase the protein content (from 3.5% in the control up to 19.7% for the 40% soy fortification), there were accompanying significant unfavorable effects on the physicochemical and organoleptic properties of the fermented dough (Achi 1999).

While often used as a fortificant, due to their high protein quality, soybean fermented products (for example, soy yogurt) have also been investigated for fortification, but with micro- rather than macronutrients. Soy yogurt has a high nutritional value and low production cost and has thus been considered as a suitable vehicle for iron fortification. However, due to its having a lower calcium content than cow's milk yogurt, fortification with calcium is also necessary. The selection of appropriate sources of iron and calcium is important, as they must offer high bioavailability without reducing product acceptability and shelf life. The use of microencapsulation technology presents a potential means of meeting these requirements, as it separates sensitive or incompatible ingredients, thus avoiding nutritional interaction and undesirable changes in the sensory characteristics. When used to fortify soy yogurt, microencapsulated ferrous sulfate has been found to provide a bioavailable source of iron, which has little impact on the color and flavor of the yogurt. Studies on calcium citrate found it suitable as a fortificant due to its bioavailability, compatibility with iron sources, and sensory properties. The use of microencapsulated ferrous sulfate and calcium citrate together as dual fortificants at levels of 1.2 mg iron/100 ml and 60 mg calcium/100 ml increased the iron and calcium content of the soy yogurt by 233.9% and 72.5%, respectively, with no significant effect on the rheological and sensory properties of the yogurt, despite there being noticeable changes in the pH, titratable acidity, and viable cell counts (Cavallini and Rossi 2009).

In areas of nutritional deficiency, all forms of indigenous foods are often considered as a means of fortification in order to increase intake of the particular nutrient of interest by the target population. In parts of Africa, the locally produced fermented sorghum beer has been investigated as a route for increasing the thiamine intake of habitual consumers of the beer, and thus counteracting Wernicke-Korsakoff syndrome (a neurological disorder that can be caused by thiamine deficiency linked to high alcohol intake). As thiamine is relatively stable in alcoholic beverages and is undetectable in beer to professional taste testers at levels theoretically sufficient to guarantee adequate absorption (Bishai and Bozzetti 1986), its use as a fortificant in sorghum beer would appear to be a practical option. Studies found that the naturally occurring thiamine concentration in sorghum beer averaged 0.33 mg/l, with more than 95% in the cellular fraction and therefore not available for absorption. However, when thiamine hydrochloride was added in concentrations exceeding 2.5 mg/l, the uptake by the cellular fraction was less than 30%, and it was concluded that 3 mg thiamine hydrochloride per liter of beer would supply approximately 2 mg of absorbable thiamine, thus meeting the recommended daily allowance in the target population (van der Westhuyzen et al. 1985, 1987). Nonetheless, despite the interest in this route of fortification during much of the 1980s, little progress appears to have occurred since, and this is possibly due to the nutritional and educational implications of fortifying a substance that is considered to be potentially harmful (Binns, Carruthers, and Howat 1989).

Fermented vegetables are a common food product in central and northern Europe, traditionally providing a source of nutrients during the winter months. Although the development of frozen fruits and vegetables and the importation of foods from southern Europe have removed the dependence on fermented vegetable products for their micronutrient content, the folate fortification of sauerkraut has been investigated. The process involved the fermentation of the cabbage with propionibacteria, which have been shown to synthesize vitamin B_{12} and folate in symbiosis with the yeast *Kluyveromyces fragilis* when in nutrient media. *Propionibacterium shermanii* was studied for the nutritive enrichment of sauerkraut, and it was found to increase the folate concentration by 287% when compared to the raw material, while maintaining the color, taste, and consistency at an acceptable level (Hozova, Batorova, and Cerna 1991).

8.6.3 FERMENTED MEAT, FISH, AND CONDIMENTS

Fermented meat has been a less common vehicle for fortification, which could be due to the high cost of animal meat in many parts of the world where fortification is considered a requirement. However, the fortification of fermented fish products, particularly those used as condiments, is an area that has been closely investigated, as has the fortification of other fermented condiment products such as soy sauce (Table 8.3).

One fermented meat that has been looked at for fortification is Thai-style fermented pork sausage, which is traditionally made from the lactic acid bacterial fermentation of minced lean pork mixed with salt, potassium / sodium nitrate / nitrate, cooked rice, and seasonings. The fortificant investigated was calcium, with the objective of producing a nondairy dietary source of the micronutrient. The proposed ingredients were a commercial calcium lactate preparation and an eggshell calcium lactate preparation, the latter chosen due to the wide availability of eggshells as a source of calcium. The calcium ingredients were added to the pork sausage mix at levels of calcium of 150, 300, and 450 mg per 100 g prior to fermentation. It was found that, at 150 mg calcium from either source, there were no significant differences in pH, lactic acid bacterial count, or color, though the shear force (texture) of the eggshell-sourced

---

TABLE 8.3

Examples of Fermented Meat and Fish Products and Other Condiments and Related Nutrients That Are Currently in Use or Have Been Studied for Possible Use for Fortification

Food	Fortified with
Thai-style fermented pork sausage	Calcium
Fermented fish, fish paste, and fish sauce	Iron, Iodine
Soy sauce	Iron

calcium-fortified product was significantly decreased when compared to the control or the commercial calcium source fortified sausage. However, despite this textural difference, the overall acceptability score in terms of sensory evaluation of the fortified fermented sausage was not significantly different between either of the calcium sources and the control. The higher levels of calcium content, at 300 and 450 mg per 100 g, caused significant differences in the physicochemical, microbiological, and color analyses for both the commercial and eggshell sources of the calcium. It was thus concluded that calcium fortification levels should be limited to 150 mg per 100 g (Daengprok, Garnjanagoonchorn, and Mine 2002).

Due to the prevalence of anemia in a number of countries, the route of iron fortification has been of great interest, and various studies have been undertaken to investigate the fortification of indigenous fermented products with this micronutrient. In many eastern and southeastern Asian countries there is a high intake of fermented fish products, including fish-based condiments such as fish sauce and fish paste, and these, due to their strong flavors, have proven to be a suitable vehicle for iron fortification. Iron fortification can lead to problems with color, taste, and odor in the chosen food, and using a highly flavored product can minimize these acceptability issues. Fish sauce has long been considered as suitable for iron fortification due to its low price, widespread consumption, fairly constant daily consumption level per person, and its overall availability in the market. In addition, due to its high salt content, overconsumption is unlikely to occur (Garby and Areekul 1974). The fortificant that is usually chosen is iron(II) sodium ethylenediaminetetraacetate (NaFeEDTA), as preliminary studies have indicated that this iron source is not precipitated in the fish sauce, does not cause storage problems, and does not affect the sensory characteristics of the product (Fidler et al. 2003; Cook and Reusser 1983; Garby and Areekul 1974). One advantage of using NaFeEDTA as the iron source is that its bioavailability is less affected by the presence of phytates, which is of considerable importance considering that fish sauce or fish paste are often consumed with rice meals (Hurrell et al. 2000). A comparison of the absorption of NaFeEDTA from fish sauce to that of ferrous sulfate in fish sauce indicated that there was no significant difference, although the latter iron source causes unacceptable precipitation in the product (Fidler et al. 2003). NaFeEDTA is similarly suitable as a fortificant for fermented fish paste, which in certain countries, such as the Philippines, is consumed by the majority of individuals in the lower socioeconomic sector of the population (Cook and Reusser 1983).

The other mineral that has been considered for fortification of fermented fish products is iodine, as iodine deficiency is prevalent in many Asian countries. The usual route of fortifying salt with iodine is not suitable in a number of these countries, as solid salt is not generally in the market in the local village shops, whereas fish sauce or fish paste are; these products are added during cooking of most types of foods (Lotfi et al. 1996). Potassium iodide was added to fish sauce in combination with various iron compounds in order to study the joint fortification of iodine and iron, and no significant effect on sensory quality was noted. In addition, there was no detected loss of iodine or iron during the three-month shelf-life testing. When commercially produced fermented fish and fish sauces were fortified with iodated rock and sea salt grains, at a level of approximately 30 mg/kg, it was found that there was a 16% loss of iodine in the fermented fish after six months of fermentation, 55% loss of iodine in fish sauce after 12 months fermentation exposed to sunlight, and 13% loss of iodine in fish sauce after 12 months fermentation in the shade. There was no significant difference in the sensory qualities of the fortified or unfortified products (Chanthilath et al. 2009).

Another condiment commonly used in many Asian countries is soy sauce, produced from the yeast fermentation of soybeans and wheat or soybeans alone. Soy sauces are available in different types and grades, making them accessible to consumers in all socioeconomic groups, and their strong flavor and daily use in many countries where iron deficiency is a problem makes them an ideal product for iron fortification. Surprisingly, as soy products in general inhibit iron absorption due to their high phytic acid concentration, soy sauce has been demonstrated to enhance iron absorption, possibly due to the substantial degradation of the phytic acid content during the fermentation process (Fidler et al. 2003). As with fish sauce and paste, the correct choice of iron source is important to avoid adversely affecting the sensory acceptability of the product, and again NaFeEDTA is a popular fortificant due to its lack of effect on the sensory characteristics of the sauce. However, it can be an expensive source of iron and thus unappealing to the policy makers and industry. Research was carried out on six alternative iron sources, and it was found that ferrous sulfate, ferric ammonium citrate, ferrous lactate, and ferrous gluconate did not significantly affect the sensory qualities of the soy sauce over a three-month period. In contrast, ferrous fumarate and ferrous bisglycinate caused unacceptable precipitation (Watanapaisantrakul, Chavasit, and Kongkachuichai 2006). However, the bioavailability of the alternative iron sources was not assessed, and this would be an important consideration if proceeding with the cheaper fortificant sources.

8.7 USE OF FERMENTED PRODUCTS TO FORTIFY OTHER FOODS

Although, as discussed previously, fermented foods are often fortified with macro- and micronutrients, the use of fermented foods as fortificants is far less common, which could be due to their relatively low pH and comparatively strong organoleptic properties. However, one fermented food product that has been used as a fortificant is the Japanese fermented soy product *natto*. Due to its specific fermentation process, *natto* naturally produces a particularly bioavailable form of the vitamin K_2 (menaquinone) known as MK-7. In recent years, this vitamin has been isolated from the *natto* product and has gained approval in the European Union as a fortificant for use in other foods (European Food Safety Authority 2008). However, prior to the isolation of MK-7, *natto* itself was being used as the fortificant source of vitamin K in other food products.

8.8 CONCLUDING REMARKS

This review of the developments in the fortification of fermented foods has shown that considerable success can be achieved with the addition of protein, fiber, or micronutrients to many indigenous staple foods. As previously discussed, the most effective foods for a mass fortification program are those that are regularly consumed by the target population. Experience has shown that this can either be a staple commodity or a commonly used condiment.

The success of a fortification program depends on the cultural acceptability of the food, and the most successful foods are those that blend into the diet of the target population and where the organoleptic properties of the fortified food are indistinguishable from the nonfortified form. It is essential that a detailed knowledge of the dietary habits, food consumption patterns, and levels of food intake is obtained before embarking on a fortification exercise. This information is needed for the assessment of fortification levels, particularly with regard to micronutrients. For example, if the dietary survey shows that a significant proportion of the food is consumed by infants and young children, the quantities of certain micronutrients such as iron and iodine in the fortified food should be adjusted to ensure that their intake is within safe levels for this subset of the population. While the collection of dietary intake data is difficult in some communities, it is an important part of any public health program that fortification should introduce a high level of benefit without also introducing unintended negative consequences.

When considering a fortification program as part of a national public health policy, the primary consideration must be cost. In developing countries, particularly, a fortified staple food must only have a minimal impact on the family food budget, and preferably the price should be at parity with its unfortified equivalent. Government subsidies may be necessary to achieve this.

Over the past two decades a number of programs have been destabilized by the rapidly increasing costs of the fortificants, particularly vitamins. Such increases must not impact the purchaser/consumer in low-income economies.

The issues surrounding food fortification are multifaceted, and all aspects must be taken into consideration in order to achieve success.

REFERENCES

Achanta, K., and K. J. Aryana. 2007. Fat free plain set yogurts fortified with various minerals. *LWT Food Science and Technology* 40 (3): 424–429.

Achi, O. K. 1999. Quality attributes of fermented yam flour supplemented with processed soy flour. *Plant Foods for Human Nutrition* 54 (2): 151–158.

Adelekan, A. O., and O. B. Oyewole. 2010. Production of Ogi from germinated sorghum supplemented with soybeans. *African Journal of Biotechnology* 9 (42): 7114–7121.

Afoakwa, E. O., K. J. Edem, A. S. Budu, and G. A. Annor. 2010. Souring and starch behaviour during co-fermentation of cassava and soybean into gari "farina." *Asian Journal of Food and Agro-Industry* 3 (3): 371–385.

Aminigo, E. R., and J. O. Akingbala. 2004. Nutritive composition and sensory properties of Ogi fortified with okra seed meal. *Journal of Applied Sciences & Environmental Management* 8 (2): 23–28.

Anal, A. K., and H. Singh. 2007. Recent advances in microencapsulation of probiotics for industrial applications and targeted delivery. *Trends in Food Science & Technology* 18 (5): 240–251.

Antunes, A. E. C., T. F. Cazetto, and H. M. A. Bolini. 2005. Viability of probiotic microorganisms during storage, postacidification and sensory analysis of fat-free yogurts with added whey protein concentrate. *International Journal of Dairy Technology* 58:169–173.

Aportela-Palacios, A., M. E. Sosa-Morales, and J. F. Velez-Ruiz. 2005. Rheological and physicochemical behavior of fortified yogurt with fiber and calcium. *Journal of Texture Studies* 36:333–349.

Baker, A. A. B. A., A. H. El Tinay, and A. E. A. Yagoub. 2010. Protein content and digestibility of sorghum (sorghum bicolor)-based fermented gruel: Effect of supplementation with soybean protein. *Electronic Journal of Environmental, Agricultural and Food Chemistry* 9 (9): 1495–1501.

Bari, M. R., R. Ashrafi, M. Alizade, and L. Rofehgarineghad. 2009. Effects of different contents of yogurt starter/probiotic bacteria, storage time and different concentration of cysteine on the microflora characteristics of bio-yogurt. *Research Journal of Biological Sciences* 4 (2): 137–142.

Barrantes, E., A. Y. Tamime, A. M. Sword, D. D. Muir, and M. Kaláb. 1996. The manufacture of set-type natural yoghurt containing different oils: 1. Compositional quality, microbiological evaluation, and sensory properties. *International Dairy Journal* 6 (8–9): 811–826.

Berry Ottaway, P. 2008. The stability of vitamins in fortified foods and supplements. In *Food fortification and supplementation*, ed. P. Berry Ottaway, 88–107. Cambridge, U.K.: Woodhead Publishing.

Binns, C.W., S. J. Carruthers, and P. A. Howat. 1989. Thiamin in beer: A health promotion perspective. *Community Health Studies* 13 (3): 301–305.

Bishai, D. M., and L. P. Bozzetti. 1986. Current progress toward the prevention of the Wernicke-Korsakoff syndrome. *Alcohol and Alcoholism* 21 (4): 315–323.

Cavallini, D. C. U., and E. A. Rossi. 2009. Soy yogurt fortified with iron and calcium: Stability during the storage. *Alimentos e Nutrição (Brazilian Journal of Food and Nutrition)* 20 (1): 7–13.

Chanthilath, B., V. Chavasit, S. Chareonkiatkul, and K. Judprasong. 2009. Iodine stability and sensory quality of fermented fish and fish sauce produced with the use of iodated salt. *Food and Nutrition Bulletin* 30 (2):183–188.

Clifton, P. M., M. Noakes, D. Sullivan, N. Erichsen, D. Ross, G. Annison, A. Fassoulakis, M. Cehun, and P. Nestel. 2004. Cholesterol-lowering effects of plant sterol esters differ in milk, yoghurt, bread, and cereal. *European Journal of Clinical Nutrition* 58:503–509.

Cook, J. D., and M. E. Reusser. 1983. Iron fortification: An update. *American Journal of Clinical Nutrition* 38:648–659.

CPHA. 2010. Food fortification with vitamins and minerals. Canadian Public Health Association (CPHA).http://cpha100.ca/12-great-achievements/food-fortification-vitamins-and-minerals.

Cueva, O., and K. J. Aryana. 2008. Quality attributes of a healthy heart yogurt. *LWT Food Science and Technology* 41 (3): 537–544.

Daengprok, W., W. Garnjanagoonchorn, and Y. Mine. 2002. Fermented pork sausage fortified with commercial or hen eggshell calcium lactate. *Meat Science* 62:199–204.

Dave, R. I., and N. P. Shah. 1997a. Effectiveness of ascorbic acid as an oxygen scavenger in improving viability of probiotic bacteria in yoghurts made with commercial starter cultures. *International Dairy Journal* 7:435–443.

———. 1997b. Effect of cysteine on the viability of yoghurt and probiotic bacteria in yoghurts made with commercial starter cultures. *International Dairy Journal* 7 (8–9): 537–545.

———. 1998. Ingredient supplementation on viability of probiotic bacteria in yoghurt. *Journal of Dairy Science* 81:2804–2816.

Donkor, O. N., A. Henriksson, T. Vasiljevic, and N. P. Shah. 2006. Effect of acidification on the activity of probiotics in yoghurt during cold storage. *International Dairy Journal* 16 (10): 1181–1189.

Donkor, O. N., S. L. I. Nilmini, P. Stolic, T. Vasiljevic, and N. P. Shah. 2007. Survival and activity of selected probiotic organisms in set-type yoghurt during cold storage. *International Dairy Journal* 17 (6): 657–665.

European Food Safety Authority (EFSA). 2008. Scientific opinion of the Panel on Dietetic Products Nutrition and Allergies on a request from the European Commission on the Safety of "Vitamin K₂." *EFSA Journal* 822:1–32.

———. 2009. Plant stanols and plant sterols and blood LDL-cholesterol. *EFSA Journal* 1175:1–9.

FAO/WHO. 2002. Guidelines for the evaluation of probiotics in food. Report of a joint FAO/ WHO working group on drafting guidelines for the evaluation of probiotics in food. London, Ontario, Canada.

Farvin, K. H. S., C. P. Barona, N. S. Nielsen, and C. Jacobsen. 2010. Antioxidant activity of yoghurt peptides: Part 1: In vitro assays and evaluation in ω-3 enriched milk. *Food Chemistry* 123 (4): 1081–1089.

Fasasi, O. S., I. A. Adeyemi, and O. A. Fagbenro. 2007. Functional and pasting characteristics of fermented maize and Nile tilapia (*Oreochromis niloticus*) flour diet. *Pakistan Journal of Nutrition* 6 (4): 304–309.

Fernández-Garcia, E., J. U. McGregor, and S. Taylor. 1998. The addition of oat fiber and natural alternative sweeteners in the manufacture of plain yoghurt. *Journal of Dairy Science* 81:655–663.

FFI. 2010. Flour Fortification Initiative. http://www.sph.emory.edu/wheatflour/index.php.

Fidler, M. C., L. Davidsson, T. Walczyk, and R. F. Hurrell. 2003. Iron absorption from fish sauce and soy sauce fortified with sodium iron EDTA. *American Journal of Clinical Nutrition* 78:274–278.

Garby, L., and S. Areekul. 1974. Iron supplementation in Thai fish sauce. *Annals of Tropical Medicine and Parasitology* 68 (4): 467–476.

Granado-Lorencio, F., C. Herrero-Barbudo, B. Olmedilla-Alonso, I. Blanco-Navarro, and B. Pérez-Sacristán. 2010. Lutein bioavailability from lutein ester-fortified fermented milk: In vivo and in vitro study. *Journal of Nutritional Biochemistry* 21 (2): 133–139.

Hanson, A. L., and L. E. Metzger. 2010. Evaluation of increased vitamin D fortification in high-temperature, short-time-processed 2% milk, UHT-processed 2% fat chocolate milk, and low-fat strawberry yogurt. *Journal of Dairy Science* 93 (2): 801–807.

Hashim, I. B., A. H. Khalil, and H. S. Afifi. 2009. Quality characteristics and consumer acceptance of yogurt fortified with date fiber. *Journal of Dairy Science* 92:5403–5407.

Hekmat, S., and D. J. McMahon. 1997. Manufacture and quality of iron-fortified yogurt. *Journal of Dairy Science* 80:3114–3122.

Herzallah, S. M. 2005. Effect of dried raisins and apricots extract on the growth of *Bifidobacteria* in cows and goats milk. *Pakistan Journal of Nutrition* 4 (3): 170–174.

Hozova, B., R. Batorova, and J. Cerna. 1991. Using *Propionibacteria* cultures for folacine fortification of sauerkraut. *Potravinarske Vedy UVTIZ* 9 (3): 211–218.

Hurrell, R. F., M. B. Reddy, J. Burri, and J. D. Cook. 2000. An evaluation of EDTA compounds for iron fortification of cereal-based foods. *British Journal of Nutrition* 84:903–910.

IFST. 1989. Nutritional enhancement of food. Institute of Food Science and Technology Technical Monograph No. 5, London.

Ilic, D. B., and S. H. Ashoor. 1988. Stability of vitamins A and C in fortified yogurt. *Journal of Dairy Science* 71:1492–1498.

ISAPP. 2009. Clarification of the definition of a probiotic. International Scientific Association for Probiotics and Prebiotics. http://www.isapp.net/.

Katan, M. B., S. M. Grundy, P. Jones, M. Law, T. Miettinen, R. Paoletti, and Stresa Workshop participants. 2003. Efficacy and safety of plant stanols and sterols in the management of blood cholesterol levels. *Mayo Clinic Proceedings* 78 (8): 965–978.

Khurana, H. K., and S. K. Kanawjia. 2007. Recent trends in development of fermented milks. *Current Nutrition and Food Science* 3:91–108.

Kohajdová, Z., and J. Karovičová. 2007. Fermentation of cereals for specific purpose. *Journal of Food and Nutrition Research* 46 (2): 51–57.

Krasaekoopt, W., B. Bhandari, and H. C. Deeth. 2006. Survival of probiotics encapsulated in chitosan-coated alginate beads in yoghurt from UHT- and conventionally treated milk during storage. *LWT Food Science and Technology* 39 (2): 177–183.

Lelana, I. Y. B., L. Purnomosari, and A. Husni. 2003. Fortification of plain cracker with fish flour. *Indonesian Food and Nutrition Progress* 10 (1): 26–28.

Lotfi, M., M. G. Venkatesh Mannar, R. J. H. M. Merx, and P. Naber-van den Heuvel. 1996. Micronutrient fortification of foods: Current practices, research, and opportunities. Ottawa: The Micronutrient Initiative, International Development Research Centre.

Maage, A., J. Toppe, M. Steiner-Asiedu, E. Asibey-Berko, and E. Lied. 2008. Inclusion of marine fish in traditional meals improved iodine status of children in an iodine deficient area. *African Journal of Food Science* 2:45–53.

Martín-Diana, A. B., C. Janer, C. Peláez, and T. Requena. 2004. Effect of milk fat replacement by polyunsaturated fatty acids on the microbiological, rheological, and sensorial properties of fermented milks. *Journal of the Science of Food and Agriculture* 84:1599–1605.

Martini, S., J. E. Thurgood, C. Brothersen, R. Ward, and D. J. McMahon. 2009. Fortification of reduced-fat cheddar cheese with n-3 fatty acids: Effect on off-flavor generation. *Journal of Dairy Science* 92:1876–1884.

Mbata, I. T., M. J. Ikenebomeh, and I. Ahonkhai. 2006. Improving the quality and nutritional status of maize fermented meal by fortification with bambara nut. *Internet Journal of Microbiology* 2 (2). http://www.ispub.com/journal/the_internet_journal_of_microbiology/volume_2_number_2_30/article/improving_the_quality_and_nutritional_status_of_maize_fermented_meal_by_fortification_with_bambara_nut.html.

Mbata, T. I., M. J. Ikenebomeh, and J. C. Alaneme. 2009. Studies on the microbiological, nutrient composition, and antinutritional contents of fermented maize flour fortified with bambara groundnut (*Vigna subterranean* L). *African Journal of Food Science* 3 (6): 165–171.

Mbata, T. I., M. J. Ikenebomeh, and S. Ezeibe. 2009. Evaluation of mineral content and functional properties of fermented maize (generic and specific) flour blended with bambara groundnut (*Vigna subterranean* L). *African Journal of Food Science* 3 (4): 107–112.

Ndangui, C. B., A. Kimbonguila, J. M. Nzikou, L. Matos, N. P. G. Pambou-Tobi, A. A. Abena, Th. Silou, J. Scher, and S. Desobry. 2010. Nutritive composition and properties physicochemical of gumbo (*Abelmoschus esculentus* L.) seed and oil. *Research Journal of Environmental and Earth Sciences* 2 (1): 49–54.

Osungbaro, T. O. 2009. Physical and nutritive properties of fermented cereal foods. *African Journal of Food Science* 3 (2): 23–27.

Papastoyiannidis, G., A. Polychroniadou, A.-M. Michaelidou, and E. Alichanidis. 2006. Fermented milks fortified with B-group vitamins: Vitamin stability and effect on resulting products. *Food Science and Technology International* 12 (5): 521–529.

Parveen, S., and F. Hafiz. 2003. Fermented cereal from indigenous raw materials. *Pakistan Journal of Nutrition* 2 (5): 289–291.

Pirkul, T., A. Temiz, and Y. K. Erdem. 1997. Fortification of yoghurt with calcium salts and its effect on starter microorganisms and yoghurt quality. *International Dairy Journal* 7 (8–9): 547–552.

Plahar, W. A., C. A. Nti, and N. T. Annan. 1997. Effect of soy fortification method on the fermentation characteristics and nutritional quality of fermented maize meal. *Plant Foods for Human Nutrition* 51:365–380.

Salem, A. S., W. A. Gafour, and E. A. Y. Eassawy. 2006. Probiotic milk beverage fortified with antioxidants as functional ingredients. *Egyptian Journal of Dairy Science* 34 (1): 23–32.

Sanz, T., A. Salvador, A. Jiménez, and S. Fiszman. 2008. Yogurt enhancement with functional asparagus fibre, effect of fibre extraction method on rheological properties, colour, and sensory acceptance. *European Food Research and Technology* 227:1515–1521.

Scala, J. 1985. *Making the vitamin connection: The food supplement story*. New York: Harper and Row.

Shakeel, M., M. Abdullah, M. Nasir, and N. Akhtar. 2009. Nutritional improvement and value-addition of buffalo milk kurut with soya protein isolate. *Pakistan Journal of Zoology* 9 (Suppl. Ser.): 379–384.

Staffolo, M. D., N. Bertola, M. Martino, and A. Bevilacqua. 2004. Influence of dietary fiber addition on sensory and rheological properties of yogurt. *International Dairy Journal* 14:263–268.

Tabatabaie, F., and A. Mortazavi. 2008. Influence of lactulose on the survival of probiotic strains in yoghurt. *World Applied Sciences Journal* 3 (1): 88–90.

UN General Assembly. 2002. Resolution adopted by the general assembly on the report of the ad hoc committee of the whole (A/S-27/19/Rev.1 and Corr.1 and 2) S-27/2. A world fit for children. A/RES/S-27/2.

UNICEF. 2009. *Tracking progress on child and maternal nutrition: A survival and development priority*. New York: United Nations Children's Fund (UNICEF), November.

Van der Westhuyzen, J., R. E. David, G. C. Icke, and J. Metz. 1985. Fortification of sorghum beer with thiamine. *International Journal for Vitamin and Nutrition Research* 55 (2): 173–179.

———. 1987. Thiamine deficiency in black male hostel-dwellers. *South African Medical Journal* 71 (4): 231–234.

Vasiljevic, T., T. Kealy, and V. Mishra. 2007. Effects of β-glucan addition to a probiotic containing yogurt. *Journal of Food Science* 72:C405–C411.

Walker, P. M. B. 1988. *Chambers science and technology dictionary*. Cambridge, U.K.: Chambers Ltd. and Cambridge University Press.

Watanapaisantrakul, R., V. Chavasit, and R. Kongkachuichai. 2006. Fortification of soy sauce using various iron sources: Sensory acceptability and shelf stability. *Food and Nutrition Bulletin* 27 (1): 19–25.

WHO. 2008. *Worldwide prevalence of anaemia 1993–2005: WHO Global Database on Anaemia*, ed. B. de Benoist, E. McLean, I. Egli, and M. Cogswell. Geneva, Switzerland: WHO Press, World Health Organization.

———. 2009. *Global prevalence of vitamin A deficiency in populations at risk 1995–2005. WHO Global Database on Vitamin A Deficiency*. Geneva, Switzerland: WHO Press, World Health Organization.

WHO/FAO Codex Alimentarius. 1987. Codex general principles for the addition of essential nutrients to foods. CAC/GL 09-1987, Codex Alimentarius Commission, Rome, Italy.

Yeganehzad, S., M. Mazaheri-Tehrani, and F. Shahidi. 2007. Studying microbial, physiochemical, and sensory properties of directly concentrated probiotic yoghurt. *African Journal of Agricultural Research* 2 (8): 366–369.

9 Fermented Cereal and Legume Products

Afaf Kamal-Eldin

CONTENTS

9.1 INTRODUCTION

Fermentation of all types of food raw materials has been a celebrated human practice since antiquity. Cereals—or the fruits of Gramineae, mainly wheat, rye, rice, maize, barley, oats, sorghum, and millet—have been fermented to various products in different parts of the world. Fermentation contributes several advantages to the food, including (a) addition of new tastes, flavors, aromas, and textures, (b) preservation, (c) enhancement of the nutritional value of food by increasing digestibility and production of vitamins, (d) elimination of toxic substances, and (e) decrease of cooking time and energy. Thus, fermentation alters food shape, texture, and flavor; increases its nutritional value; and promotes safety.

Cereal grains belonging to the family Poaceae that includes, *inter alia*, wheat (*Triticum sativum*), maize (*Zea mays*), rice (*Oryza sativa*), rye (*Secale cereale*), barley (*Hordeum vulgare*), oats (*Avena sativa*), sorghum (*Sorghum bicolor*), pearl millet (*Pennisetum glaucum*), finger millet (*Eleusine coracana*), and teff (*Eragrostis tef*). Cereals are the major staples in economically less-developed parts of the world, and the fermentation products of cereal grains and their fractions are major foods for people in these parts. Fermented products, based on different cereal grains, follow the geographic production of the grains. For example, fermented products from sorghum and millet dominate in Africa and Asia; fermented products from maize are dominant in Central and South America and Africa; fermented products from rice are common in India, China, Southeast Asia, and the Far East; and fermented products from wheat are important in the Middle East, Turkey, and the Far East (Steinkraus 1983; Hesseltine 1979; Campbell-Platt 1987; Blandino et al. 2003). The same fermentation product may have different names in different countries. The sorghum beer called *amgba* in Cameroon (Chevassus, Favier, and Orstom 1976) is *bili-bili* in Chad (Nanadoum et al. 2006), *burukutu* in Nigeria (Faparusisi, Olofinboba, and Ekundayo 1973), *dolo* in Burkina Faso (Sawadogo et al. 2007), *pito* in Ghana (Sefa-Dedeh et al. 1999), and *tchapalo* in the Ivory Coast (Marcellin et al. 2009).

The legume family, Leguminaceae (=Fabaceae) include all types of beans and peas and is second to the grass family in importance to human nutrition. While cereal grains are important sources of carbohydrates and energy, legumes are important sources of proteins. The most studied fermented products are those made of soybean (*Glycine max*), but fermentation extends to an uncountable number of legumes (Reddy et al. 1982). Fermented soy products include, but are not limited to, soybean sauce, tempeh, tofu, miso, *nattō, cheonggukjang, chunjang, doenjang, doubanjiang, gochujang,* tamari, *tauchu,* and yellow soybean paste (Liu 2004). Other fermented legume products include *amriti, dhokla, dosa, idli, papad,* and *wadi.* As for cereals, legumes are fermented as such or after malting (Zamora and Fields 2006). Legume fermentation is mainly practiced in the Far East and Southeast Asia (Japan, Korea, China, Taiwan, Thailand, Philippines, Indonesia, Burma, Malaysia, and Singapore) and to a lesser extent in Africa and South America.

In most parts of the world, fermentation is performed traditionally with different grades of grains and with mixed starter cultures, leading to variability in product quality (appearance, taste, flavor, nutritional value, and safety), while countries using advanced biotechnology use defined cultivars and fermentation culture to produce commercial products of predictable quality.

9.2 TYPES OF FERMENTED CEREAL PRODUCTS AND THEIR MANUFACTURE

Cereal fermentation can be classified into three types: lactic acid fermentation, alcoholic fermentation by baker's (*Saccharomyces cerevisiae*) and other yeasts, and acetic acid fermentation by *Acetobacter,* which converts alcohol to acetic acid in the presence of excess oxygen. The different types of fermented cereal products can broadly be classified into leavened breads, gruels and porridges, and beverages (alcoholic and nonalcoholic). The different types of bread are leavened by the baker's yeast, which ferments the sugar in the dough to produce ethanol (which evaporates during baking) and carbon dioxide as the leavening agent. These include, e.g., (a) common loafs made of wheat or mixtures of wheat with other grains worldwide, (b) tortillas made of maize and wheat in Central American countries, (c) pizza, focaccia, *ciabatta,* and pita in Italy and other southern European countries, (d) crisp breads made of rye, barley, and wheat in Scandinavian countries, and (e) *baladi* and other flat breads in North Africa and in Middle East countries. Important quality characteristics of these breads include high volume, softness, and elasticity; good shelf life; and microbiological safety (Cauvain 2003).

Breads based on sourdough, commonly rye bread, contain yeasts (usually *Candida* and *Saccharomyces*) in symbiotic combination with the *Lactobacillus* responsible for the sour taste of lactic acid. Sourdough bread is prepared mixing some 20% to 25% of previous dough, as a starting culture, with new flour and water. Depending on the fermentation time, some bacteria, including *Lactobacillus*, coexist with the yeast and contribute to bread taste and flavor. Three main types of sourdough can be recognized:

Type I sourdoughs (pH range 3.8–4.5 and fermentation temperature range 20°C–30°C) in which *Saccharomyces exiguus* leavens the dough and *Lactobacillus* species (mainly *L. sanfranciscensis*, and *L. pontis*) produce sourness

Type II sourdoughs (pH <3.5 and fermentation temperature 30°C–50°C for several days), where *Saccharomyces cerevisiae* leavens the dough and *L. pontis* and *L. panis* dominate the lactic acid–producing bacteria

Type III sourdoughs that are usually added to bread to immediately contribute flavor and taste (Vogel and Ehrmann 2008)

Other bacteria, e.g., *L. fermentum*, *L. fructivorans*, *L. brevis*, and *L. paralimentarius*, are also present in these doughs (Arendt, Ryan, and Dal Bello 2007). Sometimes other gas-producing bacteria, usually heterofermenting lactic acid bacteria (LAB), are also part of the microflora in sourdough. Besides these "plain" sourdough breads, there are other breads including other ingredients, e.g., (a) the quick-fermenting Amish friendship bread that has sugar and milk included in its sourdough starter and has baking powder and baking soda added to the dough and (b) the German pumpernickel where commercial yeasts spiked with lactic acid and/ or citric acid are used.

Special types of leavened breads are made from traditional cereals in Africa, including the Sudanese *kisra* and the Ethiopian *injera*. Sudanese *kisra* is generally made from sorghum using its natural microflora, mainly bacteria including *Lactobacillus coprophilus*, *L. cellobiosus*, and *L. brevis*, as well as *Pediococcus pentosaceus*, *Klebsiella pneumoniae*, *Enterobacter sakazakii*, *Enterobacter agglomerans*, and *Serratia ficaria*) and molds including *Aspergillus niger*, *Fusarium* spp., *Alternaria* spp., and *Penicillium* spp. (Abdel Rahman et al. 2010). Fermentation is performed using part of an old dough, called *kham'mar* added to flour to make a thick dough, or *ajin*, that is left for 12–24 hours to ferment, with pH changing from 5.5 to about 3.5. The maltose content of the dough decreases continuously, whereas glucose is accumulated as an intermediate. The *ajin* is diluted to a thin batter before baking on a hot plate rubbed with an oily cloth; the flat *kisra* bread bakes in about one minute and comes out as a thin sheet.

The fluffier, pancakelike *injera*, which is the staple food in Ethiopia, is commonly made from the indigenous small grain teff (*Eragrostis tef*), but it can also be made from sorghum and can include variable amounts of wheat, barley, or rice flour. As with *kisra*, *injera* is made by fermenting teff flour spontaneously or with an old sourdough starter. Half of the fermented slurry is cooked to gelatinize the starch, and the gruel is mixed with the uncooked half before a second fermentation for 3–5 hours (Parker, Umeta, and Faulks 1989). Fermentation of *injera* causes hydrolysis of starch to free sugars (mainly fructose), whose relative proportions change following changes in the pH and microbial population dynamics (Umeta and Faulks 1988). *Injera* is also eaten in Eritrea, and similar products known as *canjeero* or *laxoox* are eaten in Somalia and *lahoh* in Yemen (Belton and Taylor 2004).

A wide range of fermented gruels and porridges is available in Africa, including acidified thin *ambali*, *edi*, *ogi*, and *ugi* and thick *asida*, *bogobe*, *ugali*, *sankati*, and *tuwo* fermented products of sorghum and maize. *Ogi* is a fine, pastelike sour maize gruel eaten in Nigeria as a breakfast cereal by adults and as a traditional weaning food for infants. Similar foods in other African countries are *koko* or *kenkey* in Ghana and *mahewu* in South Africa. Their fermentation is dominated by lactic acid bacteria, particularly *Lb. planfreundii*, in addition to *Corynebacterium*, which hydrolyzes the corn starch, and the yeasts *Saccharomyces* and *Candida*, which contribute to flavor development.

In addition, fermented alcoholic and nonalcoholic beverages are widely made, e.g., *abrey*, *borde*, *burkutu*, *busa*, *ting*, and *obushera*. For example, *borde* is a beverage produced in Ethiopia by the fermentation of a mixture of malted and unmalted flours from single or mixed cereals including finger millet, wheat, maize, and barley. The process of *borde* production involves grinding, lactic acid fermentation, roasting, steam cooking, boiling, cooling, mashing, wet-milling, and wet-sieving (Abegaz et al. 2004). Malt is generally used in the fermentation of African and European beers, where it constitutes 25%–100% of the substrate. The brewing of *merissa*, the unclear sorghum beer in Sudan, is an exceptionally complicated and advanced process involving about 5% sorghum malt as an enzyme preparation rather than a substrate and using a caramelized sorghum product called *surij*. Parts of the substrate are half cooked, while other parts are fully cooked and overcooked to meet enzyme requirements for a mixture of raw and gelatinized starch. A clear sorghum or millet beer called *assaliya* (or *um-bilbil*) is also prepared in the Sudan by a process involving about 40 steps, which is much more complicated than the procedures used in the preparation of *otika* or *amgba* involving less than 20 steps (Dirar 1992).

Hilu-mur is an advanced fermented sorghum drink that is consumed daily in Sudan during the fasting season of Ramadan. Sorghum, particularly the red feterita cultivar, is malted and milled into flour that is finally added to unmalted sorghum previously cooked into porridge in a ratio of 1:1 or 1:2 (Bureng 1979; Agab 1985). Once the malted flour is added, amylolysis will take place, causing liquefaction of the porridge. The hot porridge is transferred from the cooking pot to a special jar in which fermentation will take place in a warm corner of the house for 2–3 days with natural flora of the sorghum, enzymes of the malt, and a starter culture of *Lactobacillus*. The dominant microorganisms in *hilu-mur* fermentation were found to be *Lactobacillus*, *Acetobacter*, *Leuconostoc*, *Gluconobacter*, and the yeasts *Saccharomyces* and *Candida* (Marhoum 1987). Other ingredients are added to the sour, sweetish, and brownish *hilu-mur* mixture one night before baking include tamarind (*Tamarindus indicum*), karkade (*Hibiscus sabdariffa*), cinnamon

(*Cinnamomum zeylanicum*), ginger (*Zingiber officinale*), *ghurunjal* (*Alpinia offici-narum*), and dates slurry (*Phoenix dactylifera*) (Dirar 1993). The *hilu-mur* batter, which is reddish and fluffy with a strong bouquet of malt and spices, is thinned and baked on a very hot plate in a process similar to that of *kisra* baking, but the *hilu-mur* bread sheet is much thinner. The *hilu-mur* sheets are folded and spread on an airy surface for complete dryness before storage. For making the sweet-sour taste and pleasantly flavored *hilu-mur* drink, the dry sheets are soaked in water and the drink is filtered, cooled, and sweetened with sugar.

9.3 TYPES OF FERMENTED LEGUME PRODUCTS AND THEIR MANUFACTURE

Several leguminous species are fermented around the world, but this practice is much limited compared to other practices such as boiling, toasting, and germination. An exception to this generality is soybean (Hesseltine 1989; Deshpande et al. 2000). There is a wide range of fermented soybean products, mainly composed only of soybean but also of mixtures of soybeans and other ingredients including cereals, other legumes, chili, and other spices, etc. Soy sauce is perhaps the most internationally known product of soybean fermentation. In the preparation of soy sauce, soybeans alone or mixed with cereal grains are fermented with *Aspergillus oryzae* and/or *A. sojae* molds to give a *moromi*, which is pressed to form a liquid (the soy sauce) and a cake usually used as animal feed. The fermentation inoculum is generally known as *koji*. Certain brands of soy sauce are made from acid-hydrolyzed soy protein and are characterized by longer shelf life and different flavor, aroma, and texture compared to brewed soy sauces. Wheat, unroasted or roasted, is commonly used in combination with soybeans in the preparation of Japanese soy sauces and may constitute up to 50% of the ingredients and contributes sweetness to the sauce.

Tempeh, or *tempe kedele*, is a traditional soy product that originated from Indonesia and spread to the neighboring Asian countries and the rest of the world (Ko Swan and Hesseltine 1979; Steinkraus et al. 1983; Hachmeister and Fung 1993; Nout and Kiers 2005). It is a soybean–grain product that looks like a vegetarian burger patty, with a firm rubbery texture and a pronounced earthy flavor that intensifies on aging. Simple tempeh, or *tempeh kedelai*, is made by fermentation of dehulled and partially cooked soybeans with a starter culture containing spores of the fungus *Rhizopus oligosporus* for 24–36 hours at about 30°C–32°C. A mild acid, usually vinegar, is usually added to neutralize generated ammonia, lower the pH, and maintain an optimal environment that favors the growth of the tempeh mold over competing microorganisms. During the process, the fermented beans are knitted by a mat of white or grayish mycelia. Different types of tempeh are made using alterations in the preparation process and/or other types of beans or mixtures of beans and whole-grain cereals. For example, *tempe mendoan* is thinly sliced tempeh, battered and quickly deep fried to form crispy fully deep-fried tempeh or a half-cooked soft tempeh having a limp texture. *Tempeh bacem* is a dark-colored, spicy, and sweet tempeh made by boiling tempeh with spices and palm sugar and frying for a few minutes to enhance taste. *Tempe kering* is made by deep frying of raw tempeh cut into little sticks and mixing in fried peanuts, anchovies, spices, and sugar. *Tempe hembus* is made from soybean curd residue, and *tempe bosok*, or *tempe busuk*, is a tempeh with a rotten flavor that is used in small amounts as a flavoring. *Tempe gódhóng* is a tempeh wrapped in banana leaves.

Besides soybean, tempeh can be made from other legumes, e.g., *tempeh bengook* (made from *Mucana pruriens*), *tempe gembus* (made from okra), *tempeh koro* (made from *Phaseolus lunatis*), *tempeh lamtoro* (made from *Leucaena leucocephala*), and *tempeh oncom* (an orange-colored tempeh made from peanut cake). *Tempeh bongkrèk*, which is made from coconut press cake, often gets contaminated with the undesirable bacterium *Burkholderia gladioli*, which produces two toxins (bongkrek acid and toxoflavin). Figure 9.1 shows the structures of these two toxins. Bongkrek acid is highly toxic because it shuts the ATP/ADP translocation, preventing ATP from leaving the mitochondria and provide energy to the rest of the body. Because of this toxicity, *tempeh bongkrèk* is banned in Indonesia.

Tofu, or soybean curd, originated in China and spread to the rest of its neighboring countries. Tofu is made by pressing of the curds produced by the coagulating soy milk proteins and oil using one or a mixture of salt or acid coagulants. Calcium sulfate coagulants are traditionally used to produce Chinese-style tofu, which is tender and calcium rich but slightly brittle in texture. Magnesium and calcium chlorides are also used as coagulants to produce tofu with smooth and tender textures. Acidic coagulants, such as acetic acid (vinegar) and citric acid (lemon juice), are used to coagulate soy milk and produce tofu with different tastes. The acid coagulant glucono-d-lactone produces fine-textured tofu that is almost jellylike. Coagulants are dissolved in water and stirred into boiled soybean milk until the mixture curdles into a soft gel of fresh tofu having different textures and other microscopic features, e.g., soft "silken" tofu, firm tofu, and Chinese dry tofu. Fresh tofu is always kept in water and eaten in soft form. Fresh tofu can further be processed, for example to produce fermented/pickled, flavored, dried, and fried tofu versions.

Traditional miso (*memo miso*) is a salty, thick-paste Japanese seasoning produced by fermentation of soybeans, barley, or rice with the fungus *kōjikin*. Miso is used in the preparation of sauces and spreads, pickling vegetables or meats, and mixing with dashi soup stock to produce the *misoshiru* soup. White (*shiromiso*) and red (*akamiso*) misos are prepared from boiled and steamed soybeans, respectively, while mixed miso *(awasemiso)* is prepared from soybeans and other ingredients. Red miso is fermented for relatively longer times, during which coloration is brought

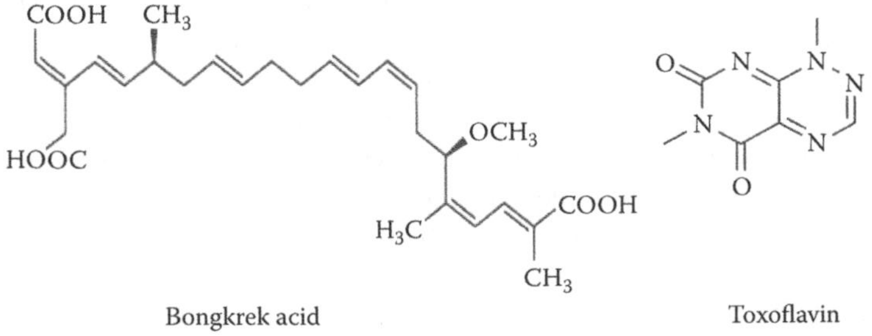

Bongkrek acid   Toxoflavin

FIGURE 9.1   Structures of bongkrek acid and toxoflavin

about by Maillard reactions. Besides soybeans, different misos are made from other ingredients such as *kome* from rice, *genmai* from brown rice, *mugi* from barley, *tsubu* from whole wheat and barley, *sobamugi* from buckwheat, *hadakamugi* from rye, and *taima* from hemp seed. The quality and flavor of the wide range of misos is dependent on processing variables such as the grain type, color, and quality, other ingredients and additives, time of fermentation (ranging from five days to several years), fermentation temperature, and salt concentration. Recently, new miso types are being produced in other countries using alternative ingredients, e.g., amaranth, adzuki beans, chickpeas, corn, and quinoa.

In Japan, soybeans are fermented with *Bacillus subtilis* (known as *nattō-kin*) to produce *nattō*, a fermented food with strong smell and flavor and slippery texture that is usually consumed as a breakfast food. Smaller beans are preferred so that the fermentation can reach their center more easily. The beans are washed, soaked in water, and steamed before mixing with the bacterium and fermented at 40°C for up to 24 hours. The *nattō* is then aged at a temperature of about 0°C to allow the bacteria to develop spores and enzymatic peptidases to hydrolyze the soybean protein. *Nattō* is believed to have several health benefits including cardioprotective anticlotting activity related to the serine protease enzyme nattokinase, which can reduce blood clotting by direct fibrinolysis of clots and/or inhibition of the plasma protein plasminogen activator inhibitor 1 (Fujita et al. 1993). In addition, *nattō* is rich in pyrazines, vitamin K_2, and phytoestrogenic soybean isoflavones with protective effects against heart disease, cancer, osteoporosis, Alzheimer's disease, and aging.

Ontjom is a fermented tempeh-like product made in Indonesia from peanut press cake, alone or in combination with coconut press cake, cassava press cake, and residues from making soybean milk or tofu. These other ingredients are added, wholly or in part, to reduce cost and/or to improve the texture. Fermentation is performed by molds, *Neurospora* and/or *Rhizopus*. The resultant fermented cake will be gray when the predominant mold is *Rhizopus* because of the abundance of blackish sporangia, and it will be pink if *Neurospora intermedia* is abundant due to the production of pink conidia. In India, for example, several fermented products are based on black gram (*Phaseolus mungo* L.) and bengal gram (= black chickpeas, *Cicer arietinum*) including, e.g., *bhallae*, *papadam*, *vadai*, and *warri* (black gram products), *idli* and *dosa* (black gram and rice products), and *dhokla* and *khaman* (bengal gram products). Other legumes such as yellow chick peas, velvet beans, winged beans, melon seeds, and broad beans are occasionally subjected to fermentation.

9.4 MICROORGANISMS USED IN FERMENTATION

The microorganisms involved in cereal fermentation include bacteria, molds, and yeasts. Lactic acid bacteria (mainly belonging to *Lactobacillus*, *Lactococcus*, *Leuconostoc*, and *Pediococcus*) are responsible for the formation of acids and flavors in these products. The type of bacterial flora in the different fermented foods depends on several factors including water activity, pH, salt concentration, temperature, and composition of the food matrix. In LAB fermentations, endogenous grain amylases generate fermentable sugars that will provide fermentable energy for the bacteria (Blandino et al. 2003). Lactic acid bacteria are divided into two groups: homofermentative (*Pediococcus*, *Streptococcus*, *Lactococcus*, and some *Lactobacilli*) and heterofermentative (*Weissella* and *Leuconostoc* and some *Lactobacilli*). While homofermentative microorganisms produce lactic acid as the major or sole end product of glucose fermentation, heterofermenters produce mixtures of lactate, CO_2, and ethanol. The production of these metabolites has significant effects on the other types of microflora that can grow in the fermentation medium.

Yeasts can grow on glucose or maltose, which they metabolize to carbon dioxide and alcohol, respectively. Alcoholic fermentation of malted grain maltose to produce beer is dominated by the brewer's yeast *Saccharomyces*. Positive symbiotic interactions between yeasts and lactic acid bacteria are commonly observed in fermented foods. Yeasts but also *Torulopsis* and homofermentative and heterofermentative lactic acid bacteria, particularly *Lactobacillus* spp., coexist in sourdough breads from wheat, rye, and other cereal grains. Yeast and LAB catalyze two types of biochemical changes: Lactobacilli cause acidification of the dough with lactic and acetic acids, and yeast causes leavening of the dough with carbon dioxide. Yeasts, especially *Saccharomyces cerevisiae* and *Candida* spp., contribute to the pleasant flavor of fermented cereal products. In addition, heterofermentative strains such as *Lb. sanfrancisco* and *Lb. brevis* are responsible for the characteristic sensory qualities of such breads.

The microorganisms involved in the fermentation of *tempe kedele* are mainly fungi belonging to the genus *Rhizopus*, typically *R. oligosporus*, that produce the mycelia-knitted compact cake final products. Other microorganisms that are less predominant in tempehs include *Mucor* spp. and *Aspergillus* spp. *Koji* is a fungus of the genus *Aspergillus*, which is used in the fermentation of some soybean products (Table 9.1). The main species involved in these fermentations include *Aspergillus oryzae* and *Aspergillus sojae* strains having high proteolytic activities. In addition to these molds, tempeh and *koji* contain *Lactobacillus* spp. that produce lactic acid and increase the acidity in the products as well as *Bacillus* spp. causing the generation of odors and ammonia.

Apart from a few products, fermentations are practiced at home using variable raw materials, which causes some instability in the microorganisms involved, but some known component cultures are given in Table 9.1. For the few commercial products, i.e., certain soybean and bread products, starter cultures have been established for the production of consistent and safe products. Examples of strains of starter culture for soybean fermentations and their NRRL (Northern Regional Research Laboratory, USDA, Peoria, IL) numbers are given in Table 9.2.

9.5   CHEMICAL CHANGES IN FERMENTED FOODS

9.5.1   CHEMICAL CHANGES INDUCED BY PREFERMENTATION PROCESSES

Several prefermentation treatments, being dependent on the type of raw material and the end product, exist for cereal and legume products. These include drying, washing, steeping, milling, fractionation, sieving, germination, and cooking. Malted or sprouted cereals and pulses are commonly used as ingredients in fermented foods. For example, when cereal flour is cooked into porridge (stiff or liquid paste) at

TABLE 9.1
Examples of Fermented Cereal and Legume Products Produced Worldwide*

Cereal/Legume	Microorganism(s)	Product name	Countries
	Fermented Cereal Products		
Maize	*Lactobacillus spp., Aerobacter; Corynebacterium*, Yeast, moulds	Agidi (solid porridge)	Nigeria, Benin
Maize	*Lactobacillus spp.*, and yeast (*Candida and Saccharomyces spp.*)	Kenkey	Ghana
Maize	*Lactobacillus fermentum, Candida krusei, Saccharomyces cerevisae*	Mawe (Liquid porridge)	South Africa, Benin
Maize	*Lactobacillus spp.*, moulds	Banku dough	Ghana
Maize	*Enterobacter cloacae, Acinetobacter spp., Lactobacillus spp., Saccharo myces cerevisiae, Candida mycoderma*	Koko (similar to Bogobe in Botswana)	Ghana
Maize, sorghum or Millet	*Lactobacillus spp.*, yeasts	Bogobe (similar to Koko in Ghana)	Botswana
Maize, sorghum or Millet	*Lactobacillus Spp.*, yeasts	Kenkey (Solid porridge)	Ghana, Botswana
Maize, sorghum or Millet	*Lactobacillus delbrukii; L.bulgarius; Streptococcus lactis*	Mahewu (Liquid porridge)	East African Countries
Maize, sorghum, or millet	*Lactobacillus spp., Aerobacter, Corynebacterium, Cephalosporium, Fusarium, Aspergillus, Penicillium spp., Saccharomyces cerevisiae, Candida mycoderma, C. valida, C. krusei, or C. vini,*	Ogi porridge (Similar to Kenyan Uji)	Nigeria Benin, Ghana, West Africa
Pearl millet (*Pennisetum glaucum*)	*Lactobacillus spp.* especially *L. plantarum*	Ben Saalga	Burkina Fasco
Pearl millet	Lactic acid bacteria (LAB), yeast and moulds and coliforms	Dèguè	Burkina Fasco
Ragi (*Eleusine coracana*)	*Leuconostoc mesenteroides, Lactobaci/lusjermentum,* and *Saccharomyces spp.*	Ambali	India

(Continued)

TABLE 9.1 (CONTINUED)
Examples of Fermented Cereal and Legume Products Produced Worldwide*

Cereal/Legume	Microorganism(s)	Product name	Countries
Sorghum	Lactic acid bacteria (*Lactobacillus spp.*, *Enterococcus spp.*, *Pediococcus spp.*, and *Leuconostoc spp.*) and yeasts	Amgba (same as assaliya in Sudan, bili-bili in Tchad, burukutu in Nigeria, dolo in Burkina Faso, pito in Ghana, and tchapalo in the Ivory Coast)	Cameroon
Sorghum		Bili Bili (same as, amgba in Cameroon, assaliya in Sudan, burukutu in Nigeria, dolo in Burkina Faso, pito in Ghana, and tchapalo in the Ivory Coast)	Tchad
Sorghum	*Lactobacillus fermentum, Lactobacillus buchneri, Lactobacillus* spp., *Saccharomyces cerevisae, Candida inconspicua, Candida magnolia, Candida humilis, Issatchenkia orientalis.*	Ikigage (known as *Tchoukoutou* in Benin or Togo, *Dolo* in Burkina-Faso, *Pito* in Ghana, *Burukutu or Otika* in Nigeria, *Bili bili* in Tchad and, *Mtama* in Tanzania)	Rwanda
Sorghum or millet	*Lactobacillus bulgaricus, L. brevis, L. cellobiosus, L. coprophilus, Pediococcus pentosaceus, Klebsiella pneumoniae, Enterobacter sakazakii, Enterobacter agglonerans, Serratia ficaria)* and moulds including *Aspergillus niger, Fusarium spp., Alternaria spp.* and *Pencillium spp.*	Kisra	Sudan

Substrate	Product	Microorganisms	Country
Sorghum or millet	Burukutu (same as amgba in Camerron, assaliya in Sudan, bili-bili in Tchad, dolo in Burkina Faso, pito in Ghana, and tchapalo in the Ivory Coast)	*Saccharomyces cerevisiae, S. chavelieria* and *Leuconostoc mesteroides*	Nigeria
Sorghum or millet	Bushera	*Weissella confusa, Lactobacillus. spp.. Streptoccocus spp..* and *Enterococcus spp.*	Uganda
Sorghum or millet	Dolo (same as amgba in Camerron, assaliya in Sudan, bili-bili in Tchad, burukutu in Nigeria, pito in Ghana, and tchapalo in the Ivory Coast)		Burkina Faso, Ivory coast, Mali, Niger, Togo
Sorghum or millet	Assaliya or um-bilbil (same as amgba in Cameroon, bili-bili in Tchad, burukutu in Nigeria, dolo in Burkina Faso, pito in Ghana, and tchapalo in the Ivory Coast)	Unidentified lactic acid bacteria and yeasts	Sudan
Teff (*Eragrostis tef*)	Injera	Yeasts, *Lactobacillus* spp.	Eyhiopia
Millet, sorghum, guinea-corn and maize	Kunun-zaki	*Leuconostoc* spp. *Lactobacillus* spp.. *Saccharaomyces cerevisiae, S.* spp.	Nigeria
Finger millet, wheat, barley, maize	Borde	*Weissella confusa, Lactobacillus brevis, L. viridescens, Pediococcus pentosaceus* and yeasts	Ethiopia
Millet, cooked maize, wheat, or rice semolina/flour	Boza	*Lactobacillus*	Albania, Bulgaria, Romania, Turkey, the Bulkans, Egypt

(Continued)

TABLE 9.1 (CONTINUED)
Examples of Fermented Cereal and Legume Products Produced Worldwide*

Cereal/Legume	Microorganism(s)	Product name	Countries
Maize and/or sorghum	Grain endogenous enzymes and moulds (*Penicillium* spp., *Rhizopus oryzae* and *Aspergillus* spp., *Candida* spp., *Geotrichum candidum* and *Lactobacillus* spp.	Pito (same as amgba in Camerron, assaliya in Sudan, bili-bili in Tchad, burukutu in Nigeria, dolo in Burkina Faso, and tchapalo in the Ivory Coast)	Ghana, Nigeria
Wheat and malt flours	Lactic acid bacteria (*Lactobacillus* spp., *cellobiosus*, *Leuconostoc lactis*, and *Pediococcus pentosaceus*). Coliforms (*Klebsiella pneumoniae*, *Enterobacter* spp.). Yeasts (*Saccharomyces cerevisiae*, *Candida* spp., *Kloeckera japonica*, and *Rhodotorula rubra*) and moulds (*Penicillium* spp.)	Sobia	Saudia arabia
Ragi (*Eleusine coracana* (L.) Gaern.)	*Leuconostoc mesenteroides*, *Lactobacillus jermentum*, *Saccharomyces* spp.	Ambali	India
Fermented Legume Products			
Soybeans Black gram (*Phaseolus mungo*) kidney bean (*Phaseolus vulgaris*) Broad bean (*Vicia faba*) Chick pea (*Cicer arietinum*) Cow pea (*Vigna unguiculata*)	*Rhizopus oligosporun*, *Rhizopus* spp.	Tempeh or tempe	Indonesia
Soybeans	*Aspergillus oryzae* or *A. sojae*, *Lactobacillus*, *Zygosaccharomyces rouxii*	Soybean sauce	Orient
Soybeans	*B. subtilis* (*B. natto*)	Natto	Japan, Korea

Soybeans	Mucor, Penicillium, Rhizopus, A. oryzae, B. subtilis, B. pumilies, Rhadotorula flava, Torulopsis dettilo	Meju (like koj)	Korea
Soybeans	Bacteria	Thua kab, Thua.nao	Thailand
Soybeans	Bacteria	Kenima	Nepal
Soybean curd	Mucor, Actinomucor	Tou-fu-ru, Fu-ru, Tsue-fan, Sufu, Chinese cheese	China
Soybean paste	Bacteria	Jeonkukjang, Jeonahjang	Korea
Soybean residue from Tofu making	Actinomucor elegans	Meitauza	China
Soybeans	Bacillus subtilis, B. pumilis, B. licheniformis, Leuconostoc spp., Staphylococcus saprophyticus	Daddawa	Nigeria
Black soybean	Aspergillus oryzae	Keeap	Indonesia
Black soybean	Aspergillus, Rhizopus	In-ya (Soy souce)	Taiwan
Soybean/Wheat	Aspergillus oryzae	Chiang-chu	China
Soybean/Wheat	Aspergillus oryzae	Hamanatto	Japan
Soybeans/Wheat	Aspergillus	Toyo	Philippines
Soybean/Wheat/Rice	A.oryzae, A. sojae	Koji	Japan
Soybean/Barley	A. oryzae, yeast	Mugi miso	Japan
Soybean/Rice	A. oryzae, yeast	Miso	Japan
Soybean/Rice	Bacteria	Moromi	Japan
Soybeans/Wheat	A. oryzae, A. sojae, Lactobacillus delbrueckii, Pediococcus hallophyllus, Torulopsis etchclisti, Succharomyces rouxii	Soya sauce	Orient
Black gram (Phaseolus mungo)	Candida, krusei, Saccharmyces cerevisine	Waries	India
Black gram/Rice flour	Lactic bacteria (Ln. mesenteroides, E. faecalis), Torulopsis, Candida, Trichosporon pullulans	Idli	South India
Faba beans (Vicia faba)	Aspergillus	Tou-pan-chiang	China

(Continued)

TABLE 9.1 (CONTINUED)
Examples of Fermented Cereal and Legume Products Produced Worldwide*

Cereal/Legume	Microorganism(s)	Product name	Countries
Melon seed (*Citrullus vulgaris*)	*Bacillus subtilis, B. megateruim, B. firmus E.coli,. Proteus, Pediococcus, Alcaligenes* spp., *Pseudomonas* spp., *Micrococcuss* spp.. *Streptococcus*	Ogiri	Nigeria
Castor oil seed (*Ricinus communis*) Fluted pumpkin seeds (*Telferia ocidentalis*)			
Mung beans/ Wheat/Maize	Bacteria	Phool waries	India
Lima beans (*Phaseolus lunaris*)	*Rhizopus*	Tempeh koro	Indonesia
Locust bean (*Parkia biglobosa*)	*Bacillus* spp., *Bacillus subtilis, B. licheniformis, B. pumilis, Staphylococcus* spp. *Bacillus* spp., *Staphylococcus* spp.	Dawadawa or Iru	Nigeria
Bambara groundnut (*Vigna subterranean*) Soybeans			
Peanuts	*Neurospora intermedia, Rhizopus oligosporus*	Oncom or Onjom	Indonesia
Velvet bean (*Mucuna pruriens*)	*Rhizopus*	Tempeh benguk	Indonesia

*Data collected from numerous literature sources

TABLE 9.2
Examples of Strains Used in Starter Cultures of Soybean Fermentation*

Fermented Food	Microorganism used	NRRL numbers(s)
Miso (soy paste)	Aspergillus oryzae	5593
	A. sojae	3485, 3486
	Saccharomyces rouxii	Y-11785
	Candida etchellsii	Y-7583
	Pediococcus halophilus	B-4506, B-4243, B-4244
Nattō	Bacillus subtilis var. natto	B-4008, 8-3383, 8-3010
Ontjom	Neurospora intermedia	5506, 6025
Shoyu (soy sauce)	Aspergillus sojae	6271
	A. oryzae	1988, 6270, 24551: 1989, 24556
	Saccharomyces rouxii	Y-6681
	Candida etchellsii	Y-7583
	C. versatilis	Y-7584
	Pediococcus halophilus	B-4243, B-4244
	Lactobacillus delbrueckii	8-445
Sufu	Actinomucor elegans	3104, 2242
	Mucor disperses	3103
Tempeh	Rhizopus oligosporus	2710

*Modified from Hasseltine (1983)

80°C–95°C prior to fermentation, starch gelatinization results in a bulky product with high viscosity and improves the fermentation with malt enzymes. Malted flour is added to contribute amylase enzyme activities and to reduce the viscosity of the cooked flour (Hansen et al. 1989; Gopaldas, Suneeta, and Cinnamma 1988). Many desirable changes occur during the malting of cereal grains due to the breakdown of complex compounds into simple ones (Subramanian et al. 1992; Charturvedi and Sarojini 1996) and the transformation into essential constituents. The resulting product is less viscous after cooking and is suitable for fermentation with lactic acid bacteria such as Bifidobacteria. In addition, sprouting improves the palatability, digestibility, and nutritive value of these foods. Besides beverages, malted cereals are used in recipes of breads, biscuits, breakfast cereals, other bakery products, and confectionaries.

Prefermentation treatments often lead to loss of nutritional value of the products, which is especially important for staple foods such as ogi. Efforts to improve the nutritional status of these staples have been based on fortification with legumes to provide the deficient amino acids. Various attempts have been made for nutrient restoration and fortification of ogi, and these include blending with fermented and unfermented legumes and addition of pawpaw slurry at various levels of substitution (Osungbaro 2009).

9.5.2 Chemical Changes Induced by Fermentation

9.5.2.1 Changes in Carbohydrates

Fermenting microorganisms are able to degrade digestible carbohydrate polymers, including starch and different types of fiber, into mono- and oligosaccharides. *Bacillus* species are known to possess carbohydrate-hydrolyzing enzymes such as amylases, galactanases, galactosidases, glucosidases, and fructofuranosidases (Aderibigbe and Odunfa 1990; Sarkar, Cook, and Owens 1993; Omafuvbe, Shonukan, and Abiose 2000). These enzymes hydrolyze carbohydrates into sugars that are partly consumed by the microflora itself and partially converted to other products including ethanol and pyruvate. The end products of this fermentation include gases (methane, carbon dioxide, hydrogen, and hydrogen sulfide), short-chain fatty acids (SCFA, mostly acetic, propionic, and butyric acids), and lactic acid in the case of *Lactobacillus* spp. (Kandler 1983; Corsetti and Settanni 2007).

9.5.2.2 Changes in Proteins

Fermentative yeasts and bacteria use their enzymatic activities in addition to the available endogenous enzymes to induce several changes in the fermented raw cereal and grain materials (Loponen et al. 2004; Spicher and Nierle 1988; Thiele, Gänzle, and Vogel 2002, 2003; Thiele, Grassi, and Gänzle 2004). The change in pH-induced sourdough fermentation may catalyze the action of some grain enzymes such as endogenous prolamin-degrading cereal proteases (Kawamura and Yonezawa 1982; Brijs, Bleukx, and Delcour 1999) and enhance the hydrolysis of gliadins (M'hir et al. 2008), glutenins (Gocmen et al. 2007), glutamins (Vermeulen, Gänzle, and Vogel 2007), globulins (Loponen et al. 2007), and secalins (Tuukkanen et al. 2005) as well as the depolymerization and solubilization of the gluten networks (Thiele, Grassi, and Gänzle 2004; Wieser et al. 2008; Gerez, Rollan, and Valdez 2006).

The various proteolytic activities induced by fermentation hydrolyze cereal proteins to produce free amino acids (Spicher and Nierle 1988; Thiele, Gänzle, and Vogel 2002), which act as flavor precursors (Schieberle 1990; Gobbetti, Corsetti, and J. Rossi 1994). While bacterial fermentation enhances the free amino acid contents in dough, yeasts consume free amino acids for their own metabolism (Thiele, Gänzle, and Vogel 2002). Thus, the levels of free amino acids in doughs depend on the composition of the fermentation culture, the temperature and time of fermentation, and the substrates present (Katina, Poutanen, and Autio 2004). The individual amino acids are affected variably by these conditions. For example, the release of proline is favored by pH > 5.5, while the release of phenylalanine, leucine, and cysteine occurs at lower pH. Ornithine, a major contributor to bread flavor, is found only in doughs fermented with *Lactobacillus pontis* (Thiele, Gänzle, and Vogel 2002). Higher proteolytic activities are encountered for brans and whole-grain flours compared to white flour (Loponen et al. 2004).

9.5.2.3 Changes in Lipids

Lipolytic reactions may take place, leading to the liberation of free fatty acids, particularly oleic, linoleic, and linolenic acids, which could diminish the sensory and nutritional quality of the products. However, although oils constitute up to 40% of the legumes used in food fermentations, extensive lipolysis does not take place, and only low levels of lipase activity were detected in, for example, *Parkia biglobosa* during *dawadawa* production, in *Pentaclethra macrophylla* during *ugba* production, and in melon seed fermentation. Aldehydes resulting from oxidation of lipids and causing off-flavor have been reported. On the other hand, there are reports of beneficial effects of lipase in the developments of characteristic flavor and aroma. However, efforts to eliminate lipid-soluble odorants in fermented products from locust bean, melon seed, and soybean by the use of organic solvents did not succeed, illustrating the complexity of this issue (Achi 2005).

9.5.2.4 Changes in Minerals and Vitamins

Fermentation *per se* does not add minerals to the fermented foods. However, fermentation maintains the optimum pH for cereal phytases, which catalyze the enzymatic degradation of phytates. Cereal and legume phytates have strong chelating capacities and, thus, form strong insoluble complexes with physiologically important metal ions, e.g., iron, zinc, calcium, and magnesium, and inhibit their absorption into the body (Lopez et al. 2000, Leenhardt et al. 2005). As discussed in detail in Chapter 6, fermentative microorganisms are able to enhance the contents and bioavailability of a number of vitamins, including vitamins B_1 (thiamine), B_2 (riboflavin), B_5 (pantothenic acid), B_6 (pyridoxine, pyridoxamine, and pyridoxal), B_7 (biotin), B_{12} (cyanocobalamin), folate, and K.

9.5.2.5 Volatile Compounds

In general, bacterial fermentation produces acids (such as lactic, propionic, and butyric acids), while yeast fermentation produces alcohols and more-volatile compounds (Hansen and Hansen 1994; Meignen et al. 2001). The production of volatiles by bacteria is strain-specific, with homofermentative *lactobacilli* producing high levels of diacetyl, acetaldehyde, hexanal; heterofermentative *lactobacilli* producing ethyl acetate, alcohols, and aldehydes; and yeast producing ethyl acetate and iso-alcohols such as 2-methyl-1-propanol, 3-methylbutanol, 2,3-methyl-1-butanol, and 2-phenylethanol (Gassenmeier and Schieberle 1995; Damiani et al. 1996). Acetic acid, butanoic acid, phenylacetic acid, 2- and 3-methylbutanoic acid, and pentanoic acid are important flavor compounds formed during sourdough fermentation (Czerny and Schieberle 2002).

9.6 CONCLUDING REMARKS

Currently, active research is taking place regarding the effects of sourdough fermentation on gut health. The production of short-chain fatty acids by fermentation has several health benefits including improved glycemic index of foods *via* the prolongation of the gastric emptying rate and reduced postprandial glycemia (Liljeberg, Lonner, and Bjorck 1995; Ostman et al. 2002). Moreover, the proteolytic actions induced by fermentative microorganisms enhance the availability of amino acids and may provide a new approach for the development of gluten-free functional breads for celiac patients (Lappi et al. 2010). Other benefits of fermented cereal products include the fermentation of dietary fiber to produce exopolysaccharides and prebiotics, which also can enhance gut health. Fermented cereal products can also be developed to provide antidiabetic, hypocholesterolemic, and immune-stimulating functional foods. This is a rich research area for proving of effects, elucidation of mechanisms of action, and development of novel technologies and uses.

REFERENCES

Abdel Rahman, I. E., S. H. Hamad, M. A. Osman, and H. A. Dirar. 2010. Characterization and distribution of microorganisms associated with Kisra bread preparation from three sorghum varieties in Sudan. *Current Research in Bacteriol.* 3:138–147.

Abegaz, K., T. Langsrud, F. Beyene, and J. A. Narvhus. 2004. The effects of technological modifications on the fermentation of Borde, an Ethiopian traditional fermented cereal beverage. *J. Food Technol. Africa* 9:3–12.

Achi, O. K. 2005. Traditional fermented protein condiments in Nigeria. *African J. Biotechnol.* 4:1612–1621.

Aderibigbe, E. Y., and S. A. Odunfa. 1990. Growth and extracellular enzyme production by strains of *Bacillus* species isolated from fermenting African locust bean, *iru. J. Appl. Bacteriol.* 69:662–671.

Agab, M. A., 1985. Fermented food products: "Hulu-mur" drink made from sorghum bicolor. *Food Microbiol.* 2:147–155.

Arendt, E. K., L. A. Ryan, and F. Dal Bello. 2007. Impact of sourdough on the texture of bread. *Food Microbiol.* 24:165–174.

Belton, P. S., and J. R. N. Taylor. 2004. Sorghum and millets: Protein sources for Africa. *Trends Food Sci. Technol.* 15:94–98.

Blandino, A., M. E. Al-Aseeri, S. S. Pandiella, D. Cantero, and C. Webb. 2003. Cereal-based fermented foods and beverages. *Food Res. Intern.* 36:527–543.

Brijs, K., W. Bleukx, and J. Delcour. 1999. Proteolytic activities in dominant rye (*Secale cereale* L.) grain. *J. Agric. Food Chem.* 47:3572–3578.

Bureng, P. L. 1979. A study on the malting characteristics of sorghum grain and the preparation of the fermented food hullo-mur. Ph.D. thesis, University of Reading, U.K.

Campbell-Platt, G. 1987. *Fermented foods of the world: A dictionary and guide.* London, U.K.: Butterworth.

Cauvain, S. 2003. Bread making: An overview. In *Bread making—Improving quality*, ed. S. Cauvain, 8–20. Cambridge, U.K.: Woodhead Publishing.

Charturvedi, A., and G. Sarojini. 1996. Malting of pearl millet (*Pennisetum typhoideum*): Its effect on starch and protein digestibilities. *J. Food Sci. Technol.* 33:342–344.

Chevassus, A., J. C. Favier, and A. J. Orstom. 1976. Technologie traditionnelle et valeur nutritive des bires: De sorgho du Cameroun. *Cah. Nut. Diet* 11:89–104.

Corsetti, A., and L. Settanni. 2007. *Lactobacilli* in sourdough fermentation. *Food Res. Intern.* 40:539–558.

Czerny, M., and P. Schieberle. 2002. Important aroma compounds in freshly ground whole meal and white flour: Identification and quantitative changes during sourdough fermentation. *J. Agric. Food Chem.* 50:6835–6840.

Damiani, P., M. Gobbetti, L. Cossignani, A. Corsetti, M. S. Simonetti, and J. Rossi. 1996. The sourdough microflora: Characterization of hetero- and homofermentative lactic acid bacteria, yeasts and their interactions on the basis of the volatile compounds produced. *Lebensmittel Wissenschaft und Technologie* 29:63–70.

Deshpande, S. S., D. K. Salunkhe, O. B. Oyewole, S. Azam-Ali, M. Battcock, and R. Bressani. 2000. Fermented grain legumes, seeds, and nuts: A global perspective. FAO Agricultural Services Bulletin No. 142, FAO, Rome, Italy.

Dirar, H. 1992. Sudan's fermented food heritage. In: *Applications of biotechnology to traditional fermented foods*, 27–34. Report of an ad hoc panel of the Board on Science and Technology for International Development. Washington, DC: National Academy Press.

———. *The indigenous fermented foods of the Sudan: A study in African food and nutrition.* Oxfordshire, U.K.: CAB International.

Faparusisi, S. I., M. O. Olofinboba, and J. A. Ekundayo. 1973. The microbiology of burukutu beer. *Zeit Allg. Microbiol.* 13:563–568.

Fujita, M., K. Nomura, K. Hong, Y. Ito, A. Asada, and S. Nishimuro. 1993. Purification and characterization of a strong fibrinolytic enzyme (nattokinase) in the vegetable cheese natto, a popular soybean fermented food in Japan. *Biochem. Biophys. Res. Commun.* 197:1340–1347.

Gassenmeier, K., and P. Schieberle. 1995. Potent aromatic compounds in the crumb of wheat bread (French type): Influence of pre-ferments and studies on the formation of key odorants during dough processing. *Zeitchrift Lebensmittel Unterschung Forschung* 201:241–248.

Gerez, C. L., G. C. Rollan, and G. F. Valdez. 2006. Gluten breakdown by lactobacilli and pediococci strains isolated from sourdough. *Lett. Appl. Microbiol.* 42:459–464.

Gobbetti, M., A. Corsetti, and J. Rossi. 1994. The sourdough microflora: Interactions between lactic acid bacteria and yeasts: Metabolism of amino acids. *J. Microbiol. Biotechnol.* 10:275–279.

Gocmen, D., O. Gurbuz, A. Yıldırım, K. A. F. Dagdelen, and I. Sahin. 2007. The effects of wheat sourdough on glutenin patterns, dough rheology, and bread properties. *European Food Research and Technology* 225:821–830.

Gopaldas, T., D. Sunecta, and J. Cinnamma. 1988. Studies on a wheat-based amylase rich food. *Food Nutr. Bull.* 10:55–59.

Hachmeister, K. A., and D. Y. C. Fung. 1993. Tempeh: A mold-modified indigenous fermented food made from soybeans and/or cereal grains. *Crit. Rev. Microbiol.* 19:137–188.

Hansen, B., and A. Hansen. 1994. Volatile compounds in wheat sourdoughs produced by lactic acid bacteria and sourdough yeasts. *Zeitchrift Lebensmittel Unterschung Forschung* 198:202–209.

Hansen, M., B. Pedersen, L. Munck, and B. O. Eggum. 1989. Weaning foods with improved energy and nutrient density prepared from germinated cereals. 1: Preparation and dietary bulk of gruels based on barley. *Food Nutr. Bull.* 11:40–45.

Hesseltine, C. W. 1979. Some important fermented foods of mid-Asia, the Middle East, and Africa. *J. Am. Oil Chem. Soc.* 56:367–374.

———. 1983. Microbiology of Oriental fermented foods. *Ann. Rev. Microbiol.* 37:575–601.

———. 1989. Fermented products. In *Legumes: Chemistry, technology, and human nutrition*, ed. R. H. Matthews, 161–185. New York: Marcel Dekker.

Kandler, O. 1983. Carbohydrate metabolism in lactic acid bacteria. *Antonie van Leeuwenhoek* 49:209–224.

Katina, K., K. Poutanen, and K. Autio. 2004. Influence and interactions of processing conditions and starter culture on formation of acids, volatile compounds, and amino acids in wheat sourdoughs. *Cereal Chemistry* 81:598–610.

Kawamura, Y., and D. Yonezawa. 1982. Wheat flour proteases and their action on gluten proteins in dilute acetic acid. *Agric. Biol. Chem.* 46:767–773.

Ko Swan, D., and C. W. Hesseltine. 1979. Tempe and related foods. *Econ. Microbiol.* 4:115–140.

Lappi, J., E. Selinheimo, U. Schwab, K. Katina, P. Lehtinen, H. Mykkänen, M. Kolehmainen, and K. Poutanen. 2010. Sourdough fermentation of wholemeal wheat bread increases solubility of arabinoxylan and protein and decreases postprandial glucose and insulin responses. *J. Cereal Sci.* 51:152–158.

Leenhardt, F., M.-A. Levrat-Verny, E. Chanliaud, and C. Rémésy. 2005. Moderate decrease of pH by sourdough fermentation is sufficient to reduce phytate content of whole wheat flour through endogenous phytase activity. *J. Agric. Food Chem.* 53:98–102.

Liljeberg, H. G. M., C. H. Lonner, and I. M. E. Bjorck. 1995. Sourdough fermentation or addition of organic acids or corresponding salts to bread improves nutritional properties of starch in healthy humans. *J. Nutr.* 125:1503–1511.

Liu, K. 2004. Fermented soy foods. In *Handbook of food and beverage fermentation technology*, ed. Y. H. Hui, L. Meunier-Goddik, Å. S. Hansen, J. Josephsen, W.-K. Nip, P. S. Stanfield, and F. Toldra, chap. 28. Boca Raton, FL: CRC Press.

Lopez, H. W., A. Ouvry, E. Bervas, C. Guy, A. Messager, C. Demigne, and C. Remesy. 2000. Strains of lactic acid bacteria isolated from sour doughs degrade phytate and improve calcium and magnesium solubility from whole wheat flour. *J. Agric. Food Chem.* 48:2281–2285.

Loponen, J., P. Laine, T. Sontag-Strohm, and H. Salovaara. 2007. Behaviour of oat globulins in lactic acid fermentation of oat bran. *Eur. Food Res. Technol.* 225:105–110.

Loponen, J., M. Mikola, K. Katina, T. Sontag-Strohm, and H. Salovaara. 2004. Degradation of HMW glutenins during wheat sourdough fermentations. *Cereal Chem.* 81:87–90.

Marcellin, D. J., A. K. Solange, N. Y. Zinzendorf, Y. A. Celestin, and L. Y. Guillaume. 2009. Predominant lactic acid bacteria involved in the spontaneous fermentation step of tchapalo process: A traditional sorghum beer of Cote D'Ivoire. *Res. J. Biol. Sci.* 4:789–795.

Marhoum, O. A. 1987. Biochemistry and microbiology of hulu-mur fermentation. M.Sc. thesis, University of Khartoum, Sudan.

Meignen, B., B. Onno, P. Gelinas, M. Infantes, S. Guilos, and B. Cahagnier. 2001. Optimisation of sourdough fermentation with *Lactobacillus brevis* and baker's yeast. *Food Microbiology* 18:239–245.

M'hir, S., J.-M. Aldric, T. El-Mejdoub, J. Destain, M. Mejri, M. Hamdi, and P. Thonart. 2008. Proteolytic breakdown of gliadin by *Enterococcus faecalis* isolated from Tunisian fermented dough. *World J. Microbiol. Biotechnol.* 24:2775–2781.

Nanadoum, M., M. Mbaiguinam, C. Gaillardin, and J. Pourquie. 2006. Technical, analytical, and microbiological follow-up of the bili bili, Chadian traditional beer. *Afr. Sci.* 20:69–82.

Nout, M. J. R., and J. L. Kiers. 2005. Tempe fermentation, innovation, and functionality: Update into the third millennium. *J. Appl. Microbiol.* 98:789–805.

Omafuvbe, B. O., O. O. Shonukan, and S. H. Abiose. 2000. Microbiological and biochemical changes in the traditional fermentation of soybean for soy-daddawa: A Nigerian food condiment. *Food Microbiol.* 17:469–474.

Ostman, E. M., L. Elmstahl, G. M. Helena, and I. M. E. Bjorck. 2002. Barley bread containing lactic acid improves glucose tolerance at a subsequent meal in healthy men and women. *J. Nutr.* 132:1173–1175.

Osungbaro, T. O. 2009. Physical and nutritive properties of fermented cereal foods. *African J. Food Sci.* 3:23–27.

Parker, M. L., M. Umeta, and R. M. Faulks. 1989. The contribution of flour components to the structure of injera, an Ethiopian fermented bread made from tef (*Eragrostis tef*). *J. Cereal Sci.* 10:93–104.

Reddy, N. R., M. D. Pierson, S. K. Sathe, and D. K. Salunkhe. 1982. Legume-based fermented foods: Their preparation and nutritional quality. *Crit. Rev. Food Sci. Nutr.* 17:335–370.

Sarkar, P. K., P. E. Cook, and J. D. Owens. 1993. *Bacillus* fermentation of soybeans. *World J. Microbiol. Biotechnol.* 9:295–299.

Sawadogo, L. H., V. Lei, B. Diawara, D. S. Nielsen, P. L. Moller, A. S. Traore, and M. Jakobsen. 2007. The biodiversity of predominant lactic acid bacteria in dolo and pito wort for the production of sorghum beer. *J. Applied Microbiol.* 103:765–777.

Schieberle, P. 1990. The role of free amino acids present in yeast as precursors of the odorants 2-acetyl-1-pyrroline and 2-acyltetrahyropyridine in wheat bread crust. *Zeitchcrift Lebensmittel Unterschung Forschung* 191:206–209.

Sefa-Dedeh, S., A. I. Sanni, G. Tetteh, and D. E. Sakyi. 1999. Yeasts in the traditional brewing of pito in Ghana. *World J. Microbiol. Biotechnol.* 15:983–997.

Spicher, G., and W. Nierle. 1988. Proteolytic activity of sourdough bacteria. *Appl. Microbiol. Biotechnol.* 28:487–492.

Steinkraus, K. H. 1983. *Handbook of indigenous fermented foods*. New York: Marcel Dekker.

Steinkraus, K. H., R. E. Cullen, C. S. Pederson, L. F. Nellis, and B. K. Gavitt. 1983. Indonesian tempeh and related fermentations. In *Handbook of indigenous fermented foods*, ed. K. H. Steinkraus, R. E. Cullen, C. S. Pederson, L. F. Nellis, and B. K. Gavitt, 1–94. New York: Marcel Dekker.

Subramanian, V., D. S. Murty, N. S. Rao, and R. Jambunathan. 1992. Chemical changes and diastatic activity in grains of sorghum (*Sorghum biocolor*) cultivars during germination. *J. Sci. Food Agric.* 58:35–40.

Thiele, C., M. G. Gänzle, and R. F. Vogel. 2002. Contribution of sourdough lactobacilli, yeast, and cereal enzymes to the generation of amino acids in dough: Relevant for bread flavor. *Cereal Chem.* 79:45–51.

———. 2003. Fluorescence labeling of wheat proteins for determination of gluten hydrolysis and depolymerization during dough processing and sourdough fermentation. *J. Agric. Food Chem.* 51:2745–2752.

Thiele, C., S. Grassi, and M. Gänzle. 2004. Gluten hydrolysis and depolymerization during sourdough fermentation. *J. Agric. Food Chem.* 52:1307–1314.

Tuukkanen, K., J. Loponen, M. Mikola, T. Sontag-Strohm, and H. Salovaara. 2005. Degradation of secalins during rye sourdough fermentation. *Cereal Chem.* 82:677–682.

Umeta, M., and R. M. Faulks. 1988. The effect of fermentation on the carbohydrates in tef (*Eragrostis tef*). *Food Chem.* 27:181–189.

Vermeulen, N., M. G. Gänzle, and R. F. Vogel. 2007. Glutamine deamidation by cereal-associated lactic acid bacteria. *J. Applied Microbiol.* 103:1197–1205.

Vogel, R. F., and M. A. Ehrmann. 2008. Sourdough fermentations. In *Molecular techniques in the microbial ecology of fermented foods*, ed. L. Cocolin, and D. Ercolini, 119–144. Berlin: Springer.

Wieser, H., N. Vermeulen, F. Gaertner, and R. F. Vogel. 2008. Effects of different *Lactobacillus* and *Enterococcus* strains and chemical acidification regarding degradation of gluten proteins during sourdough fermentation. *Eur. Food Res. Technol.* 226:1495–1502.

Zamora, A. F., and M. L. Fields. 2006. Production of corn and legume malts for use in home fermentation. *J. Food Sci.* 43:205–214.

10 Fermented Vegetables Products

Edyta Malinowska-Pańczyk

CONTENTS

10.1 INTRODUCTION

Preservation of vegetables by fermentation has been known and practiced by humans for 4500 years. It allowed for the consumption of vegetables out of season and during long journeys. Nowadays, vegetables are processed this way to obtain the desirable sensory properties of fermented products as well as improved digestibility and higher nutritional value. Fermented vegetable products are high in nutritive value and positively influence human health. They are rich sources of vitamins, especially vitamin C, dietary fiber, mineral salts, and antioxidants. Fermented cucumbers and sauerkraut are among the earliest fermented foods and are made commercially and at home throughout the world, especially in the West. Other important fermented vegetables include kimchi, the best-known Korean fermented vegetable, and fermented olives in the Mediterranean countries. Less popular are fermented peppers, carrots, cauliflowers, green tomatoes, and green mangoes. In this chapter, we will cover in detail the fermentation of cucumbers, cabbage (sauerkraut and kimchi), and olives. Each is discussed in terms of the characteristics of raw materials, the fermentation process, the microorganisms involved, and the chemical changes resulting from the process.

10.2   FERMENTED CUCUMBERS

10.2.1   Characteristics of the Raw Materials

Cucumbers, cannot be stored for a long time and, therefore, they must be processed directly after harvesting. They are processed by fermentation or preserved by chemical methods. Depending on the cultivar, fruit size, and maturity, cucumbers differ in their chemical composition, structural characteristics such as skin thickness and tissue texture, and softening enzyme activity (Lu, Fleming, and McFeeters 2002). Cucumbers are divided into three groups depending on diameter size: small (2.7–3.8 cm), medium (3.8–5.1 cm), and large (5.1–5.7 cm) (Fleming 1984). Raw cucumbers contain an average 1.5%–2% of saccharides on a wet weight basis (about 4%–5% of dry mass). For fermentation, immature cucumbers are used. The mature fruits possess polygalacturonases, enzymes associated with ripening, causing liquefaction of the seed area during fermentation and storage (Fleming 1984).

10.2.2   The Fermentation Process

Directly after harvesting, the cucumbers are sorted and washed using brush- or reel-type washers. Diseased, broken, and moldy fruits are discarded (Etchells et al. 1973). Fermentation of cucumbers can be performed using two different methods. In the first, cucumbers are fermented in brine (concentrated sodium chloride) at concentration of 5%–8% for the time needed to complete the conversion of saccharides to acids and other products. Then, salt is added up to 10%–16% in order to ensure product stability. This process is usually conducted in the open-faced tanks in which cucumbers are stored (Fleming 1984). If required, fermented cucumbers are removed from the brine during storage and desalted to a suitable concentration of salt by leaching in water. Whole, cut, or combined with other vegetables, fermented cucumbers are then impregnated with acetic acid and spices, and packaged. Sometimes the products are heated to increase their stability. This is how so-called pickles are produced, mainly in the United States.

In the second method, cucumbers are fermented in the presence of dill weed, garlic, blackcurrant and oak leaves, and horseradish at a final brine concentration of 3%–4%. Spices give the products a desirable flavor, and antimicrobial substances present in them prevent mold and yeast growth. The cucumbers are not desalted. Fermentation is carried out in containers, typically in 190-L barrels or glass jars.

Fermented cucumbers have to be stored at 0°C–5°C. This is the way cucumbers are fermented in the northern parts of Europe, especially in Poland. Products of good quality possess a characteristic refreshing sour taste, light green color, and firm texture. There are not many reports in the literature on the physical, chemical, and microbiological changes that take place during fermentation conducted using this method. Therefore, the information presented in this chapter mainly regards the first method.

Four distinct stages exist during the natural fermentation of cucumbers: the initiation (starting with participation of gram-positive and gram-negative bacteria), the primary fermentation (lactic acid bacteria [LAB] and yeast), the secondary fermentation (yeast), and the postfermentation if open tanks are used (surface growth of oxidative yeast, molds, and bacteria). However, these natural processes can cause many difficulties that often lead to improper products.

In order to obtain proper and predictable fermentation, an alternative method was introduced in 1973, mainly in the United States (Etchells et al. 1973). This procedure is based on controlled fermentation that eliminates two stages: the initiation and secondary fermentation. It is achieved by the acidification of brine with acetic acid to inhibit growth of acid-sensitive gram-positive and gram-negative bacteria and thus to promote the growth of LABs. Subsequently, after 24 h, the brine is buffered by sodium or calcium acetate, and can be inoculated with a desired starter culture of LABs (Fleming, McFeeters, Daeschel, Humphries, and Thompson 1988). This operation makes the initial pH favorable for growth of the added culture, ensures complete fermentation of sugars, and considerably eliminates the secondary fermentation by yeasts. The presence of calcium ions also helps to ensure a firm texture of the cucumbers. To eliminate postfermentation microbiological problems, suitable anaerobic tanks are used (Fleming et al. 1988).

The next challenge is to maintain the structural integrity of the cucumbers. Bloater damage is a result of gas accumulation inside the fruit during fermentation and is typically more severe in large cucumbers than in medium or small (Fleming 1984). Fresh cucumbers contain 5%–7% by volume of intercellular gas space. In the respiratory process, the cucumbers entrap O_2, and as a result, production CO_2 takes place. Therefore, the total content of CO_2 in fermented cucumbers stems from that process and from fermentation. As the concentration of CO_2 increases, gas pressure increases inside the cucumbers, causing internal gas pockets (bloating). This problem can be resolved by bubbling N_2 through the brine, which removes the CO_2. This operation also serves to create anaerobiosis and thereby prevents problems resulting from growth of aerobic microorganisms and oxidative changes.

10.2.3 THE MICROORGANISMS PARTICIPATING IN FERMENTATION

There are numerous and variable microflora on the surface of cucumbers, including a relatively small number of LABs. When the cucumbers are brined, the salt concentration controls the types of microorganisms that can survive and multiply. Before LABs become the dominating microflora, they must predominate over various gram-positive and gram-negative bacteria. The number of coliforms may amount to about 10^4 cells per ml at the beginning of fermentation. Cells of this group of bacteria are not detected after 5-day fermentation (Costilow and Fabian 1953). Simultaneously with the decrease of coliforms, the number of LAB increases. *Leuconostoc* and *Lactobacillus* species are detected during the early hours of fermentation, whereas *Lactobacillus* and *Pediococcus* become the dominant genera during the late stage of fermentation (Singh and Ramesh 2008). Yeasts may also proliferate during the natural fermentation of cucumbers. The first few days after brining, a heterogeneous group of yeasts exists. During the most active phase of yeast growth, the predominating species are *Torulopsis holmii* and, later, *Brettanomyces versatilis* (Costilow and Fabian 1953). Fermentative yeasts may grow together with LABs and may continue growing after the LABs are inhibited by low pH, because some amount of fermentable sugars remains. High concentration of CO_2 from yeast fermentation may then result in bloater formation.

The spent brine after fermentation is a continuing problem of wastewater with high biological oxygen demand (BOD) and salt concentrations too low to be used for recycling. Reduction of the salt content in the waste can be achieved by complete or partial elimination of sodium salts in the fermentation process. Using 100–300 mM $CaCl_2$ instead of salt in brine or a mixture of both (1.03 M NaCl and 40 mM $CaCl_2$) allows the conversion of the sugars contained in the fresh cucumbers to lactic acid and reduces the pH below 3.5. The chemical changes and sensory properties of pickles obtained in the presence of $CaCl_2$ are similar to those produced in a conventional process (McFeeters and Pérez-Díaz 2010).

L. plantarum and *P. pentosaceus* are homofermentative species that do not produce CO_2 from hexoses and, therefore, might be used as starter cultures in commercial cucumber fermentations. However, starter cultures are not currently produced commercially for this purpose (Pérez-Díaz and McFeeters 2011). The microorganisms of starter cultures should grow rapidly under the low pH and high salt concentration in commercial fermentation tanks, and should improve the efficiency of fermentations. *L. plantarum* strains are the homofermentative bacteria present in large numbers during the fermentation of brined cucumbers. However, these strains produce a high amount of CO_2 from the malic acid that is present in cucumbers (McFeeters, Fleming, and Thompson 1982). McDonald et al. (1993) proposed an *L. plantarum* mutant that is not able to decarboxylate malic acid (MDC⁻) and, therefore, can be used as a very active starter culture in fermentation of cucumbers.

10.2.4 The Chemical Changes during Processing and Fermentation

Glucose and fructose contents slightly increase with the size of the cucumbers (Lu, Fleming, and McFeeters 2002). After completion of fermentation, fructose is not detected, while residues of glucose are still present. A higher amount of glucose remains in larger fermented cucumbers than in smaller ones. LABs convert saccharides mainly to lactic and acetic acids. During fermentation ethanol is also produced. This product can arise from heterofermentative LAB activity or from anaerobic respiration of the cucumber tissue or by activities of yeasts if they are present on the fruit (Lu, Fleming, and McFeeters 2002).

The content of lipids in raw cucumbers is different in the skin, seeds, and flesh and is dependent on the size and maturity of the fruits. The skin, flesh, and seeds of large mature cucumbers constitute 25%, 65%, and 10% of the total weight of the raw cucumber, respectively. A higher amount of phospholipids is contained in the skin and seed area than in the flesh (Pederson et al. 1964). During fermentation of cucumbers, significant changes take place in the amount and proportion of particular fractions of lipids. The content of free fatty acids, neutral fat, and unsaponifiables increases after fermentation. At the same time, the phospholipid fraction decreases below 10% of the amount found in raw cucumbers. The marked decrease in the content of phospholipids during fermentation suggests their utilization by the bacteria and that the quality differences between good and bloated pickles are related, in part, to lipid metabolism (Pederson et al. 1964). Among the acids that were present in amounts greater than 0.5 mole% are $C_{14:1}$, $C_{15:0}$ *iso*, $C_{19:0}$, $C_{19:1}$, $C_{19:2}$, $C_{21:0}$, $C_{21:1}$, $C_{22:0}$, $C_{22:1}$, $C_{23:0}$, and $C_{24:1}$. It is noteworthy that the flesh of the bloated pickles contained 8.6 mole% of $C_{18:4}$ acid residues from the phospholipids, while the flesh of the good pickles contained 8.0 mole% of $C_{19:1}$ acid in the form of free fatty acid. Furthermore, the content of linoleic and linolenic acids increased in the free fatty acid and neutral fat fractions in the good pickles, whereas the oleic acid increased markedly in the same fractions of the bloated pickles (Pederson et al. 1964).

The contents of the amino acids (leucine, isoleucine, valine, tryptophan, and cysteine) increase slowly in the brine during fermentation. The brines from the tanks containing small cucumbers are richer in leucine, isoleucine, and valine than the brines coming from fermentation of big cucumbers. The concentration of tryptophan and cysteine in brines does not depend on the size of the cucumbers. The greatest reduction of tryptophan occurs during the fermentation accompanied by the greatest yeast activity, indicating that these microorganisms can utilize this amino acid (Costilow and Fabian 1953). During the first 5–10 days of fermentation, a diffusion of biotin, niacin, and pantothenic acid from cucumbers to the brine takes place, and then during days 10–19 of the process a slight decline in vitamin content in brine was observed. Simultaneously, the yeast activity was the greatest (Costilow and Fabian 1953).

The flavor of cucumber pickle depends on the spices and flavoring added to the final products (Fleming et al. 1995). Most of the volatile compounds contained in fresh cucumbers are present with little or no changes in the fermented cucumbers. In the fermented products, volatile alkanes, alkenes, alcohols, aldehydes, enals, ketones, carboxylic acids, esters, aromatics, heteroaromatics, and sulfur compounds were identified (Zhou and McFeeters 1998). Among the esters are ethyl acetate, isopropyl acetate, isopropyl propionate, butyl acetate, and isopropyl butyrate. They came from the acetic acid used in the cover brine. During days 3–10 of fermentation a 10-fold increase in the concentration of linalool also takes place. The content of this compound is stable after finishing of fermentation and during further storage. On the other hand, as the fermentation proceeds, the concentrations of hexanal, (*E*)-3,7-dimethyl-1,3,6-octatriene, (*E*,*Z*)-2,6-nondienal, and 2-nonenal decrease. Ethyl benzene, *o*-xylene, and benzaldehyde are absent in fresh cucumbers but appear after fermentation. The differences in the profile of volatile compounds are responsible for the specific aroma of fermented cucumbers (Zhou and McFeeters 1998).

10.3 FERMENTED CABBAGE (SAUERKRAUT)

10.3.1 Characteristics of the Raw Material

It is difficult to produce a good quality sauerkraut (fermented cabbage) due to the dependence of the process on many factors such as quality of the raw material, the microflora involved in fermentation, the parameters of the technological process, and storage conditions. The heads of cabbage used for producing sauerkraut should be large, with white, crumbly, compacted leaves; resistant to cracking and browning; and possess a short stump (up to 50% of head height). Raw cabbage contains above 4% of saccharides, mainly glucose, fructose, and sucrose, 3% of total fiber, 1.4% of proteins, and only 0.2% of lipids. The raw material should be rich in vitamin C (as high as 30 mg per 100 g wet mass), which prevents blackening, and contain not more than 750 mg nitrates per kg wet mass. Using cabbages with green leaves, weakly compacted, and of low concentration of saccharides leads to a product with acrid smell and bitter taste.

10.3.2 The Fermentation Process

Sauerkraut is made commercially and at home throughout the world. The cabbage heads are sorted in order to discard the diseased, defective, injured, or moldy ones. Subsequently, the top and dirty green leaves are removed to reduce the number of extraneous microorganisms. Before the cabbage is cut, the core of each head has to be drilled out using a specialized machine. Loose leaves as well as the drilled cores are collected separately. After removing the cores, the cabbage is finely sliced and shredded into 0.1–1.5-mm-wide strips. The ground cores are added to the shredded material due to the high concentration of carbohydrates, which have a positive effect on the taste and the biological value of sauerkraut.

The next step in the process is addition of NaCl. The concentration of salt should be 2%–3% in order to ensure good properties of sauerkraut. Salt plays important roles in the manufacturing of sauerkraut. It facilitates the extraction of the nutrients from the plant cells required to support the growth of LAB; it inhibits the growth of undesirable microorganisms; and it acts as a flavor ingredient in the final product. Salt should be pure, without any alkaline impurities. Sometimes a mixture of spices is added simultaneously with salt.

The shredded cabbage with additives is tightly packed into fermentation containers with capacity of up to 40 tons (under commercial conditions) or into barrels or crocks (at home conditions) and is quickly pressed. Brine is formed by osmotic release of water from the shreds that are contacted by salt. Then the upper surface of the shredded, pressed cabbage is covered with plastic sheeting that is draped over the tank wall and weighted down with water or weak brine. This covering prevents contact of the kraut with air that can lead to quality loss (Fleming, McFeeters, and Humphries 1988). Fermentation starts within a few hours to 2 days, depending on the temperature. Temperature is one of the most important factors in sauerkraut processing and should be in the range of 15°C–20°C. A variation of a few degrees beyond this range alters the microbial activity and affects the quality of the final product. The optimum for growth of heterofermentative *L. mesenteroides* is about 20°C, and it is the most desirable for initiating the fermentation. Therefore, the first phase of the process is carried out at 20°C for a couple of days, and then the temperature is lowered to 15°C–18°C. Initial temperatures above 22°C favor the growth of *Lactobacillus* species.

According to Fleming, McFeeters, and Humphries (1988), sauerkraut fermentation consists of two distinct stages: gaseous and nongaseous. In the first phase, a rapid production of CO_2, quick increase in lactic and acetic acids, and decrease in pH take place as a result of growth of the heterofermentative LAB. The second, nongaseous phase is characterized by slight or no production of CO_2, further increase in lactic acid concentration, and gradual decrease in pH. The gaseous phase continues for up to 7–9 days. The nongaseous phase ends when all carbohydrates have been metabolized (after about 3 weeks) and pH has decreased below 4.1. The final product is packed into barrels or smaller containers and stored at 3°C–8°C. Sometimes the sauerkraut is pasteurized.

10.3.3 THE MICROORGANISMS PARTICIPATING IN THE FERMENTATION

Most vegetable fermentations are spontaneous and take place as a result of growth of LAB, which are naturally present on fresh vegetables. It is known that mainly four species are involved in these processes: heterofermentative *Leuconostoc mesenteroides* and *Lactobacillus brevis* and the homofermentative *Pediococcus pentosaceus* and *Lactobacillus plantarum* (Yoon et al. 2002). However, the development of molecular techniques has allowed researchers to detect that, in many environmental conditions, a higher diversity in species composition of the microbial communities occurs than that determined by culture and biochemical methods. Breidt (2004) showed that species associated with sauerkraut fermentation during the early heterofermentative phase (1 to 7 days after initiation) belong to *L. mesenteroides* and *Weissella* spp. (those were earlier classified as members of *Leuconostoc* genera by Collins et al. [1993]), and *Leuconostoc fallax*. During the first 14 days of fermentation the presence of *Lactobacillus argentinum*, *Lactobacillus coryniformis*, *Leuconostoc citreum*, *Lactobacillus paraplantarum*, and *Lactobacillus paracasei* was evidenced. However, the roles of these species in the fermentation of cabbage require further investigation. Only small numbers of *L. brevis* and *P. pentosaceus* were confirmed.

The succession of growth of particular LAB species and their metabolic activities are responsible for the quality and safety of the product. The initial heterofermentative stage of fermentation precedes the second, homofermentative stage. Shorter generation time at 18°C and a high initial number of heterofermentative bacteria allow rapid initiation of fermentation and predominate the early stage of the process. Between the third and the seventh day after the initiation of the process, these bacteria are succeeded by the more acid-tolerant homofermentative species of *Lactobacillus* genera, due to the accumulation of lactic acid to more than 1% (w/v) and the decrease in pH below 4.5.

The typical flavor and aroma of sauerkraut and stability of the product are obtained by the correct sequence of LAB species. Microbial succession depends on the initial microbial load on the cabbage, salt and acid concentration, pH, and temperature. According to Lu et al. (2003), the sequence of LAB during fermentation may also be influenced by phages. They showed that the appearance of a new group of phages was closely correlated with bacterial succession. Phages isolated from days 1–3 did not attack host isolated on or after day 7, whereas phages isolated after day 3 did not infect host isolated on day 1 or 3. Phages isolated during the first 3 days infected *Leuconostoc* and *Weissella* strains, whereas after day 7 they all belong to *Lactobacillus*-specific phages.

It was concluded earlier that inoculation of cabbage with a starter culture does not bring a considerable benefit because the naturally occurring LAB dominate the fermentation in the proper sequence to produce a good-quality sauerkraut (Pederson and Albury 1969). This guideline is applied to commercial production of sauerkraut nowadays, but after fermentation most of the brine must be discarded. This waste is rich in undesirable chloride ions and has a high biological oxygen demand (Hang et al. 1972). However, in recent studies, many authors revealed that the quality of the final product may be improved by using starter cultures. This allows not only a significant shortening of the fermentation time, but also lowers the concentration of salt, which is important for treatment of the waste water (Tolonen et al. 2002; Johanningsmeier et al. 2007; Beganovic et al. 2011). Inoculation of shredded cabbage containing 0.9% salt with a mixed culture of *L. mesenteroides* and *Pediococcus dextrinicus* reduces the pH and completes the fermentation faster than in the process conducted without a starter culture (Tolonen et al. 2002). The application of the probiotic strain *L. plantarum* L4 and *L. mesenteroides* LMG7954 positively influenced the functional properties of the final product (Beganovic et al. 2011).

10.3.4　The Chemical Changes during Processing and Fermentation

As a result of osmotic stress due to addition of salt, the saccharides pass from the vegetable tissue to the brine and are an excellent medium for growth of microorganisms. During the gaseous phase of fermentation, fructose is depleted rapidly from the brine, whereas the concentration of glucose increases up to days 3–8 and subsequently decreases (Fleming, McFeeters, and Humphries 1988). Fructose is reduced to mannitol by heterofermentative bacteria, and its concentration does not change during the process. In the first stage of fermentation, lactic and acetic acid and ethanol are also produced (Fleming, McFeeters, and Humphries 1988; Trail et al. 1996). Sometimes, propanol and glycerol are formed (Trail et al. 1996). These reactions lead to a rapid lowering of pH to about 4.0.

$$\text{Glucose} + 2\,\text{Fructose} \rightarrow CO_2 + \text{Lactate} + \text{Acetate} + 2\,\text{Mannitol} \qquad (10.1)$$

During the second, nongaseous phase, the concentration of lactic acid increases, whereas that of other acids does not change (Fleming, McFeeters, and Humphries 1988). Titratable acidity increases to about 2%, and it corresponds to pH about 3.5. The final product also contains relatively small amounts (<2 mM) of malic, succinic, and propionic acids, which are produced during the heterofermentative stage (Trail et al. 1996).

Cabbage is a rich source of glucosinolates (GLS), which can be sorted into three different classes: aliphatic GLS, indole GLS, and aromatic GLS. The dominating compounds (about 90% of total GLS) are glucoiberin and sinigrin (aliphatic GLS) and glucobrassicin (indole GLS) (Daxenbichler, VanEtten, and Williams 1980; Ciska and Pathak 2004). The endogenous enzyme myrosinase (thioglucosidase glucohydrolase), released from the damaged plant tissue during shredding, hydrolyzes the GLS. The aglucon is usually converted to isothiocyanate, thiocyanate, and nitriles or epitionitriles and oxazolidine-thiones, depending on the properties of the environment (Figure 10.1). During fermentation, further decomposition of cabbage GLS takes place (Daxenbichler, VanEtten, and Williams 1980; Tolonen et al. 2002; Ciska and Pathak 2004). The content of GLS degradation products in fermented cabbage is affected by the initial amount of GLS in raw cabbage, and may also substantially depend on the volatility of the formed compounds, as well as the stability, including microbiological stability, and reactivity in an acidic environment. Aliphatic GLS are hydrolyzed to corresponding isothiocyanates and cyanides. Isothiocyanates are formed at the beginning of fermentation when the pH is relatively high, about 4–5, while cyanidins are the predominating products at pH below 3.7 (Gil and MacLeod 1980; Ciska and Pathak 2004). During storage of sauerkraut, the contents of isothiocyanates are gradually decreasing. Chemical and microbiological instability as well as high volatility of the degradation products released from some aliphatic GLS lead to their low content in the final products. The greatest losses of volatile compounds occur during the first stages of fermentation, which are accompanied by intensive release of gaseous products.

Unstable isothiocyanate formed from enzymatically hydrolyzed glucobrassicin, indole GLS, is further decomposed to the stable thiocyanate ion and the indolo-3-ylmethyl carbonium ions. These carbonium ions react with ascorbic acid. Ascorbic

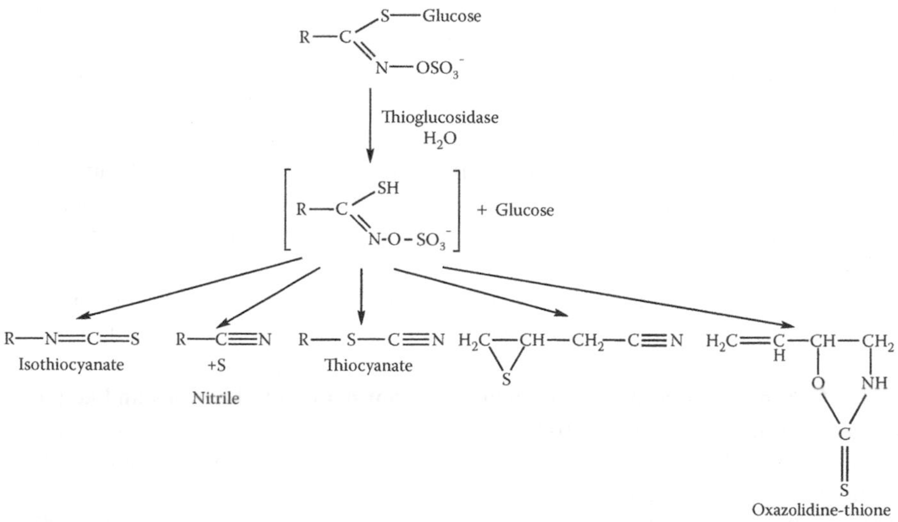

FIGURE 10.1 The enzymatic degradation products of glucosinolates (From S. F. Vaughn and M. A. Berhow. 2005. Glucosinolate hydrolysis products from various plant sources: pH effects, isolation, and purification. *Industrial Crops and Products* 21:193–202. With permission.)

acid, as a nucleophilic reagent at neutral or weakly acid conditions, favors ascorbigen formation (Buskov et al. 2000). The content of ascorbigen, which is low in raw cabbage, increases up to 20-fold in sauerkraut. Ascorbigen is stable during storage of sauerkraut up to 17 weeks (Martinez-Villaluenga et al. 2009). Ascorbigen concentration observed in low-sodium sauerkraut (0.5% NaCl) is higher than in fermented cabbage containing 1.5% salt (Martinez-Villaluenga et al. 2009). This phenomenon can be explained by a slower initiation of pH decrease in low-salted sauerkraut (Hrncirik, Valusek, and Velisek 2001; Johanningsmeier et al. 2005). At pH below 4, myrosinase can hydrolyze GLS to nitriles instead of isothiocyanate. However, the concentration of nitriles in sauerkraut remains low, and therefore they do not constitute a risk to human health (Tolonen et al. 2002).

Losses in vitamin C take place already during cutting or shredding of the cabbage (Mozafar 1994), and a further decrease in ascorbic acid concentration follows during fermentation and may amount to about 40% of the initial value (Martinez-Villaluenga et al. 2009). Salt concentration and LAB in sauerkraut seem not to affect the ascorbic acid level in sauerkraut. Loss of vitamin C in fermented cabbage is partially a result of its involvement in ascorbigen formation (Hrncirik, Valusek, and Velisek 2001). In addition, ascorbic acid is very susceptible to chemical and enzymatic oxidation by ascorbic oxidase during processing (Klieber and Franklin 2000).

Development of sauerkraut flavor depends on the cabbage cultivar, method of fermentation, and salting. The characteristic flavors of cabbage arise in part from decomposition of GLS and from degradation of S-methyl-L-cysteine sulfoxide, which is decomposed by cysteine sulfoxide lyases (C-S-lyase) during disruption of the vegetable tissues. The activity of this enzyme leads to formation of methanethiol, dimethylsulfide, dimethyldisulfide, and dimethyltrisulfide (C. Lee et al. 1974; Trail et al. 1996). The characteristic sauerkraut aroma is due to methyl methanethiosulfinate and methyl methanethiosulfonate (Chin and Lindsay 1994). Formation of these compounds during cutting of cabbage and in the early stage of fermentation as proposed by Chin and Lindsay (1994) is shown in Figure 10.2. Methanol, acetaldehyde, ethanol, n-propanol, 2-propanol, and ethyl acetate have also been identified among volatile compounds participating in the formation of the flavor of sauerkraut (Trial et al. 1996). Furthermore, low-molecular-weight fatty acids were recognized as well in the headspace over sauerkraut. Butyric acid seems to be the most important among them (Vorbeck et al. 1961).

The high number of microorganisms in many kinds of food may lead to production of toxic metabolites such as biogenic amines, which exert negative biological effects on human health. These compounds are produced as a result of decarboxylation of amino acids or by amination and transamination of aldehydes and ketones. Biogenic amines are usually produced by microorganisms possessing the activity of amino acid decarboxylase. Such activity has also been shown in some LABs participating in fermentation of sauerkraut, i.e., *Lactobacillus* spp., *L. mesenteroides*, and *Pediococcus damnosus* (Halász et al. 1994; Moreno-Arribas et al. 2003). Fermentation of cabbage caused a significant increase in the individual and total biogenic amine contents in a spontaneous process of fermentation, as well as in that initiated by a starter culture. However, some authors (Halász et al. 1994; Spicka et al. 2002) showed that the contents of biogenic amines in spontaneously fermented

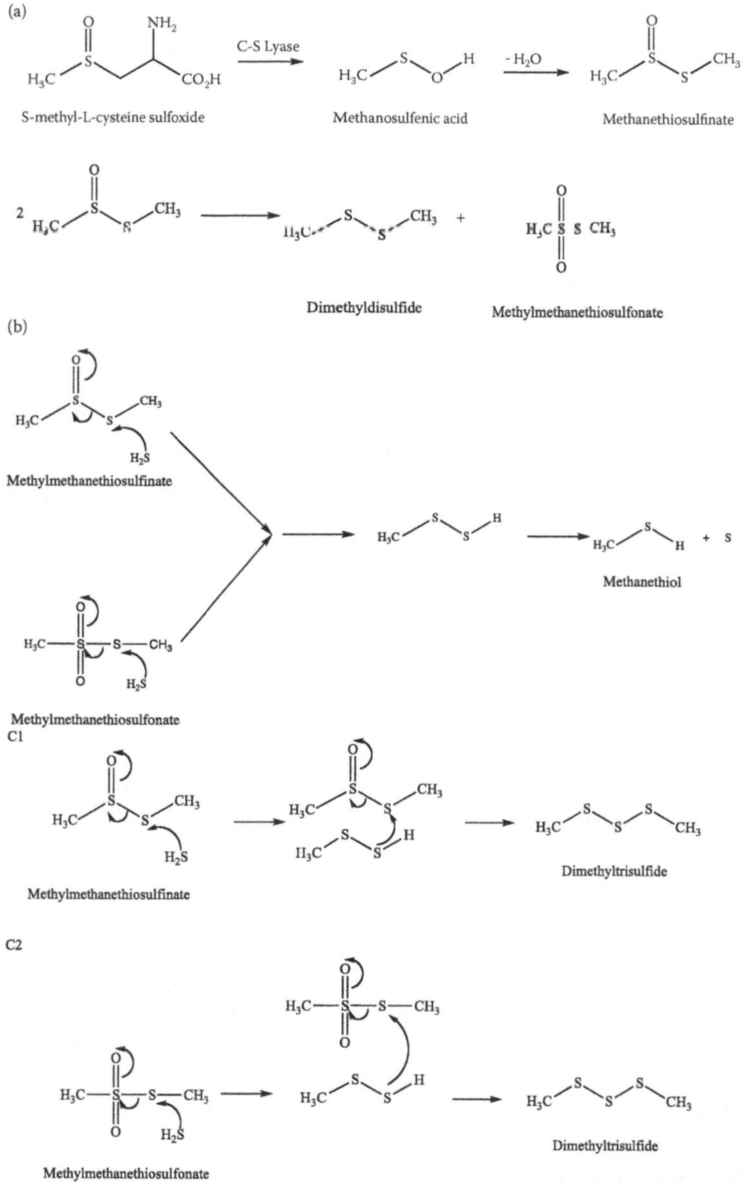

FIGURE 10.2 Suggested mechanisms for the formation of some volatile sulfur compounds following the action of C-S lyase on *S*-methyl-L-cysteine sulfoxide: (A) dimethyl disulfide and methyl methanethiosulfonate involving condensation and dehydration of methanesulfonic acid; (B) methanethiol from the reactions of methyl methanethiosulfinate and methyl methanethiosulfonate with hydrogen sulfide; (C1) dimethyl trisulfide from the reactions of methyl methanethiosulfinate with hydrogen sulfide and (C2) methyl methanethiosulfonate with hydrogen sulfide (From H. W. Chin and R. C. Lindsay. 1994. Mechanisms of formation of volatile sulfur compounds following the action of cysteine sulfoxide lyases. *Journal of Agricultural and Food Chemistry* 42:1529–1536. With permission.)

sauerkraut were higher than the ones observed when a starter culture was used. Therefore, the safety of sauerkraut may depend on the composition of the natural microflora of cabbage and can be controlled by usage of starter cultures.

The level of putrescine and spermidine in sauerkraut produced by using *L. plantarum* starter culture was four times higher than that determined in raw cabbage, and the concentration of cadaverine, tyramine, histamine, and spermine was two- or threefold higher. Sauerkraut low in salt (0.5%) contains less bioactive amines than that obtained at higher salt content (Peñas et al. 2010; Kalač et al. 2000). The concentration of these compounds also depends on the fermentation conditions and the postfermentation treatment (pasteurization or sterilization). The recommended maximum levels of histamine, tyramine, putrescine, and cadaverine in sauerkraut are 10, 20, 50, and 25 mg/kg, respectively (Künsch, Schärer, and Temperli 1989).

10.4   FERMENTED CABBAGE (KIMCHI)

Another form of fermented cabbage is kimchi, a traditional Korean fermented vegetable product characterized by its sour/sweet and carbonated taste. It is usually served as a cold side dish with cooked rice or soup. Kimchi products may be divided into two groups depending on processing methods: ordinary (without added water) and *mul*-kimchi (with water). The first group contains, for example, *beachu* kimchi (diced Chinese cabbage), *tongbaechu* kimchi (whole Chinese cabbage), *yeolmoo* kimchi (young oriental radish), and *kakdugi* (cubed radish kimchi), while in the second group are: *baik* kimchi (*beachu* kimchi with water), *dongchimi* (whole radish kimchi with water) and *nabak* kimchi (cut radish and Chinese cabbage). They differ in the raw materials used for their preparation and in functional properties. There are about 200 types of kimchi made today in Korea (Mheen 2010).

10.4.1   CHARACTERISTICS OF THE RAW MATERIALS

The important raw materials used for preparing kimchi may be divided into three groups. The major vegetables are Chinese cabbages and radishes, while the minor ingredients (second group) that can be added during kimchi preparation are the following: cucumbers, mustard leaves, green onions, leeks, garlic, red peppers, and ginger. Sometimes fishery products such as fermented shrimp and anchovy are added. The third group contains ingredients that are available in various seasons and localities. The raw materials for making special kimchi in Korea include parsley, pear, chestnut, oyster, frozen pollock, starches, monosodium glutamate, and sweeteners (Mheen and Kwon 1984). The simplest kimchi may be made from cabbage (100 g), garlic (2 g), red pepper powder (2 g), green onion (2 g), ginger (0.5 g), and salt (2%–3%). The kinds and amounts of seasoning vegetables used in kimchi production differ between various manufacturers. Therefore, the quality and composition of major and subingredients may significantly affect the fermentation and the characteristics of kimchi.

Selection of good-quality raw materials and appropriate formulation of kimchi ingredients and seasonings is important for developing a desirable taste of kimchi. Vegetables having soft texture and high sugar content are more preferred for obtaining high quality of kimchi than those with harder texture and lower sugar content. However, using harder vegetables for preparing kimchi helps to avoid softening during long-term storage of the product (Mheen 2010).

Addition of subingredients can affect the growth of microorganisms. Garlic and leek retard the fermentation (M. Kim et al. 1987), as do mustard seeds or leaves (H. Park and Han 1994). Especially allicin, one of the constituents of garlic, is very effective in inhibiting the growth of various microorganisms in the initial fermentation period, which leads to improvement in the storage capacity and lesser acidification of the product (S. K. Lee et al. 1989; N. Cho et al. 1988). During fermentation of kimchi, the intensive, specific taste of garlic is slowly changed to a harmonized flavor (Mheen 2010). Ginger is also used as an ingredient in preparing kimchi. It contains many antibacterial components such as citral and linalool, gingerol, and shogaol, which additionally give a hot sensation to the product. Therefore, ginger slows down the fermentation; however, it does not significantly change the sour taste of the product (S. H. Lee and Kim 1988; Jang et al. 1991). Contrary to garlic, ginger, and other subingredients, hot pepper stimulates the fermentation of kimchi (W. S. Park et al. 1996).

Fermented anchovy and shrimp are a rich source of proteins and amino acids, and possess unique flavor. Therefore, they affect not only the nutritional balance but also contribute to the sensory quality of kimchi (Ryu et al. 1996). Starch and sugar are utilized as a carbon source by various microorganisms present in kimchi, and in consequence they promote the fermentation and contribute greatly to the harmony of taste (Jang et al. 1991).

10.4.2 THE FERMENTATION PROCESS

The quality of kimchi depends on the groups of microorganisms involved in the fermentation, temperature, oxidoreduction potential, salt, saccharides, and other available nutrients as well as inhibitory compounds. The fresh cabbage or radish are cut into halves or shredded, salted overnight, washed, and drained. The minor ingredients are cut up and mixed with the major components. The mixture is packed in earthen jars and pressed in order to immerse the vegetables in the juice and to remove oxygen present in the internal void spaces. The surface is covered (using large salted cabbage leaves) and weighed down. Sometimes water or weak brine is added to the top to prevent air access. In most cases, the kimchi fermentation process includes several stages based on the changes of pH and reducing-sugar content. In the first stage, a rapid decrease in pH, an increase of acidity, and a decrease in reducing-sugar content take place. During the second stage, the pH decreases gradually. Fermentation is finished when the pH and acidity (expressed as a percentage of lactic acid) of kimchi reach about 4.0%–4.5% and 0.5%–0.8%, respectively. Ripening of kimchi occurs in the last stage that is eventually accompanied by small changes in pH, acidity, and reducing sugars (Mheen 2010).

Salting, washing, fermentation temperature, and storage conditions are important factors affecting the quality of kimchi. Salt plays an important role in fermentation and preservation of good quality of kimchi. The best salt for making kimchi is sea salt or salt without iodine or other chemical impurities. The optimum salt concentration in kimchi amounts to 2%–3%. If the salt concentration is below the optimum, fermentation may proceed too fast, and quick acidification and softening of the product usually occur. On the other hand, the color and flavor of kimchi are not acceptable when the salt concentration is over 6% (Mheen 2010). Two types of salting procedures are used: adding salt dry (typically used conventionally at the household level) or in brine. Because of the difficulty in controlling the final salt content in kimchi salted with dry salt, the brine is preferable for commercial production because it reduces the variability in the quality of the final product (Shim et al. 2008). Salting is carried out for 3 to 15 hours, depending on the attempted salt concentration, temperature, and the cutting and size of the cabbages. During salting, the decrease in the moisture content (10%–12%), relative volume, and weight as well as the internal void space of the cabbage take place. The texture of the vegetable, especially the flexibility and firmness, are changed (Choe et al. 1991). Brining inhibits aerobic microbial activity and stimulates the growth of LAB.

Washing of salted Chinese cabbages or radishes with clean water is the next critical point during preparation of kimchi (W. P. Park et al. 2000). Minimally processed brined Chinese cabbage is susceptible to spoilage during storage. Furthermore, the metabolic reactions are accelerated by the physical damage of the tissue. Therefore, in the case of minimally processed brined Chinese cabbages, it is important to ensure the required shelf life of the product. Washing of brined Chinese cabbages with calcium hypochlorite positively affects the pH and microbial growth compared with washing in tap water (W. P. Park et al. 2000). Shim et al. (2008) also showed that pretreatment of brined raw material with calcium hypochlorite was effective in maintaining a good product quality.

Fermentation of kimchi is mainly dependent on the microorganisms naturally present in the raw material. Therefore, a significant factor in the correct succession of microflora is not only salt concentration, but also temperature. Usually, under household conditions, the fermentation proceeds at ambient temperature. In industrial conditions, different temperatures can be used. The optimum pH of kimchi fermentation of about 4.2 and acidity 0.6% (as lactic acid) are achieved after different times depending on temperature. Faster decline of pH takes place as the temperature increases and salt concentration decreases. At salt concentration of 2.75%, the acidity reaches 0.75% within 8 days at 15°C, whereas at 0°C only 0.35%–0.43% acidity is achieved after 24 days of fermentation (M. Kim and Chang 2000). Higher concentration of salt and refrigerated temperature prolong the fermentation. At 5°C and above 5.0% salt content, kimchi ripens very slowly, and at 7.0%, this process is not accomplished even after 180 days (Mheen and Kwon 1984).

The LABs involved in kimchi fermentation continue to produce organic acids after the ripening and cause changes in the composition of kimchi. These changes are called the overripening or acid-deterioration of kimchi and lead to softening of the product, excessive sourness, and inflation during storage. These phenomena are the most serious defects observed in kimchi stored for an extended time. The best way to avoid quality deterioration is to control the LAB growth (Mheen 2010). Heat processing has been used to extend the shelf life of kimchi, but undesirable changes in taste and texture can occur (N. Lee and Chun 1982). Other methods, such as the use of natural preservatives (Choi and Beuchat 1994) and gamma irradiation

(Song et al. 2004) were also investigated as ways of extending the shelf life, but none of them was particularly successful. High-pressure processing was also used to control undesirable changes in fermented kimchi products (Sohn and Lee 1998). High-pressure treatment at 400 MPa minimized the changes in pH and effectively prevented the production of CO_2 during storage. Pressure-treated kimchi maintained a palatable taste, and its shelf life was doubled. However, the negative color change induced by high-pressure treatment limits the application of this method (Sohn and Lee 1998).

10.4.3 THE MICROORGANISMS PARTICIPATING IN THE FERMENTATION

LABs are important microorganisms in kimchi fermentation (Mheen and Kwon 1984; J. Lee et al. 2005). Furthermore, archaea and yeasts were detected in fermented kimchi (Chang et al. 2008). The fermentation temperature and seasonal variations in raw material have a significant impact on the population dynamics inherent to kimchi (J. Cho et al. 2006). The following bacteria predominate among the LABs: *Leuconostoc citreum, Leuconostoc carnosum, Leuconostoc gasicomitatum, Leuconostoc kimchii, Leuconostoc lactis, Leuconostoc mesenteroides, Weissella cibaria, Weissella confusa, Lactobacillus sakei, Lactobacillus curvatus,* and *Lactobacillus brevis* (J. Cho et al. 2006; J. Lee et al. 2005; Chang et al. 2008). The composition of the microflora may differ depending on the raw material and temperature of processing. For example, some eubacteria that occur initially disappear later during the fermentation (J. Lee et al. 2005). *Weissella confusa* and *Leuconostoc citreum, Leuconostoc gasicomitatum, Leuconostoc mesenteroides, Lactobacillus sakei, Lactobacillus curvatus,* and *Lactobacillus brevis* remain throughout the fermentation process. This indicates their importance in kimchi fermentation. Also, the *Weissella koreensis* and *Lactobacillus pentosus* existing in the largest numbers among the microbial flora in the later phase of the process demonstrate their significant role in ripening of kimchi (J. Cho et al. 2006).

The composition of archaea changes throughout fermentation. Chang et al. (2008) showed that in the first stage of fermentation (pH ~ 5.3), halophilic archaea *Halococcus dombrowskii, Halococcus thermotolerans,* and *Halorubrum trapanicum* were present. When pH decreased to about 4.1, the predominant species was *H. dombrowskii.*

Yeasts identified in kimchi by Chang et al. (2008) belong to *Lodderomyces elongisporus, Trichosporon brassicae, Candida sake, Saccharomyces castellii,* and *Kluyveromyces marxianus.* No significant changes were observed in the yeast population during kimchi fermentation. This indicates that the growth of yeasts is suppressed during fermentation and that indigenous yeasts are rather not involved in the process. However, H. J. Kim, Kang, and Yang (1997) showed that exogenous yeasts could affect the properties of kimchi.

The combinations of various strains such as *Leuconostoc mesenteroides, Lactobacillus brevis, Lactobacillus plantarum,* and *Pediococcus cerevisiae,* which were isolated from kimchi, could be used as starters for kimchi fermentation. This starter culture increases the fermentation rate, and better sensory quality of kimchi is obtained than when a single strain is used (S. H. Lee and Kim 1988). Kimchi fermentation is carried out at relatively low temperature. Therefore, psychrotrophic LAB starters isolated from kimchi fermented at 5°C help to shorten the time of fermentation (So and Kim 1995). Psychrotrophic yeast, *Saccharomyces fermentati* YK-19 isolated from kimchi and used as a starter, can prevent overacidification (H. J. Kim, Kang, and Yang 1997). The addition of *S. fermentati* YK-19 prolongs the period of optimum fermentation because it inhibits the growth of *Lactobacillus* species. Furthermore, sensory indices such as acidic and moldy flavor were reduced by starter addition, while flavor scores for freshness were increased (H. J. Kim, Kang, and Yang 1997).

10.4.4 THE CHEMICAL CHANGES DURING PROCESSING ON FERMENTATION

The sugars contained in raw materials are converted to organic acids, carbon dioxide, ethanol, and other flavoring compounds by hetero- and homofermentative LABs. Lactic, acetic, citric, malic, fumaric, succinic, oxalic, tartaric, malonic, maleic, and glycolic acid were identified in kimchi samples. The kimchi fermented at low temperature (6°C–7°C) contained more lactic, acetic, and succinic acid, and less oxalic, malic, tartaric, malonic, maleic and glycolic acid than that fermented at higher temperature (22°C–23°C), while no difference was noted in citric acid level (H. O. Kim and Rhee 1975). Higher temperatures stimulated the production of carbon dioxide during kimchi fermentation (Hong and Park 2000).

The vegetables used in the production of kimchi are rich in vitamin B group, vitamin C, and precursors of other vitamins. During fermentation, the content of particular vitamins changes. Cheigh (1995), cited by Kwon and Kim (2003), showed that the ascorbic acid content of *beachu* kimchi decreased gradually during 12 days of fermentation, and subsequently increased on day 18. Storage of this kimchi at 7°C caused again a decline in ascorbic acid. B vitamins and β-carotene contents decrease to about 50% of the initial concentration during ripening.

The content of free amino acid in kimchi depends on the composition of the raw materials used. Products containing mustard leaves are a rich source of proline, glutamic acid, alanine, and histidine (S. Park et al. 1995). Fermented anchovy and shrimp are rich in proteins. Therefore, the content of amino acids is higher in final products than at the beginning of fermentation (H. Lee, Ko, and Lim 1984). Free amino acids are important for the formation of kimchi flavor (Hawer et al. 1988).

The compounds responsible for flavor of kimchi depend on the composition of vegetables and, generally, their amount increases at the early time of storage and then decreases (Kang et al. 2003; Cha, Kim, and Cadwallader 1998). Volatile compounds can be formed through chemical or enzymatic reactions. Enzymes participating in these reactions come from the vegetables and ingredients or they are produced by the microorganisms. The substrates comprise carbohydrates, free amino acids, fatty acids, sulfur compounds such as allyl glucosinolate and propenyl-cysteine sulfoxide. Among the volatile compounds found are alcohols, aldehydes, ketones, sulfur compounds, and hydrocarbons. During further storage, they can react with volatiles and nonvolatiles and form nonvolatile compounds (Kang et al. 2003). The mechanisms for the formation of volatile and nonvolatile compounds in kimchi are shown in Figure 10.3.

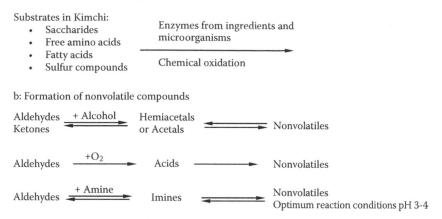

a: Formation of volatile compounds

Substrates in Kimchi:
- Saccharides
- Free amino acids
- Fatty acids
- Sulfur compounds

Enzymes from ingredients and microorganisms

Chemical oxidation

b: Formation of nonvolatile compounds

Aldehydes   + Alcohol   Hemiacetals
Ketones   or Acetals   Nonvolatiles

Aldehydes   +O₂   Acids   Nonvolatiles

Aldehydes   + Amine   Imines   Nonvolatiles
                                         Optimum reaction conditions pH 3-4

FIGURE 10.3 Probable mechanism of formation of volatiles (A), and nonvolatiles (B) during kimchi storage (From J. H. Kang et al. 2003. Changes of volatile compounds, lactic acid bacteria, pH, and headspace gases in kimchi, a traditional Korean fermented vegetable product. *Journal of Food Science* 68:849–854. With permission.)

As in the case of sauerkraut, sulfur compounds play an important role in the quality and formation of the characteristic strong flavor of kimchi (Hawer 1994). The sources of sulfur compounds in kimchi are Chinese cabbage, radish, red pepper, garlic, ginger, and green onion (Kang et al. 2003). Formation of methanethiol, dimethyl disulfide, and dimethyl trisulfide during fermentation is described in Section 10.3.4 regarding volatile compounds in sauerkraut.

The volatile compounds in kimchi also contain terpenes such as β-myrcene, α-phellandrene, β-phellandrene, α-copaene, germacrene B, germacrene D, β-sesquiphellandrene, and α-farnesene (that may come from garlic and ginger); alcohols, mainly ethanol; esters; aldehydes; ketones as well as aromatic compounds; and hydrocarbons (Kang et al. 2003).

10.5 FERMENTED OLIVES

10.5.1 CHARACTERISTICS OF THE RAW MATERIALS

Olive fruits are the raw material for the production of olive oil or table olives. They are prevalently cultivated in the Mediterranean area and in other regions with suitable climate. Olives are characteristic for low sugar (2.6%–6.0%) and high oil content (12%–30%). They are not edible during all stages of maturation because the flesh is usually acrid and bitter. Therefore, they have to be processed to obtain edible, appetizing, and well-preserved products.

Olives used on the table should have a proper size, weight, and shell shape. It must be easy to separate the stone from the flesh (Bianchi 2003). The olive's skin should be thin, elastic, and resistant to blows as well as to the action of alkaline solution or brine. High content of oil is undesirable, because it damages the consistency and shortens the shelf life of the product. However, in some cases, moderate or high oil content is acceptable, especially for making special black olives. To produce green olives, the fruits are harvested when they are green or straw-yellow green and reach a normal size. Varicolored olives are picked exactly before maturity, when the color is pale rose or wine rose. If black olives are produced, the harvesting takes place when the fruits have a deep black, wine red, or dark red skin color and achieve ripeness (IOOC 1982).

10.5.2 The Fermentation Process

Table olives can be produced using three methods, depending on the ripeness stage of the fruit: the Spanish (or Sevillian) method, especially used for producing green olives and varicolored olives; the Californian method, for black oxidized olives; and the Greek method, particularly for naturally black olives. In the Spanish and Californian methods, a lye treatment precedes fermentation in order to destroy the bitter glucoside–oleuropein in the olives.

Olives are harvested by hand, as they are delicate and require careful handling. In the Spanish method, the olives after cleaning are soaked in 2%–2.5% NaOH solution for 8–12 h, depending on the variety, maturation stage, and temperature. The solutions should cover the fruit. During this process, most of the waxy material is dissolved. The external layer of the fruit becomes more permeable, and NaOH penetrates the pulp down to stone (Bianchi 2003). Subsequently, the olives are repeatedly washed with water to remove most of the lye residue and to neutralize the medium. Sometimes, an appropriate amount of HCl solution is added or bubbling with CO_2 is applied for this purpose. This step affects the sugar and salt content but does not change the oil content. The next step is fermentation. Washed fruits are immersed in 9%–10% brine. The final concentration of brine amounts to 4%–5% due to the high water content in the olives. Spontaneous or controlled fermentation lasts several months. During that time pH decreases below 4.5.

In the Californian method, the olives are immersed in 1%–2% NaOH solution that penetrates the fruit to the pit. At the end, olives are washed with water and aerated. This treatment causes darkening of the olives and softening of the flesh. The skin becomes brown to black as a result of oxidation and polymerization of phenolic compounds. The black color is fixed by immersion of the fruits in a solution of ferrous gluconate or lactate. Then the olives are canned in light brine (Bianchi 2003).

In the Greek method, natural ripe black olives are transported to the factory as quickly as possible, sorted, and washed. Then olives are kept in 4%–8% brine solution for a suitable time. At the beginning of fermentation, the vats are closed tightly to prevent the contact of fruits with air. The brine stimulates the microbial action for fermentation and reduces the bitterness of the oleuropein. The pH value decreases rapidly during about 5 days and then slowly but continually up to the 10th day. After this time, pH is stabilized (Leal-Sánchez et al. 2003). The fermentation can be finished when the bitterness is reduced enough. Sometimes the olives are conditioned with vinegar (25% of the volume of the brine) and heat processed. Fermentation takes place in large vats (8–10-tonne capacity) or in wooden barrels.

Temperature and salt concentration are two of the most important parameters in the fermentation process. These factors influence the number and succession of the microbial populations and prevailing species of LABs. Salt in higher concentration (>8%) causes retardation of LAB growth and favors the growth of salt-tolerant yeast, giving a less sour product. Lower salt content (4%–6%) and temperature 18°C–25°C favor the role of LABs in fermentation. Therefore, the control of salt concentration and fermentation temperature is crucial to obtain high-quality final products (Tassou, Panagou, and Katsaboxakis 2002). Tassou et al. (2007) showed that brine supplemented with $CaCl_2$ (4%) caused the flesh of olives to be stronger and stiffer than those fermented without this additive.

10.5.3 THE MICROORGANISMS PARTICIPATING IN THE FERMENTATION

The composition and number of natural microflora of olives depend on the cultivar and hygienic condition during processing. The autochthonous microflora usually comprises yeasts, about 10^3 cells/g; LABs, 10^3 cells/g; enterobacteria, 10^2 cells/g; and pseudomonads, 10^2 cells/g (Tassou et al. 2007). During the first week of fermentation, the number of Enterobacteriaceae and *Pseudomonas* spp. increases quickly, but after 12 and 15 days, respectively, they are not detected (Tassou et al. 2007). At the same time, the number of LABs and yeast populations increases (Tassou, Panagou, and Katsaboxakis 2002; Panagou et al. 2008). The yeasts and bacteria are present in brine, on the surface of the fermented black olives, and within them. The yeasts occur mainly on the skin and in the stomal openings, whereas bacteria predominated in the intercellular spaces of the substomal cells (Tassou, Panagou, and Katsaboxakis 2002).

During fermentation of naturally black olives, among the LABs, *Lactobacillus mesenteroides*, *L. brevis*, *L. plantarum*, and *Lactobacillus pentosus* species were identified (Tassou, Panagou, and Katsaboxakis 2002). In the case of debittered green table olives, within the LAB population prior to brining, enterococci are the predominant species, particularly *Enterococcus casseliflavus* (De Bellis et al. 2010). *Lactobacillus coryniformis*, *Lactobacillus paracasei*, *L. plantarum*, *L. rhamnosus*, *L. brevis*, *Lactobacillus mali*, *Lactobacillus vaccinostercus*, *Lactobacillus casei*, *L. mesenteroides*, *Leuconostoc pseudomesenteroides*, *Lactococcus lactis*, *Weissella paramesenteroides*, *W. cibaria*, and *Enterococcus italicus* were present during the whole process (De Bellis et al. 2010).

Although fermentation of olives is carried out by LABs, the yeasts are present throughout the process and during storage (Ruiz-Barba et al. 1994). The number of yeast populations can range from 10^4 to 10^6 cells/g (Garrido Fernandez, Fernández Díaz, and Adams 1997) in the first 10 days and remain almost unchanged thereafter (Tassou et al. 2007). *Saccharomyces cerevisiae*, *Pichia membranifaciens*, *Pichia anomala*, and *Rhodotorula glutinis* are the most often isolated yeasts from different kinds of olives and fermentations (Arroyo-López et al. 2008). Nisiotou et al. (2010) observed a relatively broad biodiversity of yeast population in spontaneous fermentation of black Conservolea olives. At the beginning of the process, *Metschnikowia pulcherrima* are a dominant yeast species, followed by *Debaryomyces hansenii*, *Aureobasidium pullulans*, *Rhodotorula mucilaginosa*, and *Rhodosporidium diobovatum.* At 17 to 35 days of fermentation, the prevailing species are *Pichia membranifaciens, Pichia anomala,* and *Candida boidinii,* while at the end of fermentation, *P. anomala, P. membranifaciens,* and *Pichia kluyveri* species (Nisiotou et al. 2010).

When yeasts become the dominant group of microorganisms during fermentation, a product with milder taste but with shorter shelf life is produced. Excessive growth of yeasts during fermentation can also lead to a vigorous production of CO_2 that may penetrate the olives and damage the fruit. Some yeasts possess polysaccharolytic activity and can degrade polysaccharides of the olive cell wall (Hernández et al. 2007). On the other hand, yeasts produce many compounds with an important role in flavor generation and texture maintenance. Additionally, these groups of microorganisms synthesize vitamins, amino acids, and purines as well as breakdown products of carbohydrates that are indispensable for the growth of *Lactobacillus* species (Arroyo-Lopez et al. 2008).

Natural fermentation of olives sometimes fails due to excessive growth of spoilage microorganisms, which is a critical point during production of table olives. It seems to be necessary to control the fermentation process by adding a LAB starter in order to obtain high-quality products (Campaniello et al. 2005). An additional benefit from using a starter culture is a reduction of fermentation time and spoilage risk, as well as improving the process control and sensory properties of the product (Panagou et al. 2008). The decrease of pH in spontaneous fermentations is slower than that in the control process with *L. plantarum* as a starter culture (Leal-Sánchez et al. 2003).

De Bellis et al. (2010) showed that the human isolate *Lactobacillus paracasei* IMPC2 can be considered as a probiotic strain suitable for industrial processing of table olives leading to a final product with functional appeal. This strain rapidly colonizes the olive surface and is effective in accelerating the fermentation process. As a result, a reduction of the survival period of potential spoilage microorganisms (*Enterobacteriaceae* and yeasts) was observed (De Bellis et al. 2010).

Hurtado et al. (2010) revealed that co-inoculations give a better microbial development profile than single inoculations. The yeast *C. diddensiae* C6B19 associated with *L. pentosus* 5E3A18 as an adjunct starter effectively inhibits undesirable contaminating yeasts and food-borne pathogens and improves LAB development (Hurtado et al. 2010).

10.5.4 THE CHEMICAL CHANGES DURING PROCESSING ON FERMENTATION

Raw olives contain mainly glucose and fructose and lower amounts of saccharose and mannitol (3.5%–6% of the flesh). Saccharide concentrations decrease with fruit maturation. During fermentation, the saccharides in olives and those diffusing to the brine are fermented by microorganisms, mainly to lactic acid. The D-isomer of lactic acid is produced in higher amounts than the L-isomer (Tassou, Panagou, and Katsaboxakis 2002). Apart from lactic acid, acetic and propionic acids are formed in considerable amounts, whereas citric and malic acids are present at low levels and are degraded completely during the process (Panagou et al. 2008).

Olive fruits are a rich source of phenolic compounds that constitute 1%–14% weight of dry pulp (Bianchi 2003). The concentration of these compounds depends on the fruit maturity. Oleuropein, hydroxytyrosol (β-(3-4-dihydroxyphenyl)ethanol), tyrosol (β-(4-hydroxy- phenyl)ethanol) and verbascoside are the main phenols identified in olives. The most abundant is oleuropein, but its concentration decreases during ripening (Amiot, Fleuriet, and Macheix 1989). In black olives, the concentration of oleuropein can fall to zero. The other phenolic compounds detected in fresh olives are protocatechuic acid, p-hydroxyphenylacetic acid, p-coumaric acid, and ferulic and benzoic acid. In fresh green olives, additionally present are gallic, p-hydroxybenzoic, and syringic acids, whereas vanillin, m-coumaric acid, and o-coumaric acid are detected only in fresh varicolored and black olives.

During fermentation, the phenols diffuse from the pulp into the brine. During spontaneous and controlled fermentation of varicolored olives, the phenolic contents are reduced by 55% and 46%, respectively, and black olives by only 43% (spontaneous fermentation) and 32% (controlled fermentation) (Othman et al. 2009). During processing of olives, the phenolic compounds undergo changes. The lye treatment of green and varicolored olives causes rapid hydrolysis of oleuropein into β(3-4-dihydroxy-phenyl)ethanol, oleoside 11-methylester, and oleoside (Figure 10.4). In the case of black olives, this glucoside is slowly hydrolyzed during fermentation by enzymes from oleuropeinolytic *L. plantarum* strains. Oleuropein is first hydrolyzed to aglycone by β-glucosidase and then to hydroxytyrosol and elenolic acid by esterase.

Degradation of verbascoside leads to formation of caffeic acid that is not present in fresh olive, and its concentration increases during fermentation (Brenes-Balbuena, García-García, and Garrido-Fernandez 1992). The changes of other phenolic compounds depend on the olive maturation stage. The concentrations of gallic, p-hydroxyphenylacetic, vanillic, and benzoic acids decreased after fermentation of green olives. However, their concentrations increased during processing of varicolored and black olives.

Processing of olives according to the Spanish method causes loss of a part of chlorophylls a and b due to their dissolution in water. The remaining part is transformed into the corresponding pheophytins. In the case of black olives, anthocyanins are the main substances responsible for this color. The anthocyanins content also increases in the brine (Romero et al. 2004). Lye treatment causes loss in the

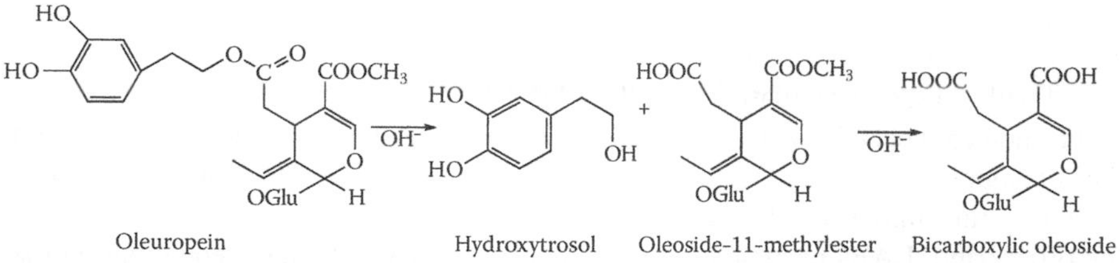

FIGURE 10.4 Hydrolysis of oleuropein in alkaline conditions (From G. Bianchi. 2003. Lipids and phenols in table olives. *Journal of Lipid Science and Technology* 105:229–242. With permission.)

oil content of the fruit. During processing of olives, the lipids are protected against oxidation due to the presence of phenolic compounds, and the quality of oil is almost wholly maintained (Bianchi 2003).

The concentration of volatile compounds varies depending on the treatments and microbial activity in the brines (Panagou and Tassou 2006). Among the volatile compounds identified during fermentation are ethanol, methanol, 4-methyl-1-pentanol, 1-pentanol, 2-pentanol, acetaldehyde, ethyl acetate, isobutyl acetate, hexyl acetate, isobutyric acid, isovaleric acid, and propionic acid. The most important sensory compound formed during fermentation of olive is acetaldehyde, which comes from yeast activity (Osborne et al. 2000). The concentration of this compound is reduced gradually throughout the process due to its high volatility (Panagou and Tassou 2006). The content of ethanol also depends on the activity of yeasts and additionally on heterofermentative LABs. The controlled fermentation (using starter cultures) leads to lower population of yeasts and results in low ethanol concentration. Ethanol is a precursor of ethyl acetate that is formed at the beginning of fermentation, and its concentration increases gradually throughout the process (Panagou and Tassou 2006). The concentration of this product should not exceed 200 mg/L to prevent unpleasant odors (Roza et al. 2003). Methanol detected in the headspace of fermented olive brines probably is attributed to the activity of pectinolytic enzymes that cleave the methoxyl group of pectin present in fruit (Montaño et al. 1992).

10.6 CONCLUDING REMARKS

Fermentation of vegetables is one of the better ways of extending the shelf life of vegetables. Products obtained by fermentation are palatable and possess a sour and carbonated taste. Additionally, such products have high nutritious value. Fermented vegetables are rich sources of vitamins, particularly vitamin C, vitamin B group, and β-carotene, dietary fiber, mineral salts, and phytochemicals, especially flavonoids and glucosinolates. These last components are decomposed by myrosinase to isothiocyanates and indole compounds during fermentation. These compounds demonstrate high biological activity, crucial from the point of view of chemoprevention of cancer. Many mechanisms are proposed in the literature to explain the antitumor properties of isothiocyanates. The best known is a mechanism based on retardation of metabolic activation of carcinogenic compounds by cytochrome P450 (phase I) connected with induction of phase II detoxification enzymes (Śmiechowska, Bartoszek, and Namieśnik 2008). Nevertheless, it is well known that products of glucosinolate hydrolysis affect some cell processes such as regulation of transcription factor levels, signal pathways, and regulation of cell cycle and apoptosis.

Fermented vegetables also contain LABs, which show a beneficial effect on the intestinal ecosystem of the consumer. Additionally, using probiotic strains as a starter culture may serve to increase the nutritional value of the final product. On the other hand, LABs participating in fermentation of vegetables, depending on the specific properties of the strain, produce the lactic acid that consists of two isomers: L-(+) and D-(-) forms. L-(+) lactic acid is the physiological form because it is easily and completely metabolized by humans due to the presence of L-lactate dehydrogenase, whereas the D-(-) form is metabolized significantly more slowly and may lead to lactate acidosis. Nowadays, it is considered that D-(-) lactic acid does not cause gastric problems for healthy adults.

REFERENCES

Amiot, M. J., A. Fleuriet, and J. J. Macheix. 1989. Accumulation of oleuropein derivatives during olive maturation. *Phytochemistry* 28:67–69.

Arroyo-López, F. N., A. Querol, J. Bautista-Gallego, and A. Garrido-Fernández. 2008. Role of yeasts in table olive production. *International Journal of Food Microbiology* 128:189–196.

Beganovic, J., A. L. Pavunc, K. Gjuračić, M. Špoljarec, J. Šušković, and B. Kos. 2011. Improved sauerkraut production with probiotic strain *Lactobacillus plantarum* L4 and *Leuconostoc mesenteroides* LMG 7954. *Journal of Food Science* 76:124–129.

Bianchi, G. 2003. Lipids and phenols in table olives. *Journal of Lipid Science and Technology* 105:229–242.

Breidt Jr., F. 2004. A genomic study of *Leuconostoc mesenteroides* and the molecular ecology of sauerkraut fermentation. *Journal of Food Science* 69:30–32.

Brenes-Balbuena, M., P. García-García, and A. Garrido-Fernandez. 1992. Phenolic compounds related to the black color formed during the processing of ripe olives. *Journal of Agricultural and Food Chemistry* 40:1192–1196.

Buskov, S., C. E. Olsen, H. Sørensen, and S. Sørensen. 2000. Supercritical fluid chromatography as basis for identification and quantitative determination of indol-3-ylmethyl oligomers and ascorbigens. *Journal of Biochemical and Biophysical Methods* 43:175–195.

Campaniello, D., A. Bevilacqua, D. D'Amato, M. R. Corbo, C. Altieri, and M. Sinigaglia. 2005. Microbial characterization of table olives processed according to Spanish and natural styles. *Food Technology and Biotechnology* 43:289–294.

Cha, Y. J., H. Kim, and K. R. Cadwallader. 1998. Aroma-active compounds in kimchi during fermentation. *Journal of Agricultural and Food Chemistry* 46: 1944–1953.

Chang, H. W., K. H. Kim, Y. D. Nam, S. W. Roh, M. S. Kim, C. O. Jeon, H. M. Oh, and J. W. Bae. 2008. Analysis of yeast and archaeal population dynamics in kimchi using denaturing gradient gel electrophoresis. *International Journal of Food Microbiology* 126:159–166.

Cheigh, H. S. 1995. Critical review on biochemical characteristics of kimchi. *Journal of East Asian Society of Dietary Life* 5:89–101.

Chin, H. W., and R. C. Lindsay. 1994. Mechanisms of formation of volatile sulfur compounds following the action of cysteine sulfoxide lyases. *Journal of Agricultural and Food Chemistry* 42:1529–1536.

Cho, J., D. Lee, C. Yang, J. Jeon, J. Kim, and H. Han. 2006. Microbial population dynamics of kimchi, a fermented cabbage product. *FEMS Microbiology Letters* 257:262–267.

Cho, N. C., D. Y. Jhon, M. S. Shin, Y. H. Hong, and H. S. Lim. 1988. Effect of garlic concentrations on growth of microorganisms during kimchi fermentation. *Food Science and Technology* 20:231–235.

Choe, S. M., Y. S. Jun, K. Y. Park, and H. S. Cheigh. 1991. Changes in the contents of moisture, reducing sugar, microorganisms, NO_2, and NO_3 during salting in various varieties of Chinese cabbage for kimchi fermentation. *Bulletin of Home Economics Pusan National University* 17:25–30.

Choi, S. Y., and L. R. Beuchat. 1994. Growth inhibition of *Listeria monocytogenes* by a bacteriocin of *Pediococcus acidilacti* M during fermentation of kimchi. *Food Microbiology* 11:301–307.

Ciska, E., and D. R. Pathak. 2004. Glucosinolate derivatives in stored fermented cabbage. *Journal of Agricultural and Food Chemistry* 52:7938–7943.

Collins, M. D., J. Samelis, J. Metaxopoulos, and S. Wallbanks. 1993. Taxonomic studies on some Leuconostoc-like organisms from fermented sausages: Description of a new genus *Weissella* for the *Leuconostoc paramesenteroides* group of species. *Journal of Applied Bacteriology* 75:595–603.

Costilow, R. N., and F. W. Fabian. 1953. Availability of essential vitamins and amino acids for *Lactobacillus plantarum* in cucumber fermentations. *Applied Microbiology* 1:320–326.

Daxenbichler, M. E., C. H. VanEtten, and P. H. Williams. 1980. Glucosinolate products in commercial sauerkraut. *Journal of Agricultural and Food Chemistry* 28:809–811.

De Bellis, P., F. Valerio, A. Sisto, S. L. Lonigro, and P. Lavermicocca. 2010. Probiotic table olives: Microbial populations adhering on olive surface in fermentation sets inoculated with the probiotic strain *Lactobacillus paracasei* IMPC2.1 in an industrial plant. *International Journal of Food Microbiology* 140:6–13.

Etchells, J. L., T. A. Bell, H. P. Fleming, R. E. Kelling, and R. L. Thompson. 1973. Suggested procedure for the controlled fermentation of commercially brined pickling cucumbers: The use of starter cultures and reduction of carbon dioxide accumulation. *Pickle Pak Science* 3:4–14.

Fleming, H. P. 1984. Developments in cucumber fermentation. *Journal of Chemical Technology and Biotechnology* 34:241–252.

Fleming, H. P., L. C. McDonald, R. F. McFeeters, R. L. Thompson, and E. G. Humphries. 1995. Fermentation of cucumbers without sodium chloride. *Journal of Food Science* 60:312–319.

Fleming, H. P., R. F. McFeeters, M. A. Daeschel, E. G. Humphries, and R. L. Thompson. 1988. Fermentation of cucumbers in anaerobic tanks. *Journal of Food Science* 53:127–133.

Fleming, H. P., R. F. McFeeters, and E. G. Humphries. 1988. A fermentor study of sauerkraut fermentation. *Biotechnology and Bioengineering* 31:189–197.

Garrido Fernández, A., M. J. Fernández Díaz, and R. M. Adams. 1997. *Table olives: Production and processing*. London: Chapman and Hall.

Gil, V., and A. J. MacLeod. 1980. The effects of pH on glucosinolate degradation by a thioglucoside glucohydrolase preparation. *Phytochemistry* 19:2547–2551.

Halász, A., A. Bárath, L. Simon-Sarkadi, and W. Holzapfel. 1994. Biogenic amines and their production by microorganisms in food. *Trends in Food Science and Technology* 51:42–49.

Hang, Y. D., D. L. Downing, J. R. Stamer, and D. F. Splittstoesser. 1972. Wastes generated in the manufacture of sauerkraut. *Journal of Milk and Food Technology* 35:432–435.

Hawer, W. D. 1994. A study on the analysis of volatile flavor of kimchi. *Analytical Science and Technology* 7:125–132.

Hawer, W. D., J. H. Ha, H. M. Seog, Y. J. Nam, and D. H. Shin. 1988. Changes in the taste and flavour compounds of kimchi during fermentation. *Korean Journal of Food Science and Technology* 20:511–517.

Hernández, A., A. Martin, E. Aranda, F. Pérez-Nevado, and M. G. Córdoba. 2007. Identification and characterization of yeast isolated from the elaboration of seasoned green table olives. *Food Microbiology* 24:346–351.

Hong, S. I., and W. S. Park. 2000. Use of color indicators as an active packaging system for evaluating kimchi fermentation. *Journal of Food Engineering* 46:67–72.

Hrncirik, K., J. Valusek, and J. Velisek. 2001. Investigation of ascorbigen as a breakdown product of glucobrassicin autolysis in Brassica vegetables. *European Food Research and Technology* 212:576–581.

Hurtado, A., C. Reguant, A. Bordons, and N. Rozès. 2010. Evaluation of a single and combined inoculation of a *Lactobacillus pentosus* starter for processing *cv. Arbequina* natural green olives. *Food Microbiology* 27:731–740.

IOOC. 1982. Unified qualitative standard applying to table olives in international trade. International Olive Oil Council, Madrid.

Jang, K. S., M. J. Kim, Y. A. Oh, I. D. Kim, H. K. No, and S. D. Kim. 1991. Effects of various sub-ingredients on sensory quality of Korean cabbage kimchi. *Journal of the Korean Society of Food Science and Nutrition* 20:233–240.

Johanningsmeier, S. D., H. P. Fleming, R. L. Thompson, and R. F. McFeeters. 2005. Chemical and sensory properties of sauerkraut produced with *Leuconostoc mesenteroides* starter cultures of differing malolactic phenotypes. *Journal of Food Science* 70:343–349.

Johanningsmeier, S. D., R. F. McFeeters, H. P. Fleming, and R. L. Thompson. 2007. Effects of *Leuconostoc mesenteroides* starter culture on fermentation of cabbage with reduced salt concentration. *Journal of Food Science* 72:166–172.

Kalač, P., J. Špička, M. Křížek, and T. Pelikánová. 2000. Changes in biogenic amine concentrations during sauerkraut storage. *Food Chemistry* 69:309–314.

Kang, J. H., J. H. Lee, S. Min, and D. B. Min. 2003. Changes of volatile compounds, lactic acid bacteria, pH, and headspace gases in kimchi, a traditional Korean fermented vegetable product. *Journal of Food Science* 68:849–854.

Kim, H. J., S. M. Kang, and C. B. Yang. 1997. Effects of yeast addition as starter on fermentation of kimchi. *Korean Journal of Food Science and Technology* 29:790–799.

Kim, H. O., and H. S. Rhee. 1975. Studies on the nonvolatile organic acids in kimchis fermented at different temperatures. *Korean Journal of Food Science and Technology* 7:74–81.

Kim, M. H., and M. J. Chang. 2000. Fermentation property of Chinese cabbage kimchi by fermentation temperature and salt concentration. *Journal of the Korean Society of Agricultural Chemistry and Biotechnology* 43:7–11.

Kim, M. H., M. S. Shin, D. Y. Jhon, Y. H. Hong, and H. S. Lim. 1987. Quality characteristics of kimchis with different ingredients. *Journal of the Korean Society of Food and Nutrition* 16:268–277.

Klieber, A., and B. Franklin. 2000. Ascorbic acid content of minimally processed Chinese cabbage. *Acta Horticulturae* 518:201–204.

Künsch, U., H. Schärer, and A. Temperli. 1989. Biogene amine als qualitätsindikatoren von sauerkraut. In *XXIV Vortragstagung der Deutschen Gesellschaft für Qualitätsforschung. Qualitätsaspekte von Obst und Gemüse*. Ahrensburg, Germany.

Kwon, H., and Y. K. L. Kim. 2003. Korean fermented foods: Kimchi and doenjang. In *Handbook of Fermented Functional Foods*, ed. E. R. Farnworth, 287–304. Boca Raton, FL: CRC Press.

Leal-Sánchez, M. V., J. L. Ruiz-Barba, A. H. Sánchez, L. Rejano, R. Jiménez-Díaz, and A. Garrido. 2003. Fermentation profile and optimization of green olive fermentation using *Lactobacillus plantarum* LPCO10 as a starter culture. *Food Microbiology* 20:421–430.

Lee, C. Y., T. E. Acree, R. M. Butts, and J. R. Stamer. 1974. Flavor constituents of fermented cabbage. *Proceedings of the 4th International Congress of Food Science and Technology* 1:175–178.

Lee, H. S., Y. T. Ko, and S. J. Lim. 1984. Effects of protein sources on kimchi fermentation and on stability of ascorbic acid. *Korean Journal of Nutrition* 17:101–107.

Lee, J. S., G. J. Heo, J. W. Lee, Y. J. Oh, Y. A. Park, J. H. Park, Y. R. Pyun, and Y. S. Ahn. 2005. Analysis of kimchi microflora using denaturing gradient gel electrophoresis. *International Journal of Food Microbiology* 102:143–150.

Lee, N. J., and J. K. Chun. 1982. Studies on the kimchi pasteurization. II: Effects of kimchi pasteurization conditions on shelf life of kimchi. *Journal of the Korean Agricultural Chemical Society* 25:197–200.

Lee, S. H., and S. D. Kim. 1988. Effect of starters on fermentation of kimchi. *Journal of the Korean Society of Food and Nutrition* 17:342–347.

Lee, S. K., M. S. Shin, D. Y. Jhung, Y. H. Hong, and H. S. Lim. 1989. Changes during fermentation of kimchis containing various levels of garlic. *Korean Journal of Food Science and Technology* 21:68–74.

Lu, Z., F. Breidt, V. Plengvidhya, and H. P. Fleming. 2003. Bacteriophage ecology in commercial sauerkraut fermentations. *Applied and Environmental Microbiology* 69:3192–3202.

Lu, Z., H. P. Fleming, and R. F. McFeeters. 2002. Effects of fruit size on fresh cucumber composition and the chemical and physical consequences of fermentation. *Food Science* 8:2934–2939.

Martinez-Villaluenga, C., E. Peñas, J. Frais, E. Ciska, J. Honke, M. K. Piskula, H. Kozlowska, and C. Vidal-Valverde. 2009. Influence of fermentation conditions on glucosinolates, ascorbigen, and ascorbic acid content in white cabbage (*Brassica oleracea* var. *capitata* cv. Taler) cultivated in different seasons. *Journal of Food Science* 74:62–67.

McDonald, L. C., D. H. Shieh, F. P. Fleming, R. F. McFeeters, and R. L. Thompson. 1993. Evaluation of malolactic-deficient strains of *Lactobacillus plantarum* for use in cucumber fermentation. *Food Microbiology* 10:489–499.

McFeeters, R. F., H. P. Fleming, and R. L. Thompson. 1982. Malic acid as a source of carbon dioxide in cucumber juice fermentations. *Journal of Food Science* 47:1862–1865.

McFeeters, R. F., and I. Pérez-Díaz. 2010. Fermentation of cucumbers brined with calcium chloride instead of sodium chloride. *Journal of Food Science* 75:291–296.

Mheen, T. I. 2010. Kimchi fermentation and characteristics of the related lactic acid bacteria. *Korea Institute of Science and Technology Information* 1:1–34.

Mheen, T. I., and T. W. Kwon. 1984. Effect of temperature and salt concentration on kimchi fermentation. *Korean Journal of Food Science and Technology* 16:443–450.

Montaño, A., A. de Castro, L. Rejano, and A.-H. Sánchez. 1992. Analysis of zapatera olives by gas and high-performance liquid chromatography. *Journal of Chromatography A* 594:259–267.

Moreno-Arribas, M. V., M. C. Polo, F. Jorganes, and R. Muñoz. 2003. Screening of biogenic amine production by acid lactic bacteria isolated from grape must and wine. *International Journal of Food Microbiology* 84:117–123.

Mozafar, A. 1994. *Plant vitamins: Agronomic, physiological, and nutritional aspects.* Boca Raton, FL: CRC Press.

Nisiotou, A. A., N. Chorianopoulos, G.-J. E. Nychas, and E. Z. Panagou. 2010. Yeast heterogeneity during spontaneous fermentation of black Conservolea olives in different brine solutions. *Journal of Applied Microbiology* 109:396–405.

Osborne, J. P., R. Mira de Orduña, G. J. Pilone, and S. Q. Liu. 2000. Acetaldehyde metabolism by wine lactic acid bacteria. *FEMS Microbiology Letters* 191:51–55.

Othman, N. D., D. Roblain, N. Chammen, P. Thonart, and M. Hamdi. 2009. Antioxidant phenolic compounds loss during the fermentation of Chétoui olives. *Food Chemistry* 116:662–669.

Panagou, E. Z., U. Schillinger, C. M. A. P. Franz, and G.-J. E. Nychas. 2008. Microbiological and biochemical profile of cv. Conservolea naturally black olives during controlled fermentation with selected strains of lactic acid bacteria. *Food Microbiology* 25:348–358.

Panagou, E. Z., and C. C. Tassou. 2006. Changes in volatile compounds and related biochemical profile during controlled fermentation of cv. Conservolea green olives. *Food Microbiology* 23:738–746.

Park, H. J., and Y. S. Han. 1994. Effect of mustard leaf on quality and sensory characteristics of kimchi. *Journal of the Korean Society of Food and Nutrition* 23:618–624.

Park, S. K., Y. S. Cho, J. R. Park, and J. S. Moon. 1995. Changes in the contents of sugar, acid, free amino acid, and nucleic acid-related compounds during fermentation of leaf mustard-kimchi. *Journal of the Korean Society of Food and Nutrition* 24:48–53.

Park, W. P., K. D. Park, J. H. Kim, Y. B. Cho, and M. J. Lee. 2000. Effect of washing conditions in salted Chinese cabbage on the quality of kimchi. *Journal of the Korean Society of Food Science and Nutrition* 29: 30–34.

Park, W. S., S. W. Moon, M. K. Lee, B. H. Ahn, and Y. J. Koo. 1996. Comparison of fermentation characteristics of the main types of Chinese cabbage kimchi. *Food Biotechnology* 5:128–135.

Pederson, C. S., and M. N. Albury. 1969. The sauerkraut fermentation. New York State Agricultural Experiment Station bulletin 824.

Pederson, C. S., L. R. Mattick, F. A. Lee, and R. M. Butts. 1964. Lipid alterations during the fermentation of dill pickles. *Applied Microbiology* 12:513–516.

Peñas, E., J. Frias, B. Sidro, and C. Vidal-Valverde. 2010. Impact of fermentation conditions and refrigerated storage on microbial quality and biogenic amine content of sauerkraut. *Food Chemistry* 123:143–150.

Pérez-Díaz, I., and R. F. McFeeters. 2011 preparation of a *Lactobacillus plantarum* starter culture for cucumber fermentations that can meet kosher guidelines. *Journal of Food Science* 76:120–123.

Romero, C., M. Brenes, K. Yousfi, P. Garcia, A. Garcia, and A. Garrido. 2004. Effect of cultivar and processing method on the contents of polyphenols in table olives. *Journal of Agricultural and Food Chemistry* 52:479–484.

Roza, C., A. Laca, L. A. Garcia, and M. Díaz. 2003. Ethanol and ethyl acetate production during the cider fermentation from laboratory to industrial scale. *Process Biochemistry* 38:1451–1456.

Ruiz-Barba, J. L., D. P. Cathcart, P. J. Warner, and R. Jiménez-Díaz. 1994. Use of *Lactobacillus plantarum* LPCO10, a bacteriocin producer, as a starter culture in Spanish style green olive fermentation. *Applied and Environmental Microbiology* 60:2059–2064.

Ryu, B. M., Y. S. Jeon, Y. S. Song, and K. S. Moon. 1996. Physicochemical and sensory characteristics of anchovy added kimchi. *Journal of the Korean Society of Food and Nutrition* 25:460–469.

Shim, S. M., M. J. Kim, B. S. Kim, and G. H. Kim. 2008. Effects of washing solution and LDPE package on the quality changes of brined Chinese cabbages at different storage temperatures. *Food Science and Technology Research* 14:395–402.

Singh, A. K., and A. Ramesh. 2008. Succession of dominant and antagonistic lactic acid bacteria in fermented cucumber: Insights from a PCR-based approach. *Food Microbiology* 25:278–287.

Śmiechowska, A., A. Bartoszek, and J. Namieśnik. 2008. Cancer chemopreventive agents: Glucosinolates and their decomposition products in white cabbage (*Brassica oleracea* var. *capitata*). *Postępy Higieny i Medycyny Doświadczalnej* 62:125–140.

So, M. H., and Y. B., Kim. 1995. Identification of psychrotrophic lactic acid bacteria isolated from kimchi. *Korean Journal of Food Science and Technology* 27:495–505.

Sohn, K. H., and H. J. Lee. 1998. Effects of high pressure treatment on the quality and storage of kimchi. *International Journal of Food Science and Technology* 33:359–365.

Song, H. P., D. H. Kim, H. S. Yook, M. R. Kim, K. S. Kim, and M. W. Byun. 2004. Nutritional, physiological, physicochemical, and sensory stability of gamma irradiated kimchi (Korean fermented vegetables). *Radiation Physics and Chemistry* 69:85–90.

Spicka, J., P. Kalac, S. Bover-Cid, and M. Krizek. 2002. Application of lactic acid bacteria starter cultures for decreasing the biogenic amine levels in sauerkraut. *European Food Research and Technology* 215:509–514.

Tassou C. C., C. Z. Katsaboxakis, D. M. R. Georget, M. L. Parker, K. W. Waldron, A. C. Smith, and E. Z. Panagou. 2007. Effect of calcium chloride on mechanical properties and microbiological characteristics of cv. Conservolea naturally black olives fermented at different sodium chloride levels. *Journal of the Science of Food and Agriculture* 87:1123–1131.

Tassou, C. C., E. Z. Panagou, and K. Z. Katsaboxakis. 2002. Microbiological and physicochemical changes of naturally black olives fermented at different temperatures and NaCl levels in the brines. *Food Microbiology* 19:605–615.

Tolonen, M., M. Taipale, B. Viander, J. M. Pihlava, H. Korhonen, and E. L. Ryhänen. 2002. Plant-derived biomolecules in fermented cabbage. *Journal of Agricultural and Food Chemistry* 50:6798–6803.

Trail, A. C., H. P. Fleming, C. T. Young, and R. F. McFeeters. 1996. Chemical and sensory characterization of commercial sauerkraut. *Journal of Food Quality* 19:15–30.

Vaughn, S. F., and M. A. Berhow. 2005. Glucosinolate hydrolysis products from various plant sources: pH effects, isolation, and purification. *Industrial Crops and Products* 21:193–202.

Vorbeck, M. L., F. R. Mattick, F. A. Lee, and C. S. Pederson. 1961. Volatile flavor of sauerkraut: Gas chromatographic identification of a volatile acidic off-odor. *Journal of Food Science* 26:569–572.

Yoon, S. S., R. Barrangou-Poueys, F. Breidt Jr., T. R. Klaenhammer, and H. P. Fleming. 2002. Isolation and characterization of bacteriophages from fermenting sauerkraut. *Applied and Environmental Microbiology* 68:973–976.

Zhou, A., and R. F. McFeeters. 1998. Volatile compounds in cucumbers fermented in low-salt conditions. *Journal of Agricultural and Food Chemistry* 46:2117–2122.

11 Fermented Dairy Products

*Bhavbhuti M. Mehta and Maricê
Nogueira de Oliveira*

CONTENTS

11.1 INTRODUCTION

Fermented dairy products have been consumed for centuries. In the Indian sub-continent, they are considered an important part of the diet as a highly nutritious, therapeutic, and healthy food as proven by Ayurveda, the old science of medicine (Thirtha 1998). The microbial action and spoilage of raw milk during storage is the basis for the origin of fermented milks. In the earliest days of mankind, spontaneous acidification processes relied on lactic acid bacteria naturally present in milk as adventitious contaminants that grow and produce the lactic acid required to coagulate the milk and improve the keeping qualities and the hygienic status of processed foods. Sometimes the lactic acid bacteria are accompanied by yeasts or molds that give special features to the fermented products (Kandler 1983; Wouters et al. 2002).

The process of fermentation gradually improved as civilization proceeded. Selected strains of microorganisms commonly known as lactic acid bacteria are used for the production of various fermented dairy products of uniform quality. Thus, fermented dairy products generally refer to dairy products that employ selected microorganisms to develop characteristic flavor and/or body and texture. Fermented dairy products such as yogurt, *dahi* (curd), *shrikhand*, Bulgarian butter milk, acidophilus milk, kefir, koumiss, and varieties of cheese are commonly manufactured throughout the world.

Prescott and Dunn (1957) defined fermentation in a broad sense as

a process in which chemical changes are brought about in an organic substrate, whether carbohydrate or protein or fat or some other type of organic material, through the action of biochemical catalysts known as "enzymes" elaborated by specific types of living microorganisms.

The definition is a perfect fit in the field of dairy science and technology. Fermentations can be categorized as desirable or undesirable. Cultures of well-characterized pure or mixed bacterial strains, when used as inoculums in milk, produce desirable intermediate or end products, whereas undesirable substances may be produced in products in the presence of natural bacteria or contaminants in milk (Marth 1974).

Thus, during fermentation, various chemical/biochemical changes take place. These changes primarily depend on the quality of raw milk, various processing steps, the microorganisms used, and the type of the products manufactured. Compounds formed during fermentation, including lactic acid, acetic acid, propionic acid, diacetyl, carbon dioxide, ethyl alcohol, exopolysaccharides, and bacteriocins, affect the flavor, texture, and consistency of the product and inhibit spoilage and pathogenic microorganisms (Walstra, Wouters, and Geurts 2006a). In addition to these, fermentation causes predigestion of the nutrients, which can then be easily metabolized by the human body. Fermented dairy products are useful in the treatment of gastrointestinal disorders, as these products have anticarcinogenic, hypocholesterolemic, and immunostimulating properties (Prajapati 1995).

11.2 MICROORGANISMS USED IN FERMENTATION

Selected microorganisms, known as starter cultures, are deliberately added to milk and, under specific conditions, cause desirable changes to the milk components in the production of fermented dairy products. These dairy starter cultures are generally lactic acid bacteria (*Streptococcus*, *Lactococcus*, *Lactobacillus*, and *Leuconostoc*). However, nonlactic starters (bacteria, yeast, and mold) are also used in combination with lactic starters in the manufacture of kefir, koumiss, and mold-ripened cheeses (e.g., Camembert and Roquefort). The lactic acid bacteria are divided into two groups on the basis of their growth optimum: mesophilic bacteria, which have an optimum growth temperature between 20°C and 30°C, and thermophilic bacteria with optimum growth temperatures of between 30°C and 45°C (Wouters et al. 2002).

Mesophilic lactic acid bacteria include *Lactococcus lactis* ssp. *lactis*, *L. lactis* ssp. *cremoris*, and *Leuconostoc mesenteroides* ssp. *cremoris*. The first two species are known as *acid producers*, as they produce mainly lactic acid. The *Leuconostoc* bacteria are considered to be *aroma producers*, which also ferment citric acid to important metabolites, such as acetaldehyde, CO_2, and especially diacetyl. The balance between aroma- and acid-producers is very important (Walstra, Wouters, and Geurts 2006b). Thermophilic starter cultures are microaerophilic and include the genera of *Streptococcus* and *Lactobacillus*. The important metabolic activities of thermophilic cultures in development of fermented dairy products are production of lactic acid, acetaldehyde, polysaccharides, fatty acids, peptides, and amino acids (IDF 1988; Edward 2003).

Probiotic or therapeutic cultures include *Enterococci*, *Lactococci*, *Propionibacteria*, *Leuconostoc*, and *Pediococci*, but the principal organisms are of the genera *Lactobacillus* and *Bifidobacterium*. Mesophilic and thermophilic lactic acid bacteria, and yeasts, such as *Candida kefir*, *Saccharomyces*, *Kluyveromyces marxianus*, and *Torula koumiss,* are the main microorganisms used in yeast-lactic fermentations, while in mold-lactic fermentations mesophilic lactic acid bacteria and mold (*Geotrichum candidum*) are mostly used (Hammer and Babel 1993; Marth and Steele 2001; Robinson 2002; Edward 2003; Tamime 2006).

Most ripened cheeses are the product of metabolic activities of the lactic acid bacteria. A mixed culture of *L. delbrueckii* ssp. *bulgaricus* and *S. thermophilus* is usually employed along with a culture of *Propionibacterium shermanii*, which is added to function during the ripening process in flavor development and eye formation in Swiss cheese. The spores of *Penicillium roquefortii* are used in the formation of the blue-veined appearance in blue cheeses (e.g., Roquefort). Similarly, Camembert cheese is inoculated with spores of *Penicillium camemberti* (Jay 1992; Kosikowski and Mistry 1997).

11.3 TYPES OF FERMENTED DAIRY PRODUCTS AND THEIR MANUFACTURE

Fermented milks have been produced by traditional methods for many centuries, and there are several hundred such products recorded around the world. They are produced as a result of microbial "souring" of milk. Most are very similar, both in terms of their characteristics and in the technology used to produce them. Many fermented milk products are distinguished only by their region of origin (Campbell-Platt 1987; Kurmann, Rasic, and Kroger 1992).

Fermented milks can be classified on the basis of the type of fermentation they undergo, such as lactic (e.g., sour cream), yeast-lactic (e.g., kefir, koumiss, acidophilus-yeast milk), and mold-lactic (e.g., villi). Lactic fermentation products can be further classified according to the characteristics of the lactic microflora: mesophilic (e.g., *filmjolk*, Nordic ropy milk), thermophilic (e.g., yogurt, *shrikhand*, *skyr*, Bulgarian buttermilk), and probiotic or therapeutic (e.g., "bio"-fermented milks, acidophilus milk) (Robinson and Tamime 1990; Walstra, Wouters, and Geurts 2006b). Examples of each type of product are given Table 11.1.

TABLE 11.1
Microorganisms Involved in Fermented Dairy Products

	Microorganisms used	Fermented dairy products
1	*Lactobacillus acidophilus*	Acidophilus
2	*L. acidophilus + S. thermophilus*	Acidophilus yogurt
3	*L. acidophilus + L. bulgaricus + S. thermophilus, Bifidobacterium* spp.	Biogurt
4	*Bifidobacterium bifidum*	Bifidus milk
5	*L. delbrueckii* subsp. *Bulgaricus*	Bulgarian milk/acid buttermilk
6	*L. lactis* subsp. *lactis, L. lactis* subsp. *cremoris,* and *L. lactis* biovar. *diacetylactis* and *L. mesenteroides* subsp. *Cremoris*	Cultured buttermilk, cultured (sour) cream
7	Lactococci, *Leuconostoc* spp., *L. delbrueckii* subsp. *bulgaricus, L. acidophilus, S. thermophilus*	*Dahi, shrikhand*
8	*L. lactis* biovar. *diacetylactis, Leuconostoc lactis, L. mesenteroides* subsp. *cremoris, Leuconostoc mesenteroides* subsp. *dextranicum, S. thermophilus, L. bulgaricus,* lactose-fermenting yeasts	*Leben/labneh*
9	*Bacterium lacticus login* (synonym of *Lactococcus* spp.)	Nordic sour milks such as *filmjolk* and Nordic ropy milk
10	*Lactobacillus casei* subsp. *casei*	*Yakult*
11	*S. thermophilus + L. delbrueckii* subsp. *Bulgaricus*	Yogurt
12	*Lactobacillus kefir, Lactobacillus brevis, L. acidophilus, Lactococcus lactis* ssp. *lactis, Leuconostoc* spp., yeast (*Candida kefir*), acetic acid bacteria (*Acetobacter aceti* and *Acetobacter rasens*)	*Kefir*
13	*L. delbrueckii* subsp. *bulgaricus, L. acidophilus, Kluyveromyces fragilis, Saccharomyces lactis, Torula koumiss*	Koumiss
14	*L. lactis* subsp. *lactis* biovar. *diacetylactis, L. mesenteroides* subsp. *cremoris, Geotrichum candidum*	Villi
15	Lactococci, *Brevibacterium linens*	Brick cheese
16	*Penicillium camemberti*	Camembert, Brie cheeses
17	Lactococci, *Leu. mesenteroides* subsp. *cremoris, L. casei* (occasionally)	Cheddar, cottage cheeses
18	*Streptococcus thermophilus, Lactobacillus delbrueckii* subsp. *bulgaricus, Lb. helveticus, Lb. Lactis*	Emmental, Gruyere, Parmesan, Romano cheeses
19	*Penicillium roquefortii*	Roquefort cheese
20	Thermophilus Lactobacilli, *S. thermophilus, Prop. freudenreichii* subsp. *Shermanii*	Swiss cheese

The production of fermented milk products may vary slightly, but always includes the steps of preparation of milk and standardization, homogenization, heat treatment of milk or milk-base, and lactic acid fermentation. During the steps before fermentation, chemical changes occur that could affect fermentation. The common steps in production of fermented dairy products are given in Figure 11.1 and for specific products, for example, yogurt and cultured buttermilk, are given in Figure 11.2.

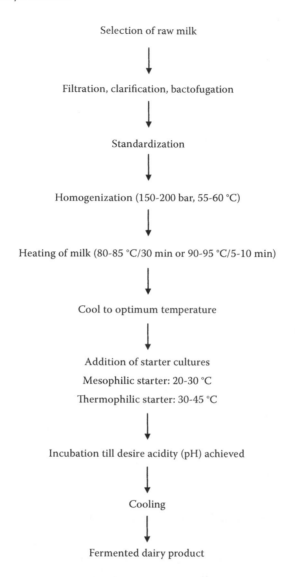

Selection of raw milk

↓

Filtration, clarification, bactofugation

↓

Standardization

↓

Homogenization (150-200 bar, 55-60 °C)

↓

Heating of milk (80-85 °C/30 min or 90-95 °C/5-10 min)

↓

Cool to optimum temperature

↓

Addition of starter cultures
Mesophilic starter: 20-30 °C
Thermophilic starter: 30-45 °C

↓

Incubation till desire acidity (pH) achieved

↓

Cooling

↓

Fermented dairy product

FIGURE 11.1 General flow diagram in manufactured of fermented dairy product(s)

11.4 CHEMICAL CHANGES DURING PROCESSING THAT HAVE AN EFFECT ON FERMENTED DAIRY PRODUCTS

The production of fermented dairy products is based on a combination of chemistry/ biochemistry, microbiology, enzymology, and physics/engineering (technology) knowledge, which makes it a very complex process (Tamime and Marshall 1997). The final quality of the product generally depends on the processing steps. During processing of milk for production of fermented dairy products, various physicochemical changes take place that ultimately affect the body, texture, and flavor of the final products.

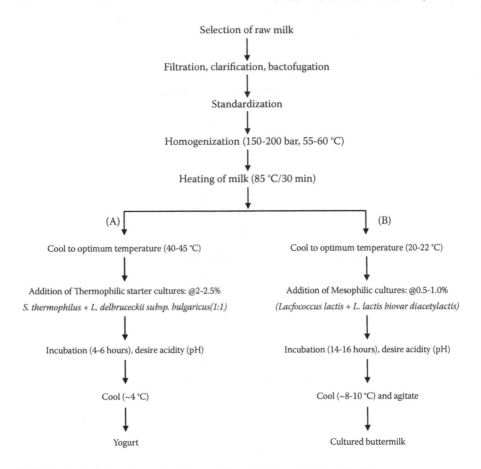

FIGURE 11.2 Manufactured of Yogurt (A) and Cultured buttermilk (B)

11.4.1 Selection of Raw Milk for Manufacture of Fermented Milk

The milk intended for the manufacture of fermented products should be fresh, clean, and free from developed acidity and any off-flavors. The initial microbial load in the raw milk should also be low, and it should not have been stored for a long time before processing. It should be free from antibiotic residues and inhibitory substances not destroyable by heat treatment. Heating of mastitic milk does not improve it as a growth medium, and even after heating, such milk may prevent normal acid development by culture organisms. The presence of leucocytes in mastitic milk leads to phagocytosis of the streptococci and affects the performance of starters and ultimately the final quality of the products (Prouty 1940; Rice 1948; Prajapati 1995).

11.4.2 Standardization and Fortification

The initial quality of milk depends on the diet and stage of lactation of the animal as well as the time of year in which it is collected. These factors are the main reasons why milk should be standardized in order to regulate the content of fat and protein for the production of fermented milk products.

The best options to achieve the desired levels of protein and solids is by the addition of powdered ingredients in the form of skim milk, concentrates, and isolates or hydrolyzed protein from milk or whey, such as caseinates (Lucey, Munro, and Singh 1999; Oliveira et al. 2001; Tamime, Robinson, and Latrille 2001; Remeuf et al. 2003; Sodini et al. 2005).

Fortification of milk protein content of 5% improves yogurt rheological properties (Fox 2001; Tamime, Robinson, and Latrille 2001). Tamime and Robinson (1999) reported that consistency was improved when the solids were in the range 12%–20%, although the major effect was observed from 12% to 14%; levels above 16% resulted in only small changes. The increase in total solids has little detrimental effect on the metabolism of starter cultures, but it increases the buffering capacity of milk, and may promote growth of some microorganisms such as *Lactobacillus acidophilus* and *S. thermophilus* (Marshall 1986; Lucey et al. 1998).

According to Prentice (1992), the increase in protein content is the main factor that influences texture. Supplementation of milk with milk powder results in the development of networks and clusters of casein micelles. Finally, the type and concentration of proteins employed significantly influence the texture and rheology of the final product (Puvanenthiran, Williams, and Augustin 2002; Sodini et al. 2004; Penna, Converti, and De Oliveira 2006; Damin et al. 2009). Fortification with whey protein powders should be limited to 1%–2% because of the increased production of sulfhydryl from cysteine residues of β-lactoglobulin upon heat treatment, which gives a cooked flavor (Tamime and Robinson 1985; Marshall 1986). Moreover, heating at high temperatures coagulates whey protein, and hence milk/whey mixtures are heated to only~ 80°C for 30 min for the production of yogurt (Abd Rabo, Partridge, and Furtado 1988; Sodini, Montella, and Tong 2005). A firmer gel is obtained after fermentation by either mesophilic or thermophilic lactic acid bacteria when the basic milk is fortified with 4% whey protein concentrates (Jelen 1993; Lucas et al. 2004).

Although the number of publications concerning supplementation of milk for manufacture of fermented milk products is high, the results are not conclusive regarding the optimal level of supplementation by the different ingredients to be used. Hence, standardization and fortification should be done in such a way that it maximizes the quality of the final products.

11.4.3 HOMOGENIZATION

Homogenization is carried out under pressure (150–200 bar) at 55°C–65°C by passing the milk through a small orifice. This results in reduction of the average diameter of the fat globules to <2 μm, which prevents the formation of a layer of fat on the surface of milk during fermentation. The increased number of small fat globules enhances the ability of homogenized milk to reflect light, and as a result the fermented dairy products appear relatively whiter than unhomogenized milk. The fat globules in raw milk are covered by a membrane that is made of protein, lipids, and carbohydrates (Mulder and Walstra 1974). In addition to increasing the surface area of the fat globules, homogenization also changes the composition of the membrane. Nearly 2% of the casein in unhomogenized milk is adsorbed on the fat globules; however, due to the turbulent effects of homogenization, about 25% of the casein micelles are adsorbed on newly created fat globules (Wiegner 1914; Trout 1950) and cover some 75% of the surface area of the fat globules (Walstra et al. 1999). Some serum proteins (whey proteins viz. α-lactalbumin (α-La) and β-lactoglobulin (β-lg) are also adsorbed on fat globules 5% of the surface area) (Figure 11.3). The fat globules in homogenized milk now act almost as large "casein micelles" as casein is fixed; as its mobility is lost, it is coagulated more easily. The risk of free whey separation onto the surface of set fermented dairy products (known as syneresis) is reduced, and the firmness of the end product is increased by homogenization. During heating (80°–85°C/30 min or 90°C/5 min) of homogenized milk, the adsorbed as well as unabsorbed whey proteins and k-casein interact with each other via disulfide (-S-S) linkages and form rigid interprotein interactions compared to those in unhomogenized heated milk. These interactions ultimately make firmer coagula during fermentation (Mulder and Walstra 1974; Dalgleish and Sharma 1993; Sharma and Dalgleish 1994; van Boekel and Walstra 1995; IDF 2007; Tamime et al. 2007).

11.4.4 HEAT TREATMENT

Milk used for fermentation is heated to 80°C–85°C/30 min or 90°C/5 min (Robinson 2003). Heat treatment kills vegetative cells of pathogenic and spoilage microorganisms present in raw milk, and reduces the competition and improves the growth medium for the starter cultures. Moreover, heat-labile indigenous milk enzymes such as lipase are inactivated. Partial hydrolysis of casein by the heat treatment, producing increased amounts of soluble nitrogen, is considered to be responsible for increased acid production by the starter organisms (Gilmour and Rowe 1990; Fox 1991; Farkye and Imafidon 1995; Law 1997).

The stability of casein micelles may be markedly affected by changes in the concentration of ions and salts, particularly by changes of the hydrogen ion concentration, during processing. Serum proteins in the native state are not sensitive to changes in the hydrogen ion concentration. They are, however, affected considerably

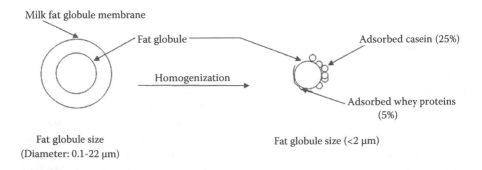

FIGURE 11.3 Effect of homogenization on fat globules

by heat treatment of milk. In particular, the specific structure of the major serum protein, β-lactoglobulin, is disrupted and unfolded in what is commonly referred to as denaturation, and denatured β-Lg forms larger aggregates by self-interaction (Xiong, Dawson, and Wan 1993). A significant effect of denaturation is a considerable decrease of serum protein solubility at the pH of coagulation, 4.6–4.7, which is important in fermented dairy products (IDF 2007; W. Lee and Lucey 2010).

Another effect is the opening up of the serum proteins, leading to a number of chemical changes and interactions such as exposure of the concealed sulfhydryl (-SH) group of β-Lg, lowering of the oxidation-reduction potential, interactions among denatured serum proteins, and a specific (hydrophobic) interaction between κ-casein and β-Lg (Haque and Kinsella 1988; Dalgleish 1990). The last reaction affects the coagulum structure and viscosity, with a corresponding development of soft-curd characteristics. β-Lg reacts with other milk components when denatured, while α-La undergoes heat-induced interactions only after severe heat treatment (Dalgleish and Sharma 1993; Sharma and Dalgleish 1994). The interaction between the κ-casein and the fat globule membrane proteins is a consequence of adsorption of the former to the fat globule surface during homogenization; this may also result in changes in the properties of the milk fat (Houlihan et al. 1992; Singh 1993; van Boekel and Walstra 1995).

Heating of milk expels oxygen and decreases the redox potential, which encourages growth of starter bacteria (Law 1997). High heat treatment slightly degrades the major milk constituents, lactose, fat, and proteins, producing various volatile compounds such as acetone, delta-caprolactone, nonanone-2, benzaldehyde, benzyl alcohol, furfural, dimethyl sulfide, dimethyl sulfone, and hydroxyl-2-pentanone-3 (Viani and Horman 1973; Rasic and Kurmann 1978), and also partially destroys some heat-labile vitamins (Puhan 1988). However, heating of milk at <100°C reduces syneresis and improves the smoothness of yogurt coagulum (Robinson and Tamime 1990).

11.4.5 INOCULATION OF STARTER CULTURE

Various types of starter culture are inoculated at 0.5%–2.5%, depending on the type of product to be manufactured, and coagulate milk by production of acids. Milk inoculated with culture commonly coagulates first at the bottom of the container.

Casein micelles are composed of α_s, β-, and κ-caseins, and colloidal calcium phosphate (acting as a cementing material). During fermentation, lactic acid is produced from degradation of lactose, which reduces the pH. This causes a gradual removal of calcium and phosphorus from the casein particles and their transfer to the soluble phase (Pouliot, Boulet, and Paquin 1989; Le Graet and Brule 1993; IDF 2007). The mean size of casein particles remains fairly constant in a pH range 6.6–5.3, and precipitation of casein is initiated at pH 5.2–5.3. Further reduction in pH leads to a reduction in the net negative charge on casein, which decreases electrostatic repulsion between the charged particles; electrostatic attraction increases, and protein-protein attraction also increases through enhanced hydrophobic interactions (Lucey and Singh 1998; Lucas et al. 2004; Lucey 2004; W. Lee and Lucey 2010). The size of the casein micellelike particles increases, and ultimately complete precipitation occurs at pH 4.6 (the isoelectric point of casein). Dissociation of β-casein from casein particles are pH and temperature dependent (Roefs 1987; Dalgleish and Law 1988; Law 1997).

Homogenization and high-heat treatment of the milk increases the hydrophilic properties of the coagulum and stability of the gel due to denaturation of β-Lg and α-La and their association with κ-casein (Dannenberg and Kessler 1988; Parnell-Clunies et al. 1988; Mottar et al. 1989). The coagulated proteins are actually a coprecipitate of the casein and denatured serum proteins, while those in the serum that remain uncoagulated are undenatured whey proteins, proteose-peptones, nonprotein nitrogen compounds such as amino acids, urea, creatin, creatinine, and other dissolved constituents. Alteration of the physical nature of the casein micelles plays a major role in acid-induced gelation of milk. Hence, desirable gel formation in fermented milk products is due to the combined effect of acid and heat, and sometimes also salt (Aoki, Kako, and Imamura 1986; Holt, Davies, and Law 1986).

Acidification to pH 4.6 and below increases the hydration of proteins and influences the consistency of the final products. The heat treatment given to the milk also considerably affects the chemical nature and structure of the coagulum. The protein coagulation is the formation of a three-dimensional network entrapping fat globules and serum with dissolved constituents. The resulting semisolid gel, opaque white in appearance, is distinguished by its smooth texture and light-custard consistency (Modler and Kalab 1983; Lucas et al. 2004; Lucey 2004; IDF 2007; W. Lee and Lucey 2010).

11.4.6 COOLING

Cooling is one of the most popular methods used to control the metabolic activity of the starter culture and its enzymes. Incorporation of air during the cooling stage leads to reduced viscosity and whey syneresis. Rapid cooling is not desirable, since it arrests acid accumulation and causes reduced diacetyl flavor (Vedamuthu 1985). Cooling that is too rapid can bring unfavorable changes in the structure of coagulum with corresponding whey separation. This defect is probably due to the very rapid contractions of the protein filaments and their disturbed hydration.

11.5 CHEMICAL CHANGES DURING FERMENTATION

Tamime and Deeth (1980) published an excellent review concerning all the technological and biochemical aspects of yogurt, followed by a second article dealing with its nutritional and therapeutic properties (Deeth and Tamime 1981). These reviews include a detailed presentation of *S. thermophilus* and *L. bulgaricus* (now *L. delbrueckii* subsp. *bulgaricus*). Since 1981, both microorganisms have been extensively studied, and reviews dealing with particular aspects of their metabolism and genetics have been published (Hutkins and Morris 1987; Mercenier and Lemoine 1989; Mercenier 1990; Zourari, Accolas, and Desmazeaud 1992).

Lactic bacteria are gram-positive, nonsporogenic microorganisms that accumulate lactic acid as a product of the primary metabolism (Ferreira 2003) of lactose. They can be subdivided into two classes: the *heterofermentative* bacteria, which utilize the Entner-Doudoroff pathway, ferment the carbon source to a mixture of lactic acid, ethanol, and acetic acid; and the *homofermentative*, which utilize the Embden-Meyerhof-Parnas pathway that leads only to lactic acid. Lactic fermentation of milk utilizes lactose and citrate and leads to the accumulation of lactic acid, the main product, and other less significant products that are useful for improving the organoleptic quality and texture of the product (de Vos and Hugenholtz 2004).

Diacetyl is partially reduced to acetoin, responsible for flavor. These two compounds accumulate in the fermentation, depending on the source of carbohydrate (lactose and glucose) and citrate. Lactic acid, diacetyl, and acetoin are derived from pyruvate, which is produced by a carbon source such as glucose and citrate fermented in the culture medium. Glucose produces two units of pyruvate, while citrate produces only one. So, either a carbohydrate or citrate can be involved in the biosynthesis of lactic acid and flavored products as diacetyl and acetoin (Ostlie, Helland, and Narvhus 2003). Many studies have shown that the co-metabolism of citrate and glucose changes the metabolic profile during the fermentation by lactic acid bacteria: Lactic acid is derived mainly from glucose, but diacetyl and acetoin are produced primarily from citrate (de Vos and Hugenholtz 2004).

Streptococcus thermophilus has cocci morphology, generally disposed in short chains. It grows at temperatures between 37°C and 45°C, and can tolerate up to 50°C. It presents high sensitivity to NaCl (<2%), and has the important bio-adjusted function in fermentative processes in the presence of probiotic bacteria (Ferreira 2003). Some strains produce exopolysaccharides, which can be useful in the production of firmer yogurts, maintaining the texture and viscosity during storage (Collet 2005). *L. delbrueckii* subsp. *bulgaricus* exists as small rods, forms long chains, and preferentially grows at temperatures between 45°C and 50°C. Both of these microorganisms are known as yogurt cultures; they are homofermentative, and their optimum fermentation temperature is 42°C.

S. thermophilus (ST) and *L. delbrueckii* subsp. *bulgaricus* (LB) exhibits in the milk an interaction that is mutually favorable and not obligatorily characterized by the fact that each bacterium produces one or more substances that stimulate the growth of the other. The symbiosis phenomenon has been studied by several authors who have observed a positive effect of the mixed culture in comparison with the corresponding pure cultures in terms of growth, acidification, production of flavors and exopolysaccharides, and proteolysis (Béal and Corrieu 1994). ST and LB (1:1) are inoculated into milk and incubated at 37°C–45°C in the manufacture of yogurt. ST grows optimally at these temperatures and hydrolyzes lactose (Driessen, Kingma, and Stadhouders 1982). LB is more proteolytic than ST. LB is stimulated by formic acid and by CO_2 produced by the *Streptococcus*, while ST is stimulated by the amino acids and small peptides from the metabolic activity of LB. Formic acid is likely to be produced by the thermal treatment of milk or by ST. Carbon dioxide is equally a stimulating factor of LB and comes from the decarboxylation of urea by urease elaborated by ST. Several amino acids are mentioned in the literature as stimulating growth factors for ST, namely, histidine, glycine, glutamic acid, methionine, phenylalanine, arginine, cysteine, valine, leucine, tryptophan, tyrosine, lysine, and serine.

11.5.1 Changes in Proteins

Cow's milk contains about 3.5% protein; however, the concentration varies significantly according to factors such as species, breed, stage of lactation, feed, season, and climate. Milk proteins can be fractionated into two well-defined groups. Under acidification at pH 4.6, at 30°C, about 80% of total proteins in milk precipitate from solution. This fraction is called casein, which is the main class of milk proteins. The protein that remains soluble under these conditions is called whey protein. Casein is in the form of colloidal micelles that, along with fat, give milk its white color. The free amino acid content of cow's milk generally does not exceed 0.01% (Rasic and Kurmann 1978; Alm 1982b).

In yogurt, proteolysis does not determine its organoleptic properties, but proteolytic activity is greatly involved in both nutrition and interactions of yogurt bacteria, since lactic acid bacteria cannot synthesize essential amino acids. Therefore, they require an exogenous nitrogen source and utilize peptides and proteins in their growth medium by more or less complex enzyme systems.

Proteolysis in cheese making during ripening plays a vital role in the development of texture as well as flavor and has been the subject of several reviews. However, proteolytic activity of strains used in the manufacture of fermented milks may be of secondary importance (Sousa, Ardo, and McSweeney 2001). Nevertheless, although the yogurt cultures are considered to be only weakly proteolytic, *S. thermophilus* and *L. delbrueckii* subsp. *bulgaricus* may, during the fermentation, cause a significant degree of proteolysis, which may be important for the following reasons:

- The enzymatic hydrolysis of milk proteins results in the liberation of peptides of varying sizes and free amino acids, and these may become involved during the formation of the gel and affect the physical structure of yogurt.
- The liberation of amino acids into the milk is essential to the growth of *S. thermophilus*.
- Although amino acids and peptides may not contribute directly to the flavor of fermented products, they do act as precursors for the multitude of reactions that produce flavor compounds.
- Important nutritional considerations apply through the release of so-called bioactive peptides (Vasiljevic and Shah 2008; Swaisgood 2010).

The profile of nitrogenous compounds in yogurt, compared with milk, changes due to the proteolytic activity of *S. thermophilus* and *L. delbrueckii* subsp. *bulgaricus*, both during the fermentation period and, to a lesser degree, during the cold storage of the product. Basically, the change amounts to an increase in the level of soluble nitrogenous compounds, which also includes amino acids and peptides from the milk proteins.

11.5.2 CHANGES IN CARBOHYDRATES

Lactose is the predominant carbohydrate in bovine milk, accounting for 50% of the milk solids. Lactose occurs in both α and β forms, with an equilibrium ratio $\beta/\alpha = 1.68$ at 20°C. Lactose, with a sweetness of about one-fifth that of sucrose, contributes to the characteristic flavor of milk (Swaisgood 1996). Although the dairy lactic acid bacteria utilize lactose as the major energy source, they have the capacity to use a number of other mono- and disaccharides.

Lactic acid is produced in the form of l(+) or d(-) lactic acid or its racemic mixture. Organisms that form the l(+) form or d(-) form have two lactate dehydrogenases (LDH), which differ in their stereospecificity. Some lactobacilli produce l(+) form that, upon accumulation, induces a racemase, which converts it into d(-) lactic acid until equilibrium is obtained (Narayanan, Roychoudhury, and Srivastava 2004). The detailed pathways, physiology, and biochemistry of sugar metabolism in LAB are available in Tamime and Robinson (2007).

Lactic acid production may also occur during yogurt storage at low temperature. It may lead to an excessive acidification known as postacidification, which affects the organoleptic properties of the product. This activity depends on the strains used and especially of the lactobacilli (Accolas et al. 1977; Béal and Corrieu 1994).

The significance of lactose fermentation to lactate is well recognized, but little research has been conducted on the metabolism of lactate (Fox 2001). The conversion of carbohydrates to lactate by LAB has been reviewed by Liu (2003), but lactate degradation in the fermentation of yogurt and milk is not known, nor is lactate racemization. The metabolism of pyruvate and lactate by lactic acid bacteria involved in food and beverage fermentations—with an emphasis on practical implications, and the formation of pyruvate and lactate from a range of substrates, including carbohydrates, organic acids, and amino acids—was also described by Liu (2003).

Polyols are sugar alcohols largely used as sweeteners and are claimed to have several health-promoting effects (low-caloric, low-glycemic, low-insulinemic, anticarcinogenic, and prebiotic). At present, chemical synthesis is the only strategy to provide the polyol market, but LAB are suited for polyol production, as they display a fermentative metabolism associated with an important redox modulation and a limited biosynthetic capacity. In addition, LAB participate in food fermentation processes, where *in situ* production of polyols during fermentation may be useful in the development of novel functional foods (Monedero, Perez-Martinez, and Yebra 2010).

11.5.3 CHANGES IN LIPIDS

Bovine milk contains the most complex lipids known. For detailed characteristics of the various lipids and discussion of their biosynthesis, see Fox and McSweeney (2006). Triacylglycerols (triglycerides) represent the greatest proportion of the lipids, comprising 96%–98% of the total. In milk, triacylglycerols are present as globules of 0.1–22 µm in diameter surrounded by a membrane, the milk fat globule membrane. A lipoprotein lipase present in raw milk causes rapid hydrolysis of milk fat. However, the lipase is prevented from accessing the fat by this membrane. The physical damage of the membrane causes lipolysis. Some lipolysis occurs during storage of raw milk, giving elevated concentrations of free fatty acids as well as mono- and diacylglycerols that produces rancid off-flavors in the milk and milk products (Bills et al. 1969; Deeth 2006).

Another important component of milk, which may have a beneficial role in human health, is conjugated linoleic acid (CLA); this fatty acid occurs as several positional and geometric isomers of octadecadienoic acid ($C_{18:2}$) with conjugated double bonds. The conjugated double bonds exist in either cis or trans configuration, mainly at positions 8 and 10, 9 and 11, 10 and 12, or 11 and 13, with the most abundant isomer in nature being the cis-9, trans-11 isomer (Fritsche and Steinhart 1999; Kim and Liu 2002; Bergamo et al. 2003; Florence et al. 2009).

CLA formation by bacterial strains has been documented since the early 1960s. For example, propionibacteria (i.e., as adjunct cultures), possibly lactic acid bacteria (Jiang, Bjorck, and Fonden 1998), and some yogurt starter cultures (Lin, Lin, and Wang 2003; Xu et al. 2005) have potential to be used for CLA enhancement in dairy products. In yogurt, there is potential to increase the CLA content by adding adjunct cultures together with free linoleic acid or a linoleic acid-containing oil and a suitable lipase. However, this approach to increasing the CLA level in the fermented milk has its limitations (Xu, Boylston, and Glatz 2005). The substrate level in the growth media may be a key factor for CLA production, since a high level of unsaturated long-chain fatty acids is inhibitory not only to cell growth, but also to the overall bio-hydrogenation steps; this applies especially to gram-positive bacteria, which lack an outer membrane (Florence et al. 2009).

The interest in dairy starter cultures for increasing the CLA content in dairy products is vast. Many researchers have investigated the synthesis of CLA by selected bacterial strains under controlled conditions in laboratory media or model systems (Ekinci et al. 2008; Jiang, Bjorck, and Fonden 1998; Rainio et al. 2002; Xu, Boylston, and Glatz 2005). Strains of lactobacilli, bifidobacteria, and propionibacteria have been found to convert efficiently linoleic acid to CLA (Florence et al. 2009).

11.5.4 CHANGES IN MINERALS

Milk is an excellent source of calcium and contains other minerals and trace elements (Valle et al. 2007). Some researchers have demonstrated that, in addition to the effects of calcium in humans, trace elements are also important and exert synergistic actions in scavenging free radicals in the body. Furthermore, copper, zinc, and certain amino acids have been shown to be essential for the function of superoxide dismutase (SOD) (Johnson and Giulivi 2005).

In fermented milks, the mineral content is similar to that of milk unless milk is supplemented. Drago and Valencia (2002) studied the mineral profile of milk, acidified milk, and yogurt fortified with ferrous sulfate (FS) or iron bis-glycinate (FBG), with or without ascorbic acid addition. Milk fermentation or acidification caused an increase in iron availability from both iron sources. Highest availability values were obtained for fermented products with added ascorbic acid (18-fold increase compared to milk).

11.5.5 CHANGES IN VITAMINS

Milk is a source of fat-soluble vitamins: A (in the form of precursor b-carotene), D, E, and K; and water-soluble vitamins: thiamine (B_1), riboflavin (B_2), niacin (B_3), pantothenic acid (B_5), pyridoxine (B_6), folic acid (B_{11}), biotin (vitamin H or B_7), cobalamin (B_{12}), and ascorbic acid (vitamin C). The fat-soluble vitamins appear to function as integral parts of cell membranes while, generally speaking, the water-soluble vitamins act as coenzymes, often as carriers of a particular chemical group (Varnam and Sutherland 1994; Baku and Dickerson 1996).

Many prokaryotes have nutritional requirements for water-soluble vitamins (Snell 1993). Moreover, some microorganisms have biosynthetic capabilities for producing B vitamins. Burgess, Smid, and Van Sinderen (2009) reported how vitamin overproduction strategies have been developed, some of which have successfully been tested in animal models. According to the authors, such innovative strategies could be relatively easily adapted by the food industry to develop novel vitamin-enhanced functional foods with enhanced consumer appeal.

11.6 OTHER CHANGES

Other changes result from fermentation, such as formation of new structures, improvement of texture, and formation of flavor compounds. In addition, fortification of milk to increase solids nonfat and the addition of stabilizers affect the rheological properties of the yogurt gel (Penna, Baruffaldi, and De Oliveira 1997; Oliveira et al. 2001; Damin et al. 2009; Marafon et al. 2010).

Polymers from plant, animal, and microbial origin play an important role in food formulations. Food polymers are long-chain, high-molecular-mass molecules that dissolve or disperse in water to alter textural properties. Most of the biopolymers used by the food industry are polysaccharides from crop plants (e.g., starch) or seaweeds (e.g., carrageenan) and animal proteins like caseinate and gelatin.

Alternative biothickeners are the microbial exopolysaccharides (EPS). These are extracellular polysaccharides that are either associated with, and covalently bound to, the cell surface in the form of capsules, or secreted into the environment in the form of slime. They are referred to as capsular (CPS) or slime (EPS) exopolysaccharides, respectively (Sutherland 1972). According to Sutherland (1999) and Crescenzi et al. (1995), EPS occur widely among bacteria and microalgae and less commonly among yeasts and fungi. A detailed review of the developments in the biosynthesis and applications of heteropolysaccharides from lactic acid bacteria was published by De Vuyst et al. (2001).

The amounts of aromatic substances produced during fermentation of milk depend on the production conditions such as fermentation time, temperature, inoculum concentration, type of milk, and microbial composition (Zajsek and Gorsek 2010). For example, the influence of kefir grains' activation time on kefir pH changes and ethanol production was investigated. The proposed exponential pH model was used successfully to describe the decrease in pH during 24-h fermentation. Furthermore, longer-activated grains produce more ethanol and have a larger number of yeasts than grains that are activated for only a few days. The number of yeasts in a kefir product also increased over the grains' activation time.

Lactic acid bacteria influence the flavor of fermented foods in a variety of ways. In many cases, the most obvious change in lactic acid fermentation is the production of acid that results in a sour flavor. Since most of the acid produced in fermentations is produced by the metabolism of sugars, sweetness decreases as sourness increases. The production of volatile flavor compounds during fermentation tends to be the first mechanism considered for the development of flavor specific to a particular fermented food. In addition to this direct mechanism, however, there are less direct ways in which fermentation microorganisms affect flavor. Lowering the pH in lactic acid fermentations may reduce the activity or completely inactivate enzymes that generate either flavor components or flavor precursor compounds. Finally, the fermentation microorganisms may directly metabolize precursor flavor compounds or flavor components themselves (McFeeters 2004).

Food-fermentation processes often result in profound changes in flavor relative to the starting ingredients. However, fermenting foods are typically very complex ecosystems with active enzyme systems from the ingredient materials interacting with the metabolic activities of the fermentation organisms. Factors such as added salt, size of particles, temperature, and oxygen levels have important effects on the chemical reactions that occur during fermentation. This is a brief review of recent research on flavor changes in food fermentations.

Acetaldehyde, a product of the metabolism of microorganisms used in the manufacture of cultured dairy products, has attracted considerable interest because of its association with the desirable flavor and flavor defects in these products. These microorganisms produce enzymes that catalyze the formation of acetaldehyde from carbohydrate, protein, or nucleic acid sources. The enzyme activities of the lactic acid bacteria have been reviewed in relation to their role in intermediate metabolism by Lees and Jago (1978a, 1978b).

Esters of short-chain fatty acids are aroma-impact compounds found in fermented dairy products. These esters are responsible for fruity flavors that can be regarded either as a defect or as an attribute by the consumer. An understanding of the mechanisms of ester biosynthesis will enable control of the development of fruity flavors in fermented dairy products, as described by Liu, Holland, and Crow (2004). The biosynthesis of flavor-active esters in dairy systems proceeds through two enzymatic mechanisms—esterification and alcoholysis. Esterification is the formation of esters from alcohols and carboxylic acids, whereas alcoholysis is the production of esters from alcohols and acylglycerols or from alcohols and fatty acyl-coenzyme A (CoA) derived from the metabolism of fatty acids, amino acids, and/or carbohydrates. Alcoholysis is essentially a transferase reaction in which fatty acyl groups from acylglycerols and acyl-CoA derivatives are directly transferred to alcohols and is the major mechanism of ester biosynthesis by dairy lactic acid bacteria and yeasts.

Amino acid catabolism is a major process for flavor formation in cheese. The ability of lactic acid bacteria (LAB) and other cheese microorganisms to degrade amino acids to aroma compounds is highly strain dependent. Generally, amino acid catabolism proceeds by two different pathways. The first one, mainly observed for methionine, is initiated by an elimination reaction and leads to major sulfur aroma compounds. The second pathway is generally initiated by a transamination reaction and is the main pathway for degradation of all amino acids by LAB. The resulting α-keto acids are then degraded to various aroma compounds via one or two additional steps. The lactococcal enzymes initiating both pathways have been well characterized, and their importance in the formation of aroma compounds has been demonstrated by using isogenic strains lacking each enzyme. From this information, several applications have been successfully developed, especially for intensifying or diversifying cheese flavor by controlling amino acid transamination (Yvon and Rijnen 2001).

Flavor development in dairy fermentations, most notably cheeses, results from a series of (bio)chemical processes in which the starter cultures provide the enzymes. In particular, the enzymatic degradation of proteins (caseins) leads to the formation of key flavor components, which contribute to the sensory perception of dairy products. More specifically, caseins are degraded into peptides and amino acids, and the latter are major precursors of volatile aroma compounds. In particular, the conversions of methionine, the aromatic and the branched-chain amino acids, are crucial. A lot of research has focused on the degradation of caseins into peptides and free amino acids, and more recently, enzymes involved in the conversion of amino acids were identified. Most data have been generated on *Lactococcus lactis*, which is the predominant organism in starter cultures used for cheese making, but also *Lactobacillus, Streptococcus, Propionibacterium*, and species used for surface ripening of cheeses are characterized according to their flavor-forming capacity. Smit, Smit, and Engels (2005) highlighted various enzymes and pathways involved in flavor formation and discussed the impact of these findings on the development of industrial starter cultures.

11.7 FUNCTIONAL PROPERTIES OF FERMENTED DAIRY PRODUCTS

Fermentation improves some of the functional properties of milk. It enhances the shelf life of milk by the release of lactic acid and antimicrobial substances that inhibit the growth of potentially harmful bacteria; alters the organoleptic properties by the release of flavor compounds and extracellular polysaccharides; produces free amino acids (valine, histidine, serine, and proline) and synthesis of vitamins; and provides therapeutic or prophylactic properties that guard against cancer and help to control the level of serum cholesterol (Parvez et al. 2006). Moreover, changes in the physical properties of casein improve its digestibility and improve absorption of calcium and other minerals (Buttriss 1997; McBean 1999). Fermented milks also contain growth factors, hormones, and immune-stimulatory molecules such as peptidoglycans, polysaccharides, and teichoic acid (Gobbetti et al. 2000).

Lactic acid (0.7%–1.2%) is found in all fermented dairy products as a result of the action of homo- and heterofermentative microorganisms. Lactic acid can exist in either the l(+) or d(-) isomeric forms and as a 50/50 DL racemic mixture. l(+) lactic acid, which generally represents 50%–70% of the total lactic acid, is completely metabolized by the body. However, d(-) lactic acid is slowly metabolized in the body and can cause metabolic disturbances. Yogurt contains both l(+) and d(-) in the ratio of l(+)/d(-) of 58:42, whereas kefir exclusively contains l(+) lactic acid (Alm 1982a; Narayanan, Roychoudhury, and Srivastava 2004).

The occurrence of various bioactive peptides in fermented milks has been reported in many studies. Due to the proteolytic activity of lactic acid bacteria, protein is degraded, resulting in biologically active peptides (e.g., angiotensin-I-converting enzymes [ACE], ACE-inhibitory peptides, casokinins, immunopeptides, casomorphins/opioid peptides, lactoferrin, phosphopeptides, and antioxidative peptides). These peptides having antimicrobial, anticarcinogenic, and antithrombotic activities as well as blood pressure regulatory, and mineral- and vitamin-binding properties (Meisel 1998; Schanbacher et al. 1998; Tome and Ledoux 1998). Readers are advised to refer to the reviews by Gobbetti et al. (2002), Korhonen and Pihlanto (2003, 2006), and Haque, Chand, and Kapila (2009).

Conjugated linoleic acid (CLA) in the fermented milk and products provides biological functional aspects. CLA may decrease risk of atherosclerosis (K. Lee, Kritchevsky, and Pariza 1994; Nicolosi et al. 1997), reduce blood LDL cholesterol levels (Kritchevsky et al. 2004), improve hyperinsulinemia (Houseknecht et al. 1998), alter the low-density lipoprotein/high-density lipoprotein cholesterol ratio (K. Lee, Kritchevsky, and Pariza 1994), prevent gastrointestinal and colon cancers (anticarcinogenic effect), enhance bone formation and immune function in the body (Devery, Miller, and Stanton 2001; Belury 2002; Parodi 2002; Bhattacharya et al. 2006; Soel et al. 2007), reduce body fat content and increase muscle mass (Chin et al. 1992), and enhance the immune system (Miller et al. 1994). In fermented dairy products, certain strains of bifidobacteria and lactobacilli produce CLA isomers and enhance the health properties of fermented dairy products (Coakley et al. 2003; Akalin et al. 2007; Yadav, Jain, and Sinha 2007).

11.8 CONCLUDING REMARKS

Fermented dairy products have been produced and consumed from ancient times up to the present and are considered to be one of the highly nutritious, therapeutic, and health-promoting foods in our normal diet. The fermentation process is gradually improving with advancements in science and technology by changing the various processing steps, selecting specific types of microorganisms (i.e., lactic acid bacteria, probiotics), and fortifying milk with more nutritional ingredients. The various processing steps, *viz.* heating, homogenization, and fortification, modify the chemical properties of milk and ultimately affect the final quality of fermented dairy products. Depending on the type of fermented dairy product being produced, various types of lactic acid bacteria (mesophilic, thermophilic, therapeutics) are used. The lactic acid bacteria use different metabolic pathways to break down primary nutrients such as proteins, carbohydrates, and fats into compounds such as amino acids, lactic acid, fatty acids, and vitamins. These end products can alter the body and texture of the fermented products and improve their flavor and nutritional/functional properties. To fully understand fermented dairy products, one should have good knowledge of various disciplines such as chemistry, biochemistry, microbiology, enzymology, and technology.

REFERENCES

Abd Rabo, F. H. R., J. A. Partridge, and H. M. Furtado. 1988. Production of yoghurt utilizing ultrafiltration retentate of salted whey as a partial substitution of milk. *Egyptian Journal of Dairy Science* 16:319–329.

Accolas, J. P., R. Bloquel, R. Didiene, and J. Regnier. 1977. Acid producing properties of thermophilic lactic bacteria with respect to yogurt making. *Le Lait* 57:1–23.

Akalin, A. S., O. Tokusoglu, S. Gonc, and S. Aycan. 2007. Occurrence of conjugated linoleic acid in probiotic yoghurts supplemented with fructooligosaccharide. *Int. Dairy J.* 17:1089–1095.

Alm, L. 1982a. Effect of fermentation on L(+) and D(-) lactic acid in milk. *J. Dairy Sci.* 65:515–520.

———. 1982b. Effect of fermentation on proteins of Swedish fermented milk products. *Journal of Dairy Science* 65:1696–1704.

Aoki, T., Y. Kako, and T. Imamura. 1986. Separation of casein aggregates cross-linked by colloidal calcium phosphate from bovine casein micelles by high performance gel chromatography in the presence of urea. *Journal of Dairy Research* 53:53–59.

Baku, T. K., and J. W. T. Dickerson. 1996. *Vitamins in human health and disease.* Wallingford, U.K.: CAB International.

Béal, C., and G. Corrieu. 1994. Viability and acidification activity of pure and mixed starters of *Streptococcus salivarius* ssp. *thermophilus* 404 and *Lactobacillus delbrueckii* ssp. *bulgaricus* 398 at the different steps of their production. *Leberem. Wus. Technol.* 27:86–92.

Belury, M. A. 2002. Inhibition of carcinogenesis by conjugated linoleic acid: Potential mechanisms of action. *J. Nutr.* 132:2995–2998.

Bergamo, P., E. Fedeli, L. Iannibelli, and G. Marzillo. 2003. Fat-soluble vitamin contents and fatty acid composition in organic and conventional Italian dairy products. *Food Chemistry* 82:625–631.

Bhattacharya, A., J. Banu, M. Rahman, J. Causey, and G. Ferrandes. 2006. Biological effect of conjugated linoleic acids in health and disease. *J. Nutr. Biochem.* 17:789–810.

Bills, D. D., R. A. Scanlan, R. C. Lindsay, and L. Sather. 1969. Free fatty acids and flavor of dairy products. *Journal of Dairy Science* 52:1340–1345.

Burgess, C., E. Smid, and D. Van Sinderen. 2009. Bacterial vitamin B2, B11, and B12 overproduction: An overview. *International Journal of Food Microbiology* 133:1–7.

Buttriss, J. 1997. Nutritional properties of fermented milk products. *Int. J. Dairy Technol.* 50:21–27.

Campbell-Platt, G. 1987. *Fermented foods of the world.* London: Butterworth.

Chin, S. F., W. Liu, J. M. Storkson, Y. L. Ha, and M. W. Pariza. 1992. Dietary sources of conjugated dienoic isomers of linoleic acid, a newly recognized class of anticarcinogens. *Journal of Food Composition and Analysis* 5:185–197.

Coakley, M., R. P. Ross, M. Norgren, G. Fitzgerald, R. Devery, and C. Stanton. 2003. Conjugated linoleic acid biosynthesis by human-derived *Bifidobacterium* species. *J. Appl. Microbiol.* 94:138–145.

Collet, L. S. F. C. 2005. A Influência da adição de caseinato de sódio sobre o escoamento e posterior recuperação estrutural do iogurte batido. In *Chemical engineering department,* 105. São Paulo, Brazil: São Paulo University.

Crescenzi, V., D. Imbriaco, C. L. Velasquez, M. Dentini, and A. Ciferri. 1995. Novel types of polysaccharidic assemblies. *Macromolecular Chemistry and Physics* 196:2873–2880.

Dalgleish, D. G. 1990. Denaturation and aggregation of serum proteins and caseins in heated milk. *Journal of Agricultural and Food Chemistry* 38:1995–1999.

Dalgleish, D. G., and A. J. R. Law. 1988. pH-induced dissociation of bovine casein micelles: I. analysis of liberated caseins. *Journal of Dairy Research* 55:529–538.

Dalgleish, D. G., and S. K. Sharma. 1993. Interaction between milk fat and milk proteins: The effect of heat on the nature of the complexes formed. In *Protein and fat globule modifications by heat treatment, homogenization, and other technological means for high quality dairy products,* 7–17. IDF Special Issue No. 9303, International Dairy Federation, Brussels, Belgium.

Damin, M. R., M. R. Alcantara, A. P. Nunes, and M. N. Oliveira. 2009. Effects of milk supplementation with skim milk powder, whey protein concentrate, and sodium caseinate on acidification kinetics, rheological properties, and structure of nonfat stirred yogurt. *Lwt-Food Science and Technology* 42:1744–1750.

Dannenberg, F., and H. G. Kessler. 1988. Effect of denaturation of β-lactoglobulin on texture studies of set-style nonfat yoghurt: 1. Syneresis. *Milchwissenschaft* 43:632–635.

Deeth, H. 2006. Lipoprotein lipase and lipolysis in milk. *International Dairy Journal* 16:555–562.

Deeth, H. C., and A. Y. Tamime. 1981.Yogurt: Nutritive and therapeutic aspects. *Journal of Food Protection* 44:78–86.

Devery, R., A. Miller, and C. Stanton. 2001. Conjugated linoleic acid and oxidative behaviour in cancer cells. *Biochemical Society Transactions* 29:341–344.

De Vos, W., and J. Hugenholtz. 2004. Engineering metabolic highways in Lactococci and other lactic acid bacteria. *Trends in Biotechnology* 22:72–79.

De Vuyst, L., F. De Vin, F. Vaningelgem, and B. Degeest. 2001. Recent developments in the biosynthesis and applications of heteropolysaccharides from lactic acid bacteria. *International Dairy Journal* 11:687–707.

Drago, S., and M. Valencia. 2002. Effect of fermentation on iron, zinc, and calcium availability from iron-fortified dairy products. *Journal of Food Science* 67:3130–3134.

Driessen, F. M., F. Kingma, and J. Stadhouders. 1982. Evidence that *Lactobacillus bulgaricus* in yoghurt is stimulated by carbon dioxide produced by *Streptococcus thermophilus*. *Netherlands Milk and Dairy Journal* 36:135–144.

Edward, R. F. 2003. *Handbook of fermented functional foods*. Boca Raton, FL: CRC Press.

Ekinci, F. Y., O. D. Okur, B. Ertekin, and Z. Guzel-Seydim. 2008. Effects of probiotic bacteria and oils on fatty acid profiles of cultured cream. *Journal of Lipid Science and Technology* 110:216–224.

Farkye, N. Y., and G. I. Imafidon. 1995. Thermal denaturation of indigenous milk enzymes. In *Heat-induced changes in milk*, 2nd ed., ed. P. F. Fox, 331–348. IDF Special Issue No. 9501, International Dairy Federation, Brussels, Belgium.

Ferreira, C. L. L. F. 2003. Grupo de bactérias láticas: Caracterização e aplicação tecnológica de bactérias probióticas. 7–33. Viçosa: UFV.

Florence, A. C. R., R. C. Da Silva, A. P. D. Santo, L. A. Gioielli, A. Y. Tamime, and M. N. De Oliveira. 2009. Increased CLA content in organic milk fermented by bifidobacteria or yoghurt cultures. *Dairy Science & Technology* 89:541–553.

Fox, P. F. 1991. *Food enzymology*. London: Elsevier Applied Science.

———. 2001. Milk proteins as food ingredients. *International Journal of Dairy Technology* 54:41–55.

Fox, P. F., and P. L. H. McSweeney. 2006. *Dairy Chemistry, 2: Lipids*. New York: Springer.

Fritsche, J. R. R., and H. Steinhart. 1999. Formation, contents, and estimation of daily intake of conjugated linoleic acid isomers and trans-fatty acids in foods. *Adv. in CLA Res.* 1:378–396.

Gilmour, A., and M. T. Rowe. 1990. Microorganisms associated with milk. In *Dairy microbiology: The microbiology of milk*, vol. 1, 2nd ed., ed. R. K. Robinson, 37–75. London: Elsevier Applied Science.

Gobbetti, M., P. E. Ferranti, F. Smacchi, and A. Goffredi. 2000. Production of angiotensin-I-converting-enzyme inhibitory peptides in fermented milks started by *Lactobacillus delbrueckii* subsp. *Bulgaricus* SS1 and *Lactococcus lactis* subsp. *Cremoris* FT4. *Appl. Environment Microbiol.* 66:3898–3904.

Gobbetti, M., L. Stepaniak, M. De Angelis, A. Corsetti, and R. Di Cagno. 2002. Latent bioactive peptides in milk proteins: Proteolytic activation and significance in dairy processing. *Crit. Rev. Food Sci. Nutr.* 42:223–239.

Hammer, B. W., and F. J. Babel. 1993. *Dairy bacteriology*, 4th ed. New York: John Wiley & Sons.

Haque, E., R. Chand, and S. Kapila. 2009. Biofunctional properties of bioactive peptides of milk origin. *Food Reviews International* 25:28–43.

Haque, Z., and J. E. Kinsella. 1988. Interaction between heated k-casein and β-lactoglobulin: Predominance of hydrophobic interactions in the initial stages of complex formation. *Journal of Dairy Research* 55:67–80.

Holt, C., D. T. Davies, and A. J. R. Law. 1986. Effects of calcium phosphate content and free calcium ion concentration in the milk serum on the dissociation of bovine casein micelles. *Journal of Dairy Research* 53:557–572.

Houlihan, A. V., P. A. Goddard, B. J. Kitchen, and C. J. Masters. 1992. Changes in the structure of the bovine milk globule membrane on heating whole milk. *Journal of Dairy Research* 59:321–329.

Houseknecht, K., J. Vanden Heuvel, S. Moya-Camarena, C. P. Portocarrero, L. W. Peck, K. P. Nickel, and M. A. Belury. 1998. Dietary conjugated linoleic acid normalizes impaired glucose tolerance in the Zucker diabetic fatty fa/fa rat. *Biochemical and Biophysical Research Communications* 244:678–682.

Hutkins, R. W., and H. A. Morris. 1987. Carbohydrate metabolism by S*treptococcus thermophilus:* A review. *Journal of Food Protection* 50:876–884.

IDF. 1988. Fermented milks: Science and technology. IDF Bulletin No. 227, International Dairy Federation, Brussels, Belgium.

———. 2007. Coagulation of milk: Processes and characteristics. IDF Bulletin No. 420, International Dairy Federation, Brussels, Belgium.

Jay, M. James. 1992. Fermented foods and related products of fermentation. In *Modern food microbiology*, 371–409. New York: Chapman and Hall.

Jelen, P. 1993. Heat stability of dairy systems with modified casein-whey protein content. In *Protein and fat globule modifications by heat treatment, homogenization, and other technological means for high quality dairy products*, 259–266. IDF Special Issue No. 9303, International Dairy Federation, Brussels, Belgium.

Jiang, J., L. Bjorck, and R. Fonden. 1998. Production of conjugated linoleic acid by dairy starter cultures. *Journal of Applied Microbiology* 85:95–102.

Johnson, F., and C. Giulivi. 2005. Superoxide dismutases and their impact upon human health. *Molecular Aspects in Medicine* 26:340–352.

Kandler, O. 1983. Carbohydrate metabolism in lactic acid bacteria. *Antonie van Leeuwenhoek* 49:209–224.

Kim, Y., and R. Liu. 2002. Increase of conjugated linoleic acid content in milk by fermentation with lactic acid bacteria. *Journal of Food Science* 67:1731–1737.

Korhonen, H., and A. Pihlanto. 2003. Food-derived bioactive peptides: Opportunities for designing future foods. *Curr. Pharm. Design* 9:1297–1308.

———. 2006. Bioactive peptides: Production and functionality. *Int. Dairy J.* 16:945–960.

Kosikowski, F. V., and V. V. Mistry. 1997. *Cheese and fermented milk*, 3rd ed. Ann Arbor, MI: Edward Brothers.

Kritchevsky, D., S. A. Tepper, S. Wright, K. Czarnecki, T. A. Wilson, and R. J. Nicolosi. 2004. Conjugated linoleic acid isomer effects in atherosclerosis: Growth and regression of lesions. *Lipids* 39:611–616.

Kurmann, J. A., J. L. Rasic, and M. Kroger. 1992. *Encyclopedia of fermented fresh milk products*. New York: Van Nostrand Reinhold.

Law, B. A. 1997. *Microbiology and biotechnology of cheese and fermented milks*, 2nd ed. New York: Blackie Academic and Professional.

Lee, K. N., D. Kritchevsky, and M. W. Pariza. 1994. Conjugated linoleic-acid inhibits atherosclerosis in rabbits. *Atherosclerosis* 108:19–25.

Lee, W. J., and J. A. Lucey. 2010. Formation and physical properties of yoghurt. *Asian-Aust. J. Anim. Sci.* 23:1127–1136.

Lees, G. J., and G. R. Jago. 1978a. Role of acetaldehyde in metabolism: Review, 1: Enzymes catalyzing reactions involving acetaldehyde. *Journal of Dairy Science* 61:1205–1215.

———. 1978b. Role of acetaldehyde in metabolism: Review, 2: Metabolism of acetaldehyde in cultured dairy-products. *Journal of Dairy Science* 61:1216–1224.

Le Graet, Y., and G. Brule. 1993. The mineral equilibria in milk: Effect of pH on ionic strength. *Le Lait* 73:51–60.

Lin, T., C. Lin, and Y. Wang. 2003. Production of conjugated linoleic acid by enzyme extract of *Lactobacillus acidophilus* CCRC 14079. *Food Chemistry* 83:27–31.

Liu, S. 2003. Practical implications of lactate and pyruvate metabolism by lactic acid bacteria in food and beverage fermentations. *International Journal of Food Microbiology* 83:115–131.

Liu, S., R. Holland, and V. Crow. 2004. Esters and their biosynthesis in fermented dairy products: A review. *International Dairy Journal* 14:923–945.

Lucas, A., I. Sodini, C. Monnet, P. Jolivet, and G. Corrieu. 2004. Effect of milk base and starter culture on acidification, texture, and probiotic cell counts in fermented milk processing. *Journal of Dairy Science* 85:2479–2488.

Lucey, J. A. 2004. Formation, structural properties and rheology of acid-coagulated milk gels. In *General aspects*, 105–122. Vol. 1 of *Cheese: Chemistry, physics, and microbiology*, 3rd ed., ed. P. F. Fox, P. L. H. McSweeney, T. M. Cogan, and T. P. Guinee. London: Elsevier Academic Press.

Lucey, J. A., P. A. Munro, and H. Singh. 1999. Effects of heat treatment and whey protein addition on the rheological properties and structure of acid skim milk gels. *Int. Dairy J.*, 9 (3): 591–600.

Lucey, J. A., and H. Singh. 1998. Formation and physical properties of acid gels: A review. *Food Research International* 30:529–542.

Lucey, J. A., M. Tamehana, H. Singh, and P. A. Munro. 1998. A comparison of the formation, rheological properties, and microstructure of acid skim milk gels made with a bacterial culture or glucono-delta-lactone. *Food Research International* 31:147–155.

Marafon, A. P., A. Sumi, M. R. Alcântara, A. Y. Tamime, and M. N. Oliveira. 2010. Optimization of the rheological properties of probiotic yoghurts supplemented with milk proteins. *LWT-Food Science and Technology* 44:511–519.

Marshall, V. M. E. 1986. The microflora and production of fermented milks. In *Progress in industrial microbiology: Microorganisms in the production of food*, vol. 23, ed. M. R. Adams, 1–44. Amsterdam: Elsevier Science.

Marth, E. H. 1974. Fermentation. In *Fundamental of dairy chemistry*, 2nd ed., ed. B. H. Webb and A. H. Johnson. Westport, CN: AVI Publishing.

Marth, E. M., and J. L. Steele. 2001. *Applied dairy microbiology*, 2nd ed. New York: Marcel Dekker.

McBean, L. D. 1999. Emerging dietary benefits of dairy foods. *Nutr. Today* 34:47–53.

McFeeters, R. 2004. Fermentation microorganisms and flavor changes in fermented foods. *Journal of Food Science* 69:M35–M37.

Meisel, H. 1998. Overview of milk protein derived peptides. *Int. J. Dairy Technol.* 8:363–373.

Mercenier, A. 1990. Molecular genetics of S*treptococcus thermophilus. FEMS Microbiology Review* 87:61–78.

Mercenier, A., and Y. Lemoine. 1989. Genetics of *Streptococcus thermophilus*: A review. *Journal of Dairy Science* 72:3444–3454.

Miller, C. C., Y. Park, M. W. Pariza, and M. E. Cook. 1994. Feeding conjugated linoleic acid to animals partially overcomes catabolic responses due to endotoxin injection. *Biochemical and Biophysical Research Communications* 198:1107–1112.

Modler, H. W., and M. Kalab. 1983. Microstructure of yoghurt stabilized with milk proteins. *Journal of Dairy Science* 66:430–437.

Monedero, V., G. Perez-Martinez, and M. Yebra. 2010. Perspectives of engineering lactic acid bacteria for biotechnological polyol production. *Applied Microbiology and Biotechnology* 86:1003–1015.

Mottar, J., A. Bassier, M. Joniau, and J. Baert. 1989. Effect of heat-induced association of whey proteins and casein micelles on yoghurt texture. *Journal of Dairy Science* 72:2247–2256.

Mulder, H., and P. Walstra. 1974. *The milk fat globule: Emulsion science as applied to milk products and comparable foods.* Farnham Royal, U.K.: Commonwealth Agricultural Bureau.

Narayanan, N., P. Roychoudhury, and A. Srivastava. 2004. L(+) lactic acid fermentation and its product polymerization. *Electronic Journal of Biotechnology* 7:167–179.

Nicolosi, R. J., E. J. Rogers, D. Kritchevsky, J. A. Scimeca, and P. J. Huth. 1997. Dietary conjugated linoleic acid reduces plasma lipoproteins and early aortic atherosclerosis in hypercholesterolemic hamsters. *Artery* 22:266–277.

Oliveira, M. N., I. Sodini, F. Remeuf, and G. Corrieu. 2001. Effect of milk supplementation and culture composition on acidification, textural properties, and microbiological stability of fermented milks containing probiotic bacteria. *International Dairy Journal* 11:935–942.

Ostlie, H., M. Helland, and J. Narvhus. 2003. Growth and metabolism of selected strains of probiotic bacteria in milk. *International Journal of Food Microbiology* 87:17–27.

Parnell-Clunies, E., Y. Kakuda, D. Irvine, and K. Mullen. 1988. Heat-induced protein changes in milk processed by vat and continuous heating systems. *Journal of Dairy Science* 71:1472–1483.

Parodi, P. W. 2002. Health benefits of conjugated linoleic acid. *Food Ind. J.* 5:222–259.

Parvez, S., K. A. Malik, S. Ah. Kang, and H. Y. Kim. 2006. Probiotics and their fermented food products are beneficial for health. *Journal of Applied Microbiology* 100:1171–1185.

Penna, A. L. B., R. Baruffaldi, and M. N. De Oliveira. 1997. Optimization of yogurt production using demineralized whey. *Journal of Food Science* 62:846–850.

Penna, A. L. B., A. Converti, and M. N. De Oliveira. 2006. Simultaneous effects of total solids content, milk base, heat treatment temperature, and sample temperature on the rheological properties of plain stirred yogurt. *Food Technology and Biotechnology* 44:515–518.

Pouliot, Y., M. Boulet, and P. Paquin. 1989. Observations on heat-induced salt balance changes in milk: I. Effect of heating time between 4 and 90°C. *Journal of Dairy Research* 56:185–192.

Prajapati, J. B. 1995. *Fundamentals of dairy microbiology*. Gujarat, India: Akta Prakashan Nadiad.

Prentice, J. H. 1992. *Dairy rheology: A concise guide*. New York: VCH Publishers.

Prescott, S. C., and C. G. Dunn. 1957. *Industrial microbiology*. New York: McGraw-Hill.

Prouty, C. C. 1940. Observations on the growth responses of *S. lactis* in mastitis milk. *J. Dairy Sci.* 23:899–904.

Puhan, Z. 1988. Treatment of milk prior to fermentation. In *Fermented milks: Science and Technology*, 66–74. IDF Bulletin No. 227, International Dairy Federation, Brussels, Belgium.

Puvanenthiran, A., R. P. W. Williams, and M. A. Augustin. 2002. Structure and visco-elastic properties of set yoghurt with altered casein to whey protein ratios. *International Dairy Journal* 12:383–391.

Rainio, A., M. Vahvaselka, T. Suomalainen, and S. Laakso. 2002. Production of conjugated linoleic acid by *Propionibacterium freudenreichii* ssp. *shermanii*. *Le Lait* 82:91–101.

Rasic, J. L., and J. A. Kurmann. 1978. *Yoghurt: Scientific grounds, technology, manufacture, and preparations*. Copenhagen, Denmark: Technical Dairy Publishing House.

Remeuf, F., S. Mohammed, I. Sodini, and J. P. Tissier. 2003. Preliminary observations on the effects of milk fortification and heating on microstructure and physical properties of stirred yogurt. *International Dairy Journal* 13:773–782.

Rice, E. B. 1948. Studies on starters, V: Effect of pathological or physiologically abnormal milk on acid production by lactic acid streptococci. *Dairy Inds.* 13:983.

Robinson, R. K. 2002. *Dairy microbiology handbook: Microbiology of milk and milk products*, 3rd ed. New York: John Wiley & Sons.

———. 2003. Yogurt: Types and manufacture. In *Encyclopedia of dairy sciences*, ed. H. Roginski, J. W. Fuquay, and P. F. Fox, 1055–1058. New York: Academic Press.

Robinson, R. K., and A. Y. Tamime. 1990. Microbiology of fermented milks. In *Dairy microbiology: The microbiology of milk products*, vol. 2, 2nd ed., ed. R. K. Robinson, 291–343. London: Elsevier Applied Science Publishers.

Roefs, S. P. F. M. 1987. Structure of acid casein gels: A study of gel formed after acidification in the cold. *Netherlands Milk and Dairy Journal* 41:99–101.

Schanbacher, F. L., R. S. Talhouk, F. A. Murray, L. I. Gherman, and L. B. Willett. 1998. Milk borne bioactive peptides. *Int. Dairy Journal* 8:393–403.

Sharma, S. K., and D. G. Dalgleish. 1994. Effect of heat treatment on the incorporation of milk serum proteins into the fat globule membrane of homogenized milk. *Journal of Dairy Research* 61:375–384.

Singh, H. 1993. *Heat-induced interactions of proteins in milk in protein and fat globule modifications by heat treatment, homogenizations, and other technology means for high quality dairy products*, 191–204. IDF Special Issue No. 9303, International Dairy Federation, Brussels, Belgium.

Smit, G., B. Smit, and W. Engels. 2005. Flavour formation by lactic acid bacteria and biochemical flavour profiling of cheese products. *FEMS Microbiology Reviews* 29:591–610.

Snell, E. E. 1993. From bacterial nutrition to enzyme structure: A personal odyssey. *Annual Review of Biochemistry* 62:1–27.

Sodini, I., A. Lucas, J. Tissier, and G. Corrieu. 2005. Physical properties and microstructure of yoghurts supplemented with milk protein hydrolysates. *International Dairy Journal* 15:29–35.

Sodini, I., J. Montella, and P. S. Tong. 2005. Physical properties of yogurt fortified with various commercial whey protein concentrates. *Journal of the Science of Food and Agriculture* 85:853–859.

Sodini, I., F. Remeuf, S. Haddad, and G. Corrieu. 2004. The relative effect of milk base, starter, and process on yogurt texture: A review. *Critical Reviews in Food Science and Nutrition* 44:113–137.

Soel, S. M., O. S. Choi, M. H. Bang, J. H. Park, and W. K. Kim. 2007. Influence of conjugated linoleic acid isomers on the metastasis of colon cancer cells in vitro and in vivo. *J. Nutr. Biochem.* 18:650–657.

Sousa, M., Y. Ardo, and P. McSweeney. 2001. Advances in the study of proteolysis during cheese ripening. *International Dairy Journal* 11:327–345.

Sutherland, I. W. 1972. Microbial exopolysaccharides: Potential. *Process Biochemistry* 7: 27–30.

———. 1999. Microbial polysaccharide products. In *Biotechnology and genetic engineering reviews*, vol. 16, ed. S. E. Harding, 217–229. Andover, U.K.: Intercept.

Swaisgood, H. E. 1996. Characteristics of milk. In *Food chemistry*, ed. O. R. Fennema, 841–878. New York: Marcel Dekker.

———. 2010. Características do leite. In *Química de alimentos de fennema*, ed. S. Damodaran, K. L. Parkin, and O. R. Fennema, 689–758. São Paulo, Brazil: ArtMed.

Tamime, A. Y. 2006. *Probiotic dairy products*. Oxford, U.K.: Wiley-Blackwell.

Tamime, A. Y., and H. C. Deeth. 1980. Yogurt: Technology and biochemistry. *Journal of Food Protection* 43:939–977.

Tamime, A. Y., A. Hasan, E. R. Farnworth, and T. Toba. 2007. *Structure of fermented milks*. Oxford, U.K.: Wiley-Blackwell.

Tamime, A. Y., and V. M. E. Marshall. 1997. Microbiology and technology of fermented milks. In *Microbiology and biochemistry of cheese and fermented milk*, 2nd ed., ed. B. A. Law, 57–152. London: Blackie Academic and Professional.

Tamime, A. Y., and R. K. Robinson. 1985. *Yoghurt: Science and technology*. Oxford, U.K.: Pergamon Press.

———. 1999. *Yoghurt: Science and technology*, 2nd ed. Cambridge, U.K.: Woodhead Publishing.

———. 2007. *Yoghurt: Science and technology*. Boca Raton, FL: CRC Press.

Tamime, A. Y., R. K. Robinson, and E. Latrille. 2001. Yoghurt and other fermented milks. In *Mechanization and automation in dairy technology*, ed. A. Y. Tamime and B. A. Law, 152–203. Reading, U.K.: Sheffield Academic.

Thirtha, Sada Shiva. 1998. *The Ayurveda encyclopedia: Natural secrets to healing, prevention, and longevity*. New York: Ayurveda Holistic Center Press.

Tome, D., and N. Ledoux. 1998. Nutritional and physiological role of milk protein components. *Bulletin of the IDF* 336:11–16.

Trout, G. M. 1950. *Homogenized milk: A review and guide*, 23–58. East Lansing: Michigan State College Press.

Valle, P. S., G. Lien, O. Flaten, M. Koesling, and M. Ebbesvik. 2007. Herd health and health management in organic versus conventional dairy herds in Norway. *Livestock Science* 112:123–132.

Van Boekel, M. A. J. S., and P. Walstra. 1995. Effect of heat treatment on chemical and physical changes to milk fat globules. In *Heat-induced changes,* 2nd ed., ed. P. F. Fox, 51–65. IDF Special Issue No. 9501, International Dairy Federation, Brussels, Belgium.

Varnam, A. H., and J. P. Sutherland. 1994. *Leche y productos lácteos*. Saragossa, Spain: Editorial Acribia S.A.

Vasiljevic, T., and N. P. Shah. 2008. Probiotics—From Metchnikoff to bioactives. *International Dairy Journal* 18:714–728.

Vedamuthu, E. R. 1985. What is wrong with cultured buttermilk? *Dairy and Food Sanitation* 5:8–13.

Viani, R., and I. Horman. 1973. Composition de l'arome de yoghurt. *Travaux de Chimie Alimentaire et d'Hygiene* 64:66–70.

Walstra, P., T. J. Geurts, A. Noomen, A. Jellema, and M. A. J. S. van Boekel. 1999. *Dairy technology: Principles of milk properties and processes*, 245–264. New York: Marcel Dekker.

Walstra, P., J. T. M. Wouters, and T. J. Geurts. 2006a. Lactic fermentation. In *Dairy science and technology*, 357–398. Boca Raton, FL: Taylor & Francis Group, LLC (CRC Press).

———. 2006b. Fermented milks. In *Dairy science and technology*, 551–576. Boca Raton: Taylor & Francis Group, LLC (CRC Press).

Wiegner, G. 1914. Ueber die Aenderung einiger Physikalischer Eigenschaften der Kuhmilch mit der Zerteilung ihrer dispersen Phasen. *Kolloid Ztschr.* 15:105–123.

Wouters, J. T. M., E. H. E. Ayad, J. Hugenholtz, and G. Smit. 2002. Microbes from raw milk for fermented dairy products. *International Dairy Journal* 12:91–109.

Xiong, Y. L., K. A. Dawson, and L. Wan. 1993. Thermal aggregation of β-lactoglobulin: Effect of pH, ionic environment, and thiol reagent. *Journal of Dairy Science* 76:70–77.

Xu, S., T. D. Boylston, and B. A. Glatz. 2005. Conjugated linoleic acid content and organoleptic attributes of fermented milk products produced with probiotic bacteria. *Journal of Agriculture and Food Chemistry* 53:9064–9072.

Yadav, H., S. Jain, and P. R. Sinha. 2007. Production of free fatty acids and conjugated linoleic acid in probiotic dahi containing *Lactobacillus acidophilus* and *Lactobacillus casei* during fermentation and storage. *Int. Dairy J.* 17:1006–1010.

Yvon, M., and L. Rijnen. 2001. Cheese flavour formation by amino acid catabolism. *International Dairy Journal* 11:185–201.

Zajsek, K., and A. Gorsek. 2010. Effect of natural starter culture activity on ethanol content in fermented dairy products. *International Journal of Dairy Technology* 63:113–118.

Zourari, A., J. P. Accolas, and M. J. Desmazeaud. 1992. Metabolism and biochemical characteristics of yogurt bacteria: A review. *Le Lait* 72:1–34.

12 Fermented Seafood Products

Nilesh H. Joshi and Zulema Coppes Petricorena

CONTENTS

12.1 INTRODUCTION

Seafood products provide an important source of nutrients in diets in many countries. Among seafoods, fish and fish products have always been considered to be an excellent source of proteins and minerals in a low-fat product (Kinsella 1988) in addition to being rich sources of omega-3 fatty acids such as docosahexaenoic acid (DHA) and eicosapentaenoic acid (EPA). However, fish are highly perishable, and spoiled fish have a strong odor, soft flesh with brown traces of blood, and dark-red gills with slime on them (Ackman 1990). These deteriorations are due to autolytic spoilage by enzymes, fat oxidation, microbiological spoilage, or a combination of these. In Asia, seasonally available fish species are preserved by fermentation, and fermented fish products are consumed daily (Mizutani et al. 1987; Kimizuka et al. 1992; Fraser and Sumar 1998).

Fermented fish products have been reported as far back as ancient Greece, and trade of the fermented products was extensive in the Roman era (Beddows 1985). Fermentation is an energy-neutral process that extends the shelf life of fish and fish products and increases their palatability and nutritional value. The basic aim of the fermentation process is to transform the highly perishable substrate muscle into a stable and safe product maintaining optimum nutritive values and sensory quality. The nature of the raw material and activity of the microorganism affects the process of fermentation (Peredes-Lopez and Herry 1988).

Fermentation of fish relies on the enzymatic liquefaction of fish to produce edible soft pastes and/or sauces (Kevin and Howgate 2002). Salt is mixed with fish at ratios that vary from 1:2 to 1:9 and ferment at ambient temperature over time periods ranging from a few days (1–5) to several months (3–12). Salt-labile microorganisms are also inhibited during the process. However, salt-tolerant microorganisms (known as halophiles) as well as autolytic enzymes (located in the main tissue of the fish) lead to chemical, biochemical, and physiological changes during fermentation.

Fish fermentation produces three different types of products: those that retain the fish in their original form, those that reduce fish to pastelike products, and those that reduce fish to a liquid sauce. Lactic acid bacteria (LAB) are sometimes used in the production of Asian fish sauces, fish balls, and sausages. Fish and shrimp pastes and fish sauce (whole fish) are highly flavored, salty, and taste like cigars and malt scotch (IIF 1986; Barbara and Bledsoe 2006). Fish sauce is a popular condiment for dipping and cooking, and it is generally a clear amber to reddish-brown liquid with a mild fishy flavor and salty taste. Biochemically, fish sauce is salt-soluble protein in the form of peptides and amino acids (Lopetcharat et al. 2001).

In this chapter, the term *fermented seafood products* includes freshwater and marine fishes, shellfish, and crustaceans that are processed with salt to cause fermentation. Some of the common fish sauces and pastes are discussed in detail.

12.2 TYPES OF FERMENTED FISH PRODUCTS AND THEIR MANUFACTURE

While fish are an excellent nutritive food, they are also an extremely perishable product due to their biological and biochemical composition and their high water activity. Moreover, fish contain a wide range of nutrients that make them an ideal substrate for both spoilage microorganisms as well as endogenous enzymes (Ravipim 1991). Fermentation is one of the means used to preserve fish. Fermented fish products undergo degradative changes through microbiological or enzymatic activity in the presence or absence of salt (Essuman 1992). Some fermented fish products use a carbohydrate source, and in such products, the level of salt is <8% water-phase salt (WPS), which allows the growth of LAB and decreases the pH to <4.5. On the other hand, enzymatically hydrolyzed fish products have WPS >8% and a final pH between 5 and 7 (Paludan-Muller 2002). Fermented fish products are grouped into three categories based on the mechanism of protein breakdown (Amano 1962):

1. In the presence of low to high salt concentrations, fermentation is achieved by endogenous fish enzymes. Examples are traditional products of fish sauces and fish pastes.
2. In the presence of salt, fermentation is achieved by the combined action of endogenous fish and microbial enzymes. The microbial enzymes are

added as a starter culture, usually in the form of microorganisms growing on maize or cooked rice. Examples of traditional products are *plara* in
Thailand and *buro* in the Philippines.

3. Rapid fermentation is carried out either with enzymes or by chemical
hydrolysis to produce silage, a nontraditional fermented fish product.

According to the substrates used in fermentation, Adams, Cook, and Rattagool
(1985) divided traditional fermented fish products into two categories, *viz.*, products
made from fish and salt, and products made from fish, salt, and carbohydrate. Many
authors have tried to classify fermented fish products as per various rules as reported
by Hall (2002), but not all types of fermented fish products fit into the categories
given by Amano (1962). For further information about the classification of fermented
fish products, readers are advised to refer to papers by Saisithi (1987) and Kose and
Hall (2011).

12.2.1 Fermented Fish Sauce and Paste

Fish sauce made from fish that have been allowed to ferment is an essential ingredient in many recipes. Fish sauce is a staple ingredient in Filipino, Vietnamese, Thai,
Laotian, and Cambodian cuisine and is used in other Southeast Asian countries.
Fish sauces are known under a variety of names and vary according to both their
organoleptic and nutritive qualities (Saisithi et al. 1966; Beddows 1985; Itoh, Tachi,
and Kikuchi 1993; Ijong and Ohta 1995; Lopetcharat et al. 2001; Sanni et al. 2002;
Ruddle and Ishige 2010). The indigenous fish sauces are shown in Table 12.1. Fish
sauce can also be used in mixed form as a dipping condiment. The origin of fish
sauce is not known, but it has been suggested to have first appeared in Southeast
Asia (Ishige and Ruddle 1987). Fish pastes are made from various low-quality fish as
well as shrimp and are mainly used as a flavoring agent or in condiment preparation
(Thapa, Pal, and Tamang 2004).

Fish sauce preparation is a very simple method where fish, generally small fish,
are mixed with salt and packed in earthen pots or wooden containers and sealed to
create an anaerobic environment. The fermentation process takes place for months,
and a clear, amber-colored liquid is formed, which is separated from the residue by
pressing it out. Sometimes a fish sauce can be made during the preparation of fish
paste. The general method of preparation of fish paste is similar to that of fish sauces.
Fermentation of fish sauce takes longer than fish paste because all of the fish flesh
must be broken down to obtain a clear liquid. Fish paste must be mixed regularly to
help distribute the salt evenly. Figure 12.1 gives a general flow chart for traditional
fish sauce production.

Fish pastes are more widely produced and eaten than fish sauces, and they are
consumed with rice dishes and serve as an important source of nutrition for poor
households (Mckee 1956). There are various kinds of fish paste available in various
countries. Some examples of fish pastes are *kapi* in Thailand, *belachan* in Malaysia,
trassi-ikan in Indonesia (Irianto and Irianto 1998), *ngangapi* in Burma, and *mam-ca*
in Vietnam (Adams 2009). *Bagoong*, produced mainly in the Philippines, is a fish or
shrimp paste that has cheeselike flavor with little fishy odor and is prepared by the

TABLE 12.1
Popular Names of Fish Sauces

Country	Popular name	Salt: Fish ratio	Fermentation period
Korea	Aekjeot, myeol-chi-jeot-guk	1:3 to 4	6–12 months
France	Anchovy,	1:2	6–7 months
	Pissala	1:4	2–8 weeks
Malaysia	Budu	1:3 to 5	3–12 months
India	Colombo-cure	1:6	12 months
Pakistan	Colombo-cure	1:6	12 months
Greece	Gaross	1:9	8 days
Japan	Ishiru or shottsura, ikanago-shoyu	1:5	6 months
Indonesia	Ketjap-ikan	1:6	6 months
Ghana	Momoni	3:10	1–5 days
Thailand	Nampla, budu, thai pla	1:1 to 5	5–12 months
Burma	Ngapi		
Cambodia	Nuoc-mam or nuoc-nam	1:3 to 2:3	3 months
Philippines	Patis	1:4	3–12 months
China	Yeesu, yu lu		

fermentation of whole or ground fish, fish roe, shrimp, or shrimp roe that is partially or fully hydrolyzed. *Bagoong* can be eaten raw or cooked, and eaten also together with tomatoes or vegetables (Olympia 1992). The best quality *Bagoong* is prepared from goby fry (*Gobiidae*), herring fry (*Clupeidae*), small anchovies (*Engraulidae*), and small (less than 2.5 cm) shrimp. Salting of fish (1:2.5–3 salt:fish) is done on wooden or cement floors with the aid of wooden paddles. Fish and salt are then kept in fermenting tanks and allowed to undergo natural fermentation for several months. Occasional stirring is necessary to maintain uniform concentration of salt in the mash. When a cheeselike aroma develops, the *bagoong* is harvested and bottled for marketing. Figure 12.2 presents a flow chart for production of the fish paste *bagoong*. The best *bagoong* products are those that are sealed in tin containers. Exposure to a temperature of 45°C results in a somewhat cooked flavor and aroma (Olympia 1992). There are various kinds of *bagoong*. For example, *bagoong na alamang*, which is very rich in protein, is prepared by fermenting small shrimps with salt; *bagoong na sisi* is prepared by salting shelled oysters; and *bagoong na sida* is prepared by partial fish fermentation (Baens-Arcega 1977a). Fish pastes are also produces by fermentation in the presence of cooked/roasted rice with yeasts and molds.

For more details about fermented fish and fish products, readers are advised to refer to recent studies by Lopetcharat et al. (2001) and Panda et al. (2011).

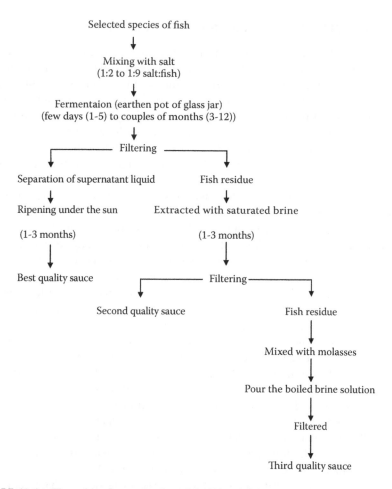

FIGURE 12.1 Flow chart for production of traditional fish sauce

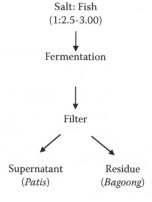

FIGURE 12.2 Flow chart for production of fish paste (*bagoong*)

12.3   MICROFLORA OF FERMENTED FISH PRODUCTS

In fermentation, raw materials are converted into products through the activity of endogenous enzymes or microorganisms (bacteria, yeasts, and molds). Fermentation in fish products degrades fish muscle proteins into smaller peptides and amino acids that are nutrients for microorganisms. Hence, fish fermentation is often combined with the addition of salt or drying to reduce water activity and eliminate proteolytic and putrefying microorganisms (Salampessy, Kailasapathy, and Thapa 2010).

A fish sauce with high salt concentration allows only halophilic and halotolerant bacteria to grow, and their presence enhances degradation of fish proteins while facilitating the development of flavor and aroma. According to Ijong and Ohta (1996), bacteria in fish sauces can be classified into those that produce proteolytic enzymes and those that produce flavor and aroma. The increase of bacteria with time corresponds to a decrease of pH that can be attributed to the production of lactic acid by lactic acid bacteria (LAB) (Taira et al. 2007). However, after days of fermentation, the number of microorganisms decreases, and *Lactobacillus* becomes the predominant microorganism. Such behavior occurs in the production of both fish sauce and fish paste, where LABs are the higher microorganisms in fish product fermentation (Thapa, Pal, and Tamang 2004). Temperature has showed no differences for the preparation of fish sauce using tuna viscera, either maintained at room temperature or refrigerated, since nearly no changes were detected in chemical, physical, and microbiological characteristics generated during fermentation (Dissaraphong et al. 2006). Kuda et al. (2009) have analyzed the microflora of 12 *aji-no-susu* products, a Japanese fermented food that is prepared from salted horse mackerel *Trachurus japonicus*. The authors found that lactic acid concentration was very high compared to *narezushi*, another typical Japanese fermented fish product.

Generally, the microbial population experiences a slight reduction in the first 10 days, and the first high initial microbial increase is probably due to fermentable sugar that promotes the growth of acid-forming bacteria (Kasankala, Xiong, and Chen 2011). Y. Xu et al. (2010) reported an increase of lactic acid bacteria (LAB) within 24 hours of fermentation and a decrease in other microbes, which can be explained by the rapid drop in pH and an increased stress against the growth of other microorganisms such as gram-negative bacteria (Xu et al. 2008).

12.4   CHEMICAL CHANGES DURING PROCESSING THAT AFFECT FERMENTATION

The fermentation process is affected by various factors such as the type of fish species, preliminary treatment (washing, storage, handling, etc.), types of salt, ratio of salt to fish, addition of additives, etc. Various types of fish such as anchovy (*Stolephorus* spp.), mackerel (*Rastrelliger* spp.), and herring (*Clupea* spp.) are generally used in the production of fish sauces (Lopetcharat et al. 2001). For small fish, the fermentation time is around 6 months and, for larger fish, up to 18 months (Rose 1919). The chemical compositions—protein, fat, minerals, fat- and water-soluble vitamins, moisture content, etc.—of these fish are qualitatively and quantitatively different (Ozden and Erkan 2011). The type of protein is a ready source for microorganisms and a substrate for enzymes. The fatty acid profile is slightly different in every species of fish, and unsaturated fatty acids decrease during fermentation (Kim et al. 2001).

The types and number of microorganisms present in fish depend on place, season, species, catching method, storage condition, etc. The washing of fish to remove mud adhering to the fish surface, proper chilling/cooling during storage, and careful handling to avoid damage to the skin of fish prevent the entry of spoilage microorganisms that can grow and effectively depress the growth of beneficial microorganisms. The initial pH of the raw material will vary, which affects the activity of LAB and other bacteria. The added cooked or roasted rice, maize, palm sugar, etc., is the carbohydrate source that is fermented by lactic acid bacteria (LAB), resulting in a decrease of pH. The low pH (4.5–5) inhibits the spoilage and pathogenic bacteria (Paludan-Muller, Huss, and Gram 1999) so that they cannot compete with beneficial bacteria.

The addition of salt inhibits the spoilage-causing bacteria and promotes the growth of halophiles and LAB by decreasing the water activity (Hall 2002). However, Adams, Cook, and Rattagool (1985) found that lower water activity inhibits the LAB and affects the fermentation process. The quality of the salt, either common salt (NaCl) or solar salt, and the salting technique have different effects on the quality of finished products. The salt present in the solid state has a greater effect than the salt present in solutions.

The presence of iron impurities in the crude solar salt accelerated the autoxidation during fermentation (Saisithi 1967). The presence of calcium and magnesium chlorides impurities slow down the penetration of salt into the flesh. These impurities give a whiter color but tend to impart a bitter taste to the product. Magnesium chloride is hygroscopic and tends to absorb water, making the fish more difficult to dry and alters the product quality (Shewan 1951; Cole and Greenwood 1965).

Distribution of salt should be uniform around the fish, and if even a small portion is deprived of salt, the fish will spoil (Saisithi 1994). The salt concentration may vary from 1% to 25% (w/w) in fermented fish, and this influences the rate of fermentation as well as growth of microorganisms (Kose and Hall 2011). Bacterial and enzymatic activity are changed at different salt concentrations, which affects the fermentation process. The concentration of salt should be optimum and neither inhibit the growth of fermenting microorganisms nor deteriorate the quality of flavor and texture of the product (Achinewhu and Oboh 2002; Paludan-Muller 2002). Salt concentrations that are too high slow down the activity of enzymes, resulting in a slower rate of fermentation. On the other hand, low salt concentration will result in the growth of unwanted bacteria, enhance rapid protein hydrolysis, and increase the development of ammonia, which is undesirable in the product (Amano 1962). The minimum salt concentration should be 25% (w/w) or in the ratio of 1:4 (salt: fish) (Rose 1918; Amano 1962; Raksakulthai 1987; Ravipim 1991). Frequent reuse of brine changes the chemical composition and increases the microbial (osmophilic molds and red halophiles) load of salt solution (Kose and Hall 2011), which ultimately affects the fermentation process and reduces the product quality. Fish that are fresh, thick, and have a higher fat content show less salt penetration (Shewan 1951; Cole and Greenwood 1965).

The presence of oxygen in the fermentation tank affects the growth of microorganisms. Low oxygen levels have a synergistic effect on selecting microorganisms for the process. The oxygen concentration is quite high on the surface and low at the bottom. Fermentation under anaerobic conditions changes the quality of fish sauce. The fermentation of fish is generally regarded as a partial aerobic and anaerobic process (Sanceda et al. 1992 ; Lopetcharat et al. 2001).

The temperature of fish fermentation varies according to types of product to be produced. The temperatures range of 25°C–30°C is suitable for fermentation and promotes tissue breakdown and favors sauce production (Hall 2002). The higher the temperature, the more rapid is the salt uptake (Shewan 1951; Cole and Greenwood 1965).

12.5 CHEMICAL CHANGES DURING FERMENTATION

Fermented fish sauce and paste product are the proteinaceous products obtained through natural hydrolysis by endogenous enzymes and microorganisms (Lopetcharat et al. 2001). Thus, the major change during the fermentation process is the conversion of proteins to small peptides and free amino acids and low-molecular-weight compounds. The chemical composition of fermented fish products has been broadly investigated, chiefly on amino acids, fatty acids, organic acids, volatile compounds, and biogenic amines like histamine. How to avoid histamine production is a great goal for the research of fermented fish products. Volatile compounds developed during fermentation contribute to a unique aroma and flavor (Lindsay 1991). Peptides, amino acids, nucleotides, and organic acids also contribute to the taste of fermented fish products. The pH of fish sauces decreases during fermentation, maybe due to the production of lactic acid by LAB (Ijong and Ohta 1996), free amino acids, and large polypeptides.

12.5.1 CHANGES IN PROTEINS

The macronutrients found in seafood include proteins, fat and oil, and water. All other nutrients are considered micronutrients (Hui et al. 2006). For most seafood species, protein content ranges between 10% and 25%, with an average of 17 g in 100 g, which accounts for the 80%–90% of the energy produced per 100 g of lean species (Nunes et al. 2006). The protein found in seafood is of good quality due to its high digestibility and the specific amounts and relative proportion and availability of essential amino acids (Nunes, Bandarra, and Batista 2011). The amount of protein is similar in pelagic and demersal fish (Hui et al. 2006). Fish is the most important source of animal protein for the people in developing countries, since approximately 53% of the world's catch is from these countries, and most of the catch is consumed within those countries. In the Southeast Asian region, people receive 60%–70% of their protein from fish (Saisithi 1994).

Protein content varies from fish to fish, either being marine or freshwater, hence a protein content variation can be expected in the different fish sauces and pastes. As proteins are the most important constituents of fish, then their degradation originates the major changes that take place during fish sauce and fish paste fermentation. During fermentation of fish, proteins, fat, and glucose are converted into peptides and amino acids, fatty acids, and lactic acid by the action of enzymes and microorganisms (Salampessy, Kailasapathy, and Thapa 2010). The hydrolysis of proteins during fermentation is mainly caused by the endogenous enzymes, a trypsin-like enzyme, and cathepsins A and C (De Vecchi and Coppes Petricorena 1996), although their actions are partially minimized by the presence of high salt concentration, slowing the process of autolysis (Orejana and Liston 1982; Raksakulthai, Lee, and Haard 1986), and hence allowing the preservation of the fermented fish product.

Due to its perishable nature, fish undergoes rapid spoilage under tropical conditions of high temperature and high humidity, like in the countries of Southeast Asia, where traditional fish fermentation to preserve fish is common because of its low cost. However, as traditionally the process lasts near 24 months, attempts have been made to shorten the duration of the process. For example, *koji* has been successfully used to accelerate fish sauce fermentation (Kasankala, Xiong, and Chen 2011).

12.5.1.1 Changes in Amino Acids

The amino acid composition of fish sauce may be nutritionally important, especially in regions where fish sauce serves as a significant source of dietary protein (Hjalmarsson, Park, and Kristbergsson 2007). Fish sauces contain 20 g/L of nitrogen, of which 16 g/L are in the form of amino acids (Sanceda et al. 1996). According to Sikorski (1994), amino acids dictate the taste of seafood. Fermented fishes have unique characteristics, umami and sour/tastes (stronger than that of soy sauce), which are thought to be caused by a high concentration of extractive components, such as free amino acids and organic acids (Itou et al. 2006). Cha and Cadwallader (1998) stated that the specific free amino acids having sweet, sour, and bitter tastes may play a prominent role in the overall taste of fish sauce.

Because of taste differences among countries, the chemical composition of fish sauces differs among factories and among regions (Park et al. 2001). Many studies have been carried out to determine the taste-active components of seafood (Hayashi, Yamaguchi, and Konosu 1981; Shirai et al. 1996; Park et al. 2002b). In general, the taste-active components in seafood include several free amino acids such as glutamic acid, glycine, alanine, arginine, and nucleotides such as inosine 5' monophosphate, adenosine 5' monophosphate, and guanosine monophosphate (J. Park et al. 2002b). Also, creatine and lactate as well as succinate were found in dried skipjack and short-necked clam (Fuke and Konosu 1991). Concerning only amino acids, lysine, alanine, glycine, serine, and threonine provide a sweet taste, while arginine, leucine, valine, phenylalanine, histidine, and isoleucine give a bitter taste, and glutamic acid and aspartic acid give a sour taste, with the typical aroma of glutamic acid contributing a meaty taste (Kato, Rhue, and Nishimura 1989; Lopetcharat et al. 2001; Jiang et al. 2007). These amino acids are generated/produced during fermentation.

Through the action of enzymes and some halotolerant and halophile microorganisms, fish proteins are hydrolyzed, resulting in free amino acids, peptides, and ammonia. According to Jiang et al. (2007), high salt concentration helps to control the growth of pathogenic microorganisms, hence resulting in a desirable taste and aroma. Interestingly, fish sauce prepared with capelin from Iceland varies in its amino acid content whether capelin being caught in summer or in winter (Hjalmarsson, Park, and Kristbergsson 2007), and similar amino acid composition from summer capelin was found compared to other fish sauces from tropical countries (Ijong and Otha 1996; Lopetcharat et al. 2001; Tungkawachara, Park, and

Choi 2003). The authors found that capelin harvested during the summer season was more suitable as raw material for fish sauce than was capelin harvested during the winter season, due to the higher proteolytic activity. Another interesting study was carried out by Osako et al. (2005), who compared the amino acid content of fish sauces prepared from raw material and from heated fish, allowing them to ferment. The amino acid content of fish sauce from fermented raw fishes was different in the three species studied (Japanese anchovy, horse mackerel, and rabbit fish), but the amino acid content of fish sauce prepared from heated meat did not vary among the species. The authors concluded that the production of industrial fish sauce of similar taste, on one hand, will depend on heat treatment of meat before fermentation, and on the other hand, the production of characteristic fish sauces will be possible on the raw material.

Predominant amino acids like alanine and glutamic acid are chiefly recognized as being important in the taste of fermented fish products (Park et al. 2002a; Park, Je, and Kim 2005). Alanine and lysine tend to be high in long-fermented fish products. However, glutamic acid was not correlated with fermentation time when analyzing amino acid content in *aji-no-susu* from salted mackerel (Kuda et al. 2009). Glutamic acid, in general closely related to umami substance, was produced as a relatively stable amino acid, without any secondary decomposition during fermentation. Also, glutamic acid content of fish sauces, whether from freshwater or marine species, is less than that from shrimp sauce.

Interestingly, Park et al. (2002b) determined that glycine was not a taste-active component of fish sauce, although many other authors have reported that glycine was a taste-active component in seafood (Hayashi, Yamaguchi, and Konosu 1981; Fuke and Konosu 1991; Shirai et al. 1996; Park et al. 2001; Park et al. 2005). Thus, Park et al. (2002b) suggest that the sweetness of glycine may be masked by large amount of alanine and/or glutamic acid. D-alanine was one of the major amino acids found in several fish sauces from Japan by Abe et al. (1999). However, in fish sauce prepared from sardine, lysine was the most abundant amino acid (Dincer et al. 2010).

12.5.1.2   Generation of Peptides

Several peptides are known to have their own tastes (umami, bitter or sour taste) in various fermented foods (Ohyama et al. 1988). The bitterness of fish sauce was attributed to the presence of some tripeptides containing the α-imino acid proline. Osako et al. (2005) also determined amino acid composition and peptides including oligopeptides, where the main amino acid constituents of peptides in fish sauces from various species (Japanese anchovy, rabbit fish, and horse mackerel) were aspartic acid, glutamic acid, and glycine, which accounted for the nearly 50% of total amino acids in all fish sauces analyzed. Peptides in fermented fish products not only account for the taste, but also their presence may help to avoid oxidation of the product. Thus, recently and as a new application for a fermented fish product, Giri et al. (2010) suggested that the hydrophobic amino acids such as leucine, valine, and alanine present in the sequences of squid miso peptides are linked to antioxidative properties, since these amino acids are effective inhibitors of linoleic acid oxidation, for example.

12.5.2 Changes in Carbohydrates

Very little to no carbohydrate content appears in the fish fermentation process because the LABs identified in fish sauces are responsible for transforming glucose into lactic acid, which occurs in higher amounts (Ijong and Ohta 1996).

12.5.3 Changes in Lipids

Fat content in fish ranges widely from 0.2% to almost 30%, and contrary to terrestrial animals, in which most lipids are generally deposited in adipose tissue, fish have lipids in the liver, muscle, and perivisceral, and subcutaneous tissues (Nunes, Bandarra, and Batista 2011). Every fish has a slightly different fatty acid profile, and unsaturated fatty acids constitute up to 40% of the total fatty acids decreasing during fermentation (Kim et al. 2001). The lipid content of fish sauce is considered low, although the lipid content of the raw material is higher. For example, the lipid content of *budu* never increases more than 0.13% although the lipid content of the raw material *kanbilis* is 10 times higher (Beddows et al. 1979; Salampessy, Kailasapathy, and Thapa 2010). *Chepashutki*, a semifermented fish product, has a lipid content that varies from 19.25% to 24.97% (Nayeem et al. 2010). The lipid content in fish sauce from clupeid fish depends on the time of fermentation, varying from 5.6% at the end of 4 weeks, to 6.54% after 8 weeks fermentation, and decreasing abruptly at 12 weeks fermentation with a fat content of 3.85% (Olubunmi et al. 2010). A similar result was obtained by Ibrahim (2010) when comparing fish lipid content of fresh fish *Gambusia affinis* with the fish sauce fermentation during 16 weeks, varying from 6.56% to 1.56%, respectively. Analyzing the chemical composition of fresh fish and fermented fish, using different processors, Koffi-Nevry et al. (2011) found that while protein, ash, and mineral contents decreased among all fermented fish samples, the lipid content was similar whether being fresh fish or fermented fish sauce. Crude lipid also varies in commercial fish pastes, ranging from 1.1% to 1.3%, which was much lower than in the corresponding fresh fish (Giri, Osako, and Ohshima 2009a). Several underutilized fish species for miso-like fermented fish production were shown to have a lipid content that varies from 5.2% to 10.8% in fresh fish, while in fish pastes the lipid content decreased abruptly, ranging from 0.6% to 2.2% (Giri, Osako, and Ohshima 2009b).

Although fish lipids are composed of saturated fatty acids, monounsaturated fatty acids, and polyunsaturated fatty acids (PUFAs), what chiefly characterize fishes are PUFAs, like eicosapentaenoic (EPA: C_{20}:5, Ω-3) and docosahexaenoic (DHA: C_{22}:6, Ω-3) acids, both making marine lipids unique compared to other lipid sources (Rustad 2003; Shahidi and Miraliakbari 2006).

As fish fermentation processes also involve the use of high salt concentration and incubation at ambient temperature, essential PUFAs could be greatly damaged during severe fermentation conditions (Peralta et al. 2008). Antioxidant activity has been found in a number of fermented fishery products such as fermented blue mussels (Jung, Rajapakse, and Kim 2005), fish sauces (Harada et al. 2003), and fermented shrimp paste (Peralta et al. 2005). The increase of antioxidant activity in salt fermented shrimp paste was suggested to depend on formation of Maillard reaction products (MRDs) having antioxidant activity (Peralta et al. 2008). PUFAs such as EPA and DHA in the shrimp paste remained almost intact during prolonged fermentation, improving antioxidant ability and nutritional value such as amino acids in salt fermented shrimp paste without loss in the PUFAs. Also, sardine fish sauce was demonstrated to be a very good source of DHA and EPA which, together with amino acid content, showed to be a very good fermented fish product (Dincer et al. 2010).

12.5.4 Changes in Minerals

Mineral content in fermented fish products, like fish sauces and pastes, is not so different from the original fresh fish, since these elements do not suffer transformation throughout the process of fermentation. However, some minerals can retard the diffusion of NaCl into fish flesh during the fermentation process: calcium sulfate ($CaSO_4$) at 0.24%, magnesium sulfate ($MgSO_4$) at 0.17%, magnesium chloride ($MgCl_2$) at 0.3%, calcium chloride ($CaCl_2$) at 0.24%, water-insoluble substance at 0.4%, water 2.4%, Mg 2%, Ca^{2+}, and other impurities (Lall 1995; Lopecharat et al. 2001). Such minerals can negatively influence the action of NaCl on microorganisms, since salt controls the type of microorganisms and retards or kills some pathogenic microbes during fermentation. Moreover, the chemical composition of salt also affects the type of microbiological flora during fermentation, which in turn affects the quality of fish sauce. Salt concentration varies from country to country, but in general Thai sea salt has a sodium chloride content of 88.3% ± 2.7%, while salt content of fermented fish products from other countries has higher concentration of NaCl (97%).

On the other hand, it is important to take into consideration the fish-to-salt ratio, since it is another factor that can affect fish sauce quality. Salt concentration affects the function of various endogenous enzymes that play a relevant role in protein degradation during fermentation (Baens-Arcegá 1977b; Orejana and Liston 1982). According to Beddows (1985), Japanese fish sauce (*shottsuru*) has a ratio of fish to salt about 5:1. However, the fish-to-salt relationship in Korean fish sauce (*aekjeot*) is of 3:1 to 4:1 (Lopetcharat et al. 2001). In contrast, the Thai fish sauce *nampla* is prepared using a 1:1 to 5:1 ratio. Lopetcharat et al. (2001) show a table that summarizes the different fish sauces together with their characteristic fish-to-salt ratios. Those relationships depend on the size of fish used; hence, the product obtained varies with the final desired taste, because bacterial and enzymatic activities change with varying salt concentration, resulting in different flavors.

12.5.5 Changes in Vitamins

Although fish products are not considered a predominant source of vitamins, their levels are enough to reinforce human nutrition of vitamin intake. As fermented fish products are consumed by millions of people as a main food, it is relevant to take into account the vitamin content of fermented fish so that people can receive the vitamins that are in fish sauces. Vitamins, chiefly antioxidant ones, suffer transformation owing to oxidation not only in fresh and frozen fish, but in fermented fish products as well. Thus, as many vitamins are fat soluble, they are protected from oxidation in fish owing to oil, which allows vitamins to avoid oxygen. The vitamin content of fish products is comparable to that of mammals, except for vitamins A and D, which are found in large amounts in the meat of fatty fishes and in the liver of lean fish, such as cod and halibut. The vitamin content of fish varies with species, age, season, sexual maturity, and geographical area (Coppes Petricorena 2008). The research carried out till present concerning vitamin content in fermented fish products, like fish sauce, is scarce, although vitamin content has been investigated for many years. Vitamin B_{12} has received the most attention lately, and it has shown to be present in considerable amounts in different fish products, including fish sauce.

Fish and shellfish are well known as sources of fat-soluble vitamins: A (in the form of retinol) and D, though they also can also provide significant amounts of the B vitamins (Silva and Chamul 2000). Vitamin A is found in high amounts in oysters as well as in the oil fish such as herring, mackerel, and sprats. Vitamin A can occur in two different forms: as retinol, which is easily absorbed by the body, or as carotenoids, which are less easily absorbed and have only 50% of the absorption rate of retinol (Coppes Petricorena 2008). The easily absorbed retinol is the type of vitamin A found concentrated in the fish viscera, especially the liver. Fish liver oils, like those from cod and shark, are excellent seafood sources of vitamin A. Most oil in fish is found in the dark flesh, hence the flesh contains higher concentration of vitamin A. Among shellfish, oysters appear to be the best source of vitamin A. For example, vitamin A concentration reaches 94 mcg/100 g in shark liver oil, 44 mcg/100 g in flesh of herring, 60 mcg/100 g in sprat, 75 mcg/100 g in oyster, and 100 mcg/100 g in the edible portion of halibut (Nunes, Bandarra, and Batista 2011).

Vitamin D is chiefly found in the form of cholecalciferol (D_3), mainly in the oil-rich pelagic fish like mackerel and herring, both of which contain higher levels of vitamin D than leaner fish like flounder and sea trout (Holland, Brown, and Bush 1993). Vitamin E is present in significant amounts in seafood, providing around 10%–20% of the average daily vitamin E intake of 5–10 mg in a 100-g portion. Vitamin E is one of the most important lipid-soluble antioxidants to occur in plants and animals, specifically for protection against lipid peroxidation in biological membranes. Four homologous pairs of vitamin E have been described (α, β, γ, and δ tocopherols and tocotrienols) with the α-forms having the greatest activity. Yamamoto et al. (2001) discovered an unusual vitamin E in fish that enhances the antioxidant protection in marine organisms adapted to cold-water environments, hence reducing lipid peroxidation and avoiding rancidity. Although fat-soluble vitamins are mainly present in fish oily liver, they also are present in flesh; for example, vitamin E in salmon flesh reaches 4 mg/100 g (Nunes, Bandarra, and Batista 2011).

Water-soluble vitamins can be found throughout a fish body rather than concentrated in the viscera like the fat-soluble vitamins. Among the important vitamins in this group are thiamine, riboflavin, pyridoxamine, niacin, folic acid, pantothenic acid, vitamins B_{12} and B_6, and vitamin C. The recommended intake of thiamine for adult males is 1.25 mg/day and for women 1.1 mg/day. Fish muscle thiamine content averages approximately 100 mcg/100 g flesh. Oyster is an especially good source of thiamine, containing about 1 mg/100 g. Niacin, like thiamine, forms part of a coenzyme essential in the production of energy, and in fish muscle ranges from 0.9 mg to 3.1 mg /100 g flesh. Adult women need about 1.0 mg/day, while men need 1.4 mg/day.

The concentration of riboflavin in fish is quite variable, with the dark meat containing more than the white meat. However, the amount of riboflavin in many fish species is comparable to that found in terrestrial animals (50–980 mcg/100 g muscle).

Vitamin B_6 (pyridoxine) helps in the conversion of one amino acid to another needed in the body, and also in the conversion of linolenic acid into arachidonic acid. Adults need 2 mg/day of vitamin B_6. Whole fish is a good source of vitamin B_6, with values ranging from 100 to 1200 mcg/100 g of fish flesh. However, in fish sauce, B_6 values range from 40,000 to 50,000 mcg/100 g. Vitamin B_{12} or cobalamin, found in significant amounts in seafood, varies from 0 in shark to 1.9 mcg/100 g in Pacific herring. Recommended dietary intake of vitamin B_{12} for adults is 3 mcg, a very small amount, so seafood can generally be considered a very good source. Fish and shellfish contribute greatly to vitamin B_{12} intake in Asians, Japanese people in particular, and this trend is spreading throughout the world (Kimura et al. 2003). Anchovies, clams, herring, oysters, pilchard, and sardines are particularly rich sources of vitamin B_{12}, containing 25–40 mcg /100 g meat. Watanabe et al. (2001) have found in oyster, mussels, and short-necked clam a B_{12} content of 46.3, 15.71, and 87.0 mcg/100 g, respectively. Certain fish like salmon, sardine, trout, and tuna have a B_{12} content that varies from 3.8 to 8.9 mcg/100 g. However, mollusks and clams have an average of 98 mcg/100 g of B_{12} (Watanabe et al. 2001). A concentration of 159 mcg/100 g was found in dark muscle of skipjack (Watanabe 2007), whereas 52 mcg/100 g was the B_{12} content for yellowfin tuna muscle (Nishioka et al. 2007). *Nampla* appears to be the major source of B_{12} in Thailand, containing considerable amounts, and a B_{12} concentration of 2.3–5.5 mcg/100 g was determined in *ishiru* (a Japanese traditional fish sauce) (Takenaka et al. 2003).

Folic acid and vitamin C are not found in any significant amount in edible portions of fish (Coppes Petricorena 2008). Flesh of Atlantic salmon, European hake, and sardine have a folic acid concentration of 10, 27, and 24 mcg/100 g, respectively (Nunes, Bandarra, and Batista 2011). It is important to take into account the technology used for vitamin determination through the years. Thus it would be very valuable to unify the criteria of measurements in order to be able to compare different data obtained to date (Coppes Petricorena 2008). Thus, relevant future research could be to determine the vitamin content in different fermented fish products using advanced technology like high-performance liquid chromatography (HPLC).

12.5.6 MISCELLANEOUS CHANGES

Biogenic amines, including histamine, are formed through the decarboxylation of specific free amino acids by exogenous decarboxylases released from microbial populations associated with the seafood (Rawles, Flick, and Martin 1996) as well as environmental conditions to promote growth of histamine-forming bacteria (Lehane and Olley 2000; Tsai et al. 2006). Thus, mainly histamine has been proposed as a marker to evaluate fish freshness (Yen and Hsieh 1991; Paleologos, Savvaidis, and Kontominas 2004) and to estimate the quality of fish sauce (Lopetcharat et al. 2001). Histamine toxicosis is caused by ingesting a high level of free histidine in fish tissue that has been decarboxylated to histamine (Sanceda, Kurata, and Arakawa 1996).

Fish sauces are known to contain high levels of histamine depending upon fish type, handling conditions, and the fermentation process (Tao et al. 2009). Fish paste could contain a high concentration of biogenic amines due to raw fish being subjected to temperature abuse, as is the case with raw tuna (Naila et al. 2011). Fardiaz and Markakis (1979) have reported biogenic amines in fermented fish paste. Generally, histamine formation by bacteria is enhanced at elevated storage temperatures (Kim et al. 2009) as well as high-temperature abuse of harvested fish. Thus, histamine formation can easily be controlled by lowering the storage temperatures and implementing hygienic practices (Stratton et al. 1991; in Lopetcharat et al. 2001).

Various acids, including volatile fatty acids, have been identified in fish sauces and pastes (Peralta et al. 1996; Cha and Cadwallader 1998; Fukami et al. 2002). The origin of these acids is from lipids (caproic and heptanoic acids), amino acids (propionic, n-butanoic, and n-pentanoic acids through bacterial action), or glucose (lactic acid) (McIver et al. 1982, in Salampessy, Kailasapathy, and Thapa 2010). The volatile compounds contributing to the flavor of fish sauce are produced by nonenzymatic reactions of various components and enzymatic reactions by endogenous enzymes of fish origin and those of microorganisms surviving during fermentation (Fukami et al. 2004). Fish sauces with less volatile fatty acids contribute to the cheesy flavor and are more ammoniacal than sauces with more volatile fatty acids, which contribute to the rancid flavor (Fukami et al. 2002). In fish pastes, the main volatile compounds consist mainly of aldehydes, ketones, alcohols, aromatic compounds, N-containing compounds, esters, and S-containing compounds (Salampessy, Kailasapathy, and Thapa 2010). Benzaldehyde is the most common among the aldehydes, and 2,3-butanedione among ketones, contributing the first to a pleasant almond, nutty, and fruity aroma, and the second to an intense buttery and desirable aroma in crustaceans (Sidwell 1981). Also acetaldehyde, propanal, 2-methylpropanal, 2-methylbutanal, 3-methylbutanal, 2-butenal, octanal, nonanal, and 2,4-heptadienal may play a major role in determining the aroma of fish paste by changing their relative abundance (Giri et al. 2010).

The volatile sulfurs and nitrogen-containing compounds are the most odorous constituents in fermented fish products. The sulfur compounds can be originated either from raw materials or during the fermentation process from the free, peptidic, and proteinic sulfur amino acids as well as the glutathione pool in the fish tissue (Giri et al. 2010). Nitrogen-containing heterocyclic pyrazines, found in small quantity, might impart roasted and nutty flavor to the several fermented fish products.

Lactic acid is the greatest organic acid component of fish sauces, found in higher levels during fermentation (Taira et al. 2007), and its accumulation contributes considerably to decreasing pH and increasing total acidity of fermented fish products, as well as to taste, flavor, and texture (Visessanguan et al. 2006). Acetic acid is also an important contributor to the flavor, aroma, and texture development of fermented fish products. Both acids indicate the presence of LAB, the level of which could be used as another safety index of a fermented fish product. Thus, both acids could serve as indicators of successful batch fermentation (Motarjemi 2002). Organic acids vary with the raw material either from freshwater or marine fishes. Thus, acetic acid content in fish paste from marine fish is higher than the content of a fish paste prepared from a freshwater fish, contrary to what happens with lactic acid content (Mizutani et al. 1992). Pyroglutamic acid and succinic acid (taste-active components) are also increased during fermentation of fish sauce. When fish sauces are rich in organic acids, the pH decreases to a low level and a sour taste may be emphasized (Taira et al. 2007). Alcohols are found in abundance in all fish pastes, with the exception of shrimp pastes, whose low content corresponds to their low aldehyde content (Salampessy, Kailasapathy, and Thapa 2010).

12.6 CONCLUDING REMARKS

Seafood products, especially fish and fish products, are an excellent source of protein, minerals, and omega-3 fatty acids such as docosahexaenoic acid and eicosapentaenoic acid. Fish are highly perishable and are susceptible to autolytic spoilage by enzymes, fat oxidation, microbiological spoilage, or their combinations. These seafood products can be preserved by fermentation.

Fermentation of fish refers to the enzymatic liquefaction of these materials to produce edible soft pastes and/or sauces. Salt is mixed with fish at a ratio varying from 1:2 to 1:9, and fermentation takes place at ambient temperature over a few days (1–5) to several months (3–12). Halophilic microorganisms and autolytic enzymes are responsible for chemical, biochemical, and physiological changes during fermentation. Fish fermentation is often combined with the addition of salt or drying to reduce water activity and eliminate proteolytic and putrefying microorganisms. There are various factors like species, preliminary treatments, types of salt, ratio of salt to fish, addition of additives, types of microorganisms, etc., that affect the various chemical changes that occur during the fish fermentation process and the ultimate the quality of the finished products. Mineral content in fish sauces and pastes do not suffer transformation throughout the process of fermentation. However, some minerals can retard the diffusion of NaCl into fish flesh during the fermentation process and influence product quality. During fermentation of fish, proteins, fat, and glucose are converted into peptides and amino acids, fatty acids, and lactic acid by the action of enzymes and microorganisms. Care must be taken during the fermentation process to avoid peroxidation of unsaturated fatty acids from fish lipids. Various chemical/biochemical changes taking place before and during fermentation lead to improved palatability and nutritive value as well as extended shelf life of the fish products.

More research must be carried out concerning mineral and vitamin content in fermented fish products, since data are scarce, chiefly about vitamin content. More research also needs to be developed to produce high-quality fish sauce, mainly: (a) how to accelerate fish sauce production, (b) how to eliminate or reduce the histamine content in fish sauce products, and (c) how to produce a low-sodium fish sauce without sacrificing flavor. Fermented fish products have several disadvantages, including unhygienic preparation and high salt content that, together with color, flavor, and aroma. And, to make the situation more serious, the histamine content and the presence of carcinogenic compounds or their precursors in fermented fish products remain a matter of concern.

ACKNOWLEDGEMENTS

Dr. Coppes Petricorena thanks Japanese scientists Dr. Hiroki Abe, from the Department of Aquatic Biosciences of the University of Tokyo, and Dr. Kazufumi Osako, from the Nagasaki Prefectural Institute of Fisheries, for all of their efforts in sending information as well as sharing results, even in the aftermath of the earthquake and tsunami that struck Japan. The coauthor also thanks her children, Dr. Karin Achaval (MD, PhD) and Lic. Federico Achaval Coppes (biologist), for their enthusiasm and encouragement. Finally, I am grateful to the editor of this book, through Dr. Joshi, for inviting me to contribute to this chapter.

REFERENCES

Abe, H., J.-N. Park, Y. Fukumoto, et al. 1999. Occurrence of D-amino acids in fish sauces and other fermented fish products. *Fisheries Science* 65 (4): 637–641.

Achinewhu, S. C., and C. A. Oboh. 2002. Chemical, microbiological, and sensory properties of fermented fish products from *Sardinella* spp. in Nigeria. *J. Aquat Food Tech.* 11:53–59.

Ackman, R. G. 1990. Seafood lipids and fatty acids. *Food Rev. Int.* 6 (4): 617–646.

Adams, M. R. 2009. Fermented fish. In *Microbiology handbook: Fish and seafood*, ed. R. Fernandes, 123–140. Cambridge, U.K.: Royal Society of Chemistry.

Adams, M. R., R. D. Cook, and P. Rattagool. 1985. Fermented fish products of Southeast Asia. *Trop. Sci.* 25:61–73.

Amano, K. 1962. The influence of fermentation on the nutritive value of fish with special reference to fermented fish products of Southeast Asia. In *Fish in nutrition*, ed. E. Heen and R. Kreuzer, 180–200. London: Fishing News (Books).

Baens-Arcega, L. 1977a. Process of making bagoong in the Philippines. Symposium on Indigenous Fermented Foods, Bangkok, Thailand.

Baens-Arcega, L. 1977b. Patis, a traditional fermented fish sauce and condiment of the Philippines. Symposium on Indigenous Fermented Foods, Bangkok, Thailand.

Barbara, R., and G. Bledsoe. 2006. Seafood products: Science and technology. Vol. 4 in *Handbook of food science, technology, and engineering*, ed. Y. H. Hui, E. Castell-Perez, L. M. Cunha, et al., 159-1–159-13. Boca Raton, FL: CRC Press, Taylor & Francis Group.

Beddows, C. G. 1985. Fermented fish and fish products. Vol. 2 in *Microbiology of fermented foods*, ed. B. J. B. Wood, 2–23. London: Elsevier Applied Science.

Beddows, C. G., A. G. Ardeshir, and W. J. B. Daud. 1979. Biochemical changes occurring during the manufacture of budu. *Journal of the Science of Food and Agriculture*, 30:1097–1103.

Cha, Y. J., and K. R. Cadwallader. 1998. Aroma-active compounds in skipjack tuna sauce. *Journal of Agricultural and Food Chemistry* 46:1123–1128.

Cole, R. C., and B. L. H. Greenwood. 1965. Problems associated with the development of fisheries in tropical countries: The preservation of the catch by simple processes. *Tropical Science* 7:165–183.

Coppes Petricorena, Z. L. 2008. Relevancia de las vitaminas de los organismos marinos para la industria pesquera. *Industria Carnica Latinoamericana* 154:48–56.

De Vecchi, S., and Z. L. Coppes Petricorena. 1996. Marine fish digestive proteases: Relevance to food industry and the Southwest Atlantic region: A review. *Journal of Food Biochemistry* 20 (3): 193–214.

Dincer, T., S. Cakli, B. Kilinc, and S. Tolasa. 2010. Amino acids and fatty acid composition content of fish sauce. *Journal of Animal Veterinary Advances* 9 (2): 311–315.

Dissaraphong, S., S. Benjakul, W. Visessanguan, and H. Kishimura. 2006. The influence of storage conditions of tuna viscera before fermentation on the chemical, physical, and microbiological changes in fish sauce during fermentation. *Bioresource Technology* 97:2032–2040.

Essuman, K. M. 1992. Fermented fish in Africa: A study on processing, marketing, and consumption. In FAO Fisheries Technical paper 329. Food and Agriculture Organization of the United Nations. Rome, Italy.

Fardiaz, D., and P. Markakis. 1979. Amines in fermented fish paste. *Journal of Food Science* 44:1562–1563.

Fraser, O., and S. Sumar. 1998. Compositional changes and spoilage in fish: An introduction. *Nutr. Food Sci.* 5:275–279.

Fukami, K., Y. Funatsu, K. Kawasaki, and S. Watabe. 2004. Improvement of fish-sauce odour by treatment with bacteria isolated from the fish-sauce mush (moromi) made from frigate mackerel. *Journal of Food Science* 69: FMS45–FMS49.

Fuke, S., and S. Konosu. 1991. Taste-active components in some foods: A review of Japanese research. *Physiological Behavior* 49:863–868.

Giri, A., K. Osako, and T. Ohshima. 2009a. Extractive components and taste aspects of fermented fish pastes and bean pastes prepared using different *koji* molds as starters. *Fisheries Science* 75:481–489.

———. 2009b. Effect of raw materials on the extractive components and taste aspects of fermented fish paste: *Sakana* miso. *Fisheries Science* 75:785–796.

Giri, A., K. Osako, A. Okamoto, E. Okazaki, and T. Ohshima. 2010. Antioxidative properties of aqueous and aroma extracts of squid miso prepared with *Aspergillus oryzae*–inoculated *koji*. *Food Research International* 44:317–325.

Hall, G. M. 2002. Lactic acid bacteria in fish preservation. In *Safety and quality issues in fish processing*, ed. A. H. Bremmer, 330–349. Cambridge, U.K.: Woodhead Publishing.

Harada, K., C. Okano, H. Kadoguchi, et al. 2003. Peroxyl radical scavenging capability of fish sauces measured by the chemiluminescence method. *International Journal of Molecular Medicine* 12:621–625.

Hayashi, T., K. Yamaguchi, and S. Konosu. 1981. Sensory analysis of taste-active components in the extract of boiled snow crab meat. *Journal of Food Science* 46:479–483.

Hjalmarsson, G. H., J. W. Park, and K. Kristbergsson. 2007. Seasonal effects on the physicochemical characteristics of fish sauce made from capelin (*Mallotus villosus*). *Food Chemistry* 103:495–504.

Holland, B., J. Brown, and D. H. Bush. 1993. Fish and fish products. In third supplement to McCance and Widdowson's *The Composition of foods*. London: Royal Society of Chemistry.

Hui, H. J., N. Cross, H. G. Kristinsson, et al. 2006. Biochemistry in seafood processing. Chap. 16 in *Food biochemistry and food processing*, ed. Y. Y. Hui, 351–378. Ames, IA: Blackwell Publishing.

Ibrahim, S. M. 2010. Utilization of gambusia (*Affinis affinis*) for fish sauce production. *Turkish Journal of Fisheries and Aquatic Sciences* 10:169–172.

IIF. 1986. *Recommandations pour la preparation et la distribution des aliments congeles* [Recommendations for the processing and handling of frozen foods], 3rd ed. Paris, France: Institut International du Froid.

Ijong, F. G., and Y. Ohta. 1995. Microflora and chemical assessment of an Indonesian traditional fermented fish sauce, *bakasang*. *J. Fac. Appl. Biol. Sci.* 34:95–100.

———. 1996. Physicochemical and microbiological changes associated with *bakasang* processing: A traditional Indonesian fermented fish sauce. *Journal of the Science of Food and Agriculture* 71:69–74.

Irianto, H. E., and G. Irianto. 1998. Traditional fermented fish products in Indonesia. *FAO RAP Publication 1998*, no. 24:67–75.

Ishige, N., and K. Ruddle. 1987. Gyosho in Southeast Asia: A study of fermented aquatic products. *Bulletin of the National Museum of Ethnology* 12 (2): 235–314.

Itoh, H., H. Tachi, and S. Kikuchi. 1993. In *Fish fermentation technology*, ed. C. H. Lee, K. H. Steinkraus, and P. J. Alan Reilly, 177. Tokyo: United Nations University Press.

Itou, K., S. Kobayashi, T. Ooizumi, and Y. Akahane. 2006. Changes of proximate composition and extractive components in *narezushi*, a fermented mackerel product, during processing. *Fisheries Science* 72:1269–1276.

Jiang, J. J., Q. X. Zeng, Z. W. Zhu, and L. Y. Zhang. 2007. Chemical and sensory changes associated Yu-lu fermentation process: A traditional Chinese fish sauce. *Food Chemistry* 104:1629–1634.

Jung, W., N. Rajapakse, and S. Kim. 2005. Antioxidative activity of a low molecular weight peptide derived from the sauce of fermented blue mussels, *Mytilus edulis*. *European Food Research and Technology* 220: 535–539.

Kasankala, L. M., Y. L. Xiong, and J. Chen. 2011. The influence of douchi starter cultures on the composition of extractive components, microbiological activity, and sensory properties of fermented fish pastes. *Journal of Food Science* 76 (1): C154–C161.

Kato, H., M. R. Rhue, and T. Nishimura. 1989. Role of free amino acids and peptides in food taste. In *Flavor chemistry: Trends and developments*, ed. R. Teranishi, B. Buttery, and R. G. Shahidi, 158–174. ACS Symposium Series 388. Washington, DC: American Chemical Society.

Kevin, J. W., and P. Howgate. 2002. Glossary of fish technology terms. Glossary prepared under contract to the Fisheries Industries Division of the Food and Agriculture Organization of the United Nations.

Kim, S.-K., Y.-T. Kim, H.-G. Byun, et al. 2001. Isolation and characterization of antioxidative peptides from gelatin hydrolysate of Alaska pollack skin. *Journal of Agricultural and Food Chemistry* 49 (4): 1984–1989.

Kimizuka, A., T. Mizutani, K. Ruddle, and N. Ishige. 1992. Chemical components of fermented fish products. *Journal of Food Composition and Analysis* 5 (2): 152–159.

Kimura, N., T. Fukuwatari, R. Sasaki, et al. 2003. Vitamin intake in Japanese women college students. *Journal of Nutritional Science and Vitaminology* 49:149–155.

Kinsella, J. E. 1988. Fish and seafoods: Nutritional implication and quality issues. *Food Technology* 42 (5): 146–150.

Koffi-Nevry, R., T. S. T. Ouina, M. Koussemon, and K. Brou. 2011. Chemical composition and lactic microflora of *adjuevan*, a traditional Ivorian fermented fish condiment. *Pakistan Journal of Nutrition* 10 (4): 332–337.

Kose, S., and G. M. Hall. 2011. Sustainability of fermented fish products. In *Fish processing: Sustainability and new opportunities*, ed. George Hall, 138–166. Ames, IA: Wiley-Blackwell.

Kuda, T., R. Tanibe, M. Mori, et al. 2009. Microbial and chemical properties of *aji-sno-suso*, a traditional fermented fish with rice product in the Noto Peninsula, Japan. *Fisheries Science* 75:1499–1506.

Lall, S. P. 1995. Macro and trace elements in fish and shellfish. In: *Fish and fishery products: Composition, nutritive properties, and stability*, ed. A. Ruiter, 787. Wallingford, CT: CAB International.

Lehane, L., and J. Olley. 2000. Histamine fish poisoning revisited. *International Journal of Food Microbiology* 58:1–37.

Lindsay, R. C. 1991. Chemical basis of the quality of seafood flavors and aromas. *Marine Technology Society of Japan* 25:16–22.

Lopetcharat, K., Y. J. Choi, J. W. Park, and M. A. Daeschel. 2001. Fish sauce products and manufacturing: A review. *Food Reviews International* 17:65–88.

McIver, R. C., R. I. Brooks, and G. A. Reineccius. 1982. Flavour of fermented fish sauce. *Journal of Agriculture and Food Chemistry* 30: 1017–1020.

McKee, H. S. 1956. Fish sauces and pastes are palatable and nourishing. *SPC Quarterly Bulletin* 6:16–17.

Mizutani, T., A. Kimizuka, K. Ruddle, and N. Ishige. 1987. A chemical analysis of fermented fish products and discussion of fermented flavors in Asian cuisines: A study of fermented fish products. *Bulletin of the National Museum of Ethnology* 12 (3): 801–864.

———. 1992. Chemical components of fermented fish products. *Journal of Food Composition and Analysis* 5:152–159.

Motarjemi, Y. (2002). Impact of small scale fermentation technology on food safety in developing countries. *Int. J. Food Microbiol.* 75:213–229.

Naila, A., S. Flint, G. C. Fletcher, et al. 2011. Biogenic amines and potential histamine forming bacteria in *rihaakuru* (a cooked fish paste). *Food Chemistry* 128 (2): 479–484.

Nayeem, M. A., K. Pervin, M. S. Reza, M. N. A. Khan, M. N. Islam, and M. Kamal. 2010. Quality assessment of traditional semi-fermented fishery product (cheap *shutki*) of Bangladesh collected from the value chain. *Bangladesh Research Publications Journal* 4 (1): 41–46.

Nishioka, M., Y. Tanioka, E. Miyamoto, T. Enomoto, and F. Watanabe. 2007. TLS analysis of a corrinoid compound from dark muscle of the yellowfin tuna (*Thunnus* albacore). *Journal of Liquid Chromatography Related Technologies* 30:1–8.

Nunes, M. L., N. M. Bandarra, and I. Batista. 2011. Health benefits associated with seafood consumption. Chap. 29 in *Handbook of seafood quality, safety, and health applications*, ed. C. Alasalvar, F. Shahidi, K. Miyashita, and V. Wancsundare, 3169–3179. Ames, IA: Wiley-Blackwell.

Nunes, M. L., N. M. Bandarra, L. Oliveira, I. Batista, and M. A. Calhau. 2006. Composition and nutritional value of fishery products consumed in Portugal. In *Seafood research from fish to dish: Quality, safety, and processing of wild and farmed fish*, ed. J. B. Luten, C. Jacobsen, K. Bekaertk, A. Saebo, and J. Oehllenschlager, 477–487. Wageningen, Netherlands: Academic Publishers.

Ohyama, S., N. Ishibashi, M. Tamura, H. Nishizaki, and H. Okay. 1988. Synthesis of bitter peptides composed of aspartic and glutamic acids. *Agricultural and Biological Chemistry* 52:871–872.

Olubunmi, F., S. Suleman, I. Uche, and B. Olumide. 2010. Preliminary production of sauce from clupeids. *New York Science Journal* 3 (3): 45–49.

Olympia, M. S. 1992. Fermented fish products in the Philippines. In *Application of biotechnology to traditional fermented foods*. Washington, DC: National Academy Press.

Orejana, F. M., and J. Liston. 1982. Agents of proteolysis and its inhibition in *patis* (fish sauce) fermentation. *Journal of Food Science* 47:198–203.

Osako, K., M. A. Hossain, K. Kuwahara, A. Okamoto, A. Yamaguchi, and Y. Nozaki. 2005. Quality aspect of fish sauce prepared from underutilized fatty Japanese anchovy and rabbit fish. *Fisheries Science* 71:1347–1355.

Ozden, Z., and N. Erkan. 2011. A preliminary study of amino acid and mineral profiles of important and estimable 21 seafood species. *British Food Journal* 113 (4): 457–469.

Paleologos, E. K., I. N. Savvaidis, and M. G. Kontominas. 2004. Biogenic amines formation and its relation to microbiology and sensory attributes in ice-stored whole, gutted, and filleted Mediterranean sea bass (*Dicentrarchus labrax*). *Food Microbiology* 21:549–557.

Paludan-Muller, C. 2002. Microbiology of fermented fish products. Ph.D. thesis, Danish Institute of Fisheries Research, Lyngby and the Royal Veterinary and Agricultural University, Copenhagen.

Paludan-Muller, C., H. H. Huss, and L. Gram. 1999. Characterization of lactic acid bacteria isolated from a Thai low-salt fermented fish product and the role of garlic as substrate for fermentation. *Int. J. Food Microbiol.* 46:219–229.

Panda, S. M., R. C. Ramesh, A. F. El Sheikha, M. Didier, and W. Wanchai. 2011. Fermented fish and fish products: An overview. Vol. 2 in *Aquaculture microbiology and biotechnology*, ed. M. Didier and R. C. Ramesh, 132–172. Boca Raton, FL: Science Publishers, CRC Press/Taylor & Francis Group.

Park, J.-N., Y. Fukumoto, E. Fijita, et al. 2001. Chemical composition of fish sauces produced in Southeast and East Asian countries. *Journal of Food Composition Analysis* 14:113–125.

Park, J.-N., K. Ishida, T. Watanabe, et al. 2002a. Taste effects of oligopeptides in a Vietnamese fish sauce. *Fisheries Science* 68:921–928.

Park, J.-N., T. Watanabe, K.-I. Endoh, H. Watanabe, and H. Abe. 2002b. Taste active components in a Vietnamese fish sauce. *Fisheries Science* 68:913–920.

Park, P. J., J. Y. Je, and S. K. Kim. 2005. Amino acid changes in the Korean traditional fermentation process for mussel *Mytilus edulis*. *Journal of Food Biochemistry* 29:108–116.

Peralta, R. R., M. Shimoda, and Y. Osajima. 1996. Further identification of volatile compounds in fish sauce. *Journal of Agriculture and Food Chemistry* 44:3606–3610.

Peralta, E. M., H. Hatate, D. Kawabe, K. Rui, W. Shinji, Y. Tamami, and M. Hisashi. 2008. Improving antioxidant activity and nutritional components of Philippine salt-fermented shrimp paste through prolonged fermentation. *Food Chemistry* 111:72–77.

Peralta, E. M., H. Hatate, D. Watanabe, D. Kawabe, H. Murata, Y. Hama, and R. Tanaka. 2005. Antioxidative activity of Philippine salt-fermented shrimp paste and variation of its content during fermentation. *Journal of Oleo Science* 54:553–558.

Peredes-Lopez, O., and G. I. Harry. 1988. Food biotechnology review: Traditional solid state fermentations of plant raw materials: Application, nutritional significance, and future prospects. *Critical Review in Food Science and Nutrition* 27: 159–187.

Phithakpol, B., W. Varanyanoud, S. Reungrnaneepaitoon, and H. Wood. 1995. *The traditional fermented food of Thailand*, 3. Kuala Lumpur, Malaysia: Asian Food Handling Bureau.

Puwastien, P., M. Raroengwichit, P. Sungpuag, and K. Judprasong. 1999. *Thai food composition tables*, 1st ed. Bangkok, Thailand: Institute of Nutrition, Mahidol University (INMO), Thailand Asean Foods, Regional Database Centre of INFOODS.

Raksakulthai, N. 1987. Role of protein degradation in fermentation of fish sauce. Ph.D. thesis, Memorial University of Newfoundland, St. John's Newfoundland, Canada.

Raksakulthai, N., Y. Z. Lee, and N. F. Haard. 1986. Influence of mincing and fermentation aids on fish sauce prepared from male, inshore capelin, *Mallotus villosus*. *Canadian Institute of Food Science and Technology Journal* 19:28–33.

Ravipim, C. 1991. Acceleration of fish sauce fermentation using proteolytic enzymes. Master of Science thesis, Department of Food Sci. and Agricultural Chemistry, Macdonald Campus of McGill University, Montreal, Canada.

Rawles, D. D., G. I. Flick, and E. E. Martin. 1996. Biogenic amines in fish and shellfish. *Advances in Food and Nutrition Research* 39:329–364.

Rose, E. 1918. Recherches sur la fabrication et la composition climique du Nuoc-mam. *Bull. Econ. Indoch.* 129:155–190.

———. 1919. *Annales De Institute Pasteur.* 33:275.

Ruddle, K., and N. Ishige. 2010. On the Origins, diffusion, and cultural context of fermented fish products in Southeast Asia. In *Globalization, food, and social identities in the Asia Pacific region*, ed. James Farrer. Tokyo: Sophia University Institute of Comparative Culture.

Rustad, T. 2003. Utilization of marine by-products. *Electronic Journal of Environmental, Agricultural, and Food Chemistry* 2 (4): 458–463.

Saisithi, P. 1967. Studies on the origin and development of the typical flavor and aroma of Thai fish sauce. Ph.D. thesis, University of Washington, Seattle.

———. 1987. Traditional fermented fish products with special reference to Thai products. *Asian Food J.* 3:3–10.

————. 1994. Traditional fermented fish: Fish sauce production. Chap. 5 in *Fisheries pro-cessing biotechnological application*, ed. A. M. Martin, 111–131. London: Chapman and Hall.

Saisithi, P., B. O. Kasemasarn, J. Liston, and A. M. Dollar. 1966. Microbiology and chemistry of fermented fish. *Journal of Food Science* 31:105–110. http://icc.fla.sophia.ac.jp/global food papers/html/ruddle_ishige.htm.

Salampessy, J., K. Kailasapathy, and N. Thapa. 2010. Fermented fish products. In *Fermented foods and beverages of the world*, ed. J. P. Tamang and K. Kailasapathy, 289–307. Boca Raton, FL: CRC Press.

Sanceda, N., T. Kurata, and N. Arakawa. 1996. Accelerated fermentation process for the man-ufacture of fish sauce using histidine. *Journal of Food Science* 61 (1): 220–225.

Sanceda, N. G., T. Kurata, Y. Suzuki, and N. Arakawa. 1992. Oxygen effect on volatile acids during fermentation in manufacture of fish sauce. *J. Food Sci.* 57:1120–1122.

Sanni, A., J. Morlon Guyot, and J. P. Guyot. 2002. New efficient amylase producing strains of *Lactobacillus plantarum* and *L. fermentum* isolatd from different Nigerian traditional fermented foods. *Int. J. Food Microbiol.* 72:53–62.

Shahidi, F., and H. Miraliakbari. 2006. Marine oils: Compositional characteristics and health effects. In *Nutraceutical and specialty lipids and their co-products*, ed. F. Shahidi, 227–250. Boca Raton, FL: CRC Press, Taylor & Francis Group.

Shewan, J. 1951. Common salt: Its varieties and their suitability for fish processing. In *World fisheries yearbook*. London: British Continental Trade Press.

Shirai, T., Y. Hirakawa, Y. Koshikawa, H. Toraishi, M. Terayama, T. Suzuki, and T. Hirano. 1996. Taste components of Japanese spiny and shovel-nosed lobsters. *Fisheries Science* 62:283–287.

Sidwell, V. D. 1981. Chemical and nutritional composition of finfishes, whales, crustaceans, mollusks, and their products. NOAA Technical Memorandum, NMFS/Sec. II. U.S. Department of Commerce, Washington, DC.

Sikorski, Z. E. 1994. The contents of proteins and other nitrogenous compounds in marine animals. In *Seafood proteins*, ed. Z. E. Sikorski, B. S. Pan, and F. Shakidi, 6–12. New York: Chapman and Hall.

Silva, J. L., and R. S. Chamul. 2000. Composition of marine and freshwater finfish and shell-fish species and their products. In *Marine and freshwater products handbook*, ed. R. E. Martin et al., 31. Lancaster, PA: Technomic.

Stratton, J. E., R. W. Hutkins, and S. L. Taylor. 1991. Biogenic amines in foods, a review. *Journal of Food Protection*, 54:460–470.

Taira, W., Y. Funatsu, M. Satomi, T. Takano, and H. Abe. 2007. Changes in extractive compo-nents acid microbial proliferation during fermentation of fish sauce from underutilized fish species and quality of final products. *Fisheries Science* 73:913–923.

Takenaka, S., T. Enomoto, S. Tsuyama, and F. Watanabe. 2003. Analysis of corrinoid com-pounds in fish sauces. *Journal of Liquid Chromatography and Related Technologies* 26:2703–2707.

Tao, Z., M. Sato, T. Yamaguchi, and T. Nakano. 2009. Formation and diffusion mechanism of histamine in the muscle of tuna fish. *Food Control* 20 (10): 923–926.

Thapa, N., J. Pal, and P. Tamang. 2004. Microbial diversity in *ngari*, *hentak*, and *tungkap*, fermented fish products of North-East India. *World Journal of Microbiology and Biotechnology* 20:599–607.

Tsai, Y.-H., C.-Y. Lin, L.-T. Chien, et al. 2006. Histamine content of fermented fish products in Taiwan and isolation of histamine forming bacteria. *Food Chemistry* 98 (1): 64–70.

Tungkawachara, S., J. W. Park, and Y. J. Choi. 2003. Biochemical properties and consumer acceptance of Pacific whiting fish sauce. *Journal of Food Science* 68 (3): 855–860.

Visessanguan, W., S. Benjakul, T. Smitinont, C. Kittikun, P. Thepkasikul, and A. Panya. 2006. Changes in microbiological, biochemical and physico-chemical properties of *Nham* inoculated with different inoculums levels of *Lactobacillus curvatus*. *LWT - Food Science and Technology* 39:814–826.

Watanabe, F. 2007. Vitamin B_{12} sources and bioavailability. *Experimental Biology and Medicine* 232:1266–1274.

Watanabe, F., H. Katsura, S. Takenaka, et al. 2001. Characterization of vitamin B_{12} compounds from edible shellfish, clam, oyster, and mussel. *International Journal of Food Science and Nutrition* 52:263–268.

Xu, W., G. Yu, C. Xue, Y. Xue, and Y. Ren. 2008. Biochemical changes associated with fast fermentation of squid processing by-products for low salt fish sauce. *Food Chemistry* 107:1597–1606.

Xu, Y., W. Xu, F. Yang, J. M. Kim, and X. Nie. 2010. Effect of fermentation temperature on the microbial and physicochemical properties of silver carp sausages inoculated with *Pediococcus pentosaceus*. *Food Chemistry* 118:512–518.

Yamamoto, Y., A. Fukisawa, A. Hara, and W. C. Dunlap. 2001. An unusual vitamin E constituent (α-tocomonoenol) provides enhanced antioxidant protection in marine organisms adapted to cold-water environments. *Proceedings of the National Academy of Sciences* 98 (23): 13144–13148.

Yen, G.-C., and C.-L. Hsieh. 1991. Simultaneous analysis of biogenic amines in canned fish by HPLC. *Journal of Food Science* 56 (1): 158–160.

13 Fermented Meat Products

Kazimierz Lachowicz, Joanna Żochowska-Kujawska, and Malgorzata Sobczak

CONTENTS

13.1 INTRODUCTION

Meat and meat products are an important part of the human diet. The origin of the manufacture of dry-cured meat sausages probably goes back to Babylonian times, when methods such as drying, salting, or fermentation of meat were already in use (Fanco et al. 2002). These fermented meat products are found in most parts of the world, but Europe remains the major producer and consumer of these products.

The dry-curing process used in meat fermentation utilizes a mixture of curing ingredients, mainly salt, nitrate, and/or nitrite and sugars. Generally, the dry cure is applied without any added water. Consequently, the curing agents are solubilized in the original moisture present in the meat, and they penetrate by diffusion (Flores 1997). The dry-curing of minced meats is used for the manufacture of fermented sausages. Thus, dry-fermented sausage (DFS) is defined as comminuted meat and fat—mixed with salt, nitrate and/or nitrite, sugar, and spices (black pepper at least)—that is stuffed into casings and subjected to a fermentation, and then to a drying process and, in some cases, smoked. The final product preserves well and has an increased shelf life as a consequence of the inhibition of pathogenic and spoilage bacteria.

During the ripening of fermented sausages, the proteins and lipids experience great changes. Proteolysis influences both texture and flavor development due to the formation of several low-molecular-mass compounds, including peptides, amino acids, aldehydes, organic acids, and amines, all of which are important flavor compounds or precursors of flavor compounds (Demeyer et al. 1995; Fadda, Vignolo, and Oliver 2001). Lipolysis plays an essential role in the development of dry-sausage flavor. Lipids are hydrolyzed by enzymes, generating free fatty acids, which are substrates for the oxidative changes that are responsible for flavor compounds (Samelis, Aggelis, and Metaxopoulos 1993; Stahnke 1995; Verplaetse 1994).

DFSs are meat products with a high fat content, which is visible in the sliced product, and when made with a normal recipe, they have fat contents around 30% directly after manufacture, and up to 40%–50% after 4 weeks of drying (Wirth 1988). Therefore they have acquired a negative image associated mainly with high fat content and the presence of sodium chloride, which have been associated with several diseases such as hypertension and obesity. It is, therefore, of interest to produce healthier meat products by modifying their composition, either by modifying the ingredients themselves or by reformulating the products, for example, by reducing their fat or sodium content, or by incorporating functional ingredients (Jiménez-Colmenero, Reig, and Toldrá 2006).

13.2 TYPES OF FERMENTED MEAT PRODUCTS AND THEIR MANUFACTURE

13.2.1 Types of Fermented Sausages

Fermented sausages are a very inhomogeneous product group, often with great variations in ingredients levels (for example the NaCl concentration) and from the various drying conditions used during ripening (Roca and Incze 1990). In Europe, natural fermented sausages have a long tradition originating from Mediterranean countries during Roman times (Lücke 1985). In the Mediterranean region, the term *cured* is normally understood to describe product that has undergone a long ripening (aging) process, where complicated biochemical, proteolytic, and lipolytic modifications take place, and which in turn are responsible for the distinctive flavors. In the north of Europe, the term *cured* has a more general meaning, and it is mainly applied to meat products treated with nitrites (Flores 1997).

The technology of dry-cured sausages allows many variations as long as the basic concepts (reduction of pH and water activity) are kept in mind. Consequently, these products vary greatly across all of the producer countries. Many different types of naturally fermented sausages are known only at the local or regional level. DFSs are classified according to the following criteria: composition, caliber, degree of mincing of the ingredients, spices and condiments added, smoked or not, duration of the ripening period, etc. Although their manufacture always involves a combination of fermentation and dehydration processes, there are clear regional differences.

Typical Mediterranean sausages include the Italian salami, Spanish chorizo, and French *saucisson* set as well as the northern European-, German-, or Hungarian-style salami. Traditional Mediterranean products have a slow curing process where nitrite is not usually used and smoking is not applied, which allows the development of molds and yeasts on the surface. Most sausages are made with spices and are dried in air. In the United States, semicured sausages are common. These are rapidly fermented at high temperatures followed by a short drying period. In northern and central Europe, which have a wetter climate, nitrite is used, and fermentation is generally combined with smoking to prevent the development of yeasts and molds, and the drying period is shorter (Cassens 1994). So these cured and smoked treatments can be considered mainly responsible for the differences in the sensory characteristics of the products. Nitrite is usually preferred in northern Europe, but nitrate, which has been reported to give an improved flavor generation compared to nitrite cured sausages (Wirth 1991), is commonly used in southern Europe (Flores 1997). Traditionally, in Europe, dry sausages are manufactured with little or no added sugar; however since the 1950s, fermented sausages contain at least 0.3%–3% added carbohydrates. These compounds are converted primarily to lactic acid by fermentation by (mainly added) lactic acid bacteria, which reduce the pH of the sausage within a few days.

Modern types of fermented dry sausage can be divided into two groups, depending on the manufacturing method: (a) slow fermented sausages that are ripened over 4 weeks (leading to a firm texture with a mildly acidic, salty taste) or (b) semidry sausages (fast or medium-fast fermented), very popular in northern Europe and the United States, that are, depending on the product diameter, ripened between 7 and 28 days and are less intensively dried, delivering a strongly acidic, salty, mild taste and a softer texture (Houben and van't Hooft 2005). Fermented sausages can also be divided into sliceable raw sausages (salami, summer sausage, pepperoni) and spreadable raw sausages (*teewurst, mettwurst*) or, depending on the amount of moisture that they contain, they can be grouped as moist (10% weight loss), semidry (20% weight loss), and dry (30% weight loss).

Fermented sausages are usually prepared from seasoned raw meat stuffed in casings and allowed to ferment and mature (Moretti et al. 2004). It is impossible to cover all of the numerous types and potential variants of fermented sausages, but it is possible to indicate some of the major raw materials for their production. The meat of different species of adult, well-fed animals is preferred in raw sausage production: pork, beef, lamb, goat, venison, poultry, or combinations thereof. In Germany, fermented sausages are often made from equal amounts of pork and beef; in Poland, pork is most popular. Hungarian and Italian sausages also contain mostly pork. Salted and fermented meat products have been made all over the world, and different geographical locations have often dictated what kind of animals will thrive in a particular climate zone. For example, goats generally prosper well in mountainous locations; lamas live in the Andes of South America, and their meat is popular in Bolivia and Chile. Ostrich is consumed in South Africa, and moose, reindeer, or caribou meats will end up as raw materials for making dry sausages in Alaska. Norwegian fermented sausages such as Faremorr, Sognemorr gilde, Stabbur, and Tiriltunga contain not only pork but beef, horse, and lamb meats (Marianski and Mariański 2009).

Traditional fermented products from the Near East, named *sujuk* or *sucuk*, are produced from beef, water buffalo, mutton, and sheep tail fat; salt; some spices such as black pepper, red pepper, cumin, and garlic; and nitrite, nitrate, or nitrite/nitrate (Bozkurt and Erkmen 2002). Whereas MUM, a Thai traditional fermented sausage, is made from three kinds of beef: Thai native beef (TN beef), Brahman fed pineapple beef (BP beef), and Kamphaengsean beef (KU beef) (Sumon and Sumon 2009). Camel meat, with its high moisture and high protein content, provides a product with good nutritional value (El-Faer et al. 1991; Elgasim and Alkanhal 1992; Dawood 1995; Kalalou, Faid, and Ahami 2004) that is comparable in taste and texture to beef meat and is also favorable for fermented sausage making.

Also in Europe, in recent years consumer interest in game meat as an alternative for pork and beef is now increasing because of their high nutritional value, and cured, fermented, and dried products from different game species have recently appeared on the market (Paleari et al. 2000). For example, production of cured sausages in Spain amounted to 191,844 tons in 2003, and game preparations and canned meat have increased notably in recent years (Soriano et al. 2006). Despite the increased popularity of game meat, there is a lack of research comparing the nutritional and sensory qualities of meat and meat products, especially fermented sausages, from different game species.

In recent years, several studies have been carried out on the physicochemical characteristics of deer or wild boar meat (Stevenson, Seman, and Littlejohn 1992; Zomborszky et al. 1996) and their histochemistry, textural, and sensory properties (Żochowska-Kujawska, Sobczak, and Lachowicz 2009; Żochowska-Kujawska et al. 2010). However, only a very few works have been found on the proteolysis, physicochemical characteristics, and free fatty acid composition of dry sausages made with meat from deer or wild boar (Vioque et al. 2003) or from other exotic animals like kudu, zebra, or camel. According to van Schalkwyk et al. (2011), meat from wild exotic animals, similar to pork or beef, could be an appropriate raw material for fermented sausage production. They showed no significant differences in the chemical composition of the game salami produced from different species, although differences were observed among salami from springbok, gemsbok, kudu, and zebra when evaluated by trained panelists and consumers. Panelists liked the salami produced from the gemsbok, kudu, and zebra, but disliked salami produced from springbok because of its gamy flavor, sour aroma, sour flavor, dullness, and less compact appearance as well as its fatty mouth feel. These results indicated that the type and quality of the raw materials are important when producing salami from different game-meat species. This is confirmed by Żochowska-Kujawska et al. (unpublished), who showed that *kindziuk* produced from pork with 80% deer or red deer meat addition as a lean meat scored the highest for tenderness and juiciness as well as taste and was superior compared to *kindziuk* with the lower and higher game-meat addition. A higher amount of deer meat in the final product was connected with a strong gamy flavor and a dark brown color, and it consequently scored low on a scale of sensory analysis.

Soriano et al. (2006) compared 10 types of sausages made with deer or wild boar meat purchased from various supermarkets and meat factories in Ciudad Real and Toledo (Spain). Their results showed that chorizos and *saucissons* made with deer meat and commercialized in the Spanish market had a higher fat content than those prepared with wild boar meat. An additional two factors that may be differentiating were the protein nitrogen and phosphorus contents, which were higher in dry sausages made with wild boar. They also pointed that diet has a great influence on game meat fatty acid composition. Game animals are mainly herbivorous, eating a great variety of indigenous plants, grains, and fruits. Consequently, it is difficult to establish the influence of diet on the lipid composition of their meat. According to Vioque et al. (2003), who studied free fatty acids in commercial *saucissons* made with deer meat, the amounts of oleic, palmitic, linoleic, and stearic acids are similar to those found in chorizos made with lean pork and beef and fat from pork (Soriano et al. 2006). Paleari et al. (2003) obtained a total saturated fatty acid (SFA) content of 35.5%–44.0%, monounsaturated (MUFA) content of 30.3%–45.7%, and polyunsaturated fatty acid (PUFA) content of 16.2%–19.6% in fat extracted from cured products of deer and wild boar from farms. On the other hand, Soriano et al. (2006) showed that the amounts of saturated acids were lower and the monounsaturated acid content was higher, and this may be attributed to the fact that the dry sausages are made with meat from wild animals bred in freedom. The free access to pasture for grazing, and thus the greater contribution of fresh grass and herbs to the diet of these species, could be responsible for such positive values (Volpelli et al. 2003). However, the results of those studies contribute to the typification of the delicatessens made with game meat, and are interesting for companies producing these types of products.

13.2.2 Dry-Fermented Sausage Manufacture

The manufacture of dry-fermented sausages, regardless of whether the dry sausage is produced from game or pork/beef meat, is usually considered to entail three main steps: formulation, fermentation, and ripening/drying (Ordóñez et al. 1999). During the manufacture of dry-fermented sausages, the high fat content (30%–40%) provides optimal sensory properties—hardness, juiciness, and flavor—as well as useful technological functions (Wirth 1988).

In the formulation stage, the meat and pork fat are minced to a specific size. The use of chilled meat with a low pH is most suitable. For beef, the lowest pH of about 5.4–5.5 is obtained within a couple of days. Pork usually acidifies a little faster, with final pH of 5.7–5.8. It is recommended to use meat that has been slightly frozen (-3°C to -5°C) for one or two days prior to processing. The same holds true for fatty tissue. The use of beef is somewhat preferred for semidry sausages, while pork seems to be more suitable for dry-sausage manufacture. Game meat is suitable for salami production where a rich, dark red color is preferred (van Schalkwyk et al. 2011). The color of game meat is darker than traditional beef and lamb due to its higher myoglobin content (Hoffman, Kritzinger, and Ferreira 2005). The low pH of fermented sausages initially destabilizes the myoglobin and increases the rate of oxidation to metmyoglobin. The heme group is dissociated, and color is primarily attributed to nitrosomyoglobin formation due to the reduction of nitrates (van Schalkwyk et al. 2011).

Several spices, mainly added as flavorings and coloring agents, are used for the production of fermented sausages in various concentrations, depending on the type of sausage. Typical ingredients of a sausage mixture are ground pepper, paprika, garlic, mace, and pimento (Verluyten, Leroy, and De Vuyst 2004). In the case of game salami, ground green cardamom, juniper berries, or mustard are also used to enhance the taste and aroma of wild animal meat. Some spices, such as pepper (present in all types of sausages at the 0.2% to 0.3% level) or garlic, added to meat have been found to accelerate lactic acid production by the lactic acid bacterial starter culture (Kumar and Berwal 1998; Nes and Skjelkvåle 1982). On the other hand, some spices (garlic, nutmeg, mace, paprika, rosemary, and sage) contain powerful antioxidants that can extend the shelf life of dry-fermented sausages (Aguirrezábal et al. 2000; Madsen and Bertelsen 1995; Nassu et al. 2003). Other spices, such as garlic, which contains an active substance identified as allicin, or rosemary display antimicrobial activities through their essential oils (Zaika 1988; Ankri and Mirelman 1999). In addition, according to Nassu et al. (2003), rosemary extract at 0.05% concentration showed an effective protection against oxidation in fermented goat-meat sausages.

During the mincing process, the muscle fiber is broken, and the myofibrillar proteins are exposed to the action of the salt. The salt facilitates electrostatic processes involved in the formation of the film of protein that surrounds the fatty particles and favors protein-fat and protein-protein interactions (Ordóñez et al. 1999). Gradually, the conditions are established for the formation of a three-dimensional lattice structure necessary to generate the texture of these products. Several fermented sausages are manufactured using traditional technologies without the addition of starter cultures, and the traditional process favors the growth of autochthonous microflora, also known as "house-flora." Nevertheless, it is not possible to ensure that the number and the strains of microorganisms present in the minced meat material will always be the same and behave in the same way. The use of starter cultures guarantees products with reproducible hygienic and organoleptic properties in a shorter ripening time (Martin et al. 2007).

The mixes are stuffed into natural or artificial casings and, after stuffing, the sausages undergo a drying–ripening process in ripening cabinets under conditions of controlled temperature, relative humidity, and air flow. During 1 to 2 days of fermentation in a controlled temperature of about 18°C–26°C and relative humidity of 90%, several critical microbiological changes take place. However, there is a tendency for lower incubation temperature in Europe (20°C–24°C) and a higher incubation temperature in the United States (about 37°C) that may go even higher with the aim of inactivation of *Listeria* species (Hammes, Rölz, and Banteon 1985). According to Casaburi et al. (2008), the fermentation is a crucial phase of the curing process of sausages, since at this stage the major physical, biochemical, and microbiological transformations take place (Lizaso, Chasco, and Beriain 1999; Villani et al. 2007). These changes can be summarized as follows: decrease in pH; changes in the initial microflora; reduction of nitrates to nitrites, and the latter to nitric oxide; formation of nitrosomyoglobin; solubilization and gelification of myofibrillar and sarcoplasmic proteins; proteolytic, lipolytic, and oxidative phenomena; and dehydration (Casaburi et al. 2007).

Smoking may or may not be utilized in the production of fermented sausages. If it has been used, there is no standardized procedure for smoking different raw sausages. Sausages are smoked from several hours to several days or weeks according to the diameter of the product: small diameter sausages, from 10 to 40 hours; medium diameter sausages, 15–45 hours; and large diameter sausages, several days and, exceptionally, 1–3 weeks. Semidry sausages are usually smoked at a higher temperature (22°C–32°C or above), and a heavier smoke is usually applied. The smoking temperatures required for different dry sausages vary from 12°C–22°C but never exceed 30°C–31°C. A temperature of 14°C is considered as optimum in smoking dry sausages. Van Schalkwyk et al. (2011) suggested for game salami cold smoke with oak wood chips for 36 h at 22°C–24°C at relative humidity of >92%. Some producers in Europe and the United States do not smoke their dry sausages at all but keep them for at least 10 days to several months in a room in which the desired air conditions are obtained by various combinations of temperature, relative humidity, and air circulation. Some sausages are lightly smoked before drying.

Drying or aging is a key operation, especially in dry-sausage production. The drying rate for dry sausages should be as low as possible. Sausages are dried at low temperatures (6°C–12°C) to avoid a rapid and intense fermentation, achieving final pH values of over 5.3 (Aymerich et al. 2003; Y. Sanz et al. 1998). The most critical point in drying is to avoid the pronounced surface coagulation of proteins and the formation of sausage surface skin. If the sausages lose moisture too rapidly during the initial stages of the drying period, the surface becomes hardened, and a crust or ring develops immediately adjacent to the casing. This hardened ring inhibits further transmission of moisture, and the sausage has an excessively moist center. Only a sufficiently wet and soft casing, a high relative humidity at the outset of the drying operation, and the use of a lower relative humidity in the advanced stages of the process will permit moisture to migrate from the interior of the sausage into the outer layer. Thus, the sausage should dry from the inside outwards. If the outer layers of the sausage become hard, the diffusion of water is inhibited, and the sausage tends to spoil. In conclusion, if the drying rate is adequately slow, the sausage casing will enable gradual drying.

Apart from microbiological modifications, chemical and physicochemical changes occur during the drying of fermented sausages, especially dehydration, fermentation of carbohydrates and acidification, development of color, lipolysis and fat autoxidation, and proteolysis (Ordonez et al. 1999). These changes are responsible for the organoleptic characteristics of the final products.

13.3 MICROORGANISMS USED IN FERMENTATION

Fermentation of sausages is a well-known microbial process, and pioneering contributions to the study of the ecology during ripening are available from the 1960s (Reuter 1972). These studies stated that lactic acid bacteria (LAB, *Lactobacillus* spp.) and coagulase-negative cocci (CNC, *Staphylococcus* and *Kocuria* spp.) are the two main groups of bacteria that are considered technologically important in the fermentation and ripening of sausages (Corbiere Morot-Bizot, Leroy, and Talon 2006). In addition, depending on the product, other groups may play a role, such as molds, enterococci, and yeasts (Casaburi et al. 2007).

Many traditional slightly fermented sausages are produced in small-scale processing units by spontaneous meat fermentation, which cannot always guarantee the product to be safe and stable (Aymerich et al. 2003). In spontaneous meat fermentation, the LAB species such as the most commonly identified in traditional fermented sausages—*Lactobacillus sakei, Lactobacillus curvatus,* and *Lactobacillus plantarum* (Urso, Comi, and Cocolin 2006; Hammes and Hertel 1996)—and other required microorganisms originate from the meat itself or from the environment and constitute a part of the so-called house flora (Santos et al. 1998). These microorganisms are responsible for lactic acid production resulting from carbohydrate utilization as well as a low pH value (5.9–4.6). Finally, the inhibition of pathogenic and spoilage bacteria is a consequence of the accumulation of lactic acid as well as acetic acid, formic acid, ethanol, ammonium, fatty acids, hydrogen peroxide, acetaldehyde, antibiotics, and bacteriocins.

Today, the modern meat industry has to ensure high quality, reduce variability, and enhance organoleptic characteristics in sausage production, which is not feasible using spontaneous fermentation methods. In Europe, starter cultures are made up of a balance between the two main groups of bacteria that are responsible for meat fermentation: LAB and CNC (Hugas and Monfort 1997). Inoculation of the sausage batter with a starter culture composed of selected LAB, i.e., homofermentative lactobacilli and/or pediococci, and gram-positive, catalase-positive cocci (GCC), i.e., nonpathogenic, coagulase-negative staphylococci and/or cocuriae, improves the quality and safety of the final product and standardizes the production process (Hugas and Monfort 1997; Lücke 2000).

The LAB species are especially well adapted to the ecological conditions of meat fermentation, inhibiting spoilage and pathogen development mainly as a result of competitive growth and acidification of the product (Hugas and Monfort 1997). They produce sufficient amounts of lactic acid to reduce the pH of the meat to values of 4.8 to 5.0 (Benito et al. 2007). Acidification promotes the formation of color and the cohesion of sausages (Lücke 2000). Lactic acid production is responsible for the tangy flavor of sausages (Demeyer 1982).

Lactic acid bacteria with important industrial functionality are being developed to produce antimicrobial substances, sugar polymers, sweeteners, aromatic compounds, and vitamins, or those having probiotic properties (Leroy and de Vuyst 2004). However, other works showed that the use of starter cultures reduces the production of some amines (Maijala et al. 1995), a finding that is connected with the fact that the effectiveness of the starter culture depends on the hygienic quality of the raw material used (Bover-Cid, Izquierdo-Pulido, and Vidal-Carou 2000). Additionally, the formation of small amounts of acetic acid, ethanol, acetoin, carbon dioxide, and pyruvic acid, which are produced during fermentation, depend on the starter applied, the carbohydrate used, and the sources of meat proteins and additives (Demeyer 1982; Thornill and Cogan 1984).

It is well known that LAB, apart from organic acids, have a potential to produce bacteriocins and thus can suppress the growth of undesirable bacteria (Hammes and Hertel 1996). The microflora that develops is closely related to the ripening technique utilized. Sausages with a short ripening time have more lactobacilli right from the early stages of fermentation. In contrast, sausages with longer maturation times contain higher numbers of Micrococcaceae (Demeyer, Verplaetse, and Gistelinck 1986).

Cocci are also very important agents of meat fermentation. The addition of CNC to dry-sausage manufacturing improves their sensory properties (Nychas and Arkoudelos 1990). According to Berdague et al. (1993) and Montel et al. (1996), they moderate the level and nature of volatiles originating from lipid oxidation. *Staphylococcus xylosus* and *Staphylococcus carnosus* are commonly used as starter cultures in fermented meat products. Montel et al. (1996) proved that staphylococci contribute to the generation of typical fermented sausage flavor. They also showed that *S. xylosus* and *S. carnosus* were able to enhance the dry-salami odor and to increase the concentration of methyl ketones, ethyl esters, and leucine degradation products. Those species can also degrade the branched-chain amino acids (BCAA) leucine, isoleucine, and valine into methyl-branched aldehydes, alcohols, and acids (Larrouture et al. 2000; Vergnais et al. 1998) as well as free fatty acids (Stahnke et al. 2002; Olesen, Meyer, Stahnke 2004), and as a consequence modulate the aroma of fermented sausages. Additionally, *S. carnosus* has been shown to decrease the maturation time of Italian dried sausages by more than 2 weeks (Stahnke et al. 2002).

In general, the strains *S. carnosus* as well as *Staphylococcus simulans* were able to hydrolyze sarcoplasmic but not myofibrillar proteins (Casaburi et al. 2005). However, in other staphylococci they detected no protease activity but low aminopeptidase and high esterase activity (Casaburi et al. 2006). *Staphylococcus* species have several other technological advantages such as nitrite and nitrate reductase activity, oxygen consumption, and catalase activity that improve color stability and decrease rancidity development in the product (Geisen, Lücke, and Kröckel 1992; Lücke and Hechelmann 1987). *Staphylococcus* and *Kocuria*, due to their proteolytic and lipolytic activities, participate in the development of flavors of dry-fermented sausages (Demeyer, Verplaetse, and Gistelinck 1986; Cai et al. 1999). They influence the composition of nonvolatile and volatile compounds mainly by degrading free amino acids and inhibiting the oxidation of unsaturated free fatty acids (Hammes and Hertel 1998; Søndergaard and Stahnke 2002). According to López-Díaz et al. (2001), molds play an important role in the manufacture and ripening of fermented meat products in many southern European countries such as Italy, Spain, France, Hungary, and southern Germany. Those fermented sausages are characterized by the presence of a white, sometimes green mold covering that appears naturally on the surface.

The use of molds on sausage surfaces can lead to both desirable and undesirable effects. According to Ludemann et al. (2004), many of these molds may lead to serious problems for both the consumer and the producer. Some of these molds are capable of producing mycotoxins (López-Díaz et al. 2001), which, besides their acute toxic effects, often also have long-term toxigenic, carcinogenic, hemorrhagic, or liver-degenerative effects (Samson et al. 1995), and they also present quality problems associated with color and mycelium appearance because they produce green, brown, or black spots that are not acceptable to most consumers. One other hand, the positive effects are the typical appearance, flavor, and taste mediated by lactate oxidation, proteolysis, degradation of amino acids, lipolysis, and β-oxidation (Grazia et al. 1986). García et al. (2001) observed that inoculating experimental sausages with molds selected for their enzymatic activity positively influenced the development of flavor and aroma. Molds have also been studied for their capacity to enhance and stabilize the cured color through the reduction of nitrites, catalase activity, oxygen consumption, and protection against light (Bruna et al. 2001; Lücke and Hechelmann 1987). Additionally, molds play an important role in protection against spontaneous colonization with unwanted molds, yeasts, and bacteria; delay of rancidity; and reducing the risk of development of a dry edge and water loss due to slower water evaporation (Lücke and Hechelmann 1987).

The first toxicologically and technologically suitable mold starter culture for meat products was a *Penicillium nalgiovense* selected by Mintzlaff and Leistner (1972) and commercially available as *edelschimmel Kulmbach*. Today, all major starter culture suppliers offer molds as a part of their product range (Sunesen and Stahnke 2003). Additionally, Glenn, Geisen, and Leistner (1989) showed that some strains of *P. nalgiovense* are able to inhibit *Listeria monocytogenes*.

According to Selgas, Ros, and García (2003), the composition of the microflora changes during the processing of dry-fermented sausages: Yeasts are the predominant microorganisms during early ripening, and after 2 weeks, the number of yeasts and molds balance. At the end of the drying process, molds may constitute more than 95% of the mycoflora (S. Andersen 1995; Encinas et al. 2000). Yeast counts in similar sausages range from 10^2–10^6 cfu/g (Lizaso, Chasco, and Beriain 1999; Osei Abunyewa et al. 2000; Selgas, Ros, and García 2003). The earliest studies on salami (Capriotti 1954) showed that *Debaryomyces hansenii* was the most commonly isolated yeast. Most recently, several researches confirmed these results, but other yeast genera were found, such as *Candida, Kluyveromyces, Pichia, Saccharomyces, Yarrowia, Rhodotorula, Debaryomyces, Cryptococcus*, and *Trichosporon* (Grazia et al. 1989). Several authors suggest that the presence of *Debaryomyces hansenii* in dry-fermented sausages (commonly used as a starter culture) enhances their sensory characteristics and may contribute to their flavor through their lipolytic and proteolytic activity (Encinas et al. 2000). Several endo- and exopeptidases have been purified from *Debaryomyces hansenii* contributing to enhance the content of free amino acids and peptides in meat products (Bolumar et al. 2003, 2008). In addition, a glutaminase was purified from *D. hansenii* that was able to neutralize the acid pH of fermented sausages and generate L-glutamate that can act as a flavor enhancer (Durá, Flores, and Toldrá 2002). However, the enzymatic activity of *D. hansenii* seems to be inhibited in dry-fermented sausages by low pH and low temperatures (Sørensen 1997).

As shown by Gardini et al. (2001b), lipolytic activity in dry-fermented sausages may also be strongly modulated by the presence of a specific yeast species, such as *Yarrowia lipolytica*. On the other hand, Lizaso, Chasco, and Beriain (1999) reported that the enzymatic action of yeasts was of secondary importance in the manufacture of Spanish *salchichon* due to its low counts compared to other microbial groups.

Moreover, the use of *D. hansenii* in combination with *P. chrysogenum* in dry-cured loins produced a high proteolytic activity of *P. chrysogenum*, while *D. hansenii* did not show proteolytic activity (Martın et al. 2001).

It is well known that dairy products such as fermented milk and yogurt are often used as carriers for probiotic cultures. Recently, attention has been directed to the use of fermented sausages as food carrier (L. Andersen 1998; Arihara et al. 1998; Erkkilä et al. 2000, 2001b; Petäjä et al. 2003; Pennacchia et al. 2004) because these products are not heated and harbor high numbers of lactic acid bacteria. According to Klingberg et al. (2005), several additional properties of the culture are demanded in order to use probiotics as starter cultures for fermented sausages, including desirable technological, sensory, and safety properties. The probiotic culture should be well adapted to the conditions of fermented sausage to become dominant in the final product, since fermented meat products naturally contain a high background of microbiota. In addition, the culture should not develop off-flavors in the product. Thus, it should be emphasized that the commercial starter cultures in Europe are mainly produced in northern European countries and are not likely to compete well with the "house flora" colonizing southern European meat plants, so their use often results in losses of desirable sensory characteristics (Leroy, Verluyten, and de Vuyst 2006; Samelis et al. 1998; Casaburi et al. 2007). For this reason, naturally fermented sausages often had higher quality compared to those inoculated with industrial starters (Moretti et al. 2004). Therefore, the use of well-selected starter cultures with lipolytic and/or proteolytic activity, able to generate high amounts of aroma components, could help in improving sensory quality (Rebecchi et al. 1998; Casaburi et al. 2007). Environmental factors possibly affecting strain selection are ripening conditions such as temperature, RH, pH, and NaCl as well as the raw meat and other ingredients.

13.4 CHEMICAL CHANGES DURING PROCESSING THAT HAVE AN EFFECT ON FERMENTATION

The quality of dry-fermented sausages are the result of biochemical, microbiological, physical, and sensory changes occurring in a meat mixture during ripening under defined conditions of temperature and relative humidity (RH). Thus, the typical characteristics of those kinds of sausages are influenced by extrinsic (temperature, relative humidity, and air circulation) and intrinsic (sugar, sausage diameter, and fat content) factors (Pezacki 1979; Dinçer 1982; Incze 1992; Lücke 1985).

The temperature directly affects pH, water activity (a_w), microbial growth, and texture of fermented sausages. There is a strong correlation between temperature and ripening as well as drying (Baumgartner, Klettner, and Rodel 1980; Landvogt and Fischer 1990). This is confirmed by Gökalp (1986), who indicated that pH values decreased very slowly at low ripening temperatures and that a longer ripening period increased pH decline in fermented sausages (Gökalp 1986; Johansson et al. 1994; Papadima and Bloukas 1999). According to Incze (1991), there is a simple relationship among temperature, a_w, and growth: The higher the temperature and a_w, the higher is the growth rate, and vice versa, yet it is hard to find an optimum solution.

Baumgartner, Klettner, and Rodel (1980) also found a strong correlation between temperature and ripening-drying: When temperature was increased by 5°C, the process became faster by a factor of 2 (Incze 1991).

According to Papadima and Bloukas (1999), storage conditions significantly affected the microflora, pH, weight loss, water activity, and yellowness of sausages. Sausages stored in the ripening room had a faster increase in lactic acid bacteria, a lower pH, and less water activity, indicating better keeping quality compared to sausages in cold storage. However, they also showed higher weight loss. They concluded that keeping traditional sausages in a ripening room improves the keeping quality and sensory attributes of the products. However, storage of traditional sausages (with less than 30% fat) in the ripening room for longer times results in significantly increased weight loss and no positive effect on quality characteristics. This is also confirmed by Leroy and de Vuyst (1999), who found that the production of biomass and sakacin K is obviously related to temperature. Maximal values are achieved at temperatures between 20°C and 25°C, which is similar to fermentation temperatures of European sausages. It is worth mentioning that most producers favor handling raw sausages, especially in warm weather, at temperatures around 18°C–22°C. This is done to avoid condensation from gathering on smoked sausages when they are transferred from a cold to a warm place, such as the ordinary meat shop. Such condensation favors rapidly forming molds.

Apart from temperatures, air humidity inside the drying chambers also needs to be adjusted carefully to support the ripening process. When temperatures, initially kept at +22°C, are slowly reduced to +19°C, the relative humidity decreases gradually from typical values of 92%–94% on the first day to 82%–84% before the sausages are transferred to the ripening room, where the relative humidity is maintained at 75%–78% during ripening. Initially, air flow at about 0.8 m/s permits fast moisture removal, but the high humidity level moisturizes the surface of the casing, preventing it from hardening. After about 48 hours the fermentation stage ends, but the drying continues to remove more moisture from the sausage. Maintaining the previous fast air flow may harden the surface of the casing, so the air speed is decreased to about 0.5 m/s. At this time, the medium-fermented sausage will be finished. Slow-fermented sausages require additional drying time, and both the humidity and air flow are decreased again (to about 0.3 m/s). At those conditions and at temperature of about 15°C, the sausage will remain in a drying chamber for an additional 4–8 weeks, depending on the diameter of the casing.

These parameters are applied to ensure controlled bacterial fermentation and dehydration. Of course the duration of the drying process mainly depends on the diameter of the sausages as well as the type of sugars and starter cultures used. Too much humidity in the drying room favors the development of mold and sliming of products. Some producers do not object to a light white surface mold at the very beginning of the drying process. When fully developed, the mold is brushed off or, in some cases, washed off. It is believed that mold contributes to the specific flavor of some products. If humidity is reduced too fast, especially in the early stages of the process, a hard and dry crust is formed at the outer layer of the sausage. Sausages produced in tropical climates should be thoroughly dried and smoked more heavily than products in the moderate climatic regions.

The water activity can be lowered more slowly in a sausage that contains more fat than a leaner sausage. Fat contains only about 10% water, and a fattier sausage having proportionally less meat also contains less water and will dry out faster (Soyer, Erta, and Üzümcüoğlu 2005). Papadima and Bloukas (1999) reported that the decrease in water activity of the sausages with 30% fat was very small compared to the sausages with 20% and 10% fat during the ripening period. The initial pH of the sausage may influence final pH and the time until pH 5.3 is attained. This means that the high pH of meat is connected with a larger buffering capacity, and consequently an extended period of time is needed until pH can drop below the critical level, and sometimes the final pH just will not go down sufficiently (Incze 1991). According to Flores et al. (1997), a rapid decrease in pH during the initial steps of sausage manufacture, as occurs when starters are used, positively affects color development, texture, and the homogeneity of drying. Nevertheless, taste may be affected negatively. An equilibrium point must be found between acid production and taste and, as shown by Demeyer et al. (1995), ammonia production must be intensified to neutralize final acidity, thus enhancing sausage taste.

13.5 CHEMICAL CHANGES DURING FERMENTATION

Many biochemical reactions involving proteins, carbohydrates, and lipids occur during fermented sausage processing, and these determine the ultimate characteristics of the product (Hierro, de la Hoz, and Ordóñez 1999). These phenomena are due to both endogenous and microbial enzymes, the contribution of which depend mainly on the type of process (Molly et al. 1996; Hierro, de la Hoz, and Ordóñez 1999; Ordóñez et al. 1999).

13.5.1 CHANGES IN PROTEINS

Proteolysis in fermented sausages is involved in the development of flavor and texture, and this is one of the main biochemical reactions, catalyzed by either endogenous enzymes present in the meat tissues or by those of microbial origin from added starter cultures (Hughes et al. 2002; Candogan, Wardlaw, and Acton 2009). Meat proteins are known to undergo hydrolysis, first to polypeptides by endogenous muscle enzymes, such as cathepsins and calpains (Toldrá, Miralles, and Flores 1992; Molly et al. 1997), and then further to smaller peptides by the action of peptidases. Free amino acid generation from peptides by aminopeptidases is the final step in proteolysis phenomena and has been attributed to protease enzymes generated by microorganisms as well as enzymes inherent in the meat itself (Casaburi et al. 2007, 2008; Hughes et al. 2002).

Low-molecular-weight peptides and free amino acids are major components of the nonprotein nitrogen (NPN) fraction in fermented meats, and these contribute, directly or indirectly, to generation of volatile and nonvolatile flavor compounds in dry and semidry sausages (De Masi et al. 1990). Flores et al. (1997) and Candogan, Wardlaw, and Acton (2009) have also reported a large increase in the NPN fraction during fermentation of dry sausages inoculated with *L. sake* and *S. carnosus* starter culture mix, and they related it to a corresponding decrease in sausage pH value. Similarly, research in fermented pork sausages (Garcia de Fernando and Fox 1991), in a variety of fermented sausages including chorizo, *saucisson*, and salami (Astiasarán, Villanueva, and Bello 1990), and in *salchichon* (Berian, Lizaso, and Chasco 2000) independently noted significant increases in NPN concentration over various processing phases.

The generation of free amino acids directly contributes to the basic taste of dry-fermented sausages and indirectly contributes to the development of their typical aroma, since they are precursors of many volatile compounds. According to Stahnke et al. (2002), the degradation of valine, leucine, and isoleucine into methyl-branched aldehydes, acids, and alcohols has been linked to the ripened aroma of fermented foods. Aldehydes and ketones with intense aroma characteristics, resulting from further catabolism of the amino acids, have an obvious role in development of flavor (Y. Sanz and Toldrá 1997).

13.5.2 CHANGES IN LIPIDS

Lipolysis plays an essential role in the development of dry-sausage flavor. Lipids are hydrolyzed by enzymes, generating free fatty acids, which are substrates for the oxidative changes that are responsible for flavor compounds (Samelis, Aggelis, and Metaxopoulos 1993; Stahnke 1995; Verplaetse 1994). The first step in the lipid breakdown is the hydrolysis of triglycerides by both microbial (B. Sanz et al. 1988; Selgas, Sanz, and Ordóñez 1988) and endogenous (muscle and adipose tissue) lipases (Molly et al. 1997). For example, Sørensen and Samuelsen (1996) indicated that the lipases produced by *S. xylosus* and *D. hansenii* are able to hydrolyze the pork fat mainly during the initial stages of processing where pH is high.

The concentration of free fatty acids in the fat depends on the hydrolytic activity of the lipases, the microbial metabolic processes, and the oxidative reactions that work on the free fatty acids released in the lipolysis and the final products. The advanced lipolysis involves the liberation of fatty acids that undergo later enzymatic and nonenzymatic oxidative processes yielding, as final products, carbonyls and other low-molecular-weight compounds (alcohol, caboxylic acids, etc.), which are the main flavor compounds of the final sausage (Toldrá 1998).

Enzymatic hydrolysis during fermentation accelerates lipid peroxidation. Due to the high fat content and low water activity of these products, lipid oxidation is the main factor responsible for loss of quality, leading to oxidative flavors and loss of pigments and vitamins (Melton 1983). For these reasons Gray, Gomaa, and Buckley (1996) have indicated that lipid oxidation in meat products can be controlled or minimized by the addition of commercial synthetic antioxidants or natural antioxidants (Nassu et al. 2003).

The development of flavor in meat products is a very complex process due to the high number of factors involved. In general, according to Toldrá (1998), Ordóñez et al. (1999), and Fernandéz et al. (2000), flavor compounds result from the enzymatic and chemical reactions mentioned previously, such as lipid oxidation, Maillard reactions and Strecker degradations, together with raw meat properties, additives, processing conditions, etc.

13.6 FUNCTIONAL PROPERTIES OF DRY-FERMENTED SAUSAGES

Meat and meat products are important sources of protein, fat, minerals, vitamins, essential free amino acids, and other nutrients (Jiménez-Colmenero 2007; Jiménez-Colmenero, Ventanas, and Toldrá 2010; W. Zhang et al. 2010; Bhat and Bhat 2011). Dry-fermented sausages (DFSs) belong to the category of high-fat meat products, containing about 30% pork back fat directly after manufacturing and about 40%–50% after 4 weeks of drying (Wirth 1989). Fat affects flavor, color, texture, and the drying process because the granulate form of fat helps loosen the DFS mixture, and this aids in the continuous release of moisture from the inner layers of product. It is absolutely necessary for undisturbed ripening and aroma–texture formation. For this reason, DFSs are the most difficult among meat products as far as fat reduction is concerned (Wirth 1988; Muguerza et al. 2003; Jiménez-Colmenero, Carballo, and Cofrades 2001; Severini, De Pilli, and Baiano 2003; W. Zhang et al. 2010).

The development of healthier DFSs can be done in a few ways: modification of carcass composition; reformation of meat products by reducing the use of some components, i.e., fat, salt, biogenic amines, cholesterol content; and addition of functional ingredients such as oils of different origin, probiotics, dietary fiber, sterols, etc. (Arihara 2006; W. Zhang et al. 2010; Fernández-Ginés et al. 2005; Jiménez-Colmenero, Carballo, and Cofrades 2001; Valencia, Ansorena, and Astiasarán 2007; Muguerza, Ansorena, and Astiasarán 2003).

13.6.1 MODIFICATION OF CARCASS COMPOSITION THROUGH DIET

Warnants, Van Oeckel, and Boucqué (1998) prepared salami with back fat and meat produced from pigs fed a diet containing 21–28 g PUFA/kg originated from different sources. They concluded that the 25-g PUFA/kg feed, corresponding with 15% PUFA in salami, results in acceptable taste if linoleic acid from soybeans is the predominant PUFA. Partial substitution of maize with rice bran in pig feed also led to the production of salami with better concentrations of n-6/n-3 fatty acid without differences in texture and sensory features compared to the control batch (De Campos et al. 2007). Manufacture of DFSs with back fat and meat from pigs fed diets enriched with either linseed oil or olive oil or both—with addition of α-tocopherol—led to a decrease in salami n-6/n-3 ratio from 12.05 to 1.79–2.25 without adverse effect on lipid stability or textural and sensory properties (Hoz et al. 2004).

13.6.2 REPLACEMENT OF BACK FAT BY DIFFERENT OILS

The addition of preemulsified and gelated low–erucic acid rapeseed oil instead of pork back fat to the formulation of dry-fermented sausages had a relevant influence on product quality. The quality of the rapeseed-containing sausage was similar to that of traditionally made DFS up to 70% oil substitution (Nowak 2008). The ready-to-eat product was characterized by a lower amount of cholesterol and a better PUFA/SFA ratio.

Olive oil is the most monounsaturated vegetable oil. It contains 56%–87% MUFA, 8%–25% SFA, 4%–22% PUFA, and a high value of vitamin E (Bloukas, Paneras, and Fournitzis 1997; Fernández-Ginés et al. 2005). Bloukas, Paneras, and Fournitzis (1997) found that up to 20% of back fat could be substituted with olive oil in Greek fermented sausages. Texture, weight losses, appearance, and taste depended of the way of incorporating the olive oil. Direct incorporation of olive oil to DFS resulted in higher oxidation, lower weight loss, softer texture, and worse taste compared to sausages with oil that was preemulsified with soybean protein isolate (SPI). Chorizo de Pamplona with 20%–25% back fat replaced with preemulsified olive oil showed a reduction in cholesterol content of about 12%–13% and a higher total MUFA and PUFA level without any visible differences in sensory acceptance, texture, or color measurements compared to traditional products (Muguerza et al. 2001). Investigation of fat modification in Greek DFSs showed that substitution of 20% of pork back fat by preemulsified olive oil in reduced- and low-fat DFS was also possible (Muguerza et al. 2002). This modification did not result in weight losses, but rather made sausages lighter with acceptable taste and odor. However, low-fat sausages (with 10% of fat in mixture) had an unacceptable appearance due to intensively wrinkled surfaces and development of case hardening (Muguerza et al. 2002). The evaluation of the volatile compounds of DFS after two months of storage revealed that reduction of fat led to an increase of the oxidation process, whereas the olive oil replacement of pork back fat resulted in a reduction of this process.

An alternative to using olive oil is to use oils of different origin—both vegetable and marine oils—as pork back-fat substitutes to improve nutritional quality of DFS (Fernández-Ginés et al. 2005; Jiménez-Colmenero, Carballo, and Cofrades 2001; Jiménez-Colmenero 2007). Partial replacement of 15%–25% of pork back fat by preemulsified soy oil in SPI in Chorizo de Pamplona did not increase oxidation. SFA and MUFA decreased and PUFA increased due to significant increase in linoleic and α-linolenic acids. No changes in texture were noticed (Muguerza, Ansorena, and Astiasarán 2003). Replacement of 25% pork back fat in DFS with 3.3% of linseed oil and 100 mg BHA (butylated hydroxyanisole) plus 100 mg BHT (butylated hydroxytoluene) increased the PUFA/SFA ratio from 0.4 to 0.7. The Ω-6/n-3 ratio decreased almost seven-fold from 14.1 to 1.7–2.1 in modified DSF as a consequence of α-linoleic acid increment. No oxidation problem was detected (Ansorena and Astiasarán 2004b). In Dutch-style DFS, up to 20% of pork back fat can be substituted with flaxseed oil or canola oil without causing manufacturing problems. The PUFA/SFA ratio increased from 0.3 in traditional DSF to 0.42–0.48 and 0.49–0.71 and n-6/n-3 ratio decreased from 11.2 to 6.94–5.12 and to 1.33–1.05 in sausages with canola oil and flaxseed oil, respectively (Meindert Pelser et al. 2007). Oils from marine products like fish or algae have been used as a source of n-3, n-6 PUFA (Muguerza, Ansorena, and Astiasarán 2004; Meindert Pelser et al. 2007; Valencia, Ansorena, and Astiasarán 2006, 2007).

Dutch-style sausages with incorporated encapsulated fish oil as replacement for 15% of back fat resembled traditional sausages according to their physicochemical properties, but with minor differences in PUFA/SFA and n-6/n-3 ratios (Meindert Pelser et al. 2007). In Chorizo de Pamplona fortified with 0.53% and 1.07% of fish oil extracts, the n-6/n-3 ratio decreased from 16.4 in the control to 7.78–5.32, but the DFS with the higher level of oil extracts showed an increase in the values of thiobarbituric acid reactive substances (TBARs) (Muguerza, Ansorena, and Astiasarán 2004). Chorizo de Pamplona with 25% substitution of pork back fat with deodorized fish oil and addition of BHA+BHT showed a PUFA,MUFA/SFA ratio of 1.76 and n-6/n-3 ratio of 2.97 without significant loss of quality and none of the secondary oxidation products (Valencia, Ansorena, and Astiasarán 2006). Substitution of 15% pork back fat in Spanish DFSs with preemulsified microalgae oil with BHA+BHT addition led to the manufacture of DFS with acceptable sensory properties. The n-6/n-3 ratio decreased from 9.41 in traditional DFS to 2.35 in modified DFS, and the PUFA/SFA ratio increased from 0.39 to 0.49 (Valencia, Ansorena, and Astiasarán 2007).

Replacement of pork back fat with oils in DFS leads to oxidation problems during storage. Besides the addition of antioxidant, vacuum packing (VP) with the proper film and in the proper time could help minimize this problem (Ansorena and Astiasarán 2004a; Koutsopoulos, Koutsimanis, and J. Bloukas 2008; Liaros, Katsanidis, and Bloukas 2009). According to Liaros, Katsanidis, and Bloukas (2009), the most appropriate time to VP was the time when the weight loss reached about 35% using a packing film with medium permeability to oxygen and water vapor.

13.6.3 Dietary Fiber Addition

Another modification in the formulation of DFSs to make them healthier is the addition of dietary fiber polysaccharides and lignin, neither of which are digested or absorbed in the human small intestine (Naveena, Sen, and Kondaiah 2010; Bhat and Bhat 2011; Jiménez-Colmenero, Carballo, and Cofrades 2001). In many cases, this dietary fiber not only has a beneficial nutritional and healthy effect, such as reducing the product's calorie content as well as the risk of colon cancer and others disorders, but they also generate important technological properties that offset the effects of fat reduction (Jiménez-Colmenero, Carballo, and Cofrades 2001; W. Zhang et al. 2010; Naveena, Sen, and Kondaiah 2010). Inulin is being used in meat products. Adding 7.5%–12.5% of inulin into a formulation of DFS led to the production of a low-fat fermented sausage not only enriched in about 10% of fiber, but with fat content reduced about 40%–50%, with 30% less calories and improved sensory and instrumental properties due to softer texture, tenderness, and springiness (Mendoza et al. 2001). Reduced-fat DFSs with acceptable sensory properties can be produced with 10% of fat fortified with no more than 1.5% wheat, oat, peach, orange, or apple dietary fiber (García et al. 2002). The best sensory results were obtained with orange fiber. Addition of 1% orange fiber to *salchichón* (Spanish DFS) was enough to attain DFSs comparable to the original DFS in all sensory attributes. In addition, the use of orange fiber evoked a decrease in residual nitrite and enhanced micrococcus growth rate. Both effects have a positive impact on sausage safety and quality (Fernández-López et al. 2008).

Incorporation of 2.5%–5% raw dehydrated lemon albedo or 7.5% cooked albedo to Spanish dry-cured sausages yielded products demonstrating sensory properties similar to conventional sausages. Addition of both forms of albedo led to a decrease in residual nitrite level and delayed oxidation (Aleson-Carbonell et al. 2003, 2004). Only 3% of carrot dietary fiber incorporated into the formulation of *sobrassada*, a traditional Mallorca DFS, yielded sausages with physicochemical and sensory properties similar to the traditional product but with additional nutritional benefits (Eim et al. 2008). Dry tomato peel in quantities of 0.6%–1.2% has also been used to produce *salchichón* with a level of 0.26–0.58 mg of lycopene/100 g of sausage with good sensory and textural acceptability (Calvo, García, and Selgas 2008). Lupin-seed isolate (sweet variety) at a level of 2% and hydrated to a protein/water (P/W) ratio of ¼ can be used as a substitute for beef and pork meat in Greek fermented sausages without negatively affecting the sensory quality of DFS (Papavergou, Bloukas, and G. Doxastakis 1999). Animal blood is a good source of protein and is used in different meat products (Pyrcz et al. 2005; Pyrcz, Uchman, and Konieczny 1998). DFS with 9% of microbiological modified porcine blood plasma (Pyrcz, Uchman, and Konieczny 1998) showed higher stability and improved overall sensory acceptability (Pyrcz et al. 2005).

13.6.4 SODIUM CHLORIDE REDUCTION

From a technological and sensory point of view, NaCl is an important ingredient that is frequently used in the meat industry because of its effect on taste, texture, and shelf life of meat products. From a health point of view, it has been established that salt intake should not exceed 6 g NaCl per day, because sodium has been linked with hypertension and cardiovascular diseases (Ruusunen and Puolanne 2005; Desmond 2006). Average salt content in meat products can be as much as 2% in heat-treated and as much as 6%, or even more in dry-cured and/or fermented products (Jiménez-Colmenero, Carballo, and Cofrades 2001). A lot of work has been done to lower the salt content in meat products by reducing or partially substituting NaCl with other salts or ingredients.

In DSF, simply reducing the level of salt is difficult because of the fermentation process, and 2.5% is the lower limit for good-quality salami-type products (Petäjä, Kukkonen, and Puolanne 1985). Gimeno, Astiasarán, and Bello (1998, 1999, 2001a) replaced NaCl with $CaCl_2$, KCl, and $MgCl_2$ and were able to reduce NaCl content by about 50%. The only relevant difference from the traditional product was the lower consistency, probably because the salts used were not effectively reacting with myofilaments (Gimeno, Astiasarán, and Bello 2001a). A 25% reduction in NaCl content, from 3% to 2.5%, and partial (about 66%) replacement of NaCl with KCl in DFS inoculated with mixtures of *L. plantarum* and *S. carnosus* led to a slight lowering of hardness as well as better salted taste (Ibañez et al. 1996, 1997), microbiological stability and higher lipolytic activity (Quintanilla et al. 1996), and a lower degree of pigment oxidation (Ibañez et al. 1996). However, substitution of NaCl with KCl, K-lactate, and glycine in DFS (Gou et al. 1996) showed important flavor defects with substitution above 40% for the three substances compared to the original sausages.

The possible reduction of NaCl with the same constituents has also been studied by Gelabert et al. (2003). They found that the critical level regarding flavor and textural aspects of NaCl substitution by KCl, K-lactate, or glycine was 40%, 30%, and 20%, respectively. Reduced NaCl content of DFS with the mixture of 1% NaCl, 0.55% KCl, and 0.75% $CaCl_2$ compared to the control with 2.6% NaCl did not affect the development of starter culture and guaranteed the hygienic quality of the products (Gimeno et al. 2001b). Partial replacement of NaCl with 15%–45% of Ca-ascorbate caused higher acidification of DFS related to higher LAB development without any hygienic quality problems (Gimeno et al. 2001a). Concurrently, DFSs were also enriched in ascorbate and calcium, with 26%–50% of the recommended daily allowance (RDA) (Gimeno et al. 2001a).

Fortification of dry-fermented products with calcium provides an excellent opportunity for increasing Ca intake. Thai and Spanish DFS fortified with Ca-citrate, Ca-gluconate, Ca-lactate, and calcium from egg shell (Thai sausage) showed that the Ca supplementation should be limited to 19%–20% of RDA without significant differences in overall acceptability compared to the control (Daengprok, Garnjanagoonchorn, and Mine 2002; Selgas, Salazar, and García 2009).

The results presented here suggest that viable DFS can be produced with decreased NaCl content. DFS produced with partial reduction of NaCl and partial replacement of NaCl with relevant salt mixtures did not suffer significant quality and technological problems. From a sensory point of view (Guàrdia et al. 2006), it is possible to reduce NaCl content in small-caliber fermented sausages by 50% compared to conventionally produced sausages.

13.6.5 Probiotics in Dry-Fermented Sausages

Probiotics are live microorganism strains, mainly from LAB and *Bifidobacterium*, that can provide some health benefits to the host when ingested at a certain level (10^6–10^7 cfu/g). These include reduction of gastrointestinal infections, reduction in serum cholesterol, stimulation of the immune system, and anticarcinogenic properties (W. Zhang et al. 2010; Arihara 2006; de Vuyst, Falony, and Leroy, 2008; Leroy, Verluyten, and de Vuyst 2006; Ruiz-Moyano et al. 2008). The technology of fermented sausages makes them potentially excellent candidate meat products for treatment with probiotic bacteria. It is also believed that the properties of the DFS matrix can enhance the survival of probiotic lactobacilli during their passage through the gastrointestinal tract (Klingberg and Budde 2006). However, the potential negative impact of the DFS environment on bacterial viability must be taken into account, in particular with respect to its high salt content, low pH, and a_w due to acidification and drying, which complicate the choice of appropriate probiotic strains in DFS (de Vuyst, Falony, and Leroy 2008; Pennacchia, Vaughan, and Villani 2006; Ruiz-Moyano et al. 2011; Leroy, Verluyten, and de Vuyst 2006; Muthukumarasamy and Holley 2006).

Probiotic bacteria should not alter the technological and sensory properties of DFS (Ammor and Mayo 2007). The most critical aspects of probiotic strains considered for use in DFS applications are acid tolerance to pH 2.5 for 4 h, salt tolerance to 4%–6% of NaCl, bile salt resistance to 0.3% w/v ox-gall, and colonization of the human gut (W. Zhang et al. 2010; Ammor and Mayo 2007; Klingberg et al. 2005). There are mainly two ways of choosing probiotic strains relevant to DFS: (a) screening the bacteria commonly associated with DFS environments by screening natural sausage isolates and (b) screening the existing commercial starter cultures (de Vuyst, Falony, and Leroy 2008; Leroy, Verluyten, and de Vuyst 2006).

Isolation of 25 *Lactobacillus* strains from dry-fermented Italian sausages showed that only one strain, *L. paracasei* EL7, was able to ferment all of the probiotic substances tested (Pennacchia, Vaughan, and Villani 2006). Screening of potential probiotics from 22 Scandinavian DFS showed that only three strains belonging to the group of *L. plantarum/pentosus* (MF1291, MF1298, MF1300) were in conformance with the definition of a probiotic substance (Klingberg et al. 2005). Ruiz-Moyano et al. (2008) showed that among the 1,000 strains isolated from Iberian DFS, human and pig feces 51 LAB were found to have a particularly marked capacity for surviving both the ripening process and conditions simulating those in the human gastrointestinal tract. Their further studies (Ruiz-Moyano et al. 2008, 2009a, 2009b, 2011) showed that isolated strains such as *Enterococcus faecium* L. *reuteri* PL519, PL592, and *L. fermentum* HL57 were able to establish themselves on the intestinal epithelium and inhibit *L. monocytogenes*.

For example, the commercial meat starter strains *L. sakei* LB-3 and *Pediococcus acidilactici* PA-2 may be regarded as potential probiotics because of their survival under simulated gastrointestinal conditions (Erkkilä and Petäjä 2000). It was also shown (Erkkilä et al. 2001a, 2001b) that *L. rhamnosus* strains GG, LC-705, E-97800 as well as *P. pentosaceus* E-90390 and *L. plantarum* E-98098 seemed to be suitable as a commercial probiotic culture for production of DFS of northern European origin due to their lack of harmful effect to the technological and sensory properties. Given the specific technological condition during the processing and storage of DFS, the survival of probiotic strains at an appropriate level is an important consideration. Microencapsulation has been suggested as a promising tool to increase the survivability of probiotic strains in a DFS environment (W. Zhang et al. 2010; Muthukumarasamy and Holley 2006). A significantly lower reduction of encapsulated *L. reuteri* probiotic strains compared to unencapsulated strains has been demonstrated in DFS environments (Muthukumarasamy and Holley 2006). A full assessment of the probiotic effects of fermented sausages on health is not yet fully possible. Further studies are required to screen for strains that are compliant with the criteria of functional foods as well as their technological requirements.

13.6.6 ANTIBACTERIAL COMPOUNDS PRODUCTION

It is well known that LAB, apart from organic acids, have a potential to produce bacteriocins and thus can suppress the growth of undesirable bacteria (Hammes and Hertel 1996). For example, all sakacins (sakacin A, M, P, B, P, K, T, and sakacin 674) produced by *Lactobacillus sakei* strains possess strong antilisterial activity (Schillinger and Lücke 1989; Sobrino et al. 1992; Tichaczek et al. 1992; Holck et al. 1994; Samelis et al. 1994; Hugas et al. 1995; Aymerich et al. 2000a). Other bacteriocins produced by *Lactobacillus* strains—curvacins (Tichaczek et al. 1992; Sudirman et al. 1993; Garver and Muriana 1994), plantaricins (Lewus and Montville 1992; Aymerich et al. 2000b), acidocin B and salivaricin B (Brink et al. 1994), and enterocins A and B (Herranz et al. 2001)—also inhibit a large number of gram-positive bacteria and food-borne pathogens. According to Messens et al. (2003), a bacteriocin-producing strain suitable as a meat starter culture should exhibit the following characteristics during the sausage manufacturing process: bacteriocin formation during the fermentation stage, stability of the bacteriocin in the meat matrix, no inactivation of the bacteriocin by food compounds, and no interference of the bacteriocin with other useful starter organisms. Lately, it has been shown that the combination of enterocins A and B with high pressure (400 MPa) applied after ripening of low-acid-fermented sausages is viable and can reduce counts of *Salmonella* and *L. monocytogenes* below 1 log cfu/g (Jofré, Aymerich, and M. Garriga 2009).

13.6.7 BIOACTIVE PEPTIDES

Peptides that can exert different biological functions or physiological effects are known as bioactive peptides (BP). Meats used as aged and dry-cured-ripened products have been a valuable protein source for the production of BP, especially angiotensin-converting enzyme (ACE) inhibitors, because of enzymatic activity (W. Zhang et al. 2010; Jiménez-Colmenero, Ventanas, and Toldrá 2010). Although information on bioactive peptides generated from DSF is limited (W. Zhang et al. 2010; Naveena, Sen, and Kondaiah 2010; Arihara 2006), there is a possibility that BP can be formed during fermentation due to proteolytic reactions. Arihara (2006) found that the ACE inhibitory activities of extracts of European fermented sausages were higher than those obtained from nonfermented pork products. In addition, it has been shown that LAB were able to generate ACE inhibitory activities (Arihara et al. 2004).

13.6.8 OTHER NUTRITIONAL FACTORS

Most knowledge about the nutritional composition of DFS is connected with different products of proteolysis as free amino acids, free fatty acids, peptides, volatile compounds, and other components. Information about micronutrients such as vitamins and mineral salts seems to be limited. Given the fact that the meaty part of DFS formulations (50%–70%) mainly consist of pork and/or beef, it can be assumed that DFSs represent a nutritional value similar to fresh meats in respect to vitamins as well as micro and macro elements (Zanardi et al. 2010). DFSs are more protein rich than meat because of the drying process (average about 25%–35% reduction in moisture). The fermentation process may also increase digestibility. According to Dainty and Blom (1995), the process generation of peptides and amino acids reaches about 1% of dry matter during fermentation. Free amino acids are largely bioavailable due to their absorption through intestinal mucosa (Jiménez-Colmenero, Ventanas, and Toldrá 2010). For example, in Italian PDO (protected designation of origin) fermented sausages, the content of total amino acids was assessed between 1,700 and 2,160 mg/kg (Di Cagno et al. 2008), but there are great differences of amino acids contents in DFS due to the many factors that influence proteolysis. It has also been shown (W. Zhang et al. 2010) that conjugated linoleic acid (CLA) content in salami (4.2 mg/g FA methyl ester) was higher than other meat products with the exception of corned beef and German spreadable sausage. Selenium is also important for human health (RDA for Se is 55 ng/day). The average amount of Se in Spanish DFS was 35.5 µg/100 g, i.e., about 60% of RDA, and was higher compared to such other products as veal and lamb meat, hams, and *mortadella* Italian sausages (W. Zhang et al. 2010).

Consumers have become more aware of the medical benefits of meat food not only in terms of providing necessary nutrition, but also in terms of additional functions. Consequently, DFS manufacturers are developing strategies to fortify their products with functional compounds to increase their nutritional value while limiting undesirable constituents. These strategies are focused on the raw materials and technological processes used in production of DFS.

13.6.9 BIOGENIC AMINES IN DRY-FERMENTED SAUSAGES

Biogenic amines (BA) are low-molecular-weight organic bases formed mainly through decarboxylation of amino acids. Besides their undesirable physiological effects, they are also considered to be a potential precursor in the formation of carcinogenic N-nitroso compounds (Suzzi and Gardini 2003; Lu et al. 2010; Komprda, Sladkova, and Dohnal 2009). DFSs are a potential source of BA formation due to the presence of free amino acids, microorganisms, and conditions that favor decarboxylation (Komprda, Sladkova, and Dohnal 2009; Suzzi and Gardini 2003; Ansorena et al. 2002). The major BAs investigated in DFS are histamine, tyramine, putrescine, and cadaverine. The first two are most important from a toxicology point of view (Komprda, Sladkova, and Dohnal 2009; Suzzi and Gardini 2003; Latorre-Moratalla et al. 2010; Lu et al. 2010). The recommended upper limit of histamine is 10 mg/kg.

A great variability characterizes the different BA content in DFS. For example, southern European types of fermented sausages had higher contents of amines then the northern type due mainly to the presence of tyramine and phenylethylamine (Ansorena et al. 2002). Tyramine as a major amine, followed by putrescine and cadaverine, were found in artisanal DFS from different European countries. The production of BA is extremely complex and depends on several variables, including the type of microorganisms, their activities, and their interaction with processing conditions (Suzzi and Gardini 2003). DFS of Spanish origin called *fuet* had a lower amount of tyramine when starter cultures were used than the sausages with natural strains (Bover-Cid et al. 1999). Enterobacteriaceae, LAB, and Micrococcaceae strains have different abilities to produce BA (Suzzi and Gardini 2003; Bover-Cid, Izquierdo-Pulido, and Vidal-Carou 1999; Roseiro et al. 2010; Komprda, Sladkova, and Dohnal 2009). Enterobacteriaceae are generally considered to have a higher decarboxylase activity, particularly in relation to the production of cadaverine and putrescine. These microbes are present in DFS in low numbers. Histidine decarboxylase activity and histamine production were observed in a few strains belonging to Micrococcaceae. LAB from DSF are generally considered not to be toxigenic or pathogenic, but some species, e.g., *L. reuteri*, *L. sakei*, and *L. plantarum*, can produce BA, mainly tyramine. A histidine decarboxylase activity was also found among some yeasts belonging to the genera *Debaryomyces* and *Candida*.

The most important physicochemical factors influencing decarboxylation activity in DFS are pH, salt content, and temperature. Some authors (Suzzi and Gardini 2003; Bover-Cid et al. 1999b) found a correlation between increased BA production and pH decrease, while others did not find a significant relation, especially at the beginning of ripening (Lu et al. 2010). An increase of NaCl concentration between 0% and 6% markedly decreased BA production (Chander et al. 1989; Gardini et al.

2001a; Suzzi and Gardini 2003). Lower processing temperature resulted in higher level of BA (Maijala, Nurmi, and Fischer 1995). There is an increase of BA when DFSs are stored at room temperature (Komprda et al. 2004). Finally, experimental results indicate that BA content increases with sausage diameter (Bover-Cid et al. 1999; Komprda, Sladkova, and Dohnal 2009).

13.7 CONCLUDING REMARKS

In recent years, much attention has been paid to the development of meat products with physiological functions to promote health conditions and prevent the risk of diseases. It is well known that meat (and thus meat products) contains protein, fat, minerals, vitamins, essential free amino acids, and many important nutrients, including bioactive compounds such as carnosine, anserine, L-carnitine, taurine, creatine, conjugated linoleic acid (CLA), and endogenous antioxidants. However, meat products like fermented sausages can be made more functional with some modifications.

These strategies include improving the functional value of those products, and this can be realized first by adding functional compounds including CLA, vitamin E, n-3 fatty acids, and selenium in animal diets to improve carcass composition and, in this way, the quality of fresh meat for sausage production. Also game meats, as an alternative for pork and beef, with their natural high nutritional value, could be an appropriate raw material to augment the health benefits of dry sausage.

On the other hand, some functional ingredients such as vegetable proteins, dietary fibers (like inulin, orange or carrot fiber, lemon albedo), herbs and spices, oils of different origin (fish oil, olive), sterols, and some microorganism can be directly incorporated into meat products during processing to improve their functional value for consumers. The use of microorganisms from traditional sausages in large-scale fermentations could accelerate the fermentation process and improve the safety and standardization of such products and hence reduce economic losses, in this way reinforcing their position in a market. But this approach remains controversial.

Genetic engineering of starters can be performed for a variety of purposes where a suitable starter cannot easily be found in nature. One of the most promising applications is self-cloning; however, the use of genetically modified microorganisms in food and food processing remains controversial due to a lack of acceptance by consumers, especially in Europe. On the other hand, the high load of endogenous bacteria in meat raw material and the inoculation with starters may represent a problem concerning the spreading of antibiotic resistance.

The development of healthier dry-fermented sausages can also be advanced by reducing the use of certain components (e.g., pork back fat and NaCl) or replacing them with suitable alternatives. Functional compounds, especially peptides, can also be generated from meat and meat products during processing such as fermentation, curing and aging, and enzymatic hydrolysis.

The research results presented in this chapter suggest that the production of dry-fermented sausages can be altered to provide consumers with a product that aligns with a healthy diet as defined by nutrition experts. From a health standpoint, these reinvented fermented sausages seem to offer a better alternative than traditional sausages.

REFERENCES

Aguirrezábal, M. M., J. Mateo, M. C. Dominguez, and J. M. Zumalacárregui. 2000. The effect of paprika, garlic, and salt on rancidity in dry sausages. *Meat Sci.* 54:77–81.

Aleson-Carbonell, L., J. Fernández-López, E. Sayas-Barberá, E. Sendra, and J. A. Pérez-Alvarez. 2003. Utilization of lemon albedo in dry-cured sausages. *J. Food Sci. Sensory and Nutritive Qualities of Food* 68:1826–1830.

Aleson-Carbonell, L., J. Fernández-López, E. Sendra, E. Sayas-Barberá, and J.A. Pérez-Alvarez. 2004. Quality characteristics of a non-fermented dry-cured sausage formulated with lemon albedo. *J. Sci. Food and Agric.* 84:2077–2084.

Ammor, M. S., and B. Mayo. 2007. Selection criteria for lactic acid bacteria to be used as functional starter cultures in dry sausage production: An update. *Meat Sci.* 76:138–146.

Andersen, L. 1998. Fermented dry sausages produced with the admixture of probiotic cultures. In *Proc. 44th International Congress of Meat Science and Technology*, 826–827, Barcelona, Spain.

Andersen, S. J. 1995. Compositional changes in surface mycoflora during ripening of naturally fermented sausages. *J. Food Protect.* 58:426–429.

Ankri, S., and D. Mirelman. 1999. Antimicrobial properties of allicin from garlic. *Microbes and Infection* 2:125–129.

Ansorena, D., and I. Astiasarán. 2004a. Effect of storage and packaging on fatty acid composition and oxidation in dry fermented sausages made with added olive oil and antioxidants. *Meat Sci.* 67:237–244.

———. 2004b. The use of linseed oil improves nutritional quality of the lipid fraction of dry-fermented sausages. *Food Chem.* 87:69–74.

Ansorena, D., M. C. Montel, M. Rokka, R. Talon, S. Eerola, A. Rizzo, M. Raemaekers, and D. Demeyer. 2002. Analysis of biogenic amines in northern and southern European sausages and role of flora in amine production. *Meat Sci.* 61:141–147.

Arihara, K. 2006. Strategies for designing novel functional meat products. *Meat Sci.* 74:219–229.

Arihara, K., Y. Nakashima, S. Ishikawa, and M. Itoh. 2004. Antihypertensive activities generated from porcine skeletal muscle proteins by lactic acid bacteria. In *Abstracts of 50th International Congress of Meat Science and Technology*, 236, 8–13 August, Helsinki, Finland.

Arihara, K., H. Ota, M. Itoh, Y. Kondo, T. Sameshima, H. Yamanaka, M. Akimoto, S. Kanai, and T. Miki. 1998. *Lactobacillus acidophilus* group lactic acid bacteria applied to meat fermentation. *J. Food Sci.* 63:544–547.

Astiasarán, I., R. Villanueva, and J. Bello. 1990. Analysis of proteolysis and protein insolubility during the manufacture of some varieties of dry sausage. *Meat Sci.* 28:111–117.

Aymerich, M. T., M. G. Artigas, M. Garriga, J. M. Montfort, and M. Hugas. 2000a. Effect of sausage ingredients and additives on the production of enterocins A and B by *Enterococcus faecium* CTC 492: Optimization of in vitro production and anti-listerial effect in dry fermented sausages. *J. Appl. Microbiol.* 88:686–694.

Aymerich, M. T., M. Garriga, M. Monfort, I. Nes, and M. Hugas. 2000b. Bacteriocin-producing lactobacilli in Spanish-style fermented sausages: Characterization of bacteriocins. *Food Microbiol.* 17:33–45.

Aymerich, T., B. Martin, M. Garriga, and M. Hugas. 2003. Microbial quality and direct PCR identification of lactic acid bacteria and nonpathogenic staphylococci from artisanal low-acid sausages. *Appl. Environ. Microbiol.* 69:4583–4594.

Baumgartner, P. A., P. G. Klettner, and W. Rodel. 1980. The influence of temperature on some parameters for dry sausage during ripening. *Meat Sci.* 4:191–201.

Benito, M. J., A. Martín, E. Aranda, F. Pérez-Nevado, S. Ruiz-Moyano, and M. G. Córdoba. 2007. Characterization and selection of autochthonous lactic acid bacteria isolated from traditional Iberian dry-fermented salchichón and chorizo sausages. *J. Food Sci.* 72:193–201.

Berdague, J. L., P. Monteil, M. C. Montel, and R. Talon. 1993. Effects of starter cultures on the formation of flavour compounds in dry sausage. *Meat Sci.* 35:275–287.

Berian, M. J., G. Lizaso, and J. Chasco. 2000. Free amino acids and proteolysis involved in "salchichon" processing. *Food Control* 11:41–47.

Bhat, Z. F., and H. Bhat. 2011. Functional meat products: A review. *Int. J. Meat Sci.* 1 (1): 1–14.

Bloukas, J. G., E. D. Paneras, and G. C. Fournitzis. 1997. Effect of replacing pork backfat with olive oil on processing and quality characteristics of fermented sausages. *Meat Sci.* 45 (2): 133–144.

Bolumar, T., Y. Sanz, M. C. Aristoy, and F. Toldrá. 2003. Purification and properties of an arginyl aminopeptidase from *Debaryomyces hansenii*. *Int. J. Food Microbiol.* 86:141–151.

———. 2008. Purification and characterization of proteases A and D from *Debaryomyces hansenii*. *Int. J. Food Microbiol.* 124:135–141.

Bover-Cid, S., M. Izquierdo-Pulido, and M. C. Vidal-Carou. 1999. Effect of proteolytic starter cultures of *Staphylococcus* spp. on biogenic amine formation during the ripening of dry fermented sausages. *Int. J. Food Microbiol.* 46:95–104.

———. 2000. Mixed starter cultures to control biogenic amine production in dry fermented sausages. *J. Food Protect.* 63:1556–1562.

Bover-Cid, S., S. Schoppen, M. Izquierdo-Pulido, and M. C. Vidal-Carou. 1999. Relationship between biogenic amine contents and the size of dry fermented sausages. *Meat Sci.* 51:305–311.

Bozkurt, H., and O. Erkmen. 2002. Effects of starter cultures and additives on the quality of Turkish style sausages (sucuk). *Meat Sci.* 61:149–156.

Brink, B., M. Minekns, J. M. B. M. Vander Vossen, R. J. Leer, and J. H. J. Huis in't Veld. 1994. Antimicrobial activity of lactobacilli. *J. Appl. Bacteriol.* 77:140–148.

Bruna, J. M., J. A. Ordóñez, M. Fernández, B. Herranz, and L. de la Hoz. 2001. Microbial and physico-chemical sausages superficially inoculated with or having added and intracellular cell-free extract of *Penicillium aurantiogriseum*. *Meat Sci.* 59:87–96.

Cai, Y., H. Okada, H. Mori, Y. Benno, and T. Nakase. 1999. *Lactobacillus paralimentarius* sp. *nov.* isolated from sourdough. *Int. J. Syst. Bacteriol.* 49:1451–1455.

Calvo, M. M., M. L. García, and M. D. Selgas. 2008. Dry fermented sausages enriched lycopene from tomato peel. *Meat Sci.* 80:167–172.

Candogan, K., F. B. Wardlaw, and J. C. Acton. 2009. Effect of starter culture on proteolytic changes during processing of fermented beef sausages. *Food Chem.* 116:731–737.

Capriotti, A. 1954. Indagini microbiologiche sulle carni insaccate. Nota I: *I lieviti. Arch. Vet. It.* 5:113–115.

Casaburi A., M. C. Aristoy, S. Cavella, R. Di Monaco, D. Ercolini, F. Toldrá, and F. Villani. 2007. Biochemical and sensory characteristics of traditional fermented sausages of Vallo di Diano (Southern Italy) as affected by the use of starter cultures. *Meat Sci.* 76:295–307.

Casaburi, A., G. Blaiotta, G. Mauriello, O. Pepe, and F. Villani. 2005. Technological activities of *Staphylococcus carnosus* and *Staphylococcus simulans* strains isolated from fermented sausages. *Meat Sci.* 71:643–650.

Casaburi, A., R. D. Monaco, S. Cavella, F. Toldrá, D. Ercolini, and F. Villani. 2008. Proteolytic and lipolytic starter cultures and their effect on traditional fermented sausages ripening and sensory traits. *Food Microbiol.* 25:335–347.

Casaburi, A., F. Villani, F. Toldrá, and Y. Sanz. 2006. Protease and esterase activity of staphylococci. *Int. J. Food Microbiol.* 112:223–229.

Cassens, R. G. 1994. In *Meat preservation: Preventing losses and assuring safety.* Trumbull, CT: Food & Nutrition Press.

Chander, H., V. H. Batish, S. Babu, and R. S. Singh. 1989. Factors affecting amine production by a selected strain of *Lactobacillus bulguricus*. *J. Food Sci.* 54:940–942.

Corbiere Morot-Bizot, S., S. Leroy, and R. Talon. 2006. Staphylococcal community of a small unit manufacturing traditional dry fermented sausages. *Int. J. Food Microbiol.* 108:210–217.

Daengprok, W., W. Garnjanagoonchorn, and Y. Mine. 2002. Fermented pork sausage fortified with commercial or hen eggshell calcium lactate. *Meat Sci.* 62:199–204.

Dainty, R., and H. Blom. 1995. Flavor chemistry of fermented sausages. In *Fermented meats*, ed. G. Campbell-Platt and P. E. Cook, 176–193. Glasgow, Scotland: Blackie Academic & Professional.

Dawood, A. A. 1995. Physical and sensory characteristics of Najdi-camel meat. *Meat Sci.* 3:59–69.

De Campos, R. M. L., E. Hierro, J. A. Ordóñez, T. M. Bertol, N. N. Terra, and L. de la Hoz. 2007. Fatty acid and volatile compounds from salami manufactured with yerba mate (*Ilex paraguariensis*) extract and pork back fat and meat from pigs fed on diets with partial replacement of maize with rice bran. *Food Chem.* 103:1159–1167.

De Masi, T. W., F. B. Wardlaw, R. L. Dick, and J. C. Acton. 1990. Non protein nitrogen (NPN) and free amino acid contents of dry fermented and non fermented sausages. *Meat Sci.* 27:1–12.

Demeyer, D. 1982. Stoichiometry of dry sausage fermentation. *Antonie van Leeuwenhoek* 48:414–416.

Demeyer, D., H. Blom, L. Hinrichsen, G. Johansson, K. Molly, and M. C. Montel. 1995. Interaction of lactic acid bacteria with muscle enzymes for safety and quality of fermented meat products. In *Proceedings of Lactic Acid Bacteria Conference*, 1–18, Cork, Ireland.

Demeyer, D. I., A. Verplaetse, and M. Gistelinck. 1986. Fermentation of meat: An integrated process. In *Proceedings of the 32nd International Congress of Meat Science and Technology*, 241– 247, Ghent, Belgium.

Desmond, E. 2006. Reducing salt: A challenge for the meat industry. *Meat Sci.* 74:188–196.

De Vuyst, L., G. Falony, and F. Leroy. 2008. Probiotics in fermented sausages. *Meat Sci.* 80:75–78.

Di Cagno, R., C. C. Lòpez, R. Tofalo, G. Gallo, M. De Angelis, A. Paparella, W. P. Hammes, and M. Gobbetti. 2008. Comparison of the compositional, microbiological, biochemical, and volatile profile characteristics of three Italian PDO fermented sausages. *Meat Sci.* 79:224–235.

Dinçer, B. 1982. Studies on the compositional, lipolytic, and sensorial changes during the ripening of Turkish fermented sausage. *Turk. J. Vet. Anim. Sci.* 6:41–53.

Durá, M. A., M. Flores, and F. Toldrá. 2002. Purification and characterisation of a glutaminase from *Debaryomyces* spp. *Int. J. Food Microbiol.* 76:117–126.

Eim, V. S., S. Simal, C. Rosselló, and A. Femenia. 2008. Effects of addition of carrot dietary fibre on the ripening process of a dry fermented sausages (sobrassada). *Meat Sci.* 80:173–182.

El-Faer, M. Z., T. N. Rawdah, K. M. Attar, and M. V. Dawson. 1991. Mineral and proximate composition of meat of the one humped camel (*Camelus dromedarius*). *Food Chem.* 42:139–43.

Elgasim, E. A., and M. A. Alkanhal. 1992. Proximate composition, amino acids, and inorganic mineral content of Arabian camel meat: Comparative study. *Food Chem.* 45:1–4.

Encinas, J. P., T. M. López-Díaz, M. L. García-López, A. Otero, and B. Moreno. 2000. Yeast populations on Spanish fermented sausages. *Meat Sci.* 54:203–208.

Erkkilä, S., and E. Petäjä. 2000. Screening of commercial meat starter cultures at low pH and in the presence of bile salts for potential probiotic use. *Meat Sci.* 55:297–300.

Erkkilä, S., E. Petäjä, S. Eerola, L. Lilleberg, T. Mattila-Sandholm, and M.-L. Suihko. 2001a. Flavour profiles of dry sausages fermented by selected novel meat starter cultures. *Meat Sci.* 58:111–116.

Erkkilä, S., M.-L. Suihko, S. Eerola, E. Petäjä, and T. Mattila-Sandholm. 2001b. Dry sausage fermented by *Lactobacillus rhamnosus* strains. *Int. J. Food Microbiol.* 64:205–210.

Erkkilä, S., M. Venäläinen, S. Hielm, E. Petäjä, E. Puolanne, and T. Mattila-Sandholm. 2000. Survival of *Escherichia coli* O157:H7 in dry fermented sausage fermented by probiotic lactic acid bacteria. *J. Sci. Food Agric.* 80:2101–2104.

Fadda, S., G. Vignolo, and G. Oliver. 2001. Tyramine degradation and tyramine/histamine production by lactic acid bacteria and *Kocuria* strains. *Biotechnology Letters* 23:2015–2019.

Fanco, I., B. Prieto, J. M. Cruz, M. López, and J. Caraballo. 2002. Study of the biochemical changes during the processing of androlla, a Spanish dry-cured pork sausage. *Food Chem.* 78:339–345.

Fernandéz, M., J. A. Ordóñez, J. M. Bruna, B. Herranz, and L. de la Hoz. 2000. Accelerated ripening of dry fermented sausages. *Trends Food Sci. Technol.* 11:201–209.

Fernández-Ginés, J. M., J. Fernández-López, E. Sayas-Barberá, and J. A. Pérez-Alvarez. 2005. Meat products as functional food: A review. *J. Food Sci.* 70:37–43.

Fernández-López, J., E. Sendra, E. Sayas-Barberá, C. Navarro, and J. A. Pérez-Alvarez. 2008. Physico-chemical and microbiological profiles of "salchichón" (Spanish dry-fermented sausage) enriched with orange fiber. *Meat Sci.* 80:410–417.

Flores, J. 1997. Mediterranean vs. northern European meat products. Processing technologies and main differences. *Food Chem.* 59:505–510.

Flores, J., J. R. Marcus, P. Nieto, J. L. Navarro, and P. Lorenzo. 1997. Effect of processing conditions on proteolysis and taste of dry-cured sausages. *Z Lebens Unters Forsch A.* 204:168–172.

García, M. L., C. Casas, V. M. Toledo, and M. D. Selgas. 2001. Effect of selected mould strains on the sensory properties of dry fermented sausages. *Europ. Food Res. Technol.* 212:287–291.

García, M. L., R. Domínguez, M. D. Gálvez, C. Casas, and M. D. Selgas. 2002. Utilization of cereal and fruit fibres in low fat dry fermented sausages. *Meat Sci.* 60:227–236.

García de Fernando, D. G., and P. F. Fox. 1991. Study of proteolysis during the ripening of a dry fermented pork sausage. *Meat Sci.* 30:367–383.

Gardini, F., M. Martuscelli, M. C. Caruso, F. Galgano, M. A. Crudele, F. Favati, M. E. Guerzoni, and G. Suzzi. 2001a. Effects of pH, temperature, and NaCl concentration on the growth kinetics, proteolytic activity, and biogenic amine production of *Enterococcus faecalis*. *Int. J. Food Microbiol.* 64:105–117.

Gardini, F., G. Suzzi, A. Lombardi, F. Galgano, M. A. Crudele, Ch. Andrighetto, M. Schirone, and R. Tofalo. 2001b. A survey of yeasts in traditional sausages of southern Italy. *FEMS Yeast Res.* 1:161–167.

Garver, K. I., and P. M. Muriana. 1994. Purification and partial amino acid sequence of curvaticin FS47, a heat stable bacteriocin produced by *Lactobacillus curvatus* FS47. *Appl. Environ. Microbiol.* 60:2191–2195.

Geisen, R., F. K. Lücke, and L. Kröckel. 1992. Starter and protective cultures for meat and meat products. *Fleischwirtsch.* 72:894–898.

Gelabert, J., P. Gou, L. Guerrero, and J. Arnau. 2003. Effect of sodium chloride replacement on some characteristics of fermented sausages. *Meat Sci.* 65:833–839.

Gimeno, O., I. Astiasarán, and J. Bello. 1998. A mixture of potassium, magnesium, and calcium chlorides as partial replacement of sodium chloride in dry fermented sausages. *J. Agricult. Food Chem.* 46:4372–4375.

———. 1999. Influence of partial replacement of NaCl with KCl and CaCl₂ on texture and colour of dry fermented sausages. *J. Agricult. Food Chem.* 47:873–877.

———. 2001a. Calcium ascorbate as a potential partial substitute for NaCl in dry fermented sausages: Effect on colour, texture, and hygienic quality at different concentrations. *Meat Sci.* 57:23–29.

————. 2001b. Influence of partial replacement of NaCl with KCl and CaCl$_2$ on texture and colour of dry fermented sausages. *Food Microbiol.* 18:329–334.

Glenn, E., R. Geisen, and L. Leistner. 1989. Control of *Listeria monocytogenes* in mould ripened raw sausages by strains of *Penicillium nalgiovense*. *Mitteilungsbl. Bundesanstalt Fleishforch. Kulmbach* 105:317–324.

Gökalp, H. Y. 1986. Turkish style fermented sausage (soudjouk) manufactured by adding different starter cultures and using different ripening temperatures. *Fleischwirtschaft* 66:573–575.

Gou, P., L. Guerrero, J. Gelabert, and J. Arnau. 1996. Potassium chloride, potassium lactate, and glycine as sodium chloride substitutes in fermented sausages and in dry-cured pork loin. *Meat Sci.* 42:37–48.

Gray, J. J., E. A. Gomaa, and D. J. Buckley. 1996. Oxidative quality and shelf life of meats. *Meat Sci.* 43 (Suppl.): S111–S123.

Grazia, L., P. Romano, A. Bagni, D. Roggiani, and G. Guglielmi. 1986. The role of moulds in the ripening process of salami. *Food Microbiol.* 3:19–25.

Grazia, L., G. Suzzi, P. Romano, and P. Giudici. 1989. The yeasts of meat products. *Yeast* 5:S495–S499.

Guàrdia, M. D., L. Guerrero, J. Gelabert, P. Gou, and J. Arnau. 2006. Consumer attitude towards sodium reduction in meat products and acceptability of fermented sausages with reduced sodium content. *Meat Sci.* 73:484–490.

Hammes, W. P., and C. Hertel. 1996. Selection and improvement of lactic acid bacteria used in meat and sausage fermentation. *Lait* 76:159–168.

————. 1998. New developments in meat starter cultures. *Meat Sci.* 49:125–138.

Hammes, W., I. Rölz, and A. Banteon. 1985. Microbiologische Untersuchung der auf dem deutschen Markt vorhandenen Starterkulturpräparate für die Rohwurstbereitung. *Fleischwirtschaft* 65:629–636.

Herranz, C., P. Casaus, S. Mukhopadhyay, J. M. Martínez, J. M. Rodríguez, I. F. Nes, P. E. Hernández, and L. M. Cintas. 2001. *Enterococcus faecium* P21: A strain occurring naturally in dry-fermented sausages producing the class II bacteriocins enterocin A and enterocin B. *Food Microbiol.* 18:115–131.

Hierro, E., L. de la Hoz, and J. A. Ordóñez. 1999. Contribution of the microbial endogenous enzymes to the free amino acid and amine contents of dry fermented sausages. *J. Agric. Food. Chem.* 47:1156–1161.

Hoffman, L. C., B. Kritzinger, and A. V. Ferreira. 2005. The effects of region and gender on the fatty acid, amino acid, mineral, myoglobin, and collagen contents of impala. *Meat Sci.* 69:551–558.

Holck, A., L. Axelsson, K. Hühne, and L. Kröckel. 1994. Purification and cloning of sacasin 674: A bacteriocin from *Lactobacillus sake* Lb674. *FEMS Microbiol. Lett.* 115:143–150.

Houben, J. H., and B. J. van't Hooft. 2005. Variations in product-related parameters during standardised manufacture of a semi-dry fermented sausage. *Meat Sci.* 69:283–287.

Hoz, L., M. D'Arrigo, I. Cambero, and J. A. Ordóñez. 2004. Development of an *n-3* fatty acid and α-tocopherol enriched dry fermented sausages. *Meat Sci.* 67:485–495.

Hugas, M., M. Garriga, M. T. Aymerich, and J. M. Monfort. 1995. Inhibition of Listeria in dry fermented sausages by the bateriocinogenic *Lactobacillus sakei* CTC494. *J. Appl. Bacteriol.* 79:322–330.

Hugas, M., and J. M. Monfort. 1997. Bacterial starter cultures for meat fermentation. *Food Chem.* 59:547–554.

Hughes, M. C., J. P. Kerry, E. K. Arendt, P. M. Kenneally, P. L. H. McSweeney, and E. E. O'Neill. 2002. Characterization of proteolysis during the ripening of semi-dry fermented sausages. *Meat Sci.* 62:205–216.

Ibañez, C., L. Quintanilla, C. Cid, I. Astiasarán, and J. Bello. 1996. Dry fermented sausages elaborated with *Lactobacillus plantarum-Staphylococcus carnosus*. Part I: Effect of partial replacement of NaCl with KCl on the stability and the nitrosation process. *Meat Sci.* 44:227–234.

———. 1997. Dry fermented sausages elaborated with *Lactobacillus plantarum-Staphylococcus carnosus*. Part II: Effect of partial replacement of NaCl with KCl on the proteolytic and insolubilization processes. *Meat Sci.* 46:277–284.

Incze, K. 1991. Raw fermented and dried meat products. In *Proc. 37th International Congress of Meat Science and Technology*, 829–842, Kulmbach, Germany.

———. 1992. Raw fermented and dried meat products. *Fleischwirtschaft.* 12:1–5.

Jiménez-Colmenero, F. 2007. Healthier lipid formulation approaches in meat-based functional foods. Technological options for replacement of meat fats by non-meat fats. *Trends Food Sci. Technol.* 18:567–578.

Jiménez-Colmenero, F., J. Carballo, and S. Cofrades. 2001. Healthier meat and meat products: Their role as functional foods. *Meat Sci.* 59:5–13.

Jiménez-Colmenero, F., M. Reig, and F. Toldrá. 2006. *New approaches for the development of functional meat products.* In Advanced technologies for meat processing, ed. L. M. L. Nollet and F. Toldrá. Boca Raton, FL: CRC Press.

Jiménez-Colmenero, F., J. Ventanas, and F. Toldrá. 2010. Nutritional composition of dry-cured ham and its role in a healthy diet. *Meat Sci.* 84:585–593.

Jofré, A., T. Aymerich, and M. Garriga. 2009. Improvement of the food safety of low acid fermented sausages by enterocins A and B and high pressure. *Food Control* 20:179–184.

Johansson, G., J. L. Berdagué, M. Larsson, N. Tran, and E. Borch. 1994. Lipolysis proteolysis and formation of volatile components during ripening of a fermented sausage with *Pediococcus pentosaceus* and *Staphylococcus xylosus* as starter cultures. *Meat Sci.* 38:203–218.

Kalalou, I., M. Faid, and T. A. Ahami. 2004. Improving the quality of fermented camel sausage by controlling undesirable microorganisms with selected lactic acid bacteria. *Int. J. Agric. Biol.* 6:447–451.

Klingberg, T. D., L. Axelsson, K. Naterstad, D. Elsser, and B. B. Budde. 2005. Identification of potential probiotic starter cultures for Scandinavian-type fermented sausages. *Int. J. Food Microbiol.* 105:419–431.

Klingberg, T. D., and B. B. Budde. 2006. The survival and persistence in the human gastrointestinal tract of five potential probiotic lactobacilli consumed as freeze-dried cultures or as probiotic sausages. *Int. J. Food Microbiol.* 109:157–159.

Komprda, T., P. Sladkova, and V. Dohnal. 2009. Biogenic amine content in dry fermented sausages as influenced by a producer, spice mix, starter culture, sausage diameter, and time of ripening. *Meat Sci.* 83:534–542.

Komprda, T., D. Smělà, P. Pechovà, L. Kalhotka, J. Štencl, and B. Klejdus. 2004. Effect of starter culture, spice mix, and storage time and temperature on biogenic amine content of dry fermented sausages. *Meat Sci.* 67:607–616.

Koutsopoulos, D. A., G. E. Koutsimanis, and J. G. Bloukas. 2008. Effect of carrageenan level and packaging during ripening on processing and quality characteristics of low-fat fermented sausages produced with olive oil. *Meat Sci.* 79:188–197.

Kumar, M., and J. S. Berwal. 1998. Sensitivity of food pathogens to garlic (*Allium sativum*). *J. Appl. Microbiol.* 84:213–215.

Landvogt, A., and A. Fischer. 1990. Rohwurstreifung. Gezielte Steuerung der Säuerungleistung von Starterkulturen. *Fleischwirtschaft* 70:1134–1140.

Larrouture, C., V. Ardaillon, M. Pépin, and M. C. Montel. 2000. Ability of meat starter cultures to catabolize leucine and evaluation of the degradation products by using an HPLC method. *Food Microbiol.* 17:563–570.

Latorre-Moratalla, M. L., S. Bover-Cid, R. Talon, M. Garriga, E. Zanardi, A. Ianieri, M. J. Fraqueza, M. Elias, E. H. Drosinos, and M. C. Vidal-Carou. 2010. Strategies to reduce biogenic amine accumulation in traditional sausage manufacturing. *LWT-Food Science and Technology* 43:20–25.

Leroy, F., and L. de Vuyst. 1999. Temperature and pH conditions that prevail during fermentation of sausages are optimal for production of the antilisterial bacteriocin sakacin K. *Appl. Environ. Microbiol.* 65:974–981.

———. 2004. Lactic acid bacteria as functional starter cultures for the food industry. *Trends Food Sci. Technol.* 15:67–78.

Leroy, F., J. Verluyten, and L. de Vuyst. 2006. Functional meat starter cultures for improved sausage fermentation. *Meat Sci.* 106:270–285.

Lewus, C. B., and T. J. Montville. 1992. Further characterization of bacteriocins plantaricin BN, bavaricin MN, and pediocin A. *Food Biotechnol.* 6:153S –174S.

Liaros, N. G., E. Katsanidis, and J. G. Bloukas. 2009. Effect of the ripening time under vacuum and packing film permeability on processing and quality characteristics of low-fat fermented sausages. *Meat Sci.* 83:589–598.

Lizaso, G., J. Chasco, and J. Beriain. 1999. Microbiological and biochemical changes during ripening of salchichon, a Spanish dry cured sausage. *Food Microbiol.* 16:219–228.

López-Díaz, T. M., J. A. Santos, M. L. García-López, and A. Otero. 2001. Surface mycoflora of a Spanish fermented meat sausage and toxigenicity of *Penicillium* isolates. *Int. J. Food Microbiol.* 68:69–74.

Lu, S., X. Xu, G. Zhou, Z. Zhu, Y. Meng, and Y. Sun. 2010. Effect of starter cultures on microbiological ecosystem and biogenic amines in fermented sausages. *Food Control*, 21:444–449.

Lücke, F. K. 1985. The microbiology of fermented meats. *J. Sci. Food Agric.* 36:1342.

———. 2000. Utilization of microbes to process and preserve meat. *Meat Sci.* 56:105–115.

Lücke, F. K., and H. Hechelmann. 1987. Starter cultures for dry sausages and raw ham. *Fleischwirtsch.* 67:307–314.

Ludemann, V., G. Pose, M. L. Pollio, and J. Segura. 2004. Determination of growth characteristics and lipolytic and proteolytic activities of Penicillium strains isolated from Argentinean salami. *Int. J. Food Microbiol.* 96:13–18.

Madsen, H. L., and G. Bertelsen. 1995. Spices as antioxidants. *Trends Food Sci. Technol.* 6:271–277.

Maijala, R., S. Eerola, S. Lievonen, P. Hill, and T. Hirvi. 1995. Formation of biogenic amines during ripening of dry sausages as affected by starter culture and thawing time of raw material. *J. Food Sci.* 60:1187–1190.

Maijala, R., E. Nurmi, and A. Fischer. 1995. Influence of processing temperature on the formation of biogenic amines in dry sausages. *Meat Sci.* 39:9–22.

Marianski, S., and A. Mariański. 2009. *Art of making fermented sausages*. Seminole, FL: Bookmagic LLC.

Martin, A., B. Colin, E. Aranda, M. J. Benito, and M. G. Cordoba. 2007. Characterization of Micrococcaceae isolated from Iberian dry-cured sausages. *Meat Sci.* 75:696–708.

Martin, A., J. J. Cordoba, M. M. Rodriguez, F. Nunez, and M. A. Asensio. 2001. Evaluation of microbial proteolysis in meat products by capillary electrophoresis. *J. Appl. Microbiol.* 90:163–171.

Meindert Pelser, W., J. P. H. Linssen, A. Legger, and J. H. Houben. 2007. Lipid oxidation in *n-3* fatty acid enriched Dutch style fermented sausages. *Meat Sci.* 75:1–11.

Melton, S. L. 1983. Methodology for following lipid oxidation in muscle foods. *Food Technol.* 37:105–111, 116.

Mendoza, E., M. L. García, C. Casas, and M. D. Selgas. 2001. Inulin as fat substitute in low fat, dry fermented sausages. *Meat Sci.* 57:387–393.

Messens, W., J. Verluyten, F. Leroy, and L. De Vuyst. 2003. Modeling growth and bacteriocin production by *Lactobacillus curvatus* LTH 1174 in response to temperature and pH values used for European sausage fermentation processes. *Int. J. Food Microbiol.* 81:41–52.

Mintzlaff, H.-J., and L. Leistner. 1972. Untersuchungen zur Selektion eines technologisch geeigneten und toxiologish unbedenklichen Schimmelpilz-stammes für die Rohwurst-Herstellung. *Zbl.Vet.-Med. B* 19:291–300.

Molly, K., D. Demeyer, T. Civera, and A. Verplaetse. 1996. Lipolysis in a Belgian sausage: Relative importance of endogenous and bacterial enzymes. *Meat Sci.* 43:235–244.

Molly, K., D. Demeyer, G. Johansson, M. Raemaekers, M. Ghistelinck, and I. Geenen. 1997. The importance of meat enzymes in ripening and flavour generation in dry fermented sausages: First results of a European project. *Food Chem.* 59:539–545.

Montel, M. C., J. Reitz, R. Talon, J. L. Bernagué, and A. S. Rousset. 1996. Biochemical activities of Micrococcaceae and their effects on the aromatic profiles and odors of a dry sausage model. *Food Microbiol.* 13:489–499.

Moretti, V. A., G. Madonia, C. Diaferia, T. Mentasti, M. A. Paleari, S. Panseri, G. Pirone, and G. Gandini. 2004. Chemical and microbiological parameters and sensory attributes of a typical Sicilian salami ripened in different conditions. *Meat Sci.* 66:845–854.

Muguerza, E., D. Ansorena, and I. Astiasarán. 2003. Improvement of nutritional properties of Chorizo de Pamplona by replacement of pork backfat with soy oil. *Meat Sci.* 65:1361–1367.

———. 2004. Functional dry fermented sausages manufactured with high levels of n-3 fatty acids: Nutritional benefits and evaluation of oxidation. *J. Sci. Food and Agric.* 84:1061–1068.

Muguerza, E., D. Ansorena, J. G. Bloukas, and I. Astiasarán. 2003. Effect of fat level and partial replacement of pork backfat with olive oil on the lipid oxidation and volatile compounds of Greek dry fermented sausages. *J. Food Sci.: Sensory and Nutritive Qualities of Food* 68:1531–1536.

Muguerza, E., G. Fista, D. Ansorena, I. Astiasarán, and J. G. Bloukas. 2002. Effect of fat level and partial replacement of pork backfat with olive oil on processing and quality characteristics of fermented sausages. *Meat Sci.* 61:397–404.

Muguerza, E., O. Gimeno, D. Ansorena, J. G. Bloukas, and I. Astiasarán. 2001. Effect of replacing pork backfat with pre-emulsified olive oil on lipid fraction and sensory quality of Chorizo de Pamplona: A traditional Spanish fermented sausages. *Meat Sci.* 59:251–258.

Muthukumarasamy, P., and R. A. Holley. 2006. Microbiological and sensory quality of dry fermented sausages containing alginate-microencapsulated *Lactobacillus reuteri*. *Int. J. Food Microbiol.* 111:164–169.

Nassu, R. T., L. A. G. Gonçalves, M. A. A. P. da Silva, and F. J. Beserra. 2003. Oxidative stability of fermented goat meat sausage with different levels of natural antioxidant. *Meat Sci.* 63:43–49.

Naveena, B. M., A. R. Sen, and N. Kondaiah. 2010. Ensuring activity and bioavailability. *Fleischwirtschaft Int.* 4:21–28.

Nes, I. F., and R. Skjelkvåle. 1982. Effect of natural spices and oleoresins on *Lactobacillus plantarum* in the fermentation of dry sausages. *J. Food Sci.* 47:1618–1625.

Nowak, A., 2008. Produkcyjne i przechowalnicze zmiany jakości fermentowanych kiełbas surowych o zróżnicowanej zawartości tłuszczu (Post-production and storage changes of dry fermented sausages of different fat content). Praca doktorska, AR Poznań (in Polish).

Nychas, G. J. E., and J. S. Arkoudelos. 1990. Staphylococci: Their role in fermented sausages. *J. Appl. Microbiol.* 69 (Suppl. S19): 167S–188S.

Olesen, P. T., A. S. Meyer, L. H. Stahnke. 2004. Generation of flavour compounds in fermented sausages: The influence of curing ingredients, *Staphylococcus* starter culture and ripening time. *Meat Sci.* 66:675–687.

Ordóñez, J. A., E. M. Hierro, J. M. Bruna, and L. de la Hoz. 1999. Changes in the components of dry-fermented sausages during ripening. *Crit. Rev. Food Sci. Nutrit.* 39:329–367.

Osei Abunyewa, A. A., E. Laing, A. Hugo, and B. C. Viljoen. 2000. The population change of yeasts in commercial salami. *Food Microbiol.* 17:429–438.

Paleari, M. A., G. Beretta, F. Colombo, S. Foschini, G. Bertolo, and S. Camisasca. 2000. Buffalo meat as a salted and cured product. *Meat Sci.* 54:365–367.

Paleari, M. A., V. M. Moretti, G. Beretta, T. Mentasti, and C. Bersani. 2003. Cured products from different animal species. *Meat Sci.* 63:485–489.

Papadima, S. N., and J. G. Bloukas. 1999. Effects of fat level and storage conditions on quality characteristics of traditional Greek sausages. *Meat Sci.* 51:103–113.

Papavergou, E. J., J. G. Bloukas, and G. Doxastakis. 1999. Effect of lupin seed proteins on quality characteristics of fermented sausages. *Meat Sci.* 52:421–427.

Pennacchia, C., D. Ercolini, G. Blaiotta, O. Pepe, G. Mauriello, and F. Villani. 2004. Selection of *Lactobacillus* strains from fermented sausages for their potential use as probiotics. *Meat Sci.* 67:309–317.

Pennacchia, C., E. E. Vaughan, and F. Villani. 2006. Potential probiotic *Lactobacillus* strains from fermented sausages: Further investigations on their probiotic properties. *Meat Sci.* 73:90–101.

Petäjä, E., E. Kukkonen, and E. Puolanne. 1985. Einfluss des Salzgehaltes auf die Reifung von Rohwurst. *Fleischwirtsch.* 65:189–193.

Petäjä, E., T. Manninen, P. Smidtslund, and K. Sipila. 2003. Probiotic lactic acid bacteria as starters: Applicability in raw ham and in fermented meat products made from coarsely ground pork. *Fleischwirtsch.* 83:97–102.

Pezacki, W. 1979. Some basic facts about dry sausage. *Fleischwirtsch.* 59 (2): 218–220.

Pyrcz, J., R. Kowalski, P. Konieczny, and B. Danyluk. 2005. The quality of fermented raw sausages manufactured using porcine blood plasma. *EJPAU: Food Science and Technology* 8 (3): #07. www.ejpau.media.pl/volume8/issue3/art-07.html.

Pyrcz, J., W. Uchman, and P. Konieczny. 1998. The influence of microbiological modified blood plasma on quality of the "Polish salami" type raw sausages. *EJPAU: Food Science and Technology* 1 (1): #01. www.ejpau.media.pl/volume1/issue1/food/art-03.html.

Quintanilla, L., C. Ibañez, C. Cid, I. Astiasarán, and J. Bello. 1996. Influence of partial replacement of NaCl with KCl on lipid fraction of dry fermented sausages inoculated with a mixture of *Lactobacillus plantarum* and *Staphylococcus carnosus. Meat Sci.* 43:225–234.

Rebecchi, A., S. Crivori, P. G. Sarra, and P. S. Cocconcelli. 1998. Physiological and molecular techniques for the study of bacterial community development in sausage fermentation. *J. Appl. Microbiol.* 84:1043–1049.

Reuter, G. 1972. Experimental ripening of dry sausages using lactobacilli and micrococci starter cultures. *Fleischwirtsch.* 52:465–468, 471–473.

Roca, M., and K. Incze. 1990. Fermented sausages. *Food Rev. Int.* 6:91–118.

Roseiro, L. C., A. Gomes, H. Gonçalves, M. Sol, R. Cercas, and C. Santos. 2010. Effect of processing on proteolysis and biogenic amines formation in a Portuguese traditional dry-fermented ripened sausage "Chouriço Grosso de Estremoz e Borba PGI." *Meat Sci.* 84:172–179.

Ruiz-Moyano, S., A. Martín, M. J. Benito, E. Aranda, R. Casquete, and M. de Guía Córdoba. 2009a. Safety and functional aspects of preselected enterococci for probiotic use in Iberian dry-fermented sausages. *J. Food Sci.: Food Microbiology and Safety* 74:398–404.

Ruiz-Moyano, S., A. Martín, M. J. Benito, R. Casquete, M. J. Serradilla, and M. de Guía Córdoba. 2009b. Safety and functional aspects of pre-selected lactobacilli for probiotic use in Iberian dry-fermented sausages. *Meat Sci.* 83:460–467.

Ruiz-Moyano, S., A. Martín, M. J. Benito, A. Hernández, R. Casquete, and M. de Guía Córdoba. 2011. Application of *Lactobacillus fermentum* HL57 and *Pediococcus acidilactici* SP979 as potential probiotics in the manufacture of traditional Iberian dry-fermented sausages. *Food Microbiol.* 28:839–847.

Ruiz-Moyano, S., A. Martín, M. J. Benito, F. Pérez Nevado, and M. de Guía Córdoba. 2008. Screening of lactic acid bacteria and bifidobacteria for potential probiotic use in Iberian dry fermented sausages. *Meat Sci.* 80:715–721.

Ruusunen, M., and E. Puolanne. 2005. Reducing sodium intake from meat products. *Meat Sci.* 70:531–541.

Samelis, J., G. Aggelis, and J. Metaxopoulos. 1993. Lipolytic and microbial changes during the natural fermentation and ripening of Greek dry sausages. *Meat Sci.* 35:371–385.

Samelis, J., J. Metaxopoulos, M. Vlassi, and A. Pappa. 1998. Stability and safety of traditional Greek salami: A microbiological ecology study. *Int. J. Food Microbiol.* 44:69–82.

Samelis, J., S. Stavropoulos, A. Kakouri, and J. Metaxopoulos. 1994 Quantification and characterization of microbial population associated with naturally fermented Greek dry salami. *Food Microbiol.* 11:447–460.

Samson, R. A., E. S. Hoekstra, J. C. Frisvad, and O. Filtenborg. 1995. *Introduction to foodborne fungi.* Baarn, Netherlands: Centraalbureau voor Schimmelcultures.

Santos, E. M., C. Gonzalez-Fernandez, I. Jaime, and J. Rovira. 1998. Comparative study of lactic acid bacteria house flora isolated in different varieties of chorizo. *Int. J. Food Microbiol.* 39:123–128.

Sanz, B., D. Selgas, I. Parejo, and J. A. Ordóñez. 1988. Characteristics of lactobacilli isolated from dry-fermented sausages. *Int. J. Food Microbiol.* 6:199–205.

Sanz, Y., and F. Toldrá. 1997. Purification and characterization of an aminopeptidase from *Lactobacillus sake. J. Agricul. Food Chem.* 45:1552–1558.

Sanz, Y., R. Vila, F. Toldrá, and J. Flores. 1998. Effect of nitrate and nitrite curing salts on microbial changes and sensory quality of nonfermented sausages. *Int. J. Food Microbiol.* 42:213–217.

Schillinger, U., and F. K. Lücke. 1989. Antibacterial activity of *Lactobacillus sake* isolated from meat. *Appl. Environ. Microbiol.* 55:1901–1906.

Selgas, M. D., J. Ros, and M. L. García. 2003. Effect of selected yeast strains on the sensory properties of dry fermented sausages. *Eur. Food Res. Technol.* 217:475–480.

Selgas, M. D., P. Salazar, and M. L. García. 2009. Usefulness of calcium lactate, citrate, and gluconate for calcium enrichment of dry fermented sausages. *Meat Sci.* 82:478–480.

Selgas, M. D., B. Sanz, and J. A. Ordóñez. 1988. Selected characteristics of micrococci isolated from Spanish dry-fermented sausages. *Food Microbiol.* 5:185–193.

Severini, C., T. De Pilli, and A. Baiano. 2003. Partial substitution of pork backfat with extra-virgin olive oil in "salami" products: Effects on chemical, physical, and sensorial quality. *Meat Sci.* 64:323–331.

Sobrino, O. J., J. M. Rodríguez, W. L. Moreira, L. M. Cintas, M. F. Fernández, B. Sanz, and P. E. Hernández. 1992. Sakacin M, a bacteriocin-like substance from *Lactobacillus sake* 148. *Int. J. Food Microbiol.* 16:215–225.

Søndergaard, A. K., and L. H. Stahnke. 2002. Growth and aroma production by *Staphylococcus xylosus, S. carnosus,* and *S. equorum*: A comparative study in model systems. *Int. J. Food Microbiol.* 75:99–109.

Sørensen, B. B. 1997. Lipolysis of pork fat by the meat starter culture *Debaryomyces hansenii* at various environmental conditions. *Int. J. Food Microbiol.* 34:187–193.

Sørensen, B. B., and H. Samuelsen. 1996. The combined effects of environmental conditions on lipolysis of pork fat by lipases of the meat starter culture organisms *Staphylococcus xylosus* and *Debaryomyces hansenii*. *Int. J. Food Microbiol.* 32 (1–2): 59–71.

Soriano, A., B. Cruz, L. Gómez, C. Mariscal, and A. García Ruiz. 2006. Proteolysis, physicochemical characteristics, and free fatty acid composition of dry sausages made with deer (*Cervus elaphus*) or wild boar (*Sus scrofa*) meat: A preliminary study. *Food Chem.* 96:173–184.

Soyer, A., A. H. Erta, and U. Üzümcüoğlu. 2005. Effect of processing conditions on the quality of naturally fermented Turkish sausages (sucuks). *Meat Sci.* 69:135–141.

Stahnke, L. H. 1995. Dried sausages fermented with *Staphylococcus xylosus* at different temperatures and with different ingredient levels. *Meat Sci.* 41:179–223.

Stahnke, L. H., A. Holck, A. Jensen, A. Nilsen, and E. Zanardi. 2002. Maturity acceleration of Italian dried sausage by *Staphylococcus carnosus*: Relationship between maturity and flavor compounds. *J. Food Sci.* 67:1914–1921.

Stevenson, J. M., D. L. Seman, and R. P. Littlejohn. 1992. Seasonal variation in venison quality of mature, farmed red deer stags in New Zealand. *J. Anim. Sci.* 70:1389–1396.

Sudirman, I., F. Mathier, M. Michel, and G. Lefebvre. 1993. Detection and properties of curvaticin 13, a bacteriocin-like substance produced by *Lactobacillus curvatus* SB13. *Curr. Microbiol.* 27:35–40.

Sumon, W., and N. Sumon. 2009. Quality of "MUM" from beef, Thai traditional fermented sausages. In *Proceedings of the 47th Kasetsart University Annual Conference*, 79–86. Agricultural Extension and Home Economics, Kasetsart, Thailand, 17–20 March.

Sunesen, L. O., and L. H. Stahnke. 2003. Mould starter cultures for dry sausages: Selection, application, and effects. *Meat Sci.* 65:935–948.

Suzzi, G., and F. Gardini. 2003. Biogenic amines in dry fermented sausages: A review. *Int. J. Food Microbiol.* 88:41–54.

Thornill, P. J., and T. M. Cogan. 1984. Use of gas-liquid chromatography to determine the end-products of growth of lactic acid bacteria. *Appl. Environ. Microbiol.* 47:1250–1254.

Tichaczek, P., J. Nissen-Meyer, I. Nes, R. Vogel, and W. Hammes. 1992. Characterization of the bacteriocins curvacin A from *Lactobacillus curvatus* LTH1174 and sakacin P from *Lactobacillus sake* LTH673. *Syst. Appl. Microbiol.* 15:460–468.

Toldrá, F. 1998. Proteolysis and lipolysis in flavour development of dry-cured meat products. *Meat Sci.* 49:S101–S110.

Toldrá, F., M. C. Miralles, and J. Flores. 1992. Protein extractability in dry-cured ham. *Food Chem.* 44:391–394.

Urso, R., G. Comi, and L. Cocolin. 2006. Ecology of lactic acid bacteria in Italian fermented sausages: Isolation, identification, and molecular characterization. *Syst. Appl. Microbiol.* 29:671–680.

Valencia, I., D. Ansorena, and I. Astiasarán. 2006. Nutritional and sensory properties of dry fermented sausages enriched with n-3 PUFAs. *Meat Sci.* 72:727–733.

———. 2007. Development of dry fermented sausages rich docosahexaenoic acid with oil from microalgae *Schizochytrium* sp.: Influence on nutritional properties, sensorial quality, and oxidation stability. *Food Chem.* 104:1087–1096.

Van Schalkwyk, D. L., K. W. McMillin, M. Booyse, R. C. Witthuhn, and L. C. Hoffman. 2011. Physico-chemical, microbiological, textural, and sensory attributes of matured game salami produced from springbok (*Antidorcas marsupialis*), gemsbok (*Oryx gazella*), kudu (*Tragelaphus strepsiceros*), and zebra (*Equus burchelli*) harvested in Namibia. *Meat Sci.* 88: 36–44.

Vergnais, L., F. Masson, M. C. Montel, J. L. Berdagué, and R. Talon. 1998. Evaluation of solid-phase microextraction for analysis of volatile metabolites produced by staphylococci. *J. Agricul. Food Chem.* 46:228–234.

Verluyten, J., F. Leroy, and L. De Vuyst. 2004. Effects of different spices used in the production of fermented sausages on growth of and curvacin A production by *Lactobacillus curvatus* LTH 1174. *Appl. Environ. Microbiol.* 70:4807– 4813.

Verplaetse, A. 1994. Influence of raw meat properties and processing technology on aroma quality of raw fermented meat products. *Proceedings of International Congress of Meat Science and Technology* 40:45–65.

Villani, F., A. Casaburi, C. Pennacchia, L. Filosa, R. Russo, and D. Ercolini. 2007. Microbial ecology of the soppressata of Vallo di Diano, a traditional dry fermented sausage from Southern Italy, and in vitro and in situ selection of autochthonous starter cultures. *Appl. Environ. Microbiol.* 73:5453 -5463.

Vioque, M., F. Prados, A. Pino, J. Fernández-Salguero, and R. Gómez. 2003. Embutidos crudos curados elaborados con carne de venado: Características físico-químicas y composición de ácidos grasos. *Eurocarne* 122:51–56.

Volpelli, L. A., R. Valusso, M. Morgante, P. Pittia, and E. Piasentier. 2003. Meat quality in male fallow deer (*Dama dama*): Effects of age and supplementary feeding. *Meat Sci.* 65:555–562.

Warnants, N., M. J. Van Oeckel, and Ch. V. Boucqué. 1998. Effect of incorporation of dietary polyunsaturated fatty acids in pork backfat on the quality of salami. *Meat Sci.* 49:435–445.

Wirth, F. 1988. Technologies for making fat-reduced meat products. *Fleischwirtsch.* 68: 1153–1156.

———. 1989. Reducing the common salt content of meat products: Possible methods and their limitations. *Fleischwirtsch.* 69:589–593.

———. 1991. Restricting and dispensing with curing agents in meat products. *Fleischwirtsch.* 71:1051–1054.

Zaika, L. L. 1988. Spices and herbs: Their antimicrobial activity and its determination. *J. Food Safety* 9:97–118.

Zanardi, E., S. Ghidini, M. Conter, and A. Ianieri. 2010. Mineral composition of Italian salami and effect of NaCl partial replacement on compositional, physico-chemical, and sensory parameters. *Meat Sci.* 86:742–747.

Zhang, J., Z. Liu, Z. Hu, Z. Fang, J. Chen, D. Wu, and H. Ye. 2010. Effect of sucrose on the generation of free amino acids and biogenic amines in Chinese traditional dry-cured fish during processing and storage. *J. Food Sci. Technol.* 48:69–75.

Żochowska-Kujawska, J., K. Lachowicz, M. Sobczak, and G. Bienkiewicz. 2010. Utility for production of massaged products of selected wild boar muscles originating from wetlands and arable area. *Meat Sci.* 85:461–466.

Żochowska-Kujawska, J., M. Sobczak, and K. Lachowicz. 2009. Comparison of the texture, rheological properties, and myofibre characteristics of SM (*semimembranosus*) muscle of selected species of game animals. *Pol. J. Food Nutrit. Sci.* 59:243–246.

Zomborszky, G., I. Szentmihályi, I. Sarudi, P. Horn, and C. S. Szabó. 1996. Nutrient composition of muscles in deer and boar. *J. Food Sci.* 61:625–635.

14 Process Control in Food Fermentation

Robert Tylingo

CONTENTS

14.1 INTRODUCTION

Food quality is a very complicated and complex problem. Safety is an important aspect often associated with the quality of food. Numerous physically, biologically, or chemically measurable properties indicate to what extent the product meets various requirements and standards and finally contributes to the consumer's satisfaction. The food production process, manufacturing, handling, packing, labeling, storage, trade, and environmental impact—all must meet international standards and requirements. It is evident that we can talk about high quality of food only when minimum conditions exist to ensure that the food is safe.

In many countries, the policies related to the food industry seek to establish legal requirements to minimize the risks that cause food poisoning. Such impact should be mandatorily controlled. Very serious are microbiological threats, which can cause diseases transmitted by food. Over the past few decades, many countries have seen a significant rise in the incidence of diseases caused by *Salmonella* spp., *Campylobacter* spp., and *Escherichia coli*. In addition, prion diseases that cause spongiform encephalopathies may be classified as a new risk. Chemical hazards are a further group of threats. Some substances, which occur naturally in foods, like mycotoxins, as well as environmental pollutants such as polycyclic aromatic hydrocarbons (PAHs) or heavy metals are those of the chemical pollution group. Another group of chemical food contaminants includes pollution introduced by unsound agricultural cultivation and farming processes—for example, excessive use of pesticides and fertilizers (polychlorinated biphenyls, PCBs), nitrates, and nitrites; drying by combustion gases that introduce PAHs; or the lack of safety control in animal feeds, which can include dioxins. More and more frequently, new technologies such as sterilization using microwaves, genetic engineering, irradiation, and modified-atmosphere packaging are used in food production. In all of these cases, the assessment of risks and threats should be carried out objectively, taking into account socioeconomic factors as well as ethical and environmental issues (WHO 2002; Gorris 2005; Alinorm 2009; Codex Alimentarius 2003).

Information sharing between supply chain partners is considered essential for food safety and traceability. Some of these systems are obligatory, such as Hazard Analysis & Critical Control Points (HACCP) in Europe, or entail optional certification, such as ISO 22000 or the BRC (British Retail Consortium) Global Standard for International Food. Although the sometimes-cited number of more than 380 certification schemes in the European Union (EU) is presumably somewhat exaggerated, in Germany alone about 40 different schemes are used for certifying farms and firms in the agribusiness (Theuvsen 2009).

14.2   GOOD MANUFACTURING PRACTICES (GMP)

The previous chapters have described a wide range of fermented food products. The food is produced using advanced technology or traditional techniques, which affect the degree of automation of production. In the brewing industry, which produces millions of hectoliters of beer per year, advanced technologies are used, and all technological processes are carried out in closed computer-controlled systems. This type of solution affects the application of good manufacturing practices differently than in, for example, production of regional ripened cheese. The principles of good manufacturing practices (GMP) are closely related to good hygienic practices (GHP). GMP and GHP should be considered as an absolute basis for all activities related to ensuring the overall safety of the foods produced. Applying the principles of GMP and GHP allows introducing such quality management systems as HACCP, ISO 22000, among others, at subsequent stages. Basic requirements for GMP relate to equipment, design, construction, location, and size of the rooms in which the food is produced. Floors, walls, and ceilings should be constructed of durable material, resistant to mechanical damage, nonabsorbent and impermeable, easy to wash and disinfect, not slippery, and resistant to acids (acetic acid) and salt. Various materials are used to construct the floors, mostly concrete with a usable area made of polyacrylic, polyester, and epoxy resin. Floor structure should prevent the formation of puddles, and all water must be drained into the sewers.

Drains must be formed from nonabsorbent elements with smooth, easy-to-clean surfaces. The doorways should be sufficiently wide and made of material that is easy to wash and clean and be resistant to corrosion and impact. Interior doors between the rooms should be swinging, with an eye-level peephole. Just as doors, the windows also must be made of materials that are easy to keep clean, and their structure should protect against the accumulation of dust. Sills should be constructed of smooth, waterproof material and have the least possible dimensions, their inner slope should be 45°, and the distance to the floor must not be less than 1 m. Ventilation systems must be equipped with appropriate filters and constructed so as to secure the area against cross-contamination. One of the interesting technologies is the use of air ventilation supply systems that produce a slight overpressure in the clean sections and thereby direct the flow of air. Properly functioning ventilation must ensure the flow of air from clean to dirty parts of the industrial plant, and in no event expose the rooms to infection caused by microorganisms, dust.

Internal paths in the production halls, which are frequented by both the staff and raw materials, and waste, should be as short as possible. The applicable rule is to avoid crossing the movement paths of raw materials, waste, packing, and finished products. Also, there must be an adequate number of social rooms for workers employed in food plants, e.g., cloakrooms, toilets, showers, washrooms, dining rooms, and laundries. GMP rules also describe general requirements for construction of equipment used for food production. All machine parts in contact with food must be indifferent to it. They cannot penetrate or be absorbed by it. In the case of pipelines, the splices must be smooth. Places with disjoint connections (collars) should be minimized to prevent retention of the product in dead spaces. Pumps used in the food industry cannot have dead spaces, should be easily disassembled, and consist of as few parts as possible. The outflow pipe stubs must be installed in such a way that the tank is self-emptying. The same applies to lids and sight glasses. All of the surfaces of equipment and connections must be smooth, without sharp fractures and cracks in which the remnants of raw materials or products might accumulate. In the machine construction, sharp bends as well as dead and hardly accessible spaces should be avoided. Easy and effective cleaning and disinfection must be possible. The internal surfaces of the equipment should be rounded for the same reasons. The machine construction should be such that all liquids—including leaks from raw materials and washing liquids—could easily flow down the machine components without creating bottlenecks inside. All machine parts that have contact with food must be clearly visible, and easily and quickly disassembled, so that they can be checked and repaired. All internal surfaces of the machine structure should be designed so that the machine could do self-cleaning (Codex Alimentarius 2003; Bata et al. 2006).

14.3 SANITATION

In this era of emphasis on food safety and security, the high-volume food-processing and preparation operations have increased the need for improved sanitary practices from processing to consumption. The word *sanitation* is derived from the Latin word *sanitas*, meaning "health." Applied to the food industry, sanitation is "the creation and maintenance of hygienic and healthful conditions." It is the application of science to provide wholesome food processed, prepared, merchandised, and sold in a clean environment by healthy workers; to prevent contamination with microorganisms that cause food-borne illnesses; and to minimize the proliferation of food-spoilage microorganisms. Effective sanitation refers to all the procedures that help accomplish these goals.

Sanitation is a critical component of any successful food-processing operation. Standards for hygienic practices continue to rise, and they have never been higher than they are today. Not only is it essential to have proper procedures and equipment for cleaning in place, but installing such products as components that are designed to facilitate easy and efficient sanitation is critical in successfully maintaining high standards of hygiene. Food handlers are potential sources of microorganisms that cause illness and food spoilage. *Hygiene* is a word used to describe sanitary principles for the preservation of health. Personal hygiene refers to the cleanliness of a person's body. Parts of the body that contribute to the contamination of food include the skin, hands, hair, eyes, mouth, nose, nasopharynx, respiratory tract, and excretory organs. These parts are contamination sources as carriers, through direct or indirect transmission, of detrimental microorganisms.

Management must select clean and healthy employees and ensure that they follow hygienic practices. Employees must be held responsible for personal hygiene so that the food that they handle remains wholesome (Marriott and Gravani 2006).

14.4 HAZARD ANALYSIS CRITICAL CONTROL POINTS (HACCP)

The acceptance of HACCP as the food safety control system of choice by the food industry, governments, and regulatory bodies with responsibility for food safety has led to the proliferation of HACCP plans on a worldwide basis. There are a number of excellent texts outlining practical approaches to the application of HACCP that are being used as guidelines, such as Alinorm 97/13A (Codex Alimentarius Commission 1996) and Regulation (EC) no. 852/2004 (European Parliament 2004).

The seven HACCP principles are as follows:

1. *Conduct hazard analysis.* Identify any hazards that must be prevented, eliminated, or reduced to acceptable levels. This principle involves the identification of all potential biological, chemical, and physical properties entering into the production process with raw materials and auxiliary materials at all stages of the manufacturing process until the final product. Hazard analysis is to determine the risks associated with every identified potential hazard, namely the determination of the probability of its occurrence in certain conditions and its implications for the consumer. It also includes the specification of parameters and/or characteristics and measures of hazard control.
2. *Identify critical control points (CCPs)* at the step or steps at which control is essential to prevent or eliminate hazard or to reduce it to acceptable levels. Hazard analysis gives a basis to determine CCPs, i.e., places, elements, or steps in the process that must be controlled because of the significant risks identified and that, if not controlled, can cause excessive risk unacceptable for the health quality of a food product.
3. *Establish critical limits* at CCPs, which separate acceptability from unacceptability for the prevention, elimination, or reduction of identified hazards.

According to this principle, control parameters or indicators (features) are defined for each CCP, i.e., the critical limits for accepted target values that should be maintained in the production process and that, if exceeded, can cause a hazard or increase it to an unacceptable level, so consequently the product may not meet the required health quality.

4. *Establish and implement effective monitoring procedures* at CCPs. For each CCP, it is necessary to specify the method of monitoring the set of control parameters, i.e., methods; testing and measuring instruments; the person responsible for the control; frequency of measurements, testing, or observation; and how to save the results.

5. *Establish corrective actions* when monitoring indicates that a CCP is not under control. The HACCP system must have a corrective action plan, specific to each CCP, i.e., actions taken during the monitoring of CCP when critical limits are exceeded. The corrective actions must restore control parameters monitored in all CCPs with acceptable values that guarantee the required health quality of products and secure the product that is produced when CCPs have exceeded the critical value.

6. *Establish verification procedures* that should be carried out regularly to confirm that the measures outlined in principles 1 to 5 are working effectively. According to this principle, it is necessary to identify the ways of verifying the HACCP system in order to ensure that the assumptions are correct, whether they are implemented in practice, and whether the activities carried out within this system are effective, i.e., provide safe products. Verification procedures include an evaluation system in the case of changes adopted in materials, the manufacturing process, or the procedures.

7. *Establish documents and records* commensurate with the nature and size of the food business to demonstrate the effective application of the measures outlined in principles 1 to 6. The principle imposes an obligation to document the HACCP system. The documentation should include a description of all the operations and activities on the basis of which the adopted HACCP program is implemented. The previously mentioned rules should specify the items that must be covered by the HACCP system and indicate the sequence of procedures in developing the system. In practice, this is usually carried out in various stages, taking into account the specific nature and conditions of individual plants, always in compliance with the general principles of the HACCP system.

Food business operators are to ensure:

1. That food handlers are supervised, instructed, and trained in food hygiene matters commensurate with their work activity.
2. That those responsible for the development and maintenance of the HACCP procedure or for the operation of relevant guides have received adequate training in the application of the HACCP principles.
3. Compliance with any requirements of national law concerning training programs for persons working in certain food sectors.

Many countries have undertaken a comprehensive evaluation and reorganization of their food inspection and control systems in order to improve efficiency and harmonize approaches. Such evaluation of food control systems has resulted in convergence toward the necessity to implement a preventive approach based on the HACCP principles and away from the traditional approach that relied heavily on end-product sampling and inspection. In a company in which the HACCP system is implemented, the responsibilities of employees are very clear, which streamlines work organization and the use of resources (raw materials, machines, and people) and finally leads to a reduction in overall production costs.

HACCP can be implemented at minimal cost to traditional operations relying on very simple techniques and instruments such as visual inspections, the use of pH strips, thermometers, and timing of unit operations. Though such a system is sustainable, its effectiveness will require commitment and vigilant supervision of the entrepreneur. (Lee and Hathaway 2000; Azanza and Zamora-Luna 2005; Jin, Zhou, and Ye 2008; Merican 2000; Ababouch 2000; Codex Alimentarius 2003; Mayes 1999; Lee and Hathaway 1999; Suwanrangsi 2000; Ropkins and Beck 2003; Christaki and Tzia 2002; Asefa et al. 2011; Martin et al. 2003; Orriss and Whitehead 2000).

14.5 MICROORGANISMS IN FOOD PROCESSING—QUALITY

In most fermented foods, the processes used in their production are inhibitory to many microorganisms. The antimicrobial effects of fermentation are not confined to spoilage organisms alone and can also affect pathogens that might be present. As a result, fermented products generally have a longer shelf life than their original substrates, and their ultimate spoilage is different in character. Fermented foods can be produced from all raw food materials, both of plant and animal origin. Apart from the desired fermentation cultures, there may also be pathogens that can survive in an optimal environment for fermentation cultures. Large-scale microbiological hazards can be presented on the basis of cheese production. The manufactured varieties include those made from pasteurized and unpasteurized cow, sheep, and goat milk. When basic GHP requirements are not met concerning, for example, preparation of milk pasteurization, a potential source of pathogenic cultures such as *Listeria monocytogenes*, *Escherichia coli*, *Clostridium* spp., *Salmonella* spp., and *Staphylococcus* spp. can survive.

14.5.1 *LISTERIA MONOCYTOGENES*

Since the early 1980s, when *Listeria monocytogenes* was first recognized as an emerging food-borne pathogen, a considerable number of well-documented outbreaks of human listeriosis have been reported: Most of them were associated with the consumption of dairy and plant foods (Samelis and Metaxopoulos 1999). *Listeria monocytogenes* has been associated with raw milk, cheeses (particularly soft-surface ripened varieties), ice cream, raw fruits and vegetables, raw and cooked poultry, raw meat, and raw and smoked fish. Listeriosis is a potentially lethal disease with a mortality rate of about 30%. *Listeria monocytogenes* can proliferate at refrigeration temperatures and can also be relatively resistant to pasteurization, is more acid tolerant than most food-borne pathogens, and is able to grow at relatively high NaCl concentrations, which facilitates its survival in the kinds of food containing NaCl and organic acids as preservatives. Some cheeses may be risk products from a listeriosis point of view, since they constitute a suitable medium for the growth of *Listeria monocytogenes*. Soft cheeses especially provide excellent growth conditions and have often been implicated in outbreaks of listeriosis. With respect to soft cheese, the contamination is localized almost exclusively on the cheese surface, the rind. Numerous recent studies have demonstrated that *L. monocytogenes* is appreciably more heat resistant than other non-spore-forming food pathogens. In practice, heat-resistant strains arise when bulk food, like ham, is given a mild heat treatment under slowly increasing temperatures (Samelis and Metaxopoulos 1999). The presence of nonpathogenic listeriae in a product indicates an unsatisfactory process at risk of contamination with *L. monocytogenes*. It is therefore recommended that food samples in general should be tested for the occurrence of the genus *Listeria* (Rudolf and Scherer 2001; Reissbrodt 2004; Margolles, Mayo, and Reyes-Gavilan 2000; O'Brien et al. 2009; Gormley et al. 2010).

14.5.2 *Escherichia coli*

The gram-negative members of Enterobacteriaceae have not previously exhibited a real problem with fermented food, unless serious mistakes of a hygienic and/or technological nature have been made. *Escherichia coli*, being common contaminants of food raw materials, usually disappear from the fermented food as a result of the combined effect of low pH, low temperature, and low water activity. This effect is enhanced to some extent by the presence of nitrite and also by other metabolites produced by starter cultures. In this case, these microorganisms are an indicator of clean technology. The pathogenic impact of *Escherichia coli* occurs mainly in fermented, unpasteurized food such as raw fermented sausages. To eliminate this danger, it is necessary to heat the product to at least 60°C and significantly reduce water activity (a_w). It is evident that such products lose their probiotic properties after heating (Adams and Mitchell 2002; Moore 2004; Incze 1998; O'Brien et al. 2009).

14.5.3 *Clostridium* spp.

Butyric fermentation is a process that deteriorates the quality of many food products. It is caused by anaerobic bacteria of the genus *Clostridium*, which live mainly in soil and have different sensitivity to oxygen exposure. These microorganisms, in adverse conditions, produce spores resistant to heating and hydrolases for assimilation of dextrin, cellulose, hemicelluloses, and pectin. Some species are mesophiles, and the others are thermophiles, with optimum temperature of growth from 60°C to 75°C. Depending on the species of microorganisms and culture conditions, the following by-products are formed: ethyl and butyl alcohols, acetic acid, acetone, carbon dioxide, and hydrogen. It is also the type of *Clostridium* that reduces acetone and produces isopropyl alcohol. Butyric fermentation bacteria may cause bloating of ripened rennet cheeses and spoilage of pasteurized milk and food silages. The harmful effects of the butyric fermentation bacteria include the production of food with

352   Fermentation: Effects on Food Properties

an intense, unpleasant smell and the reduction of yeast growth in distillery mashes. *Clostridium botulinum* produces the best-known potent toxins, with seven distinct serotypes currently defined (A–G). These toxins can cause life-threatening systemic toxicity through natural causes such as food poisoning. To eliminate the hazard caused by *Clostridium botulinum*, it is necessary to apply proper heat-sterilization regimes or to use nitrites as food additives, especially in meat products. Nitrate III effectively inhibits the growth of *Clostridium* spp. (Adams and Mitchell 2002; Ross, Morgan, and Hill 2002).

14.5.4 SALMONELLA SPP.

Salmonellae are a group of bacteria that reside in the intestinal tract of human beings and warm-blooded animals and are capable of causing disease. They are the second most common cause of bacterial food-borne illness. They are facultative anaerobic gram-negative rods. *Salmonella* spp. are members of the Enterobacteriaceae family. The *Salmonella* genus contains two species: *Salmonella enterica* and *Salmonella bongori*.

Salmonella spp. are not particularly heat resistant, and most serotypes are killed by normal cooking conditions (i.e., cooking to a core temperature of 75°C instantaneously or an equivalent time–temperature combination, e.g., 70°C for 2 min). However, a few highly heat-resistant serotypes have been reported (e.g., *S. Senftenberg* 775W and *S. Irumu*). Heat resistance is influenced by a_w, the nature of the solutes, and the pH of the suspending medium. Greater heat resistance is observed for cells in sucrose compared with NaCl at the same a_w values.

Salmonellae are shed in the feces. Poor hygiene practices can result in the dissemination of this pathogen to (a) the hands of humans and (b) the feet, hair, and skin of animals as they walk, sit, or lie in fecally contaminated ground or litter. Spread can also occur through fecal contamination of streams, rivers, and coastal waters or the use of improperly treated sewage for agricultural purposes.

The human infection caused by *Salmonella* spp. is referred to as salmonellosis. Although salmonellosis can arise from contact with infected animals, consumption of contaminated food is the most usual cause. Many kinds of food, particularly those of animal origin and others that may be subject to fecal contamination, have been identified as vehicles for the transmission of this pathogen to humans. Those of particular importance include meat, poultry, eggs, milk, chocolate, fruit, and vegetables. Spread of this pathogen may occur in the food-processing environment through cross-contamination from raw food or infected food handlers. Although salmonellae do not form spores, they can survive for long periods in food (Commission of European Communities 2005; Adams and Mitchell 2002; Moore 2004; Gormley et al. 2010; Todd 2003; Barbuti and Parolari 2002).

14.5.5 STAPHYLOCOCCUS AUREUS

Staphylococcus aureus is a bacterium belonging to the genus *Staphylococcus*. Staphylococci are toxin-producing, gram-positive, catalase-positive cocci that grow aerobically but are capable of facultative anaerobic metabolism. Staphylococcal food-borne intoxication is a common cause of bacterial food poisoning. *Staphylococcus aureus* is an ubiquitous organism occurring on the skin and mucous membranes of most warm-blooded animals, including humans. It is commonly detected in food of animal origin, such as raw meat and raw bulk milk; however, it is a poor competitor and rarely causes food poisoning in raw products (an exception being milk from a mastitic cow). Approximately 50% of humans are carriers of this organism, and food handlers are frequently implicated in the transmission of this pathogen to food. In addition, the organism survives well in the environment of food factories, where it may become part of the flora of the processing equipment and act as a source of contamination or recontamination.

Staphylococcal enterotoxins (SE) are produced in food by many strains of *S. aureus* and by some other coagulase-positive staphylococci (e.g., *S. intermedius*, *S. hyicus*, *S. delphini*). The enterotoxigenic strain needs to grow to levels >10^5 cfu/g before the toxin is produced at detectable levels. In addition, SE formation is influenced by temperature, pH, a_w, redox potential, and bacterial antagonisms (e.g., starter cultures used in the production of fermented milk products can prevent *S. aureus* growth and SE production). Once formed, SE are extremely difficult to eliminate from food. They are resistant to heat, freezing, and irradiation. They will survive commercial pasteurization and may even survive processes used for the sterilization of canned foods. Currently, 16 types of SE have been identified (A, B, C_1, C_2, C_3, D, E, G, H, I, J, K, L, M, N, and O). Outbreaks and sporadic cases of staphylococcal food poisoning have been linked with foods such as cheese, salami, bakery products, pasta, canned meat, canned fish, and canned vegetable products. In relation to cheese, failure of the starter culture provided an opportunity for *S. aureus* to grow and produce the SE (Gormley et al. 2010; Barbuti and Parolari 2002).

14.6 MICROORGANISMS IN FOOD PROCESSING—CONTROL

The Regulation on Microbiological Criteria for Foodstuffs (Commission of European Communities 2005) contains microbiological criteria for specific food/microorganism combinations and the implementing rules to be complied with by food business operators at all stages of the food chain. These criteria should be used by food business operators when validating and verifying the correct functioning of their HACCP-based procedures and other hygiene-control measures. This regulation differentiates microbiological criteria into:

1. *Process hygiene criteria*: These criteria indicate if the production process is operating in a hygienic manner. They are applicable to foodstuffs at various stages throughout their production processes. The regulation lays down process hygiene criteria for coagulase-positive staphylococci in dairy products and fishery products. In relation to dairy products, testing for the SE must be undertaken if coagulase-positive staphylococci are detected at levels >10^5 cfu/g.
2. *Food safety criteria*: These criteria define the acceptability of foodstuffs in terms of their microbiological safety. They are applicable to foodstuffs placed on the market and throughout their shelf life. The regulation lays down a food-safety criterion for the SE in dairy products.

To control *S. aureus* in the food chain:

- Avoid the use of raw materials that may be contaminated with high numbers of *S. aureus*. (Although the organism may be killed by further processing, SE may be present and is unlikely to be removed or destroyed.)
- Ensure that food handlers are aware of the importance of GHP, particularly hand washing and the need to report skin infections.
- Implement a food-safety management system based on the principles of HACCP, which includes good process control (e.g., temperature control during cooking and storage). Since low numbers of microorganisms can cause illness, it is important that control measures be taken at all stages in the food chain. These are essential to protect the health of consumers and the integrity of the business. Examples of control measures include:
  - Implementation of GHP and GMP at all stages in the food chain, i.e., at farm level, manufacturing, processing, catering, and retail. Particular attention should be paid to the prevention of cross contamination between raw and ready-to-eat food.
  - Implementation of a food-safety management system based on the principles of HACCP. This includes good process control (e.g., temperature control during cooking and storage).
- Validation and verification of HACCP-based procedures through the use of microbiological criteria.

14.7   ISO 22000

The ISO 22000 international standard specifies requirements for food-safety management systems, which involve interactive communication, a quality-management system, prerequisite programs, and the HACCP principles. The standard applies to the general concept of food safety and relates to food-borne hazards present at the time of consumer consumption. The introduction of food-safety hazards can occur at any stage of the food supply chain; therefore, it is necessary to conduct appropriate control over the food supply chain. Thus, food safety is ensured by a combination of efforts made by all sides involved in the food supply chain. Organizations in the food chain range from food producers to primary producers through food manufacturers, transport and storage service operators, as well as subcontractors of retail trade and food service retailers (including organizations internally related, such as producers of equipment, packaging materials, cleaning products, components, and additives). These organizations also include suppliers of services.

ISO 22000 combines the principles of HACCP and the stages of implementation developed by the Codex Alimentarius Commission (Table 14.1). Using requirements that can be audited, it combines the HACCP plan with the preliminary program (PRP), i.e., application of GMP and GHP. Hazard analysis is the key to an effective food-safety management system and helps to gain the knowledge necessary to establish an effective combination of hazard control measures. As shown in Table 14.1, the ISO 22000 standard requires several additional steps prior to implementation of the HACCP system. These preceding steps are well defined in Codex Alimentarius.

TABLE 14.1

References between HACCP Principles, Implementation Stages, and Chapters of ISO 22000

HACCP principles	Stages of implementation of HACCP system			ISO 22000
Logistics	Create the HACCP team	step 1	7.3.2	Food-safety team
	Describe the product	step 2	7.3.3	Product characteristics
			7.3.5.2	Process steps and control measures
	Identify the purpose	step 3	7.3.4	Purpose
	Develop scheme of the process	step 4	7.3.5.1	Scheme of the process
	Verify scheme of the process	step 5		
Rule 1				
Conduct hazard analysis	Develop list of potential hazards	step 6	7.4	Hazard analysis
	Conduct hazard analysis		7.4.2	Hazard identification and determination of acceptable levels
			7.4.3	Hazard assessment
	Evaluate surveillance measures		7.4.4	Select and evaluate surveillance measures
Rule 2				
Identify CCPs	Identify CCPs	step 7	7.6.2	Identify CCPs
Rule 3				
Establish critical limits	Establish critical limits for each CCP	step 8	7.6.3	Establish critical limits for CCPs
Rule 4				
Establish monitoring for CCPs	Establish a monitoring system for each CCP	step 9	7.6.4	System of monitoring CCPs
Rule 5				
Establish corrective actions taken in situations where monitoring indicates that the CCP is not controlled	Establish corrective actions	step 10	7.6.5	Actions in situations where monitoring results exceed critical limits

(Continued)

TABLE 14.1 (CONTINUED)
References between HACCP Principles, Implementation Stages, and Chapters of ISO 22000

HACCP principles	Stages of implementation of HACCP system		ISO 22000
Rule 6			
Establish verification procedures to confirm that HACCP system is working effectively	Establish verification procedures	step 11	7.8	Planning verification
Rule 7			
Establish documentation concerning all procedures and records appropriate for these principles and their implementation	Prepare documentation and maintain records of the surveillance measures	step 12	4.2/7.7	Requirements concerning documentation, updating initial information and documents defining initial programs and HACCP plan

Activities preceding hazard analysis should incorporate establishment of a food-safety team. The food-safety team should possess a combination of multidisciplinary knowledge and experience in developing and implementing the food-safety management system. Records should be maintained to show that the food-safety team has the required knowledge and experience. Then it is necessary to characterize the raw materials and products. Such characterization should consider:

- Biological, chemical, and physical properties
- Composition of the ingredients, including additives and auxiliary materials used in production
- Origin
- Production method
- Methods of packing and delivery
- Storage conditions and storage time
- Preparation and/or procedures before using or processing
- Eligibility criteria related to food safety or specifications of purchased materials and components proper for their intended use
- Product name or similar identification
- Marking/labeling for food safety and/or instructions
- Distribution methods

In addition, the expected treatment of the final product, its purpose, and all unintentional (but likely to occur) inappropriate handlings—as well as the use of the final product—should be described in documents at the extent necessary to conduct hazard analysis. For each product, a group of consumers should be identified. The consumer groups who are particularly vulnerable to food safety hazards must be taken into account. Each process should be presented in the form of a schematic diagram, which would take into account interactions at all stages of the process. In the schemes, attention should be paid to the entry of raw materials, packaging materials, additives, and the output of waste. Each stage of production must be fully described in regard to the surveillance measures and external requirements, as specified in the legislation. Further investigation has already been described in the seventh HACCP principle.

Today, certification systems play various roles in food supply chains. Some systems aim at informing consumers, whereas others are used to reduce quality uncertainties in business-to-business relationships. Whereas minimum requirement standards mainly contribute to the reduction of food safety risks, differentiation schemes aim at triggering consumers' willingness to pay for higher product or process qualities. To what extent this goal is accomplished depends on the criteria the certification system sets and its reliability and credibility (Theuvsen 2009; Gaaloul, Riabi, and Ghorbel 2011; Massoud et al. 2010; ISO 22000 2005).

14.8 CONCLUDING REMARKS

Nowadays, food production is closely related to the control of all stages of production, which facilitates the elimination of various hazards. It is necessary to control the history of raw materials, their production technologies, and the history of the final product until it reaches the consumer's table. This approach is complicated, but it facilitates the production of safe foods of specific, consistent quality. To meet these requirements it is necessary to use the systems that are often imposed by the legislature of the given country (for example, the HACCP system in the European Union). Appropriate involvement of management and staff in activities related to the HACCP system allows a fully controlled production of safe food. A wider approach to the control of the whole food chain (as recommended by ISO 22000) facilitates selecting suppliers and ensuring that the consumer obtains consistent quality of the final product. Certified systems also ensure the manufacturer's compliance with regulatory and legal requirements as well as contract law. However, the use of quality control systems in food production has some drawbacks. The main problems are a large number of audits, inadequate qualifications of the auditors, and the generation of vast amounts of documents. It is believed that food control in the future will become increasingly stringent as the penetration of hazards (e.g., problems with dioxins in various foods, pathogenic microorganisms, etc.) come about more and more often. It is necessary to control every step of production, not only the manufacturing processes, but also the primary production of raw materials and distribution routes for the final products. Increasing the scope of the control will certainly generate additional costs.

REFERENCES

Ababouch, L. 2000. The role of government agencies in assessing HACCP. *Food Control* 11:137–142.

Adams, M., and R. Mitchell. 2002. Fermentation and pathogen control: A risk assessment approach. *International Journal of Food Microbiology* 79:75– 83.

Alinorm. 2009. Joint FAO/WHO food standards programme. Report 09/32/Rep. Codex Alimentarius Commission. Thirty-Second Session FAO Headquarters, Rome, Italy, 29 June–4 July.

Asefa, D., C. Kure, R. Gjerde, S. Langsrud, M. Omer, T. Nesbakken, and I. Skaar. 2011. A HACCP plan for mycotoxigenic hazards associated with dry-cured meat production processes. *Food Control* 22:831–837.

Azanza, M., and M. Zamora-Luna. 2005. Barriers of HACCP team members to guideline adherence. *Food Control* 16:15–22.

Barbuti, S., and G. Parolari. 2002. Validation of manufacturing process to control pathogenic bacteria in typical dry fermented products. *Meat Science* 62:323–329.

Bata, D., E. Drosinos, P. Athanasopoulos, and P. Spathis. 2006. Cost of GHP improvement and HACCP adoption of an airline catering company. *Food Control* 17:414–419.

Christaki, T., and C. Tzia. 2002. Quality and safety assurance in winemaking. *Food Control* 13:503–517.

Codex Alimentarius. 2003. Recommended international code of practice. General principles of food hygiene. CAC/RCP 1-1969, Rev. 4-2003. Rome, Italy.

Codex Alimentarius Commission. 1996. Joint FAO/WHO food standards programme. Alinorm 97/13A. Rome, Italy.

Commission of European Communities. 2005. Commission Regulation (EC) No. 2073/2005 on microbiological criteria for food stuffs. *Official Journal of the EU.L.* 338:1–26. http://eur-lex.europa.eu/LexUriServ/LexUriServ.do?uri=CELEX:32005R2073:EN:HTML.

European Parliament. 2004. Regulation (EC) No. 852/2004 of the European Parliament and of the council of 29 April 2004 on the hygiene of foodstuffs. http://eur-lex.europa.eu/LexUriServ/LexUriServ.do?uri=CELEX:32004R0852:en:NOT.

Gaaloul, I., S. Riabi, and R. Ghorbel. 2011. Implementation of ISO 22000 in cereal food industry "SMID" in Tunisia. *Food Control* 22:59–66.

Gormley, F., C. Little, K. Grant, E. Pinna, and J. McLauchlin. 2010. The microbiological safety of ready-to-eat specialty meats from markets and specialty food shops: A UK wide study with a focus on *Salmonella* and *Listeria monocytogenes*. *Food Microbiology* 27:243–249.

Gorris, L. 2005. Food safety objective: An integral part of food chain management. *Food Control* 16:801–809.

Incze, K. 1998. Dry fermented sausages. *Meat Science* 49:169–177.

ISO 22000. 2005. Food safety management systems—Requirements for any organization in the food chain. Edition: 1, Stage: 90.93, TC 34/SC 17 ICS: 67.020.

Jin, S., J. Zhou, and J. Ye. 2008. Adoption of HACCP system in the Chinese food industry: A comparative analysis. *Food Control* 19:823–828.

Lee, J., and S. Hathaway. 1999. Experiences with HACCP as a tool to assure the export of food. *Food Control* 10:321–323.

———. 2000. New Zealand approaches to HACCP systems. *Food Control* 11:373–376.

Margolles, A., B. Mayo, and C. Reyes-Gavilan. 2000. Phenotypic characterization of *Listeria monocytogenes* and *Listeria innocua* strains isolated from short-ripened cheeses. *Food Microbiology* 17:461–467.

Marriott, N. G., and R. B. Gravani. 2006. *Principles of food sanitation*, 1–15. New York: Springer.

Martin, T., E. Dean, B. Hardy, T. Johnson, F. Jolly, F. Matthews, I. McKay, R. Souness, and J. Williams. 2003. A new era for food safety regulation in Australia. *Food Control* 14:429–438.

Massoud, M., R. Fayad, M. Fadel, and R. Kamleh. 2010. Drivers, barriers, and incentives to implementing environmental management systems in the food industry: A case of Lebanon. *Journal of Cleaner Production* 18:200–209.

Mayes, T. 1999. How can the principles of validation and verification be applied to hazard analysis? *Food Control* 10:277–279.

Merican, Z. 2000. The role of government agencies in assessing HACCP: The Malaysian procedure. *Food Control* 11:371–372.

Moore, J. 2004. Gastrointestinal outbreaks associated with fermented meats. *Meat Science* 67:565–568.

O'Brien, M., K. Hunt, S. McSweeney, and K. Jordan. 2009. Occurrence of foodborne pathogens in Irish farmhouse cheese. *Food Microbiology* 26:910–914.

Orriss, G., and A. Whitehead. 2000. Hazard analysis and critical control point (HACCP) as a part of an overall quality assurance system in international food trade. *Food Control* 11:345–351.

Reissbrodt, R. 2004. New chromogenic plating media for detection and enumeration of pathogenic *Listeria* spp.—An overview. *International Journal of Food Microbiology* 95:1–9.

Ropkins, K., and A. Beck. 2003. Using HACCP to control organic chemical hazards in food wholesale, distribution, storage, and retail. *Trends in Food Science & Technology* 14:374–389.

Ross, R., S. Morgan, and C. Hill. 2002. Preservation and fermentation: Past, present, and future. *International Journal of Food Microbiology* 79:3–16.

Rudolf, M., and S. Scherer. 2001. High incidence of *Listeria monocytogenes* in European red smear cheese. *International Journal of Food Microbiology* 63:91–98.

Samelis, J., and J. Metaxopoulos. 1999. Incidence and principal sources of *Listeria* spp. and *Listeria monocytogenes* contamination in processed meats and a meat processing plant. *Food Microbiology* 16:465–477.

Suwanrangsi, S. 2000. HACCP implementation in the Thai fisheries industry. *Food Control* 11:377–382

Theuvsen, L. 2009. Food quality and safety: Role, dissemination, and assessment of certification systems. In *Food Quality and Safety*, ed. G. Krasnowska and A. Pęksa, 14–26. Wroclaw, Poland: Wydawnictwo Uniwersytetu Przyrodniczego we Wroclawiu.

Todd, E. 2003. Microbiological safety standards and public health goals to reduce foodborne disease. *Meat Science* 66:33–43.

WHO. 2002. *WHO global strategy for food safety: Safer food for a better health*. Geneva, Switzerland: World Health Organization.

15 Final Remarks

Bhavbhuti M. Mehta and Afaf Kamal-Eldin

Complex microbial ecosystems are useful in the production of fermented products from almost all types of food raw materials, including cereals, legumes, vegetables and fruits, milk, meat, and seafoods. The products such as bread, vinegar, alcoholic beverages, sauerkraut, olives, fermented milk products, sausages, seafoods, etc., are different in structures, tastes, flavors, rheological properties, keeping quality, and nutritional value when compared to the raw materials. Different compounds are generated in the fermented foods as products of the catabolic reactions of the carbohydrates, proteins, and lipids, and the majority of these reactions are enzyme catalyzed. A wide range of bioactive compounds like polysaccharide polymers, peptides, and low-molecular-weight vitamins, antioxidants, etc., are produced, mostly by bacteria and fungi. There is a need to identify the most potent microorganisms, determine the molecular structures of unexplored bioactive compounds, and optimize yields. The different chapters of this book mainly focus on the chemical and property changes caused by fermentation and prefermentation processes.

Lactic acid fermentation deserves certain considerations and special mention. The production of acids affects the protein structure and ultimately the texture and final porosity of the products. Lactic acid bacteria fermentation has long been regarded with special favor as a technology to produce safe (by elimination of harmful compounds present in raw foods) and nutritious foods that elicit positive effects on human health and well-being. The exopolysaccharides (EPS) produced by lactic acid bacteria have a positive effect on water absorption as well as the rheology and stability of the fermenting dough effecting final porosity and improve resistance to retrogradation. Of special importance are the probiotic effects related to the production of lactic acid in the intestines by *Lactobacillus* bacteria, particularly *L. acidophilus*. The favorable acidic environment, buffered by lactic acid, inhibits the growth of undesirable bacteria (e.g., *Streptococcus*, *Staphylococcus*, *Salmonella*, *Clostridium botulinum*, and *E. coli*) and facilitates the absorption of proteins and a number of minerals, including calcium, copper, iron, magnesium, and manganese. Lactic acid bacteria produce two isomers of the lactic acid, namely l-(+) and d-(-) forms in fermented vegetables. The l-(+) lactic acid is easily and completely metabolized, whereas the d-(-) form is metabolized significantly more slowly and may lead to lactate acidosis. However, it is nowadays accepted that d-(-) lactic acid does not cause gastric problems for healthy adults. Compared to bacterial fermentations, other fermentations utilizing yeasts, other fungi, and/or a combination of microorganisms are less well understood and deserve in-depth investigations. An outstanding example is sourdough fermentation of cereals leading to notable health benefits, including antidiabetic, hypocholesterolemic, and immune-stimulating effects.

Areas that particularly need further attention are the understanding of enzymatic processes taking action during fermentation and the design of mixed raw materials and starter cultures. Various prefermentation processes such as steeping, germination, homogenization, heating, and fortification modify the chemical properties of the raw materials and ultimately affect the quality of the final fermented products. Several enzymes are involved in the fermentation processes, including a wide range of proteases, amylases, galactosidases, xylanases, cellulases, pectinases, lipases, tannases, etc. The reactions catalyzed by these enzymes include hydrolysis, dehydration/condensation, oxidation, hydroxylation, dehydrogenation, decarboxylation, amination/deamination, and isomerization. The enzymatic activities are modulated not only by the producing microorganisms and the substrate, but also by other factors including temperature, salt concentrations, pH, etc., making fermentation technologies a complex art. In times where standard product quality is a determinant criterion for product success, it is important to understand the chemistry/physics and design the processes in ways that ensure consistent product quality. Development of starter cultures becomes vital, especially for industrial processes.

The elimination of various hazards is essential during all stages of food production. To ensure the quality and safety of fermented food products, it is necessary to use the systems that are often imposed by the legislatures of the countries where foods are fermented and follow the Good Manufacturing Practice (GMP) and Hazard Analysis Critical Control Point (HACCP) systems. A wider approach to the control of the whole food chain (as recommended by ISO 22000) allows selecting suppliers and ensuring that the consumer obtains consistent quality and a safe final product. Finally, to fully understand the fermentation processes, one should have proficient knowledge of various disciplines such as chemistry, biochemistry, microbiology, enzymology, and technology. Given the breadth of knowledge needed to achieve these objectives, well-coordinated multidisciplinary teams will be needed to orchestrate this art.

Index

Printed and bound by CPI Group (UK) Ltd, Croydon, CR0 4YY

21/10/2024

01777103-0006